Advances in understanding insect pests affecting wheat and other cereals

It is widely recognised that agriculture is a significant contributor to global warming and climate change. Agriculture needs to reduce its environmental impact and adapt to current climate change whilst still feeding a growing population, i.e. become more 'climate-smart'. Burleigh Dodds Science Publishing is playing its part in achieving this by bringing together key research on making the production of the world's most important crops and livestock products more sustainable.

Based on extensive research, our publications specifically target the challenge of climate-smart agriculture. In this way we are using 'smart publishing' to help achieve climate-smart agriculture.

Burleigh Dodds Science Publishing is an independent and innovative publisher delivering high quality customer-focused agricultural science content in both print and online formats for the academic and research communities. Our aim is to build a foundation of knowledge on which researchers can build to meet the challenge of climate-smart agriculture.

For more information about Burleigh Dodds Science Publishing simply call us on +44 (0) 1223 839365, email info@bdspublishing.com or alternatively please visit our website at www.bdspublishing.com.

Related titles:

Achieving sustainable cultivation of maize Volume 2: Cultivation techniques, pest and disease control
Print (ISBN 978-1-78676-028-9); Online (ISBN 978-1-78676-031-9, 978-1-78676-030-2)

Achieving sustainable cultivation of rice Volume 2: Cultivation, pest and disease management
Print (ISBN 978-1-78676-012-8); Online (ISBN 978-1-78676-014-2, 978-1-78676-015-9)

Achieving sustainable cultivation of wheat Volume 1: Breeding, quality traits, pests and diseases
Print (ISBN 978-1-78676-016-6); Online (ISBN 978-1-78676-018-0; 978-1-78676-019-7)

Integrated management of insect pests: Current and future developments
Print (ISBN 978-1-78676-260-3); Online (ISBN 978-1-78676-262-7, 978-1-78676-263-4)

Advances in monitoring of native and invasive insect pests of crops
Print (ISBN 978-1-80146-107-8); Online (ISBN 978-1-80146-110-8, 978-1-80146-109-2)

Chapters are available individually from our online bookshop:
https://shop.bdspublishing.com

climate•SMART•publishing
IN AGRICULTURAL SCIENCE

BURLEIGH DODDS SERIES IN AGRICULTURAL SCIENCE
NUMBER 129

Advances in understanding insect pests affecting wheat and other cereals

Edited by Professor Sanford Eigenbrode, University of Idaho, USA and Dr Arash Rashed, Virginia Tech, USA

burleigh dodds
SCIENCE PUBLISHING

Published by Burleigh Dodds Science Publishing Limited
82 High Street, Sawston, Cambridge CB22 3HJ, UK
www.bdspublishing.com

Burleigh Dodds Science Publishing, 1518 Walnut Street, Suite 900, Philadelphia, PA 19102-3406, USA

First published 2023 by Burleigh Dodds Science Publishing Limited

Library of Congress Control Number: 2022949426

British Library Cataloguing in Publication Data
A catalogue record for this book is available from the British Library

ISBN 978-1-80146-113-9 (print)
ISBN 978-1-80146-116-0 (PDF)
ISBN 978-1-80146-115-3 (ePub)
ISSN 2059-6936 (print)
ISSN 2059-6944 (online)

DOI 10.19103/AS.2022.0114

Typeset by Deanta Global Publishing Services, Dublin, Ireland

Contents

Series list

Title	Series number
Achieving sustainable cultivation of maize - Vol 1 From improved varieties to local applications Edited by: Dr Dave Watson, CGIAR Maize Research Program Manager, CIMMYT, Mexico	001
Achieving sustainable cultivation of maize - Vol 2 Cultivation techniques, pest and disease control Edited by: Dr Dave Watson, CGIAR Maize Research Program Manager, CIMMYT, Mexico	002
Achieving sustainable cultivation of rice - Vol 1 Breeding for higher yield and quality Edited by: Prof. Takuji Sasaki, Tokyo University of Agriculture, Japan	003
Achieving sustainable cultivation of rice - Vol 2 Cultivation, pest and disease management Edited by: Prof. Takuji Sasaki, Tokyo University of Agriculture, Japan	004
Achieving sustainable cultivation of wheat - Vol 1 Breeding, quality traits, pests and diseases Edited by: Prof. Peter Langridge, The University of Adelaide, Australia	005
Achieving sustainable cultivation of wheat - Vol 2 Cultivation techniques Edited by: Prof. Peter Langridge, The University of Adelaide, Australia	006
Achieving sustainable cultivation of tomatoes Edited by: Dr Autar Mattoo, USDA-ARS, USA and Prof. Avtar Handa, Purdue University, USA	007
Achieving sustainable production of milk - Vol 1 Milk composition, genetics and breeding Edited by: Dr Nico van Belzen, International Dairy Federation (IDF), Belgium	008
Achieving sustainable production of milk - Vol 2 Safety, quality and sustainability Edited by: Dr Nico van Belzen, International Dairy Federation (IDF), Belgium	009
Achieving sustainable production of milk - Vol 3 Dairy herd management and welfare Edited by: Prof. John Webster, University of Bristol, UK	010
Ensuring safety and quality in the production of beef - Vol 1 Safety Edited by: Prof. Gary Acuff, Texas A&M University, USA and Prof. James Dickson, Iowa State University, USA	011
Ensuring safety and quality in the production of beef - Vol 2 Quality Edited by: Prof. Michael Dikeman, Kansas State University, USA	012
Achieving sustainable production of poultry meat - Vol 1 Safety, quality and sustainability Edited by: Prof. Steven C. Ricke, University of Arkansas, USA	013
Achieving sustainable production of poultry meat - Vol 2 Breeding and nutrition Edited by: Prof. Todd Applegate, University of Georgia, USA	014

Introduction

Insect pests remain a significant threat to cereal production due to the direct damage they can cause to crops, but also due to their ability to act as vectors of wheat pathogens, particularly plant viruses. With growing concerns surrounding the impact of climate change on both native and invasive pest invasions, coupled with the rising threat of global food insecurity, more research is required to understand the major insect pests of cereals, including how best to control and monitor them. This volume provides a comprehensive review of the wealth of research that addresses this challenge and discusses the major insect pests affecting key cereal crops. Each chapter covers the background and most recent developments in fundamental and applied research on major pests and shows how a better understanding of these pests can be used to improve integrated pest management strategies.

The chapters are split into four parts: Part 1 focuses on foliar feeding pests such as cereal leaf beetle, grasshoppers and other orthopteran pests. Chapters in Part 2 examine gall midges and stem feeding pests, focusing specifically on the Hessian fly, wheat midge and wheat stem sawfly. Part 3 chapters cover phloem feeding pests, mites and root feeding pests such as Russian wheat aphid, greenbug, fescue aphid, the wheat aphid, wheat curl mite and wireworms. Chapters in Part 4 highlight emerging issues involving insect pests affecting cereal crops, focusing specifically on recent invasions, the potential and challenges of using biotechnology for wheat crop protection and online decision support systems, remote sensing and artificial intelligence applications for pest management.

Part 1 Foliar feeding pests

The first chapter of the book discusses the cereal leaf beetle (*Oulema melanopus L.*). Chapter 1 begins by first examining the factors that can affect the cereal leaf beetle's ecology and severity, such as host range, spatial distribution within and among grain fields and surrounding landscape. The chapter also discusses the biological control of leaf beetle, focusing specifically on classical, conservation and augmentative biological control methods. It moves on to review insect-plant interactions, drawing specific attention to host plant resistance, host volatile organic compounds, additional induced host responses to cereal leaf beetle and the effects of endophytes. A section on thresholds and other components of integrated pest management is also included.

The subject of Chapter 2 is grasshoppers and other orthopteran pests. The chapter opens with an overview of the current grasshopper, locust and katydid pest species that affect cereal crops. This is then followed by a section that addresses ways to monitor population expansion and migration of grasshoppers, Mormon crickets and locusts. The chapter moves on to review nutrient availability and orthopteran outbreaks. This is then followed by a review of effective means of preventing and suppressing these outbreaks through methods such as chemical control, the use of microbial control agents, insect immunity, diet, migration and immunity to fungal control. The chapter also discusses insect colouration and age and how this affects immunity to fungal control, spraying and endophytic fungal control as well as diet, migration and immunity to bacteria. A section on the use of orthopteran pests for food or animal feed is also provided, followed by a review of the effect of climate change on Orthoptera populations.

Part 2 Gall midges and stem feeding pests

Part 2 opens with a chapter that focuses on the Hessian fly, a destructive pest of wheat and barley. Chapter 3 first describes the distribution and biology of Hessian fly, then examines the feeding mechanisms of the pest and how these can be used to damage crops. The chapter also provides and overview of how Hessian fly can be monitored along with the various factors that can affect the population dynamics of the pest. A section on advances in integrated pest management of Hessian fly is provided, as well as a review of the advances in the identification and mapping of resistance genes in wheat. The chapter examines current advances towards understanding the mechanism of wheat resistance to Hessian fly, the mechanisms of Hessian fly biotype differentiation, as well as the effects of climate change on Hessian fly management.

Chapter 4 reviews wheat midge (*Sitodiplosis mosellana*) and its management in the Northern Great Plains of the United States and Canada. The chapter first provides an overview of the morphology of wheat midge at each stage of its life cycle. It then moves on to discuss the life cycle of the pest in more depth. The damage that wheat midge causes to crops is also reviewed, which is then followed by an examination of the different pest management methods for the pest, such as monitoring and economic thresholds, cultural control, host plant resistance and biological and chemical control.

The final chapter of Part 2 looks at wheat stem sawfly (*Cephus cinctus* Norton) and its effect on wheat production. Chapter 5 first reviews the pest's historical distribution, then moves on to examine how it damages crops, focusing specifically on life cycle-driven losses and yield losses. A section on ways to manage the pest is also included, drawing specific attention to the use of insecticides, cultural tactics and various methods of biological control. The chapter also reviews host plant resistance then concludes by analysing future directions for wheat stem sawfly research.

Part 3 Phloem feeding pests, mites and root feeding pests

The first chapter of Part 3 provides an overview of the Russian wheat aphid (*Diuraphis noxia*). Chapter 6 begins by reviewing the global occurrence and pest status of the Russian wheat aphid, then moves on to discuss population growth and conducive circumstances of outbreaks, highlighting factors such as climate change and how the Russian wheat aphid can respond to this. A case study on controlling the Russian wheat aphid is also included, drawing specific attention to control measures used in South Africa. This is then followed by a review of how Russian wheat aphid populations can be managed using cultural control, host plant resistance, biological control and insecticides. The chapter highlights Russian wheat aphid biotypes and also examines the use of decision support for integrated pest control.

Chapter 7 provides an overview of greenbug (*Schizaphis graminum*), a major insect pest of cereal grains worldwide, especially in the southern Great Plains of the United States. The chapter summarises greenbug life history and biology, then provides a description of the direct and indirect crop injury caused by greenbug to wheat and sorghum. The chapter also reviews current research efforts concerning greenbug management, including biological control, cultural control focusing on host plant resistance, monitoring and chemical control that can be incorporated into an integrated pest management program.

Expanding on topics previously discussed in Chapter 7, Chapter 8 focuses on greenbug-wheat interactions, as well as greenbug pest management and host resistance. The chapter first introduces integrated pest management practices, then moves on to analyse the importance of host plant resistance. A section on how the environment of a wheat crop can influence the presence of *Schizaphis graminum*, focusing on aspects such as temperature, ambient CO_2 concentration and host nutritional changes due to abiotic stresses. The chapter also reviews molecular observations and draws attention to wheat breeding for *Schizaphis graminum* resistance.

Chapter 9 examines the fescue aphid (*Metopolophium festucae*). The chapter begins by describing how different fescue aphid subspecies can be identified according to their appearance, then moves on to highlight where various species of the pest are currently present. This includes a summary of the invasion of the aphid into cereal systems of the Pacific Northwest USA and where it has become abundant just within the last decade. This is then followed by an overview of the reproductive biology and host range of several fescue aphid subspecies. The chapter also reviews the pest's status and the damage it can cause, focusing specifically on its role as a direct pest and potential as a virus vector. A section on the potential for host plant resistance is also provided, followed by an overview of potential management practices for fescue aphid.

The next chapter focuses on the English grain aphid (*Sitobion avenae*). Chapter 10 first explores the distribution and economic importance of the English grain aphid, drawing specific attention to its life cycle and reproduction strategies. The chapter moves on to highlight various factors affecting the pest's population growth and spread, such as climatic factors, water availability, geographic divergence and local adaptation as well as the effect of climate change. A section on host plants is also included, followed by a review of English grain aphid biotype development, focusing on biotype discovery and mechanisms underlying their development. The chapter also reviews options for integrated management, such as host plant resistance, cultural control, biological control and chemical control.

The subject of Chapter 11 is wheat curl mite ecology and the epidemiology of its associated wheat viruses. It begins by describing ways to identify wheat curl mites, then goes on to review the pest's life cycle. The chapter moves on to discuss how wheat curl mites can damage crops and how these crops can respond to this. A section on the impact of the virus complex is also provided, focusing on wheat streak mosaic virus, High Plains wheat mosaic virus and Triticum mosaic virus. This is then followed by an overview of alternative hosts and green bridge species for the mite-virus complex. The chapter reviews mite dispersal and provides a case study on the spatial characterisation of mite dispersal and virus spread. It also examines ways to manage the mite-virus complex, such as green bridge management, adjusting planting date as well as mite and virus resistance.

The final chapter of Part 3 focuses on advances in managing wireworms in cereal crops, specifically the challenges and future directions for this pest. Chapter 12 provides a review of what is known about wireworm ecology and management, focusing on the most recent advances in species identification, within-species population variation, and species-specific research on their management with respect to pest monitoring, prevention, cultural, biological, and chemical control methods. Throughout, the existing knowledge gaps are highlighted and avenues for future research are proposed.

Part 4 Emerging issues

Part 4 opens with a chapter that focuses on recent invasions of insect pests of wheat and sorghum. Chapter 13 provides a general description of the invasion and management response of six insects that have affected wheat and other cereal grains: *Agriphila aeneociliella Eversmann* (Lepid: Crambidae), *Metopolophium festucae cerealium Stroyan* (Hem: Aphididae), wheat stem sawfly (*Cephus cinctus* Norton (Hymen: Cephidae)), *Sipha maydis Passerini* (Hem: Aphididae), Russian wheat aphid (*Diuraphis noxia* (Kurdjuimov) (Hemi: Aphididae)) and sugarcane aphid (Melanaphis sorghi Theobold (Hemi:

Aphididae)). A case study focused on the recent invasion of *M. sorghi* across the sorghum agroecosystem of North America is also included. This near continent-scale invasion provides perspectives on invasion progression (both ecology and management) across ecological and agricultural gradients, and on management response appropriate at local and regional scales in the context of environmental (landscape and weather) factors that influence pest-natural enemy-plant interactions. The chapter concludes with a discussion of future trends in light of climate change and lessons learned from these recent insect pest invasions.

Chapters 14 draws attention to the use of biotechnology for wheat crop protection, focusing on its potential and current challenges. The chapter first describes new breeding technologies that have been developed to improve pest resistance in the context of wheat crops. It moves on to review how the wheat genome can be considered an obstacle to precise genome editing. A section on epibreeding is also included, followed by a discussion on the importance of understanding host plant–insect pest interactions. The chapter also addresses the importance of understanding the pest beyond pest biology and reviews how RNA interference can be used as an integrated pest management tool.

The final chapter of the book looks at online decision support systems, remote sensing and AI applications for wheat pest management. Chapter 15 begins by providing an overview of current pest forecasting, observation and support tools for wheat pest management, using case studies to provide examples of these tools in action. This is then followed by a review of the use of remote sensing and artificial intelligence image analysis to inform pest identification and promote decision-making. The chapter also reviews the uptake of decision support systems amongst farmers and provides an overview of the factors that can determine the success of these support systems as well.

Acknowledgement

We wish to acknowledge the following for their help in reviewing a particular chapter:

- Chapter 9: Mr Andy Jensen, Washington State University, USA, Dr Roger Blackman, The Natural History Museum, UK and Ms Susan Halbert, FDACS, USA

Part 1

Foliar feeding pests

Chapter 1

Cereal leaf beetle (*Oulema melanopus*)

Edward W. Evans, Utah State University, USA

1 Introduction

The cereal leaf beetle (*Oulema melanopus* [L.], Coleoptera: Chrysomelidae, Criocerinae; referred to hereafter as CLB) is now widely distributed throughout temperate northern latitudes (Olfert et al., 2004; Philips et al., 2011). It feeds on many grasses, preferring young plant tissues, and can become sufficiently numerous in cultivated small grains (especially wheat, barley, and oats) to inflict economic damage (Haynes and Gage, 1981; Wenda-Piesik et al., 2018).

In its native, Palearctic range, CLB co-occurs with congeners of similar ecology and habits, including notably *Oulema gallaeciana* [Heyden], as referred to previously (e.g. by Dysart et al., 1973), and as now synonymized with *Oulema obscura* [Stephens] (Leroy et al., 2020). Members of this *Oulema* spp. complex are referred to widely as cereal leaf beetles [*sensu lato*] and typically are studied as a group in co-infesting small grain crops in the Old World (Bechini et al., 2013; Schmidtt and Uhl, 2017; Wenda-Piesik et al., 2018; Van de Vijver et al., 2019; Steinger et al., 2020). Three species in particular, CLB and its sibling species *Oulema duftschmidi* [Redtenbacher] and *O. obscura*, occur most often and pose the greatest threat in small grains in Europe (Leroy et al., 2020).

Following the identification of CLB in 1962 as a recently introduced pest of small grains in Michigan (Haynes and Gage, 1981), CLB has spread widely through North America; excellent reviews of CLB have been published, including those of Dysart et al. (1973), Haynes and Gage (1981), Kher

http://dx.doi.org/10.19103/AS.2022.0114.14

et al. (2011), and Philips et al. (2011). The present chapter builds on earlier publications addressed in these reviews, by considering recent research advances concerning CLB, as published especially over the past 10 to 15 years.

The native range of CLB encompasses most of Europe, including north to central Scandinavia; coastal regions of North Africa; and western to central Asia, including eastward into Mongolia and China, and north into Siberia (Dysart et al., 1973; Haynes and Gage, 1981; Olfert et al., 2004). In the New World, CLB has spread from initial establishment in north central North America to occur now in most of the United States east and south of North Dakota (Philips et al., 2011). Beginning in the mid-1980s, CLB was established and spread also in the northern half of the western United States (Philips et al., 2011; Roberts, 2016), occurring in arid agricultural regions, especially in irrigated fields of small grains, and appearing for the first time most recently in California in 2013 (Dowell and Pickett, 2016).

Paralleling the establishment and spread of CLB in the eastern United States, the pest similarly became well-established in eastern Canada from the 1960s on (Harcourt et al., 1984; LeSage et al., 2007). Olfert et al. (2004) predicted from climate (CLIMEX) modeling that CLB could establish in the prairie ecozone of western Canada, potentially threatening millions of acres of small grains grown there commercially. Indeed, following its appearance in British Columbia in 1998 (Kher et al., 2011), CLB subsequently was found to have been established in Alberta in 2005, Saskatchewan in 2008, and Manitoba in 2009 (Dosdall et al., 2011; Kher et al., 2011).

The first isolated appearance of CLB in Manitoba, in the northwestern part of the province, likely resulted from long-distance dispersal with anthropogenic causes (Dosdall et al., 2011). Similar 'jump dispersal' had occurred previously when CLB first appeared in the western United States in Utah in 1984, hundreds of km distant from the nearest source population in eastern North America (Evans et al., 2006).

CLB has one generation each year (Haynes and Gage, 1981; Buntin et al., 2004); as an exception, late-developing CLB larvae in mid-summer in Virginia were reported by McPherson (1983) as from a small but distinct second generation. Adults overwinter in small-grain stubble and other debris, often at the edges of fence rows and woodlots, or in protected sites within the woodlots themselves (Casagrande et al, 1977; Haynes and Gage, 1981). On warm, early spring days, the adults emerge from overwintering sites and fly to fields of small grains and other grass habitats, where they feed and mate (Helgesen and Haynes, 1972; Dysart et al., 1973; Casagrande et al., 1977).

From mid-spring through early summer, eggs are laid, singly or less frequently as end-to-end doublets or triplets, along the midvein on the upper surfaces of host plant leaves (Dysart et al., 1973; Kher et al., 2011). At constant temperatures of 18°C and 22°C, the eggs hatch about 10 and 6 days later, respectively (Guppy and Harcourt, 1978).

CLB larvae, distinctively covered with a dark, moist fecal shield, pass through four instars to complete their development in 11-16 days at temperatures of 18-22°C (Guppy and Harcourt, 1978). They skeletonize host leaves by feeding on the upper epidermis and parenchyma tissue, while leaving the basal epidermis intact as a thin, translucent membrane (Dysart et al., 1973; Philips et al., 2011). Feeding damage to leaves can intensify water stress and the risk of disease as well as reduce photosynthesis (Steinger et al., 2020), with decreased yield and grain quality (Philips et al., 2012).

Upon fully maturing, larvae drop from the host plant to pupate in the soil at a depth of approximately 5 cm (Dysart et al., 1973). New adults emerge 2-3 weeks later during mid-summer, depending on soil temperature, to feed for a short time on the foliage of small grains and other grasses (e.g. corn; Buntin et al., 2004), before departing for overwintering sites (Dysart et al., 1973; Casagrande et al., 1977).

In both the Old and New World, outbreaks of CLB are sporadic, and it is difficult to predict when and where these outbreaks will occur (Dysart et al., 1973; Philips et al., 2011). Reduced tillage and increased area of small grain monocultures over the past few decades in Europe have led to increased frequency and severity of economic damage from high populations of CLB and related *Oulema* spp. (Wenda-Piesik et al., 2018). More widespread adoption in Europe of organic farming, and integrated pest management (IPM) for a variety of pests, could also lead to increasing economic damage in particular from CLB and congeners, which heretofore often have been contained as secondary pests by broad-spectrum insecticide application (Leroy et al., 2020).

In the United States, the severity of CLB damage to small grains declined overall with time in the north-central states where it had first established, but crop damage from CLB has continued to be more problematic in mid-Atlantic and southern states (Buntin et al., 2004; Philips et al., 2012; Reisig et al., 2017). When outbreaks of CLB occur, most economic damage results from larval rather than adult feeding, with reported estimates of yield loss in some cases approaching or even exceeding 50%, both in CLB's native and introduced range (Philips et al., 2011; Kher et al., 2016; Skuhrovec et al., 2018; Wenda-Piesik et al., 2018).

In light of the continual threat from CLB to wheat and other cereals throughout much of the northern hemisphere, considerable research in recent years has focused on understanding and controlling CLB infestations in small grain crops. Attention has been directed toward factors that affect its ecology and severity; issues of implementation and promotion of biological control; aspects of plant-insect interactions, especially as related to host plant resistance against CLB; and economic thresholds and other components of IPM of the pest. Recent research advances in each of these areas are discussed below.

2 Factors that affect the cereal leaf beetle's ecology and severity

2.1 Host range

Preference by CLB for particular host plant species, including wheat, barley, and oats, has been difficult to pin down (Dysart et al., 1973; Philips et al., 2011). CLB readily feeds on both wheat and oats (Philips et al., 2012), and oats can be planted effectively as a trap crop to attract CLB adults (Roberts, 2016). Assessing host preference under field conditions is challenging in part because host plant suitability and availability, as expressed in agricultural landscapes, can depend on when precisely individual fields of small grains are planted, e.g. as affected by weather and farmer priorities (Casagrande et al., 1977; Haynes and Gage, 1981; Evans et al., 2006).

Kher et al. (2016) recently extended our understanding of the suitability of cereals for CLB, building on prior assessments that focused especially on larval survivorship and weight gain in wheat and oats. Kher and colleagues evaluated CLB performance in the laboratory on a broad set of cereals that are grown commercially in western Canada. Cultivars of wheat, barley, and oats supported the fastest development and highest larval survivorship, adult weight, and fecundity. CLB larvae survived well also on rye and triticale, but development was prolonged and adult weight and fecundity were low. Corn was a relatively poor host in all respects. Interestingly, however, CLB adults laid eggs in similar numbers on all cereals, suggesting that the less favorable cereals may serve as secondary hosts in the field. Indeed, the initial detection of CLB in central Alberta was in a field of triticale (Kher et al., 2016).

2.2 Spatial distribution within and among grain fields

CLB males produce an aggregation pheromone that attracts adults of both sexes during spring migration to cereal fields from overwintering sites (Rao et al., 2003). Pheromone-based recruitment accentuates uneven spatial distribution of CLB adults, and subsequently of eggs and larvae, among cereal fields during each growing season. Markedly uneven distribution occurs in North American agricultural landscapes, with some fields infested heavily while many others are infested only lightly (Evans et al., 2014; Riesig et al., 2017).

The key features making some grain fields much more attractive than others to dispersing CLB adults are not fully understood. Crop phenology, however, is well-recognized as one major factor; recent studies reinforce that CLB adults prefer newly planted stands, with young cereal foliage, over older stands (Hoffman and Rao, 2010; Pickett et al., 2021). In particular, CLB females preferentially oviposit on young leaves, particularly in response to reduced tissue toughness (Hoffman and Rao, 2011). Consequently, CLB adults may first colonize winter wheat fields each spring, only to migrate thereafter to newly emerging spring wheat fields. If, however, fields of spring grains are planted late enough, e.g. as the result of spring rains, these fields may largely escape CLB infestation, with few overwintering adults remaining to colonize them (e.g. Evans et al., 2006).

Tiller density may also affect CLB colonization: results from Riesig et al. (2017) suggest that CLB adults prefer to oviposit in denser wheat fields. As had been found previously by Honěk (1991) for central Europe, Riesig et al. (2017) found that more CLB eggs per m^2 generally were laid in denser wheat fields throughout the southeastern United States, with the number of eggs per tiller positively correlated with the number of tillers per m^2. The number of CLB larvae per tiller occurring subsequently, however, was not correlated with tiller density.

Recently developed geostatistical tools, including Spatial Analysis by Distance Indices (SADIE) and semivariogram modeling, have been applied to explore further how much CLB individuals are aggregated also within individual wheat fields, both in the southeastern United States (Reay-Jones, 2010, 2012, 2017) and in western Canada (Kher et al., 2014a). Adults in particular were found by Reay-Jones (2010) to occur in greatest densities at field edges versus 5-25 m into a field. In contrast, little difference was observed in adult density at 10 m versus 40 m in winter wheat fields in Belgium (Van de Vijver et al., 2019) and in larval density at 20 m versus 100 m in winter wheat fields in western Canada (Kheirodin et al., 2020a).

Peak densities of eggs and larvae were spatially associated with peak adult densities within fields in the southeastern United States, although aggregation was detected less frequently for larvae and eggs than for adults (Reay-Jones, 2012). Higher densities of all stages were positively associated with greater wheat spike density among sampling plots within fields on one or more occasions (Reay-Jones, 2012), mirroring results for CLB and tiller spatial distributions among fields (Reisig et al., 2017).

Kher et al. (2014a) found significant positive associations of CLB larval densities with several aspects of plant vigor (plant density, plant height, and large numbers of leaves per plant) within individual fields, as measured in 10 × 10 m cells of a gridded plot. Potential associations of localized CLB larval abundance with varying nutrient content of wheat leaves (including nitrogen, phosphorus, potassium, and sulfur) were investigated also, but variable results over 3 years (2010-2012) necessitate further study (Kher et al., 2014a).

2.3 Surrounding landscape

Landscape composition and complexity can affect the local densities of pest insects in individual crop fields (Karp et al., 2018). This can be especially true for highly mobile pests such as CLB, which move readily and often across landscapes (e.g. from natal fields to overwintering sites in shelter belts, and then to new fields the following spring). This topic has been addressed recently for CLB in particular (and for its principal parasitoid - *see Section 3.1 Classical biological control*) by Kheirodin et al. (2020a) for wheat fields of western Canada.

Kheirodin et al. (2020a) assessed landscape effects at varying proximity, up to 2-km distant from the center of individual wheat fields over 2 years (2014-2015). Land cover types included were major CLB host plants (wheat and barley only, as few oats were planted during the study); minor hosts (corn and rye, plus the few oat fields that were planted); non-cereal crops (including canola and alfalfa); pasture; grassland (including fallow fields); and woodlands (including trees and riparian vegetation). Increase in woodland cover was associated with an increase in CLB abundance, as predicted because woodlands can be overwintering sites (Casagrande et al., 1977). Conversely, CLB abundance was associated negatively with an increase in pasture cover (previously in Switzerland, Tschumi et al. [2015] had found CLB abundance unrelated to non-crop cover, when all non-crop habitat types were combined for analysis).

An especially strong result from Kheirodin et al. (2020a), observed at multiple spatial scales, was that CLB occurred in greatest numbers in areas with greatest coverage of the beetle's host plants, wheat or barley, in line with the classic resource concentration hypothesis (Root, 1973). Thus, the number of CLB larvae per 50 sweeps in an individual wheat field was positively related to the percentage of the landscape that was planted with wheat or barley, both in the current and previous years.

Kheirodin et al. (2020a) note that the influence of major host plantings in the previous year most likely reflected a carry-over effect in the local emergence of large numbers of overwintered CLB adults in the current year. Interestingly, a negative correlation between the extent of planting of major hosts in the current versus the previous year, likely the result of annual crop rotation, leads to opposing effects on CLB abundance in any given year. Kheirodin et al. (2020a) point out that an overall reduction in major host plant coverage with increased regional crop diversity could be effective in reducing CLB infestations in individual cereal fields.

3 Biological control of cereal leaf beetle

3.1 Classical biological control

The purposeful introduction of natural enemies to invaded regions from the pest's native range has emerged as the most successful alternative to insecticides in controlling CLB infestations in many parts of North America and has continued to garner much research interest. CLB is attacked by a number of co-evolved parasitoids in the Old World (Dysart et al., 1973). The establishment of CLB in the 1960s as a major pest in U.S. midwestern states and eastern Canada, despite efforts initially to eradicate the pest with intensive insecticide application and strict quarantine (Haynes and Gage, 1981), soon led to a large-scale program in classical biological control (Dysart et al., 1973).

Three larval parasitoids, [*Tetrastichus julis* (Walker) (Hymenoptera: Eulophidae); and *Diaparsis temporalis* Horstmann and *Lemophagus curtus* Townes (Hymenoptera: Ichneumonidae)] became established in eastern North America upon introduction from Europe (Stehr, 1970; Maltby et al., 1971; Dysart et al., 1973; Haynes and Gage, 1981). An egg parasitoid, *Anaphes flavipes* (Förster) (Hymenoptera: Mymaridae), did so as well (Maltby et al., 1971).

Widespread establishment was obtained through redistribution from field nursery sites as a practical alternative to laboratory mass rearing (Haynes and Gage, 1981), an approach still practiced widely in CLB biocontrol efforts (Evans et al., 2006; Roberts, 2016; Pickett et al., 2021). Although Haynes and Gage (1981) initially questioned the extent of impact and success of this biocontrol program, it has been credited with significantly reducing CLB populations and damage in much of eastern North America (Harcourt et al., 1984; Ellis et al., 1988; Kher et al., 2013a).

Most research recently concerning classical biological control of CLB has focused on the larval endoparasitoid, *T. julis*. This species is the most widespread, numerous, and effective across North America of the various parasitoids released (Evans et al., 2006; Kher et al., 2011). When CLB spread to western North America beginning in the 1980s, individuals of *T. julis* were collected for release from previously established, eastern North American populations. Individuals of the egg parasitoid (*A. flavipes*) were released also from eastern North American source populations, but this species failed to establish itself in the west, perhaps because of low ambient humidity where releases were made (Evans et al., 2006; Roberts, 2016).

In contrast to *A. flavipes*, *T. julis* established strongly and spread rapidly in western North America, with high rates of parasitism that now frequently exceed 50% and can even approach 100% at various times and places (Evans et al., 2006; Kher et al., 2014a; Roberts, 2016; Kheirodin et al., 2020a). Most recently, Pickett et al. (2021) reported that newly establishing CLB populations in commercial grain fields of northern California declined by nearly 90% from 2016 to 2019, as parasitism by *T. julis* climbed to nearly 80% overall following releases of the wasp and its likely immigration also from neighboring Oregon.

The success of *T. julis* in western North America mirrors earlier success with this parasitoid in north-central North America (Logan et al., 1976; Harcourt et al., 1984; Ellis et al., 1988). It is striking, in contrast, that *T. julis* is largely absent from attacking CLB in the southeastern United States (Philips et al., 2011). Climatic factors seem unlikely to account for this absence, as *T. julis* was collected widely throughout southern as well as northern Europe for initial release in North America (Dysart et al., 1973). A more likely explanation may be widespread, prophylactic application of insecticides for CLB in southeastern states (Philips et al., 2011).

Tillage practices may also influence the establishment success or failure of *T. julis*. No-till practices in Ontario were associated with a much more rapid increase in *T. julis* numbers and impact following initial release than had been recorded previously in Michigan (Harcourt et al., 1984; Ellis et al., 1988). Annual plowing or disking, as occurred at the time in Michigan, is well-recognized as highly detrimental to the overwintering, soil-dwelling immature stages of the parasitoid (Haynes et al., 1973; Leibee and Horn, 1979).

Several attributes of *T. julis* promote its suitability and effectiveness as a biocontrol agent. Hervet et al. (2016) conducted a thorough literature review of the parasitoid's documented host range, and extensive laboratory and field testing with selected non-target species of Chrysomelidae. Their studies reinforce that *T. julis* is highly host-specific; it has been documented only to parasitize CLB and its close, Palearctic congener, *O. gallaeciana*. The parasitoid is gregarious, with high reproductive potential, as confirmed recently in western North America by both Evans (2021) and Pickett et al. (2021), who report not only high percentages of CLB larvae parasitized but also high numbers of larval parasitoids (up to 28 and 21 in California and Utah, respectively) developing within individual host larvae.

As a specialist co-evolved with a very mobile host, *T. julis* also has strong dispersal capacity to track CLB populations across agricultural landscapes. When *T. julis* was first introduced to eastern North America, it was found to disperse readily between cereal fields (Gage and Haynes, 1975; Harcourt et al., 1977; Ellis et al., 1988). Similar results have been obtained in western North America, wherein adults of the parasitoid disperse hundreds of meters to new grain fields upon emerging from the soil of a given field planted with grains in the previous growing season (Evans et al., 2015; Evans, 2018).

The strong dispersal capacity of *T. julis* appears largely to prevent CLB, when it invades new stands of cereals, from escaping in space from its parasitoid even temporarily (Evans et al., 2015; Evans, 2018). This rules out, for this host-parasitoid interaction, a colonization lag as is often hypothesized to limit biological control of pests of annual or otherwise frequently disturbed crops (Kogan et al., 1999; van Driesche et al., 2008; Kaplan, 2012).

CLB can escape from parasitism over time to some degree, as its life cycle timing as a univoltine pest differs from that of the partially bivoltine *T. julis*. The parasitoid has evolved to emerge as an adult in the spring from the soil, where it has overwintered in larval diapause, to attack CLB larvae as they first begin to hatch from eggs (Gage and Haynes, 1975). Following maturation within actively feeding CLB larvae, the parasitoid larvae complete their development by killing the host larva once it has formed a pupal cocoon in the soil (Dysart et al., 1973).

A partial second generation of adult wasps emerges subsequently from the host's cocoon in the same growing season to attack host larvae hatching late in the season. The percentage of individuals forgoing larval diapause, and instead emerging as the second generation of adults in a single summer, declines as the season advances (Dysart et al., 1973; Gage and Haynes, 1975; Evans, 2021). This percentage increases, on the other hand, with the number of conspecifics sharing a host, both early and late in the season (likely reflecting nutritional stress; Evans, 2021). The net result for biological control of CLB, as occurs in western North America, is that rates of parasitism are high early and late in the season, but the rate of parasitism often drops to low levels in mid-season (Evans, 2021; Evans et al., 2006; Pickett et al., 2021).

Thus, precisely because the parasitoid has evolved as a partially bivoltine specialist attacking the univoltine CLB, a limit is placed on the biocontrol capacity of the parasitoid, wherein some host individuals every year will escape in time. Interestingly, the percentage of hosts escaping in this way appears lower in Washington (Roberts, 2016) and California (Pickett et al., 2021) than in Utah (Evans et al., 2006, 2013). It is possible that higher ratios of parasitoids to hosts occur with the relatively low-density infestations of CLB found in Washington and California compared to Utah.

Especially at high densities of CLB, the ability of *T. julis* to fully parasitize the host population may be limited by insufficient sources of carbohydrates in cereal fields (e.g. nectar or aphid honeydew) as nourishment for *T. julis* adults, resulting in foreshortened parasitoid longevity (Evans et al., 2010). Additionally, warm springs (e.g. as tied to climate change) can accentuate the mismatch in host and parasitoid phenology, resulting in reduced levels of parasitism (Evans et al., 2013).

In their landscape study of CLB and *T. julis* abundance, Kheirodin et al. (2020a) examined the association between CLB density and *T. julis* parasitism among cereal fields. These authors found, for their 2 years of study combined, that the rate of parasitism by *T. julis* was relatively weakly, negatively associated with CLB abundance among individual fields. In addition, however, Kheirodin and colleagues reported a positive association for the year in particular of lowest CLB abundance (2015), a year in which the rate of parasitism among individual fields was also positively associated at multiple spatial scales with coverage of major host crops.

Kheirodin et al. (2020a) found local parasitism to be unrelated to surrounding landscape coverage of woodlands, grasslands, and pasture but found an intriguing positive association between parasitism by *T. julis* in a given wheat field and the coverage of alfalfa and canola crops in the surrounding landscape. This suggests that adult *T. julis* may avail themselves of carbohydrate resources in these adjacent crops. The degree to which this specialist parasitoid forages in these non-CLB host crops is unknown, however, and the available evidence suggests that locally available carbohydrate resources within cereal fields themselves may be the major influence on the rate of parasitism (Evans et al., 2010).

On a smaller scale, Kher et al. (2014a) found spatial density-dependence in parasitism rate among sparse, patchily distributed, local CLB populations within individual wheat fields of western Canada. This pattern may reflect in part that adult females of *T. julis* locate hosts over short distances by responding to olfactory cues that emanate both from CLB larvae and from their fecal shields (Kher et al., 2017).

Further study will be helpful, in adding to the results of Kher et al. (2014a) and Kheirodin et al. (2020a), to illuminate the potentially variable nature of spatial density-dependence of CLB parasitism by *T. julis* as it may differ with overall levels of host and parasitoid abundances, and with spatial scale. As for temporal density-dependence among years, Evans et al. (2013) found the rate of parasitism by *T. julis* in northern Utah to be unrelated to regional CLB population size (i.e. as averaged over multiple fields) in either the current or previous year, suggesting that parasitism provides variable, density-independent suppression of CLB numbers across years.

3.2 Conservation biological control

Indigenous predators and parasitoids of CLB, in both its native and new range, potentially can contribute significantly to the pest's management. Dysart et al. (1973) reported widespread larval parasitism throughout Europe, sometimes exceeding 50%. Noordijk et al. (2016) reported a varied set of European species of Ichneumonid and Chalcidoid parasitoids with collective potential to provide natural control of CLB in the Netherlands (see also Jeloková and Gallo [2008] for parasitoids of *O. gallaeciana* in Slovakia); applications of broad-spectrum insecticides, however, currently undercut such potential.

Further study of host-parasitoid interactions will benefit both conservation and classical biological control of CLB. For example, a recent study reveals that eggs of some individuals of *O. gallaeciana* in the Czech Republic have a dark sticky covering that defends them from the egg parasitoid *A. flavipes* (Samková et al., 2020).

Many potential CLB predators forage in cereal fields. Aphid-eating lady beetles (Coccinellidae) were recognized early on in eastern North America as potentially important predators also of CLB (Shade et al., 1970; Dysart et al., 1973). Kheirodin et al. (2019, 2020b) recently assessed the consumption of

CLB eggs and larvae by lady beetles and other generalist predators in western Canada.

In Petri dish no-choice tests, Kheirodin et al. (2019) found high rates of CLB egg and larval consumption by various species of Coccinellidae, Staphylinidae, and Carabidae (damsel bugs [Nabidae] and wolf spiders also readily ate CLB larvae but not eggs). Lady beetles also consumed many CLB eggs when foraging on potted wheat plants infested as well with cereal aphids. Few eggs were consumed by rove and ground beetles on potted plants, however, which may reflect that they tended not to climb up on plants versus forage on the ground (Kheirodin et al., 2019).

PCR confirmed that generalist predators ate CLB in wheat fields; overall proportions of individual predators collected from wheat fields that tested positive for CLB consumption were 0.24 for damsel bugs, 0. 14 for lady beetles, and 0.08 for wolf spiders (Kheirodin et al., 2020b). In addition, an average of 24.5% of sentinel CLB eggs disappeared (presumably as consumed by predators) within 24 h after being placed in wheat fields in late May (Kheirodin et al., 2019).

Collectively, the results of Kheirodin et al. (2019, 2020b) illustrate the potential contribution that generalist predators may make to biological control of CLB. The importance of naturally occurring predators in reducing CLB larval numbers on infested host plants was demonstrated in a recent large-scale field exclusion experiment by Lajos et al. (2020) in the lowlands of Hungary.

Lajos et al. (2020) compared the damage to flag leaves on caged versus open plants (i.e. with or without predators excluded), after initially infesting the plants with CLB eggs, at distances up to 75 m into commercial wheat fields. CLB feeding damage was reduced significantly on open versus caged plants, equally at all studied distances into a field. The magnitude of benefit from predation for a given field (i.e. the degree to which feeding damage was reduced on open versus caged plants) was positively associated with the amount of linear woody habitat (such as shelterbelts and hedgerows) occurring close to the field, particularly within 150–200 m, suggesting a spatially extensive, large-scale movement (spillover) of natural enemies from the linear woody habitats into the wheat fields.

In contrast, increasing landscape coverage of nonlinear woody habitats (consisting of plantations in the Hungarian lowlands), especially occurring at the scale of within 200–500 m from a study field, reduced the predation benefit, presumably by diminishing the rate of predator immigration into CLB-infested fields. As Lajos et al. (2020) note, the net benefit of linear woody habitat for CLB pest management in a given cereal-growing region may depend on the balancing of increased predation pressure on the one hand, against increased availability of woodlands as overwintering habitat for the pest on the other hand (e.g. see Kheirodin et al., 2020a).

As illustrated by Lajos et al. (2020), adjacent habitat that provides key resources (e.g. shelter, alternative prey, nectar, and pollen) can bolster populations of pest natural enemies, with spillover benefits for pest suppression in crops. Recent studies have evaluated whether strips of wildflower plantings on wheat field edges can serve in this fashion. Tschumi et al. (2015) recorded many more adults of diverse predators (ground beetles, predatory bugs, lacewings, and lady beetles) in 3-m-wide strips planted with annual wildflowers than in 3-m-wide control (wheat) strips bordering wheat fields on the Swiss plateau. These adult predators (and lady beetle and lacewing larvae) tended also to be more numerous (although in most cases not significantly so), while CLB densities (of both larvae and newly matured adults thereafter) and feeding damage were strongly reduced, up to 20 m into the wheat fields from wildflower versus control strips.

In a companion study, Tschumi et al. (2016) again found strong reductions in CLB egg and larval numbers, and also in CLB crop damage, with crop yield increasing by 10%, up to 10 m into wheat fields bordered by 20-m-wide strips sown with perennial wildflowers versus in control (wheat) strips. The importance of the wildflower strips in enhancing natural enemy numbers and predation was not studied directly but rather was inferred from the results of Tschumi et al. (2015). The authors noted that at least some of the increase in wheat yield could have resulted from wildflower-enhanced natural enemy suppression of other pests as well, such as cereal aphids, which were not included as part of the study.

In a subsequent experimental field study of wildflower strips, Pollier et al. (2019) tested for depressed abundance of both CLB larvae and aphids, from both enhanced predation and parasitism, in winter wheat fields of western France. Results were suggestive of enhanced aphid predation and parasitism in wheat close to (5 m from) wildflower strips, but CLB larval abundance was not reduced by the presence of wildflower strips (Pollier et al., 2019).

Schulz-Kesting et al. (2021) examined whether the presence of adjacent fields of faba bean (*Vicia faba* L.) strengthened the biological control of pests in German wheat fields. Although the abundance of predators in wheat fields was elevated when an adjacent field was planted to faba bean (rather than also to wheat), the abundance of aphids and CLB (*Oulema* spp.) in these wheat fields, CLB damage to leaf flags, and wheat yield were not affected. Overall, CLB abundance and damage were low in the wheat fields during the study.

3.3 Augmentative biological control

Strong initial establishment of introduced CLB parasitoids negated need and interest concerning additional, augmentative releases. Another parasitic group, entomopathogenic nematodes (Steinernematidae and Heterorhabditidae), offers promise for augmentative biocontrol of CLB (Laznik et al., 2010, 2012). Laznik et al. (2012) explored the potential of applying a naturally occurring and commercially available strain of the nematode *Steinernema carpocapsae* in Slovenia to suppress field populations of CLB.

Laznik et al. (2012) applied foliar sprays of these nematodes to field plots of winter wheat in late May and early June, when most CLB eggs had hatched, and when ambient humidity and temperatures were favorable for the nematode to penetrate the insect host. Grain yields in treated plots exceeded yields in adjacent untreated, control plots, with low and high nematode concentrations producing similar gains in yield (thereby raising the prospect that even lower concentrations, at even less cost of application, might make such augmentation economically competitive). The study provides tantalizing proof of principle that under suitable environmental conditions (e.g. see Shapiro-Ilan et al., 2006), foliar applications of entomopathogenic nematodes can play a role in IPM for CLB.

4 Plant-insect interactions

4.1 Host plant resistance

Early in the North American effort to reduce losses from CLB, emphasis was placed on host plant resistance, as expressed particularly by wheat and especially through oviposition deterrence with increased leaf pubescence (Haynes and Gage, 1981; see also Konyspaevna, 2012). Resistant varieties, however, did not yield well in comparison with other wheat varieties, thereby undercutting their potential as a stand-alone, long-term solution for managing CLB (Philips et al., 2011).

Most recently, Kher et al. (2013b, 2014b) evaluated six wheat genotypes of central Asian origin for antixenosis and antibiosis against CLB, as compared with a local, susceptible (control) cultivar of wheat planted widely in western Canada. In both choice and no-choice laboratory tests for nonpreference, significant differences among the Asian genotypes were recorded in both attractiveness for CLB oviposition and extent of subsequent larval feeding damage (Kher et al., 2014b).

Similarly, significant differences among these wheat genotypes were recorded in suitability for CLB larval rate of development and survival, and resultant adult weight, although not for egg viability and rate of hatching (Kher et al., 2013b). In particular, several of the Asian genotypes were markedly less susceptible to, and suitable for, CLB than was the local Canadian cultivar of wheat. The results of Konyspaevna (2012) and Kher et al. (2013b, 2014b) highlight the potential that lies in further development and deployment of host plant resistance as an IPM tactic for CLB (Kher et al., 2011).

The genetic basis for wheat resistance to CLB was examined by Würschum et al. (2020a), who grew 876 cultivars in a common garden in Germany, with naturally occurring pest infestation. CLB larval damage to flag leaves was more severe in earlier heading cultivars, as under strong genetic control. A major quantitative trait locus (QTL) accounted for more than a third of the genotypic variance in cultivar resistance to CLB, likely acting through control of the timing of heading. But underscoring the complex genetics, many additional QTLs, each with a small effect, appear also to influence host resistance to damage from CLB. This reduces the potential of resistance selection by QTL marker assistance, versus by more traditional means of phenotypic selection (Würschum et al., 2020a).

Würschum et al. (2020a) did not assess wheat leaf pubescence (it was difficult to do so in their experimental design) but found instead that cultivars with more glaucous leaves were less heavily attacked by CLB larvae. The quality of being glaucous (i.e. silvery blue or gray), versus glossy green, depends on the scattering of light by wax crystals in the wheat leaf epicuticle and is also under strong genetic control (Würschum et al., 2020b). Given the exploratory nature of their study, Würschum et al. (2020a) are cautious in interpreting the importance of glaucous leaves for CLB resistance but suggest that the possibility merits further investigation.

Meissle et al. (2012) examined resistance to CLB of corn (maize) genetically engineered to express the gene Cry3Bb1 from Bt (*Bacillus thuringiensis*). This Bt maize was developed originally to reduce damage from corn rootworms (*Diabrotica* spp., Chrysomelidae). Survival of young larvae, but not of older instars or adults, was reduced dramatically when CLB fed on the Bt maize. Although corn is a minor host plant for CLB, the study illustrates the potential for reduced CLB damage in genetically modified cereals.

4.2 Host volatile organic compounds

One aspect of host plant resistance, researched widely and intensively among crop plants in recent decades, is plant production of volatile organic compounds (VOCs) in response to damage from insects and pathogens, and insect responses to such production. It is now well-recognized that VOCs may repel or otherwise deter many insect herbivores (i.e. they may serve as induced plant defenses), or in other cases, they may attract insect herbivores (Dicke and Baldwin, 2010).

In recent years, this general topic has been addressed for CLB and cereals. CLB adults, particularly reproductively active females, orient positively to plant volatiles from intact (undamaged) wheat, barley, and oat plants (Kher et al., 2017). Attention has focused also on plant volatiles produced by damaged cereal plants, and to CLB responses to these volatiles.

Piesik et al. (2010a,b, 2011a) found that VOCs (both green leafy volatiles [GLVs] and terpenes) were released in greater quantities by cultivars of wheat, oats, and barley, with CLB adult feeding damage, than with mechanical damage or without damage (control plants). Ratios of VOCs in blends released by the host plant differed in response to damage inflicted by *O. melanopus* versus *O. cyanella* (Piesik et al., 2011a).

The cereals also produced VOCs when infected with *Fusarium* spp. (Piesik et al., 2010b, 2011a, 2013); these fungal-induced blends differed in overall composition from those produced in response to CLB damage (Piesik et al., 2011a). In addition, Delaney et al. (2013) induced wheat and barley plants to produce eight VOCs at very high rates by exposing the plants to doses of synthetic *cis*-jasmone for periods of up to 6 h.

Wheat plants with CLB or mechanical damage, induced neighboring, uninjured plants of all three cereals (wheat, oats, and barley), placed up to 3 m away, to produce more VOCs than they did when placed with control plants (Piesik et al., 2010b). This was found also for individuals of these cereals, as well as corn, infected with *Fusarium* spp. and nearby, uninfected conspecifics (Piesik et al., 2011a,b, 2013).

In laboratory testing with Y-tube olfactometers, adult CLB were responsive to VOCs produced by cereals with CLB damage or *Fusarium* infection, but effects (and in particular, attraction versus repellency) varied with VOC concentration. Variability in responses among trials also complicates biological interpretation. For example, CLB adults were attracted to the GLVs, (Z)-3-hexenal and (Z)-3-hexenyl acetate, at low concentrations and repelled at high concentrations (125 ng/min = 7500 ng/h), as found by Piesik et al. (2010a, 2013) and Delaney et al. (2013), but were attracted also at high concentration in other testings (Piesik et al., 2011b).

These intriguing results call for further investigation as to the implications for CLB–host plant interactions that may be associated with host plant production of, and CLB response to, VOCs. This will be especially of interest as it applies to stands of cereals naturally sustaining CLB feeding damage and/or *Fusarium* infection, or even perhaps as treated with synthetic *cis*-jasmone in an effort to induce host plant resistance.

The possible role of host plant VOCs in attracting natural enemies (parasitoids and predators) as an induced defense against insect herbivores also has generated much interest and study (Turlings and Erb, 2018). This topic is yet to be explored for CLB and cereals.

4.3 Additional induced host responses to CLB damage

Another form of induced defense may lie in host plant production of proteinase inhibitors (PIs) that bind and interfere with digestive proteases in the insect gut. Wielkopolan et al. (2018) have investigated gene expression of serine and cysteine PIs in three cultivars of wheat in response to CLB larval feeding damage. The gene *SerPI* was up-regulated following CLB damage, both at the site of insect attack (local response) and even more so in intact leaves elsewhere on the host plant (systemic response). In contrast, decreased expression of the gene *CysPI* was recorded in response to CLB larval feeding damage.

Micro-organisms occurring naturally in the CLB digestive tract and saliva appear to weaken or suppress the *SerPI* response to insect attack (Wielkopolan et al., 2018). *SerPI* was most strongly and rapidly up-regulated when these micro-organisms were eliminated through the administration of antibiotics to CLB larvae. Hence, these micro-organisms (possibly species of *Wolbachia* or *Rickettsia*) in the insect gut and saliva may help their host CLB larvae in countering the wheat plants' PI-induced defense. Differences among wheat cultivars in *SerPI* response to experimental treatments (including damage from CLB larvae treated with and without antibiotics) may hold promise for selection for enhanced host plant resistance.

4.4 Endophytes

Other micro-organisms conversely may aid host grasses in defending against CLB. Clement and Elberson (2009) and Clement et al. (2010) investigated the potential effects of fungal endophytic (*Neotyphodium* sp. Glenn, Bacon, and Hanlin) infection of wild grasses on CLB adult feeding preference, and larval development and survival. In feeding readily on two accessions of wild tall fescue (*Lolium arundinaceum* [Schreb.] S.J. Dabyshire), CLB adults consumed significantly more tissue from uninfected versus infected plants of one of these accessions (Clement and Elberson, 2009).

In addition, CLB larvae survived much less well on one accession when the host plant was infected versus uninfected and failed to survive altogether on the other accession when the host plant was infected. Interestingly, however, rates of development were indistinguishable between surviving larvae that fed on infected versus uninfected hosts, and these rates were only slightly slower than the rate of development when barley served as the host (Clement and Elberson, 2009).

In contrast, CLB adults did not show feeding preference for uninfected over infected host plants of two accessions of Alpine Timothy (*Phleum alpinum* L.). Furthermore, CLB larvae that fed on uninfected versus infected host plants of either accession of this host grass did not differ in rates of either survival or development (Clement and Elberson, 2009; Clement et al., 2010).

The differing responses of CLB to infected host plants among these accessions likely reflect differences in the types and concentrations of metabolites (especially alkaloids) produced by the fungal endophytes in the different grasses (Clement et al., 2010). The results of these studies call for further investigation to explore the potential for host plant resistance of cultivated grasses to damage from CLB, as associated with fungal endophytic infection.

5 Thresholds and other components of integrated pest management

5.1 Economic injury level and economic threshold

Implementation of a successful IPM program depends on a sound estimate of the economic injury level (EIL) and the associated economic threshold (ET) at which, when necessary, insecticide application is triggered (Pedigo et al., 2021). The EIL for CLB on winter wheat in Switzerland recently has been re-assessed experimentally by Steinger et al. (2020). Previously, the EIL used commonly in Switzerland had been set at one CLB larva per stem, as has been practiced during heading throughout much of North America (i.e. one larva per flag leaf; Webster and Smith, 1983). Steinger et al. (2020), however, conclude from their field studies that the EIL for CLB in fact may be considerably higher in Switzerland, occurring at close to three larvae per flag leaf at current costs of pest control and values of the crop.

This estimate for the EIL is based on results of variable artificial defoliation (0–60% removal of flag leaf tissue) on crop yield from trials in 11 location years, spanning the years 2014–2017 at multiple locations (Steinger et al., 2020). In these trials, grain yield declined linearly by 1.14% for each 10% increase in defoliation. The authors derive the corresponding EIL using the widely accepted estimate that a single CLB larva will inflict 10% defoliation to a flag leaf. The authors found similar results of varying defoliation on yield when the damage was caused by CLB larvae rather than manually.

While the extensive trials of Steinger et al. (2020) include a wide spectrum of abiotic conditions, the authors stress that their EIL estimates remain to be tested under more extreme conditions, including both very dry and wet years (with the latter possibly promoting plant diseases that could interact with CLB damage to reduce yield even further). Intriguingly, the authors also report substantial variability among cultivars when flag leaves of eight varieties of wheat were defoliated manually by 40%, with three cultivars in particular showing complete tolerance. These results highlight the need for further investigation of the application of EIL under different growing conditions and among cultivars.

The impact of defoliation on grain yield depends on when in the plant's growth cycle the injury occurs, with greatest negative impact on yield occurring in the three weeks prior to anthesis (Buntin et al., 2004; Steinger et al., 2020).

Steinger and colleagues manually defoliated flag leaves 7-10 days after anthesis, rather than before, in seeking to mimic the timing of peak feeding damage of CLB fourth instars under typical field conditions as occurred at their study sites. Overall, the results of this study suggest that the EIL for the CLB, as widely adopted currently, may be too low (i.e. it often may encourage unnecessary insecticide application).

Webster and Smith (1983) similarly had suggested earlier that winter wheat, as grown in North America, often can tolerate damage greater than as inflicted by one CLB larva per flag leaf during heading. In contrast, however, studies for the southeastern United States have found just the opposite. In particular, Ihrig et al. (2001) and Buntin et al. (2004) reported losses in yield exceeding 10% per CLB larva per stem. Buntin et al. (2004) hence proposed a conservative EIL of 0.4 larvae per stem during the period between spike emergence and anthesis, reflecting that peak CLB egg hatch often occurs at spike emergence, in the southeastern United States. An even more conservative ET of 0.25 eggs and/or small larvae per stem has been chosen for IPM programs in Virginia and North Carolina (Philips et al., 2011).

Taken together, these various studies underscore the importance of understanding the degree of synchrony, and the factors influencing such degree, in host plant and pest population development in cereal crop fields over the growing season. This point is illustrated strikingly by the results of Würschum et al. (2020a), in which a strong correlation emerged among hundreds of wheat cultivars grown in a common garden, between timing of heading and severity of CLB damage to flag leaves. A recent study by Bechini et al. (2013) also is highly instructive. These authors report varying degree of synchrony in winter wheat and CLB population development across a large-scale transect stretching from Switzerland through northern Italy. It is thus important to recognize that the EIL for the CLB as a pest of any given crop of small grains may vary considerably among geographic regions.

5.2 Insecticidal treatments and integrated pest management

Insecticides continue to play a major role in controlling CLB and are often applied as prophylactic treatments regardless of the pest's population levels (Philips et al., 2011; Reisig et al., 2012). Insecticidal applications for CLB ideally are best used sparingly, however, as integral elements of IPM programs, only as called for when CLB outbreaks threaten economic damage, and only when the majority of eggs have hatched (Buntin et al., 2004). Initial efforts to control CLB when it first established in North America involved widespread, intensive insecticide treatments with the organophosphate malathion and the carbamate carbaryl (Haynes and Gage, 1981). In recent years, these insecticides increasingly have been replaced by pyrethroids (Reisig et al., 2012; Chen et al., 2015).

Wenda-Piesik et al. (2018) recently investigated the effectiveness of insecticide application (including cost-effectiveness) in controlling high populations of CLB (and *O. gallaeciana*) in Poland. Wenda-Piesik and colleagues applied a seed treatment of imidacloprid (a neonicotinoid) and/or a foliar insecticide application (a blend of the carbamate primicarb, the organophosphate dimethoate, and the pyrethroid cypermethrin) to wheat planted in early or late autumn, or spring. For each planting date, numbers of CLB adults and larvae, and the amount of feeding damage to host leaves, were reduced markedly in each of 3 years (2011–2013), by either seed treatment or foliar application with insecticides (and especially when both seed and foliar treatments were applied). Nonetheless, only in spring plantings (with greater control-plot feeding damage from CLB than in autumn plantings) were insecticidal treatments cost-effective (i.e. wherein the gain in yield more than covered the cost of application).

In exploring alternatives to synthetic insecticides for controlling CLB with foliar sprays, Skuhrovec et al. (2018) investigated insecticidal properties of essential oils extracted from several common herbs. Initial results indicate great potential, as high mortalities were recorded for both CLB adults and larvae following 24 h of tarsal contact upon being placed in the laboratory on moist filter sprayed with the essential oils.

In a comprehensive analysis incorporating both cost estimates and field data for 2010–2011 in the southeastern United States, Reisig et al. (2012) addressed the merits of an IPM approach versus prophylactic insecticide application for CLB to wheat early in the spring (coinciding with nitrogen application), as practiced by many growers in the region. The IPM approach incorporates weekly scouting and insecticide application later in the growing season, only if CLB larval densities exceed the ET (set at 0.25 eggs and/or small larvae per stem; Philips et al., 2011). While IPM incurs scouting costs and necessitates additional application costs (i.e. beyond those for nitrogen application), the prophylactic approach risks insecticide application either too early in the seasonal development of CLB larval populations in any given year or when the ET would not have been reached in any case.

The analysis indicated that prophylactic insecticide application on average was best financially for an individual farmer but could result in substantial financial penalty relative to IPM when CLB populations were high. Beyond these economic considerations, Reisig et al. (2012) stressed the environmental benefits of IPM, including preservation of natural enemies for biological control of CLB and other cereal pests. Chen et al. (2015) reported reductions in numbers of natural enemies in some cases following prophylactic insecticide applications in Maryland during heading to flowering of winter wheat. These late-season prophylactic sprays did not result in increased grain yield when pest pressure, including from CLB, was low.

Reisig et al. (2012) emphasized the importance of further research to address reducing monetary costs of IPM, as achieved especially through increased ability to predict when and where sporadic outbreaks of CLB are most likely to occur. One important avenue for such research lies in further development of degree-day modeling, as pursued for CLB in the southeastern United States, in particular by Philips et al. (2012), and as discussed next for CLB in general.

5.3 Degree-day modeling

Degree-day modeling of CLB appearance and population development in small grain cereals holds promise for managing the pest. Field census at predicted peak egg density, as guided by degree-day modeling, can be used to anticipate the timing and intensity of subsequent peak larval density, which in turn is positively linked with plant damage and loss in crop yield (Ihrig et al., 2001; Philips et al., 2012; Evans et al., 2014).

Van de Vijver et al. (2018) reviewed and applied previously published, simple growing degree-day (GDD) models, as based on field data from diverse locations in North America and Europe, to predict dates of first and peak appearance of CLB eggs and larvae in Belgian winter wheat fields. Predictions were derived directly from the individual models. Each of these models accumulates GDD with a start date of January 1, but they vary in their estimates of both base temperature and GDD thresholds (i.e. expected GDD accumulations for marker events in CLB field population development, such as peak egg density).

Predictions were derived also from a combined model, drawn from the individual models, with base temperature set at 9°C and threshold GDDs set at 80°C (first egg), 150°C (peak egg), 180°C (first larva), and 360°C (peak larvae). Finally, predictions were derived for optimal models for each marker event, in which the base temperature was selected (from a range of 0–10°C) to fit best the field data. For all of the models, predictions were compared with field estimates as determined by weekly sampling. The previously published models, and the combined and optimal models, were inadequate to predict with sufficient accuracy the calendar timing of seasonal development of CLB populations in the Belgian wheat fields.

These results of Van de Vijver et al. (2018) reflect that simple GDD models, by their very nature, often are limited in their predictive power for use in IPM. Limited predictive power of such models can arise in part from their use of an arbitrary start date for GDD accumulation (most often chosen as January 1). Evans et al. (2014) addressed this issue by incorporating estimates of spring heat accumulation to improve substantially GDD predictions over those based on simple accumulation from January 1, for peak densities of CLB eggs and larvae in any given year in Utah (western North America).

A simpler approach for insects such as CLB that overwinter as adults may be to select a start date for GDD accumulation as based on photoperiod. A critical spring photoperiod of sufficient daylength may serve well as a biofix to mark the cessation of reproductive diapause, for example, and thereby trigger GDD accumulation to predict seasonal timing of subsequent oviposition and population development (e.g. Nielsen et al., 2016). However, the interaction of temperature with photoperiod in influencing diapause (e.g. Bale et al., 2002; Danks, 2007), and possible evolution of local pest populations, as associated with such interaction over time (e.g. Bean et al., 2012), remains to be explored for further development of CLB phenological models.

Van de Vijver et al. (2018) applied even more sophisticated statistical approaches for GDD accumulation. These included using rainfall and relative humidity with GDD as independent variables, and application of diverse regression approaches (multiple linear, ridge, and tree) and moving time windows of specified duration that were characterized by summary environmental variables. The resultant models considerably improved predictions over those from simple GDD accumulation, but still failed to be sufficiently accurate to serve as forecast tools for CLB IPM in Belgian wheat fields. Nonetheless, Van de Vijver et al. (2018) illustrate how to approach analytically the complexities of environmental variation influencing CLB population development, in continued development of modeling for forecast purposes.

6 Future trends in research

The numerous studies reviewed here attest to much current interest and concern regarding CLB. These studies encompass multiple creative approaches toward control. CLB is among many insect pests in reaching economic levels more infrequently than not, but doing so in unpredictable and sporadic fashion. Such a pest poses many challenges; the most pressing need going forward is further development of IPM.

Under continual threat of possible damage, growers naturally turn to insecticides for CLB. Prophylactic spraying is an easy solution, despite being often unnecessary. Research is needed to strengthen IPM as the preferred alternative, including financially (e.g. Philips et al., 2011; Riesig et al., 2012; Wenda-Piesik et al., 2018), especially through further refinement of degree-day modeling (e.g. Van de Vijver et al., 2018), and EILs and ETs (e.g. Steinger et al., 2020), as specially developed and suited for individual geographic regions and growing conditions.

Opportunities lie in research to illuminate further when, where, and why CLB populations become high, at spatial scales both regional (e.g. Riesig et al, 2017) and more local (e.g. Kher et al., 2014a; Reay-Jones, 2017). Landscape analyses can provide important insights to address these questions at large spatial scales (e.g. Kheirodin et al., 2020a; Lajos et al., 2020). Landscape analyses can also address further the potential as well as limitations of biological control, as can additional individual-based, population studies of CLB-natural enemy interactions (e.g. Samková et al., 2019a,b,c). Additional field studies to determine more definitively the present status across North America of the exotic parasitoid species introduced previously, are needed to assess how classical and conservation biocontrol may be best integrated in future efforts. Further studies of host plant resistance, building on the studies and topics reviewed above, also offer much opportunity for advances in CLB pest management.

Most generally, further development of IPM will need to anticipate and account for likely climatic change, including further global warming, as it may affect both CLB itself (e.g. Bean et al., 2012; Olfert and Weiss, 2006), and the interactions of CLB with other species, including importantly its natural enemies and biocontrol (e.g. Evans et al., 2013; Eigenbrode et al., 2015; Skendžic et al., 2021). Although CLB continues to pose significant economic threat in many places across the northern hemisphere, the many recent advances in research reviewed here pave the way for further progress in understanding and effective management of this pest.

7 Conclusion

CLB continues to pose a significant economic threat to crops of small grains across much of the northern hemisphere. The sporadic and unpredictable nature of CLB outbreaks makes management of this pest challenging. Recent research advances in understanding and controlling this pest in wheat and other small grains provide many prospects for further development of integrated pest management of CLB. Further research to explore these prospects more fully thus offers excellent potential to reduce the ongoing economic threat from this pest.

8 Where to look for further information

Excellent previous reviews of CLB and its control include Dysart et al. (1973), Haynes and Gage (1981), Kher et al. (2011), and Philips et al. (2011).

Current overviews and case studies of CLB pest management can be found for Europe in Van Vijver et al. (2018), Wenda-Piesik et al. (2018), and Steinger et al. (2020); and for North America in Philips et al. (2011, 2012) and Reisig et al. (2012, 2017).

For implementation of classical biological control for CLB, as most recently pursued in western North America, see Roberts (2016) and Pickett et al. (2021).

Many additional summaries from classical biocontrol efforts in the late 1900s and early 2000s are posted at the website, Cereal Leaf Beetle Research - Western North America. [Washington State University Libraries Exchange, Pullman WA] https://research.libraries.wsu.edu/xmlui/handle/2376/1928.

9 References

Bale, J. S., Masters, G. J., Hodkinson, I. D., Awmack, C., Bezemer, T. M., Brown, V. K., Butterfield, J., Buse, A., Coulson, J. C., Farrar, J., Good, J. E. G., Harrington, R., Hartley, S., Jones, T. H., Lindroth, R. L., Press, M. C., Symrnioudis, I., Watt, A. D. and Whittaker, J. B. (2002). Herbivory in global climate change research: direct effects of rising temperature on insect herbivores, *Glob. Change Biol.* 8(1), 1-16.

Bean, D. W., Dalin, P. and Dudley, T. L. (2012). Evolution of critical day length for diapause induction enables range expansion of *Diorhabda carinulata*, a biological control agent against tamarisk (*Tamarix* spp.), *Evol. Appl.* 5(5), 511-523.

Bechini, L., Morlacchi, P. and Baumgärtner, J. (2013). Coinciding development of winter wheat and leaf beetles along an Alpine transect, *Acta Oecol.* 47, 95-104.

Buntin, G. D., Flanders, K. L., Slaughter, R. W. and DeLamar, Z. D. (2004). Damage loss assessment and control of the Cereal Leaf Beetle (Coleoptera: Chrysomelidae) in winter wheat, *J. Econ. Entomol.* 97(2), 374-382.

Casagrande, R. A., Ruesink, W. G. and Haynes, D. L. (1977). The behavior and survival of adult cereal leaf beetles, *Ann. Entomol. Soc. Am.* 70(1), 19-30.

Chen, G. H., Hinds, J., Zobel, E., Rosario-Lebron, A. and Hooks, C. R. R. (2015). Evaluation of prophylactic sprays on pest abundance, foliar damage and yield in winter wheat, *Int. J. Pest Manag.* 61(2), 161-170.

Clement, S. L. and Elberson, L. R. (2009). Variable effects of *Grass-Neotyphodium* associations on cereal leaf beetle (Coleoptera: Chrysomelidae) feeding, development and survival, *J. Entomol. Sci.* 45(3), 197-203.

Clement, S. L., Hu, J. G., Stewart, A. V., Wang, B. and Elberson, L. R. (2010). Detrimental and neutral effects of a wild grass-fungal endophyte symbiotum on insect preference and performance, *J. Insect Sci.* 11, article 77.

Danks, H. V. (2007). The elements of seasonal adaptations in insects, *Can. Entomol.* 139(1), 1-44.

Delaney, K. J., Wawrzyniak, M., Lemańczyk, G., Wrzesińska, D. and Piesik, D. (2013). Synthetic cis-jasmone exposure induces wheat and barley volatiles that repel the pest cereal leaf beetle, *Oulema melanopus* L., *J. Chem. Ecol.* 39(5), 620-629.

Dicke, M. and Baldwin, I. T. (2010). The evolutionary context for herbivore-induced plant volatiles: beyond the 'cry for help', *Trends Plant Sci.* 15(3), 167-175.

Dosdall, L. M., Cárcamo, H., Olfert, O., Meers, S., Hartley, S. and Gavloski, J. (2011). Insect invasions of agroecosystems in the Western Canadian prairies: case histories, patterns, and implications for ecosystem function, *Biol. Invas.* 13(5), 1135-1149.

Dowell, R. V. and Pickett, C. (2016). The cereal leaf beetle, *Oulema melanopus* Linnaeus, 1758 (Coleoptera: Chrysomelidae) in California: detection and response, *Pan-Pacific Entomol.* 92(3), 165-167.

Dysart, R. J., Maltby, H. L. and Brunson, M. H. (1973). Larval parasites of *Oulema melanopus* in Europe and their colonization in the United States, *Entomophaga* 18(2), 133-167.

Eigenbrode, S. D., Davis, T. S. and Crowder, D. W. (2015). Climate change and biological control in agricultural systems: principles and examples from North America. In: Björkman, C. and Niemelä, P. (Eds.), *Climate Change and Insect Pests*, CABI Publishing International, Wallingford, UK, pp. 119-135.

Ellis, C. R., Kormos, B. and Guppy, J. C. (1988). Absence of parasitism in an outbreak of the cereal leaf beetle, *Oulema melanopus* (Coleoptera: Chrysomelidae), in the central tobacco growing area of Ontario, *Proc. Entomol. Soc. Ont.* 119, 43-46.

Evans, E. W. (2018). Dispersal in host-parasitoid interactions: crop colonization by pests and specialist enemies, *Insects* 9(4), 1-14. DOI: 10.3390/insects9040134.

Evans, E. W. (2021). Partial bivoltinism in a gregarious endoparasitoid: larval diapause as influenced by season and sharing a host, *Entomol. Exp. Appl.* 169(1), 145-153.

Evans, E. W., Anderson, M. R. and Bowling, P. D. (2010). Targeted sugar provision promotes parasitism of the cereal leaf beetle, *Oulema melanopus*, *Agric. For. Entomol.* 12(1), 41-47.

Evans, E. W., Bolshakova, V. L. J. and Carlile, N. R. (2015). Parasitoid dispersal and colonization lag in disturbed habitats: biological control of cereal leaf beetle metapopulations, *J. Appl. Entomol.* 139(7), 529-538.

Evans, E. W., Carlile, N. R., Innes, M. B. and Pitigala, N. (2013). Warm springs reduce parasitism of the cereal leaf beetle through phenological mismatch, *J. Appl. Entomol.* 137(5), 383-391.

Evans, E. W., Carlile, N. R., Innes, M. B. and Pitigala, N. (2014). Infestation of grain fields and degree-day phenology of the cereal leaf beetle (Coleoptera: Chrysomelidae) in Utah: long-term patterns, *J. Econ. Entomol.* 107(1), 240-249.

Evans, E. W., Karren, J. B. and Israelsen, C. E. (2006). Interactions over time between cereal leaf beetle (Coleoptera: Chrysomelidae) and larval parasitoid *Tetrastichus julis* (Hymenoptera: Eulophidae) in Utah, *J. Econ. Entomol.* 99(6), 1967-1973.

Gage, S. H. and Haynes, D. L. (1975). Emergence under natural and manipulated conditions of *Tetrastichus julis*, an introduced larval parasite of the Cereal Leaf Beetle, with reference to regional population management, *Environ. Entomol.* 4(3), 425-434.

Guppy, J. C. and Harcourt, D. G. (1978). Effects of temperature on development of the immature stages of the cereal leaf beetle, *Oulema melanopus* (Coleoptera: Chrysomelidae), *Can. Entomol.* 110(3), 257-263.

Harcourt, D. G., Guppy, J. C. and Ellis, C. R. (1977). Establishment and spread of *Tetrastichus julis* (Hymenoptera: Eulophidae), a parasitoid of the cereal leaf beetle in Ontario, *Can. Entomol.* 109(3), 473-476.

Harcourt, D. G., Guppy, J. C. and Ellis, C. R. (1984). *Oulema melanopus* (L.), cereal leaf beetle (Coleoptera: Chrysomelidae). In: Kelleher, J. S. and Hulme, M. A. (Eds.), *Biological Control Programmes Against Insects and Weeds in Canada 1969-1980*, Commonwealth Agricultural Bureaux, Farnharm Royal, England, pp. 65-67.

Haynes, D. L., Brandenburg, R. K. and Fisher, P. D. (1973). Environmental monitoring network for pest management systems, *Environ. Entomol.* 2(5), 889-900.

Haynes, D. L. and Gage, S. H. (1981). The cereal leaf beetle in North America, *Annu. Rev. Entomol.* 26(1), 259-287.

Helgesen, R. G. and Haynes, D. L. (1972). Population dynamics of the cereal leaf beetle, *Oulema melanopus* (L.) (Coleoptera: Chrysomelidae): a model for age specific mortality, *Can. Entomol.* 104, 787-814.

Hervet, V. A. D., Cárcamo, H. A., Dosdall, L. M., Miller, T. D. and Kher, S. V. (2016). Assessment of potential non-target effects of *Tetrastichus julis*, a biological control agent of the cereal leaf beetle, *Oulema melanopus*, *BioControl* 61(4), 399–411.

Hoffman, G. D. and Rao, S. (2010). Cereal leaf beetle, *Oulema melanopus* (L.) (Coleoptera: Chrysomelidae), attraction to oat plantings of different ages, *J. Entomol. Soc. B.C.* 107, 57–66.

Hoffman, G. D. and Rao, S. (2011). Oviposition site selection on oats: the effect of plant architecture, plant and leaf age, tissue toughness, and hardness on cereal leaf beetle, *Oulema melanopus*, *Entomol. Exp. Appl.* 141(3), 232–244.

Honěk, A. (1991). Crop density and abundance of cereal leaf beetles (*Oulema* spp.) in winter wheat (Coleoptera, Chrysomelidae), *J. Plant Dis. Prot.* 98, 174–178.

Ihrig, R. A., Herbert, D. A., Van Duyn, J. W. and Bradley, J. R. (2001) Relationship between cereal leaf beetle (Coleoptera: Chrysomelidae) egg and fourth-instar populations and impact of fourth-instar defoliation of winter wheat yields in North Carolina and Virginia, *J. Econ. Entomol.* 94(3), 634–639.

Jeloková, M. and Gallo, J. (2008). Parasitoids of cereal leaf beetle, *Oulema gallaeciana* (Heyden, 1879), *Plant Protect. Sci.* 44(3), 108–113.

Kaplan, I. (2012). Attracting carnivorous arthropods with plant volatiles: the future of biocontrol or playing with fire?, *Biol. Control* 60(2), 77–89.

Karp, D. S., Chaplin-Kramer, R., Meehan, T. D., Martin, E. A., DeClerck, R., Grab, H., Gratton, C., Hunt, L., Larsen, A. E., Martínez-Salinas, A., O'Rourke. M. E. et al. (2018). Crop pests and predators exhibit inconsistent responses to surrounding landscape composition, *Proc. Natl. Acad. Sci. U. S. A.* 115(33), 7863–7870.

Kheirodin, A., Cárcamo, H. A. and Costamagna, A. C. (2020a). Contrasting effects of host crops and crop diversity on the abundance and parasitism of a specialist herbivore in agricultural landscapes, *Landsc. Ecol.* 35(5), 1073–1087.

Kheirodin, A., Sharanowski, B. J., Cárcamo, H. A. and Costamagna, A. C. (2020b). Consumption of cereal leaf beetle, *Oulema melanopus*, by generalist predators in wheat fields detected by molecular analysis, *Entomol. Exp. Appl.* 168(1), 59–69.

Kheirodin, A., Costamagna, A. C. and Cárcamo, H. A. (2019). Laboratory and field tests of predation on the cereal leaf beetle, *Oulema melanopus* (Coleoptera: Chrysomelidae), *Biocontrol Sci. Technol.* 29(5), 451–465.

Kher, S. V., Cárcamo, H. A., Evenden, M. L. and Dosdall, L. M. (2017). Olfactory host-finding behavior of *Oulema melanopus* (Coleoptera: Chrysomelidae) and its parasitoid, *Tetrastichus julis* (Hymenoptera: Eulophidae), *J. Appl. Entomol.* 141(9), 740–750.

Kher, S. V., Dosdall, L. M. and Cárcamo, H. A. (2011). The cereal leaf beetle: biology, distribution and prospects for control, *Prairie Soils Crops* 4, 32–41.

Kher, S. V., Dosdall, L. M. and Cárcamo, H. A. (2013a). *Oulema melanopus* (L.), cereal leaf beetle (Coleoptera: Chrysomelidae). In: Mason, P. G. and Gillespie, D. R. (Eds.), *Biological Control Programmes in Canada 2001–2012*, CABI Publishers, Wallingford, UK, pp. 233–237.

Kher, S. V., Dosdall, L. M., Cárcamo, H. A. and El-Bouhssini, M. (2013b). Antibiosis resistance to cereal leaf beetle, *Oulema melanopus* (L.) (Coleoptera: Chrysomelidae), in Central Asian wheat germplasm, *J. Appl. Entomol.* 137(10), 751–760.

Kher, S. V., Dosdall, L. M. and Cárcamo, H. A. (2014a). Plant vigor metrics determine spatio-temporal distribution dynamics of *Oulema melanopus* (Coleoptera: Chrysomelidae) and its larval parasitoid, *Tetrastichus julis* (Hymenoptera: Eulophidae), *Environ. Entomol.* 43(5), 1295–1308.

Kher, S. V., Dosdall, L. M., Cárcamo, H. A. and El-Bouhssini, M. (2014b). Antixenosis resistance to *Oulema melanopus* (Coleoptera: Chrysomelidae) in Central Asian wheat germplasm, *J. Econ. Entomol.* 107(1), 410–416.

Kher, S. V., Dosdall, L. M. and Cárcamo, H. A. (2016). Biology, host preferences and fitness of *Oulema melanopus* (Coleoptera: Chrysomelidae), a recent invasive pest in western Canada, *Arthropod Plant Interact.* 10(5), 365–376.

Kogan, M. D., Gerling, D. and Maddox, J. V. (1999). Enhancement of biological control in annual agricultural environments. In: Bellows, T. S. and Fisher, T. W. (Eds.), *Handbook of Biological Control: Principles and Applications of Biological Control*, Academic Press, San Diego, CA, pp. 789–818.

Konyspaevna, S. K. (2012). Spring wheat resistance against cereal leaf beetle (*Oulema melanopus* L.) in relation to leaf pubescence, *Aust. J. Basic Appl. Sci.* 6, 515–518.

Lajos, K., Császár, O., Sárospataki, M., Samu, F. and Tóth, F. (2020). Linear woody landscape elements may help to mitigate leaf surface loss caused by the cereal leaf beetle, *Landsc. Ecol.* 35(10), 2225–2238.

Laznik, Ž., Tóth, T., Lakatos, T., Vidrih, M. and Trdan, S. (2010). *Oulema melanopus* (L.) (Coleoptera: Chrysomelidae) adults are susceptible to entomopathogenic nematodes (Rhabditida) attack: results from a laboratory study, *J. Plant Dis. Prot.* 117(1), 30–32.

Laznik, Ž., Vidrih, M., Vučajnk, F. and Trdan, S. (2012). Is foliar application of entomopathogenic nematods (Rhabditida) an effective alternative to thiametoxam in controlling cereal leaf beetle (*Oulema melanopus* L.) on winter wheat?, *J. Food Agric. Environ.* 10, 716–719.

Leibee, G. L. and Horn, D. J. (1979). Influence of tillage on survivorship of cereal leaf beetle and its larval parasites, *Tetrastichus julis* and *Lemophagus curtus*, *Environ. Entomol.* 8(3), 485–486.

Leroy, J., Chapelin-Viscardi, J. D., Genson, G., Haran, J., Pierre, E. and Streito, J. C. (2020). Distinct barcodes for the cereal leaf beetles *Oulema melanopus* and *Oulema duftschmidi* (Coleoptera: Chrysomeidae), two syntopical sibling species, *Eur. J. Entomol.* 117, 490–503.

LeSage, L., Dobesberger, E. J. and Majka, C. G. (2007). Introduced leaf beetles of the Maritime Provinces, 2: the cereal leaf beetle *Oulema melanopus* (Linnaeus) (Coleoptera: Chrysomelidae), *Proc. Entomol. Soc. Wash.* 109, 286–294.

Logan, P. A., Stehr, F. W. and Sauer, R. J. (1976). The subcolonization and buildup of *Tetrastichus julis*, (Hymenoptera: Eulophidae) a larval parasitoid of the cereal leaf beetle, (Coleoptera: Chrysomelidae) in the lower peninsula of Michigan, *Gr. Lakes Entomol.* 9, 75–78.

Maltby, H. L., Stehr, F. W., Anderson, R. C., Moorehead, G. E., Barton, L. C. and Paschke, J. D. (1971). Establishment in the United States of *Anaphes flavipes*, an egg parasite of the cereal leaf beetle, *J. Econ. Entomol.* 64(3), 693–697.

McPherson, R. M. (1983). Seasonal abundance of cereal leaf beetles (Coleoptera: Chrysomelidae) in Virginia small grains and corn, *J. Econ. Entomol.* 76(6), 1269–1272.

Meissle, M., Knecht, S., Waldburger, M. and Romeis, J. (2012). Sensitivity of the cereal leaf beetle *Oulema melanopus* (Coleoptera: Chrysomelidae) to Bt maize-expressed Cry3Bb1 and Cry1Ab, *Arthropod Plant Interact.* 6(2), 203–211.

Nielsen, A. L., Chen, S. and Fleischer, S. J. (2016). Coupling developmental physiology, photoperiod, and temperature to model phenology and dynamics of an invasive heteropteran, *Halyomorpha halys*, *Front. Physiol.* 7(165), 165. DOI: 10.3389/fphys.2016.00165.

Noordijk, J., Ulenberg, S., Zwakhals, C. J., Heijerman, T. and Luske, B. (2016). Parasitic wasps of cereal leaf beetles (Coleoptera: Chrysomelidae: *Oulema*), *Entomol. Ber.* 76, 2–10.

Olfert, O. and Weiss, R. M. (2006). Impact of climate change on potential distributions and relative abundances of *Oulema melanopus*, *Meligethes viridescens* and *Ceutorhynchus obstrictus* in Canada, *Agric. Ecosyst. Environ.* 113(1–4), 295–301.

Olfert, O., Weiss, R. M., Woods, S., Philip, H. and Dosdall, L. (2004). Potential distribution and relative abundance of an invasive cereal crop pest, *Oulema melanopus* (Coleoptera: Chrysomelidae), in Canada, *Can. Entomol.* 136(2), 277–287.

Pedigo, L. P., Rice, M. E. and Krell, R. K. (2021). *Entomology and Pest Management* (7th edn.), Waveland Press, Long Grove, IL.

Philips, C. R., Herbert, D. A., Kuhar, T. P., Reisig, D. D. and Roberts, E. A. (2012). Using degree-days to predict cereal leaf beetle (Coleoptera: Chrysomelidae) egg and larval population peaks, *Environ. Entomol.* 41(4), 761–767.

Philips, C. R., Herbert, D. A., Kuhar, T. P., Reisig, D. D., Thomason, W. E. and Malone, S. (2011). Fifty years of cereal leaf beetle in the U.S.: an update on its biology, management, and current research, *J. Integr. Pest Manag.* 2(2), 1–5. DOI: 10.1603/IPM11014.

Pickett, C. H., Borkent, C. J., Wilson, R., Stadtherr, M. and Popescu, V. (2021). Introduction of the cereal leaf beetle parasitoid *Tetrastichus julis* (Hymenoptera: Eulophidae) into California, *Biocontrol Sci. Technol.* 32(1), 1–13. DOI: 10.1080/09583157.2021.1933901.

Piesik, D., Pańka, D., Delaney, K. J., Skoczek, A., Lamparski, R. and Weaver, D. K. (2011a). Cereal crop volatile organic compound induction after mechanical injury, beetle herbivory (*Oulema* spp.), or fungal infection (*Fusarium* spp.), *J. Plant Physiol.* 168(9), 878–886.

Piesik, D., Lemńczyk, G., Skoczek, A., Lamparski, R., Bocianowski, J., Kotwica, K. and Delaney, K. J. (2011b). *Fusarium* infection in maize: volatile induction of infected and neighboring uninfected plants has the potential to attract a pest cereal leaf beetle, *Oulema melanopus*, *J. Plant Physiol.* 168(13), 1534–1542.

Piesik, D., Wenda-Piesek, A., Lamparski, R., Tabaka, P., Ligor, T. and Buszewski, B. (2010a). Effects of mechanical injury and insect feeding on volatiles emitted by wheat plants, *Entomol. Fenn.* 21(2), 117–128.

Piesik, D., Lyszczarz, A., Takaka, P., Lamparski, R., Bocianowski, J. and Delaney, K. J. (2010b). Volatile induction of three cereals: influence of mechanical injury and insect herbivory on injured plants and neighbouring uninjured plants, *Ann. Appl. Biol.* 157(3), 425–434.

Piesik, D., Pańka, D., Jeske, M., Wenda-Piesek, A., Delaney, K. J. and Weaver, D. K. (2013). Volatile induction of infected and neighbouring uninfected plants potentially influence attraction/repellence of a cereal herbivore, *J. Appl. Entomol.* 137(4), 296–309.

Pollier, A., Tricault, Y., Plantegenest, M. and Bischoff, A. (2019). Sowing of margin strips rich in floral resources improves herbivore control in adjacent crop fields, *Agric. Forest Entomol.* 21(1), 119–129.

Rao, S., Cossé, A. A., Zilkowski, B. W. and Bartelt, R. J. (2003). Aggregation pheromone of the cereal leaf beetle: field evaluation and emission from males in the laboratory, *J. Chem. Ecol.* 29(9), 2165–2175.

Reay-Jones, F. P. F. (2010). Spatial distribution of the cereal leaf beetle (Coloeptera: Chrysomelidae) in wheat, *Environ. Entomol.* 39(6), 1943–1952.

Reay-Jones, F. P. F. (2012). Spatial analysis of the cereal leaf beetle (Coleoptera: Chrysomelidae) in wheat, *Environ. Entomol.* 41(6), 1516–1526.

Reay-Jones, F. P. F. (2017). Geostatistical characterization of cereal leaf beetle (Coloeptera: Chrysomelidae) distributions in wheat, *Environ. Entomol.* 46(4), 931–938.

Reisig, D. D., Bacheler, J. S., Herbert, D. A., Heiniger, R., Kuhar, T., Malone, S., Philips, C. and Tilley, M. S. (2017). Cereal leaf beetle (Coleoptera: Chrysomelidae) regional dispersion and relationship with wheat stand denseness, *Environ. Entomol.* 46(3), 425–433.

Reisig, D. D., Bacheler, J. S., Herbert, D. A., Kuhar, T., Malone, S., Philips, C. and Weisz, R. (2012). Efficacy and value of prophylactic vs. integrated pest management approaches for management of cereal leaf beetle (Coleoptera: Chrysomelidae) in wheat and ramifications for adoption by growers, *J. Econ. Entomol.* 105(5), 1612–1619.

Roberts, D. E. (2016). Classical biological control of the cereal leaf beetle, *Oulema melanopus* (Coleoptera: Chrysomelidae), in Washington State and rôle of field insectaries, a review, *Biocontrol Sci. Technol.* 26(7), 877–893.

Root, R. B. (1973). Organization of a plant-arthropod association in simple and diverse habitats: the fauna of collards (*Brassica oleracea*), *Ecol. Monogr.* 43(1), 95–124.

Samková, A., Hadrava, J., Skuhrovec, J. and Janšta, P. (2019a). Host population density and presence of predators as key factors influencing the number of gregarious parasitoid *Anaphes flavipes* offspring, *Sci. Rep.* 9(1), 6081. DOI: 10.1038/s41598-019-42503-4.

Samková, A., Hadrava, J., Skuhrovec, J. and Janšta, P. (2019b). Effect of adult feeding and timing of host exposure on the fertility and longevity of the parasitoid *Anaphes flavipes*, *Entomol. Exp. Appl.* 167(11), 932–938.

Samková, A., Hadrava, J., Skuhrovec, J. and Janšta, P. (2019c). Reproductive strategy as a major factor determining female body size and fertility of a gregarious parasitoid, *J. Appl. Entomol.* 143(4), 441–450.

Samková, A., Hadrava, J., Skuhrovec, J. and Janšta, P. (2020). Host specificity of the parasitic wasp *Anaphes flavipes* (Hymenoptera: Mymaridae) and a new defence in its hosts (Coleoptera: Chrysomelidae: *Oulema* spp.), *Insects* 11(3), 1–11.

Schmidtt, M. and Uhl, G. (2017). *Oulema septentrionis* and *O. erichsonii* are neither conspecific nor melanic variants of *O. melanopus* as assessed by micro CT analysis of their lectotypes (Insecta, Coleoptera, Chrysomelidae, Criocerinae), *ZooKeys* 720, 121–130. https://doi.org/10.3897/ zookeys.720.19760.

Schulz-Kesting, K., Thiele, J., Everwand, G. and Dauber, J. (2021). Neighbourhood effect of faba bean (*Vicia faba* L.) on density of vegetation-dwelling natural biocontrol agents in winter wheat, *Biol. Control* 160, 104673. DOI: 10.1016/j.biocontrol.2021.104673.

Shade, R. E., Hansen, H. L. and Wilson, M. C. (1970). A partial life table of the cereal leaf beetle, *Oulema melanopus*, in northern Indiana, *Ann. Entomol. Soc. Am.* 63(1), 52–59.

Shapiro-Ilan, D. I., Gouge, D. H., Piggott, S. J. and Patterson, J. P. (2006). Application technology and environmental considerations for use of entomopathogenic nematodes in biological control, *Biol. Control* 38(1), 124–133.

Skendžic, S., Zovko, M., Živkovic, I. P., Lešic, V. and Lemic, D. (2021). The impact of climate change on agricultural insect pests, *Insects* 12(5), 1-31. DOI: 10.3390/insects12050440.

Skuhrovec, J., Douda, O., Zouhar, M., Maňasová, M., Nový, P., Božik, M. and Klouček, P. (2018). Insecticidal activity of two formulations of essential oils against the cereal leaf beetle, *Acta Agric. Scand. B* 68(6), 489-495.

Stehr, F. W. (1970). Establishment in the United States of *Tetrastichus julis*, a larval parasite of the cereal leaf beetle, *J. Econ. Entomol.* 63(6), 1968-1969.

Steinger, T., Klötzli, F. and Ramseier, H. (2020). Experimental assessment of the economic injury level of the cereal leaf beetle (Coleoptera: Chrysomelidae) in winter wheat, *J. Econ. Entomol.* 113(4), 1823-1830.

Tschumi, M., Albrecht, M., Bärtschi, C., Collatz, J., Entling, M. H. and Jacot, K. (2016). Perennial, species-rich wildflower strips enhance pest control and crop yield, *Agric. Ecosyst. Environ.* 220, 97-103.

Tschumi, M., Albrecht, M., Entling, M. H. and Jacot, K. (2015). High effectiveness of tailored flower strips in reducing pests and crop plant damage, *Proc. Biol. Sci.* 282(1814), 1-8.

Turlings, T. C. J. and Erb, M. (2018). Tritrophic interactions mediated by herbivore-induced plant volativles: mechanisms, ecological relevance, and application potential, *Annu. Rev. Entomol.* 63, 433-452.

Van de Vijver, E., Landschoot, S., Smagghe, G., de Baets, B., Temmerman, F., Dillen, J. and Haesaert, G. (2018). Potentials and limitations of a growing degree day approach to predict the phenology of cereal leaf beetles, *Environ. Entomol.* 47(4), 1039-1046.

Van de Vijver, E., Landschoot, S., van Roie, M., Temmerman, F., Dillen, J., de Ceuleners, K., Smagghe, G., de Baets, B. and Haesaert, G. (2019). Inter- and intrafield distribution of cereal leaf beetle species (Coleoptera: Chrysomelidae) in Belgian winter wheat, *Environ. Entomol.* 48(2), 276-283.

van Driesche, R., Hoddle, M. and Center, T. (2008). *Control of Pests and Weeds by Natural Enemies. An Introduction to Biological Control*, Blackwell Publishing, Oxford, UK.

Webster, J. A. and Smith, Jr., D. H. (1983). Cereal leaf beetle (*Oulema melanopus* (L.)) population densities and winter wheat yields, *Crop Prot.* 2(4), 431-436.

Wenda-Piesik, A., Kazek, M. and Piesik, D. (2018). Cereal leaf beetles (*Oulema* spp., *Coleoptera: Chrysomelidae*) control following various dates of wheat sowing and insecticidal treatments, *Int. J. Pest Manag.* 64(2), 157-165.

Wielkopolan, B., Krawczyk, K. and Obrepalska-Steplowska, A. (2018). Gene expression of serine and cysteine proteinase inhibitors during cereal leaf beetle larvae feeding on wheat: the role of insect-associated microorganisms, *Arthropod Plant Interact.* 12(4), 601-612.

Würschum, T., Langer, S. M., Longin, C. F. H., Tucker, M. R. and Leiser, W. L. (2020a). Refining the genetic architecture of flag leaf glaucousness in wheat, *Theoret. Appl. Genet.* 133, 981-991.

Würschum, T., Leiser, W. L., Langer, S. M., Tucker, M. R. and Miedaner, T. (2020b). Genetic architecture of cereal leaf beetle resistance in wheat, *Plants (Basel)* 9(9), 117. DOI: 10.3390/plants9091117.

Chapter 2

Grasshoppers and other orthopteran pests

Robert B. Srygley, Pest Management Research Unit, Northern Plains Agricultural Research Laboratory, USDA-ARS, USA

1 Introduction

Cereal crops grow in the wheat belts of every continent, and every continent except Antarctica has its grasshoppers and katydids (Orthoptera: Acrididae and Tettigoniidae) that feed on pasture lands but move into cereal crops during outbreaks. The omnivorous diet and broad habitat requirements of many Orthoptera present challenges to land managers who need reliable survey data over a broad spatial area and varied habitats to detect the risk of mass migration into cereals on adjacent farmlands. Promising research on orthopteran diet has shown a link between nutrient deficiencies, population outbreaks and migration, and susceptibility to microbial control. Another challenge is understanding variation in the basic biology of the pest Orthoptera across broad habitats and dietary requirements to forecast the response of

http://dx.doi.org/10.19103/AS.2022.0114.17
Published by Burleigh Dodds Science Publishing Limited, 2023.

these species and their community as the climate changes. This chapter focuses on these challenges and research into promising solutions for the management of grasshoppers and katydids that are cereal pests.

2 Grasshoppers, locusts and katydid pests

Approximately 20 grasshopper species are polyphenic locusts with a gregarious, migratory form and a solitary form. Locusts occur on all the continents except Antarctica, and the species are distantly related phylogenetically, indicating that the phase polyphenism characterizing locusts has evolved independently in multiple lineages. Locusts are widely distributed over native grasslands and move into crops periodically (Table 1) where they can cause substantial loss of crop yield, which can be particularly devastating to the economies of subsistence farmers and developing countries.

Canada and the USA do not have a locust since the Rocky Mountain locust *Melanoplus spretus* went extinct at the beginning of the twentieth century. For these countries, four species of spur-throated grasshoppers *M. sanguinipes*, *M. bivittatus*, *M. packardii*, and *M. differentialis* (commonly called the migratory grasshopper, two-striped grasshopper, Packard's grasshopper, and differential grasshopper, respectively) in addition to the clear-winged grasshopper *Camnula pellucida* have been the recurrent pests of small grains since the conversion of native grasslands to farmland (Hinks and Olfert, 1992).

In the western USA, the Mormon cricket, *Anabrus simplex*, a shield-backed katydid, is an important crop pest. Mormon crickets are renowned for their aggregation into broad bands between 1 and 2 km wide and tens of kilometers in length which contain millions of individuals and migrate from rangeland into

Table 1 Locusts that feed on cereals and their distributions (compiled from Le Gall et al., 2019)

Locust species	Common name	Continents
Schistocerca gregaria	Desert locust	Asia, Africa
S. piceifrons	Central American locust	North and South America
S. cancellata	South American locust	South America
Dociostaurus maroccanus	Moroccan locust	Asia, Africa, and Europe
Austracris guttulosa	Spur-throated locust	Australia
Gomphocerus sibiricus	Siberian locust	Asia
Ceracris kiangsu	Bamboo locust	Asia
Chortoicetes terminifera	Australian plague locust	Australia
Locusta migratoria	Migratory locust	Asia and Africa
Locustana pardalina	Brown locust	Africa
Oedaleus senegalensis	Senegalese locust	Africa

fields of wheat and other small grains. Related katydid species that also aggregate can be important crop pests in the Mediterranean region of Africa and Europe.

3 Surveying over varied habitats to monitor population expansion and migration

Grasshoppers and katydids are generally native to the regions where they are crop pests. Indeed, the crop is often non-native, and grasshoppers, Mormon crickets, and other katydids opportunistically feed on the crops during periodic outbreaks in their population cycles. Hence, monitoring crops for Orthoptera is rarely sufficient, and the native grasslands also require surveying to provide ample warning of dense orthopteran populations before they migrate into farmland.

3.1 Grasshoppers

In the USA, the US Department of Agriculture (USDA)–Agricultural Plant Health Inspection Survey (APHIS) coordinates surveying of federal rangeland in 17 western states to annually count grasshoppers. A survey of nymphs is conducted in the spring and an adult survey in the summer. Scouts are trained to count grasshoppers in a 1 ft^2 area (0.09 m^2) with 18 ft^2 areas surveyed at each site. Surveys are not generally species-specific because all grasshoppers are visually counted in the specified area, but sweep samples are collected in some of the western states to estimate relative densities of species at some of the sites. Based on survey counts, a hazard map is created each year to forecast the grasshopper densities the following year (Fig. 1). This approach guides the allocation of resources and effort to high-risk regions of the western USA. Lead time provided by these surveys is used to mobilize aerial- and ground-based control efforts.

Although not currently possible with total grasshopper counts (Kistner-Thomas et al., 2021), the incorporation of species-specific information into the hazard forecasting is likely to improve the accuracy of the predictive modeling in the future (e.g. *M. sanguinipes,* Humphreys et al., 2022a). For example, diets of grasshopper species vary widely from specialists on grasses to species that feed on broad-leaved plants. In addition, there are mixed feeders that feed on both grasses and broad-leaved plants. If the availability of forage for a particular species affects its potential population growth, then cereal feeders and those on broad-leaved plants would have different risk forecasts depending on forecasted plant composition, which could also be incorporated into the models. In addition to grasshopper counts for outbreak species, the amount of vegetation and indices of vegetation quality, such as soil moisture, could be collected locally or remotely from satellite data.

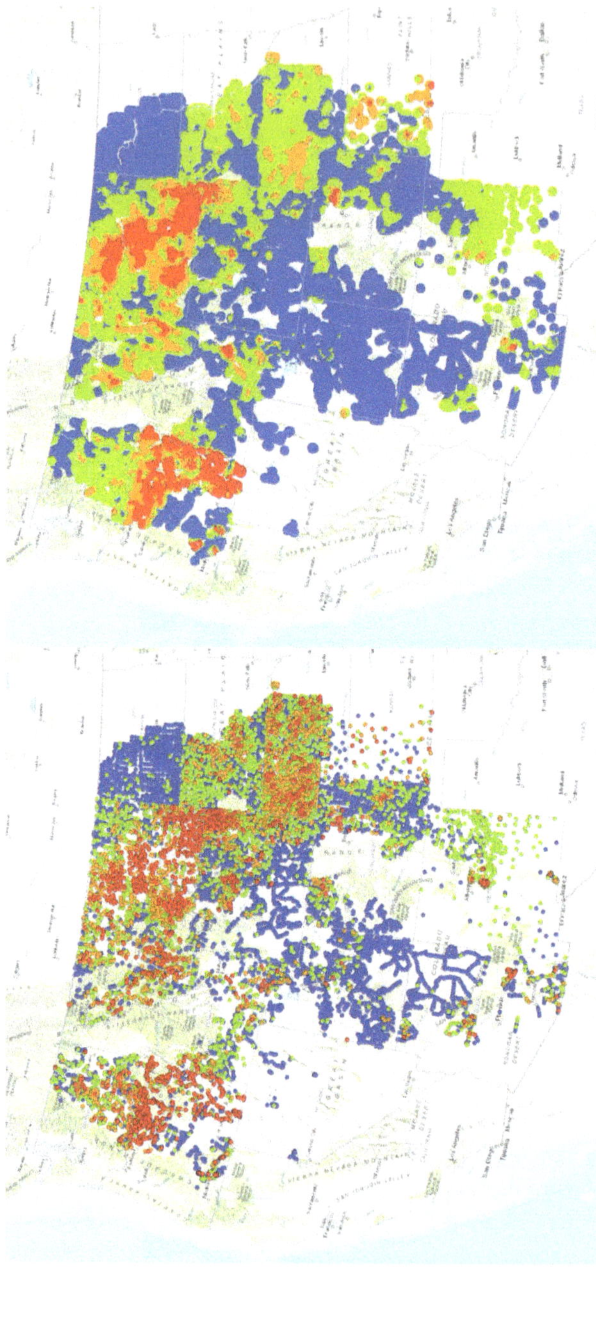

Figure 1 Adult grasshopper survey for 2020 (left panel) and 2021 nymphal hazard map (right panel) based on the rangeland survey for the western USA, as examples of maps generated by USDA-APHIS each year. Densities coded blue: less than 2.5; green: 2.5–6.7; orange: 6.7–12.6; red: more than 12.6 per m^2.

Published by Burleigh Dodds Science Publishing Limited, 2023.

3.2 Mormon crickets

Surveys of Mormon crickets are generally focused on a single species (Fig. 2). Species-specific biological information can be incorporated into forecasting, which should make the predictions more accurate than those currently used for grasshoppers. However, in some western states of the USA, only presence or absence of Mormon crickets is recorded. This approach was adopted because Mormon crickets aggregate and migrate together in bands that can be 2 km wide and contain millions of insects (Lorch et al., 2005). So density in a spot location would not represent the density in a wider area or even the density at that location the next day. The recent discovery that Mormon cricket eggs can remain in diapause underground for multiple years adds to the challenge of forecasting next year's population from nymphal and adult surveys. Indeed, Mormon crickets are notorious for disappearing suddenly from the landscape from one year to the next - a characteristic typically attributed to migration (although adults are unable to fly) but may also result from females banking eggs in the soil (Srygley, 2020).

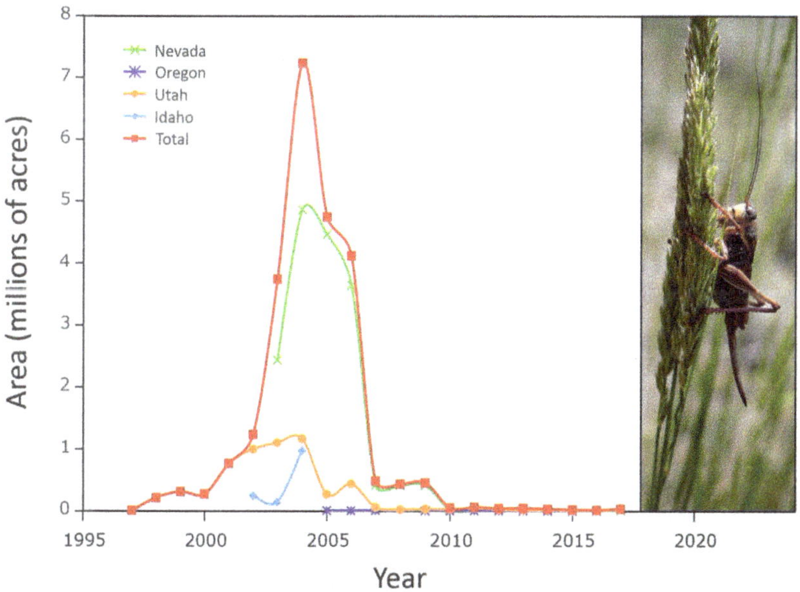

Figure 2 Total area of the western USA and that of four principal states with Mormon crickets banding and migrating during a recent outbreak (1998-2009) and relatively quiet years that followed (2010-2017). Data compiled from state presentations at annual meetings of the National Grasshopper Management Board. Data for Idaho are not shown between 2005 and 2009 because Mormon crickets were not separated from grasshoppers when reporting problem areas in those years. Photo: an adult female Mormon cricket on feral wheat in Nevada (R. Srygley, USDA-ARS).

3.3 Locusts

Because they fly large distances in swarms, adult locusts can be particularly challenging to survey and track. The desert locust *Schistocerca gregaria*, for example, is one of the most destructive insect pests on Earth. It is believed that the change from solitarious to gregarious forms occurs in several refuges (called recession zones) from which the locusts migrate to follow rain. Western Pakistan, Nepal, northwest India, southern Iran, the Arabian Peninsula, and the Horn of Africa are all included in the migratory route of the desert locust (Shrestha et al., 2021). For the desert locust and many other locusts, political instability, language barriers, or access to remote locations can all impede surveying and tracking of nymphs and adults. Remote sensing of vegetation and weather can provide important clues of population growth and movement potential (Lawton et al., 2022). In a recent review, Klein et al. (2021) summarize the application of satellite and other remotely acquired data to monitor locust populations. To date, most of the applications of remote sensing have been conducted on the desert locust, migratory locust, and the Australian plague locust. Habitat mapping, monitoring of vegetation change, and soil moisture have been important features associated with hatching, changing from solitary to gregarious phases, and outbreaks. Locust population monitoring and forecasting have benefited from improvements in spatial resolution of vegetation, open access to satellite imagery, computer processing, and weather prediction. In addition, locust management benefits from international cooperation promoted by nongovernmental organizations, such as the Food and Agriculture Organization of the United Nations and the Global Locust Initiative at Arizona State University.

4 Nutrient availability and orthopteran outbreaks

Many of the grasshopper and katydid species that are a problem on cereals persist in grasslands and only migrate into croplands during population outbreaks. For some locusts and Mormon crickets, nutrient availability on grasslands affects their consumptive and migratory behaviors. Here, we will briefly review the geometric framework of nutrition and graphically illustrate that in development and maintenance, organisms aim for an intake target corresponding to a species-specific ratio of carbohydrates to protein (Fig. 3). Because forage rarely has the same nutrient ratio as the organism's intake target, the insect must move between patches of varying nutrient ratios to achieve its target. The required nutrient patches are not always readily available in nature, however, resulting in deviation from the target. Nutritional deficiencies that result from deviation from intake targets have been shown to alter development and behavior in migratory insects, including locusts and Mormon crickets. The

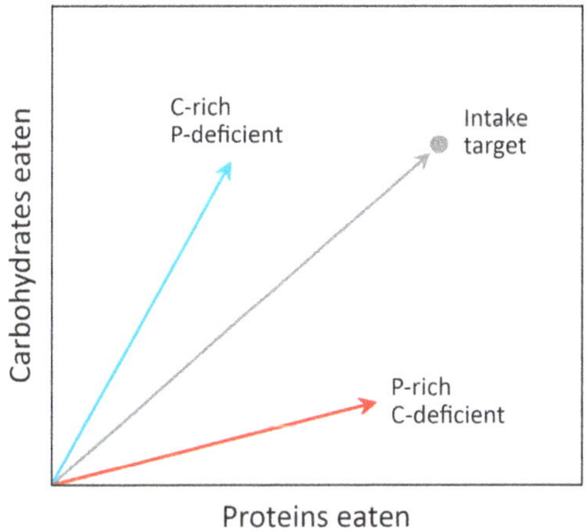

Figure 3 The geometric framework of nutrition showing the optimal diet ratio of carbohydrates and protein as the intake target (gray circle) and the diet ingested over time when balanced (gray arrow), protein deficient (blue arrow), or carbohydrate deficient (red arrow).

nutritional framework then permits further analysis of the factors that cause the nutritional imbalance in the first place and the ramifications of nutritional imbalance on the ability of the insects to defend themselves against agents of infection. This we will show in turn with examples.

Two essential macronutrients for insects are carbohydrates and proteins. Carbohydrates are required for energy, whereas proteins are typically used in the manufacture and maintenance of tissues. Although optimal foraging theory typically focuses on maximizing energy, many organisms regulate the intake of protein in addition to carbohydrates. Where insects live, nutrients vary both spatially and temporally, and the availability of protein and carbohydrates rarely matches the organism's needs. Grasshoppers and other herbivorous insects often have difficulty finding sufficient protein because plant foliage is relatively low in protein (3% on average). Some insects cope by migrating to find more suitable resources, utilizing carbohydrates and lipids to fuel their movement, which is particularly beneficial when excess carbohydrates have accumulated from feeding on plants (Zanotto et al., 1997).

The Mongolian locusts *Oedaleus asiaticus* prefer twice as much carbohydrates as proteins in their diet (dry mass), and they grow and survive best on this diet. Consequently, 2:1 carbohydrates:protein (C:P) is labeled their intake target for these macronutrients. In regions inhabited by the Mongolian locust, livestock overgrazing has resulted in loss of soil nitrogen from rangeland (Cease et al., 2015). As a result, plant tissues in heavily grazed pastures are low in nitrogen, and the Mongolian locusts, which prefer these low-protein plants, perform exceptionally well, resulting in population expansion and increased flight activity (Cease et al., 2015, 2017). Incorporation of more beef into human diet has altered regional agricultural practices, degraded rangeland, and promoted recent outbreaks of the Mongolian locust, which was not historically an outbreak pest.

Two cereal pests in North America: *M. sanguinipes* has an intake target that is nearly 1:1 C:P, whereas *M. differentialis* prefers a protein-biased diet of approximately 2:3 (C:P; Behmer and Joern, 2008). Although they feed on the same range of plants, differences in dietary requirements may result in the selection of some plant parts over others. Together with five other *Melanoplus* species in the same prairie community, the intake targets were at or near the maximum growth rate and minimum development time of the penultimate instar and, in that respect, represent the optimal diets for the species. In a dietary space defined by a carbohydrate-biased diet of 3:2 to a protein-biased diet of 3:5 (C:P), the seven species partition the niche space with very little overlap among their intake targets (Behmer and Joern, 2008). Hence, nutritional differences among species reflect physiological optima that also structure the grasshopper community.

For Mormon crickets, insects, seeds, and other items rich in protein are important constituents of their omnivorous diet. They randomly choose between diets that are equal in the total dry mass of protein and carbohydrates such that they consume equal amounts of both macronutrients (1:1 C:P). However, given a choice between a diet with greater macronutrient content and a lesser one, they choose the richer diet, although it results in a deficiency in one of the nutrients (Srygley, 2017). Spatial and temporal heterogeneity in the availability of protein or carbohydrates in nature results in Mormon crickets that are deficient in one of these nutrients migrating in aggregate. Those that are deficient in protein band together and move together in one direction to avoid contact with cannibalistic conspecifics. Likewise, those that are deficient in carbohydrates aggregate and move synchronously to avoid cannibalism (Srygley, 2016). Mormon crickets on the edge of the band that stray and become solitary have a much greater risk of predation by birds and rodents (Sword et al., 2005), whereas adult females in particular are at risk of predation by parasitic, digger wasps (Srygley and Lorch, 2016). The band moves with push-me, pull-you tension as cannibalism keeps those in front moving forward, and predation risk keeps those in the back and on the edges from straying to maintain band cohesiveness.

Similarly aggregated to provide safety in numbers, nymphal locusts migrate directionally to avoid contact with band mates (Buhl et al., 2006). Desert locusts *S. gregaria* prefer diets with equal amounts of C:P (Simpson et al., 2002), whereas migratory locusts *Locusta migratoria* prefer a diet nearer to 3:2 C:P (Chambers et al., 1995). Characterization of intake targets of outbreak species and their nutritional requirements before and during outbreak and migration will improve our understanding of the collective forces behind pest aggregation, migration, and phase change (in the case of locusts).

5 Effective means of preventing and suppressing orthopteran outbreaks

5.1 Chemical control

In the USA, managing rangeland grasshopper populations to prevent outbreaks and invasion of cropland is an active research goal. State and federal managers have readily adopted spraying of pesticide in one row and skipping the adjacent rows to enhance integrated pest management. This method serves to suppress grasshopper populations rather than eradicate them and allows the same amount of pesticide to 'protect' twice as much area as conventional application over the entire area. Reduced agent area treatments (also known as RAATs) allow grasshoppers and other insects to recolonize treated areas from adjacent untreated rows (Lockwood et al., 2002). RAATs also allow the same area to be protected for less expense incurred (Lockwood and Schell, 1997), which has accelerated its wide adoption. Controlled burning of vegetation is another means by which grasshopper populations can be reduced (Branson and Vermeire, 2016), and it is already practiced in some rangeland sites. Habitat management to promote vegetative growth and diversification not only results in more forage to reduce competition between grasshoppers and livestock but also provides shade which reduces ground temperature and the successful hatching and growth of nymphs (Branson et al., 2008).

Application of chemical insecticides remains the most prevalent means of controlling orthopterans (Zhang et al., 2019). Chemical control is expensive, but a cost-benefit assessment of chemical control of cereal pests in Australia found that of all of the pests, locusts amounted to the highest value of potential return from control measures on wheat at AUS$95 million or $7 per hectare cultivated (Table 2; Murray et al., 2013). Although the application of fenitrothion-based insecticide on cereals costs AUS$16 per hectare, not all areas are treated. The total cost of treatment for locusts was AUS$6 million for wheat, which

Table 2 Cost-benefit analysis of chemical control of locusts in Australia

Crop	Potential loss % (AUS$)	Actual loss % (AUS$)	Treatment cost	Benefit:cost
Wheat	40 (114 million)	7.4 (20 million)	$6 million	15.7:1
Barley	43 (43 million)	7.7 (1 million)	$2.5 million	16.8:1
Oat	39 (6 million)	7.5 (1 million)	$0.5 million	10.0:1
Sorghum	35 (1 million)	6.7 (0.28 million)	$0.06 million	12.0:1

amounted to a return of nearly AUS$16 for each dollar spent. In the Horn of Africa, mobilization of aerial spraying of the desert locusts with fenitrothion in 2020 protected 4 million tons of cereal (FAO, 2021). In developing countries, the benefit of control measures in local currency can often be much lower than in developed countries (Zhang et al., 2019). From desert locust control in 2020, the FAO (2021) estimates US$ 1.5 billion in protected cereal and milk production upon which 34 million people depend.

5.2 Insect-killing fungi as microbial control agents

On the continents of Africa and Australia, in particular, microbial control agents have been adopted as alternative tools in the arsenal for locust control (Hunter, 2005). On both continents, *Metarhizium acridum* (Fig. 4) is a commercially available fungus that is specific to killing Orthoptera (Lomer et al., 2001). Two strains of *M. acridum* are registered throughout sub-Sahelian Africa under the product names Novacrid and Green Muscle. *M. acridum* was applied to control the desert locust in Somalia during the recent outbreak in 2020 (Stokstad, 2020). The Australian strain is commercially known as Green Guard and has been registered in several Central Asian countries in addition to Australia. In Australia, it is applied to ecologically sensitive areas like waterways and organic cattle ranches (Hunter, 2005). Metacridum is the commercial *M. acridum* strain for Mexico.

In North America, the only fungus thus far that has been registered for application to Orthoptera on rangeland is *Beauveria bassiana*. Lacking the lethal toxins produced by *M. acridum*, *B. bassiana* kills grasshoppers much more slowly (Srygley and Jaronski, 2018). Between 2006 and 2015, a concerted

Figure 4 The pathogenic fungus *Metarhizium acridum* sporulating through the exocuticle of *Schistocerca Americana*. Photo courtesy of S. Jaronski, USDA-ARS.

attempt to discover a native strain of *M. acridum* from US soils was unsuccessful. Isolates of other *Metarhizium* species were identified, and a new species was named *M. robertsii*, after the person who led the effort. Some of these isolates, including *M. robertsii*, were found to kill Mormon cricket nymphs and adults more rapidly than *M. acridum* in the laboratory (Rangel et al., 2021), but the native isolates are less tolerant than commercialized *M. acridum* to heat and ultraviolet radiation, which may restrict their effectiveness in the field (Rangel et al., 2015).

5.3 Insect immunity

In order to improve the efficacy of microbial control, scientists at the USDA-Agricultural Research Service (ARS) in Sidney, Montana, have been studying the immune system of orthopterans to find weaknesses in their defenses to pathogens. The phenoloxidase (PO) cascade is a generalized immune response of insects that serves in clotting hemolymph to stop its loss at wound sites and walling-off foreign bodies, such as parasitoids, fungi, or bacteria, in a melanic encapsulating coat. The PO cascade releases quinones, which are not only toxic to the invaders but also to the insect. Hence, the PO cascade is tightly regulated to avoid autotoxicity. PO also occurs as an inactive zymogen prophenoloxidase (proPO), which is an order of magnitude more concentrated in the hemolymph than PO. The proPO cascade is also tightly regulated to avoid excessive or premature activation (Cerenius and Söderhäll, 2004).

5.4 Diet, migration and immunity to fungal control

Migrating Mormon crickets that are deficient in protein slow their rate of movement when their diet has been artificially supplemented with protein (Simpson et al., 2006; Srygley et al., 2009). Supplementation brings them closer to their intake target, and they slow their movement to remain in the protein patch longer. In addition, protein supplementation increases the insect's circulating PO that is spontaneously activated by wounding (Srygley et al., 2009). In a laboratory study, Mormon crickets with greater dietary protein had higher concentrations of spontaneously activated PO and greater immunity to the insect-killing fungus *B. bassiana* (Fig. 5; Srygley and Jaronski, 2018). Circulating PO is associated with attempted clearing of *B. bassiana* from the hemolymph (Srygley and Jaronski, 2011). On the other hand, Mormon crickets that are deficient in dietary protein are more susceptible to *B. bassiana* infection (Fig. 4). In addition to the invertebrates upon which they predate, Mormon crickets obtain much of their dietary protein from ovules of flowering plants and their seeds, including wheat and other cereals. Although Mormon crickets feed on vegetation throughout their development in spring, the greatest damage to

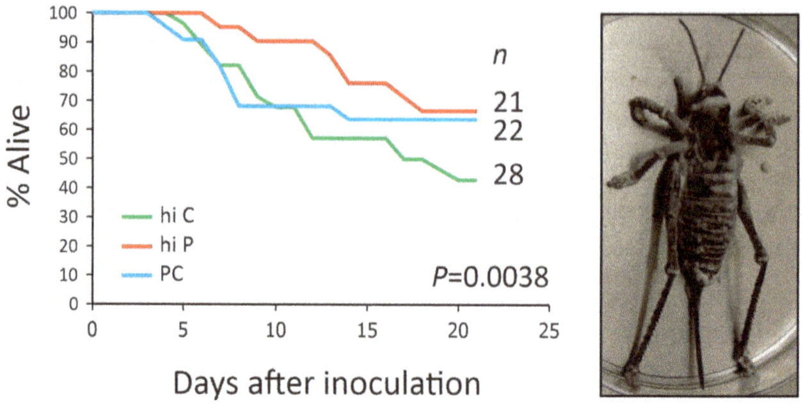

Figure 5 Mortality due to inoculation with the pathogenic fungus *Beauveria bassiana* for Mormon crickets fed a diet of 2:1 carbohydrates: protein (hi C, green line), 1:2 carbohydrates: protein (hi P, red line), or a balanced diet of 1:1 carbohydrates: protein (PC, blue line). Photo: an adult female Mormon cricket with white *B. bassiana* sporulating through the cuticle. (R. Srygley, USDA-ARS).

cereals occurs when the insects and grains are both maturing in late spring and early summer. In 2001, Utah estimated US$25 million in crop losses due to Mormon crickets, principally in wheat and alfalfa.

5.5 Insect coloration and age affect immunity to fungal control

The exoskeleton of the migratory grasshopper, *M. sanguinipes*, is paler when it develops at higher ambient temperatures to reduce excess heating from sunlight (Fielding and DeFoliart, 2005). In addition, adults have lower PO activity and proPO circulating in the hemolymph after developing at higher temperatures (Srygley and Jaronski, 2022) due to the dual role of PO in cuticular melanism and immunity. After developing at higher temperatures, *M. sanguinipes* are more susceptible to infection by *B. bassiana*. Lighter coloration is observed currently in *M. sanguinipes* at lower elevations and lower latitudes (Fielding and DeFoliart, 2005), and efficacy of *B. bassiana* is expected to increase in these regions accordingly. Lighter coloration and greater susceptibility to fungal disease are likely to become more widespread in *M. sanguinipes* with global warming. However, growth of *B. bassiana* is optimal near 28°C and has an upper limit between 34°C and 36°C (Mascarin and Jaronski, 2016).

Survivorship after *M. acridum* treatment increases as Mormon cricket adults age. This increase in survivorship is associated with exponential increases in circulating PO and proPO with adult age (Srygley, 2012). In contrast, PO and proPO increase within each nymphal molt cycle and then decrease again upon molting between instars as PO is used to melanize and harden the cuticle. As a result, the fungus is most efficacious on nymphs, which is presumably also true when applied to locusts (Stokstad, 2020). For desert locusts *S. gregaria* (Gillespie et al., 2000) and migratory locusts *L. migratoria* (Mullen and Goldsworthy, 2006), *Metarhizium* infection caused PO titers to decline until death.

5.6 Spraying and endophytic fungal control

Current control practices require inundation of the pest infestation with fungal spores. For commercially available *M. acridum*, 2.5 × 10^{12} conidia per hectare is required to achieve 80% mortality of locusts and grasshoppers, costing US$9 per hectare for the fungus alone (Jaronski, 2010). Given that it requires 5–14 days to kill infected grasshoppers, the most effective application to protect wheat and other crops is a band sprayed outside the crops so that nymphs encounter the fungus as they move to the crop. Because of the delayed mortality, little protection is afforded by treating acridids already in the field. However, in integrated pest management, microbial control agents might be introduced into cereals. For example, the Australian plague locust grew more slowly on wheat colonized by *B. bassiana* (Gurulingappa et al., 2010). The fungus deters feeding on cadavers of infected grasshoppers and Mormon crickets (Jaronski, 2013) and might similarly be a deterrent to grasshoppers feeding on *Beauveria*-colonized wheat. Colonization of wheat and other cereals with endophytic fungi that are also pathogenic to insects is an important area of future research (Vega, 2018).

5.7 Diet, migration and immunity to bacteria

In contrast to the migratory bands comprised of Mormon crickets that showed protein deficiency, other bands are comprised of Mormon crickets whose diet is deficient in carbohydrates (Srygley and Lorch, 2011, 2013). After ingesting carbohydrates, the insects slow their migratory rate relative to those that ingest protein. In addition, the ability of their hemolymph to clear bacterial infection increases. Hence, dietary deficiency plays an important role in not only motivating migration but also in the insects' defenses against bacteria, which can be exploited to improve the efficacy of bacterial control agents such as *Bacillus thuringiensis* or Bt. Currently, a Bt strain that is specific to Orthoptera is not commercially available but is sought along with other pathogenic bacteria (e.g. Quesada-Moraga et al., 2004; Oulebsir-Mohandkaci et al., 2015).

6 Use of orthopteran pests for food or animal feed

Human consumption of locusts and crickets or their use as feed for farm animals, fish, and pets can improve the output of small landowners that are otherwise negatively impacted by local outbreaks. Insects are a viable source of protein (van Huis, 2013; Dossey et al., 2016), and Orthoptera, like most insects, have high feed conversion efficiencies, low water requirements, and total protein content that rivals salmon. For grasshoppers on alfalfa, harvesting of wild populations can reduce local orthopteran populations, with results on the order of that obtained with insecticide application but with less economic and environmental costs and the benefit of acquired protein (Cerritos and Cano-Santana, 2008). Detailed quantification of the economic services and disservices that result from harvesting edible insects from cereals is needed (Payne and Van Itterbeeck, 2017). For example, the edible bush cricket *Ruspolia differens*, which arrests pollination and reduces yields of rice and millet, is harvested in East Africa, particularly when it migrates in swarms biannually (Matojo and Njau, 2010). The migratory locust and desert locust are harvested by multiple ethnic groups in Africa. Wild harvesting runs contrary to insecticidal application to the same populations which makes detailed cost-benefit analyses especially important to management decisions (Van Huis, 2021).

7 Climate and Orthoptera populations

Variation in the basic biology of pest Orthoptera across broad habitat and dietary requirements makes forecasting the response of these species and their community as the climate changes challenging. Temperature and precipitation are important drivers of orthopteran growth and reproduction and hence have major effects on population demographics that lead to pest outbreaks. However, these abiotic drivers also affect the nutritional quality of the plants on which Orthoptera feed and interactions with predators and competitors that ultimately make population dynamics less predictable (Ovadia and Schmitz, 2004). Nevertheless, understanding the role of temperature and precipitation in embryonic and nymphal development rates and adult maturation is used to model demography, improve forecasting, and optimize surveying and control efforts (Rebaudo and Rabhi, 2018). Relationships between the developmental rate of Orthoptera and temperature are curvilinear, but calculation of the lower developmental threshold and growing degree days by linear regression through the approximately straight-line portion of the curve (Campbell et al., 1974) remains the most popular method for predicting insect development for application in pest management. Sigmoidal modeling and biophysical modeling based on thermodynamics have gained in popularity and more accurately reflect development rate across the broad range of temperatures observed in nature (Rebaudo and Rabhi, 2018). The R-package DEVRATE (Rebaudo et al., 2018), for example, provides summary statistics to compare the explanatory power of more than 20 development models of varying complexity from that of Campbell et al. (1974) to Régniere et al. (2012).

Published by Burleigh Dodds Science Publishing Limited, 2023.

Developmental rates of locust, grasshopper, and katydid nymphs have been measured in laboratory and field settings. Due to active thermoregulation by the insects, field measurements are typically faster than those predicted from development in the laboratory (e.g. Lactin and Johnson, 1996). Using air temperature measured at 1.5 m and development rate relative to temperature modeled from laboratory data, cumulative development of *M. sanguinipes* in the field was underestimated by 39-53% (Fielding, 2004). Prediction of development in the field was improved by predicting the maximum elevation of body temperature with available sunlight. By including thermal elevation, Fielding (2004) improved the estimate of development rate in nature (overestimating by 5-11%).

Photoperiod, diet, and length of the growing season may also affect developmental rates. For example, developmental times of *M. sanguinipes* from populations at higher altitude were shorter than those from lower elevation at a constant temperature of 33°C (Dingle et al., 1990). Similarly, Mormon crickets *A. simplex* from high altitude populations developed faster for any given temperature than those from lower altitudes (Srygley, 2021a). Although these factors are often significant, ambient temperature and incident sunlight are by far the most important factors in estimating developmental rates of a species across broad geographic scales. Orthopteran growth rates are more closely related to soil surface temperature than air temperature at 1.5 m (Dingle et al., 1990), and niche modeling R-packages, such as microclimUS for the continental USA (Kearney, 2019), can be used to model air temperature at or near the Earth's surface. Niche modeling and the geometric framework of nutrition can be combined to link organismal traits and habitat characteristics (Kearney et al., 2010). Because the body temperatures of Orthoptera and many other insects are clearly dependent on near-surface temperatures both above and below ground, USDA-ARS is currently comparing near-surface air and soil temperatures modeled with microclimUS with temperature data collected over 5 years from locations in the western USA that vary in aspect, slope, and elevation. Accurate measurement of near-surface temperatures coupled with measurement of the effect of temperature on life-stages (e.g. Mormon crickets, Srygley, 2022) makes timing of hatching events, immature development, and adult molt more precise. Because some insecticides can only be applied to specific developmental stages, managerial decisions of timing and mode of intervention will be improved with this information. In addition, microclimate refuges from extreme weather events in the short-term (Srygley, 2021b) and climate change in the long-term (Humphreys et al., 2022a) can be mapped to improve spatial and temporal forecasting of population outbreaks and declines.

In addition to temperature, precipitation is an important abiotic factor for many orthopteran populations. For example, outbreaks of the brown locust *Locustana pardalina* in southern Africa are associated with rainfall in the prior year (Todd et al., 2002). With sufficient moisture, eggs hatch after 10–20 days, but without 15–25 mm rain, the eggs may diapause up to 15 months. Development of Mormon cricket eggs is also influenced by soil moisture (Srygley, 2014, 2020), with some populations responding to lack of moisture by entering diapause. If moisture is restored, eggs can persist in diapause for 4 or more years, and moisture restoration does not synchronize hatching of diapausing eggs - at least not in the short term (Srygley, 2020).

Abundance of the brown locust is related to the El Niño Southern Oscillation (ENSO) in the equatorial Pacific Ocean, with cool sea surface temperatures associated with La Niña events that teleconnect to bring rain followed by locust outbreaks to southern Africa (Todd et al., 2002). Like *L. pardalina*, outbreaks of the Australian plague locust *Chortoicetes terminifera* are also associated with high soil moisture (Wang et al., 2019). The ENSO is a principal driver of temperature, rainfall, and primary productivity in locust habitat of much of eastern Australia. As a result, *C. terminifera* outbreaks are largely associated with La Niña events (Deveson, 2013). The ENSO cycle teleconnects to agricultural systems in many regions of the world where it is linked to plant productivity and insect abundance and movements (e.g. Srygley et al., 2010, 2014; Humphreys et al., 2022b). The role of ENSO in orthopteran outbreaks, in particular, can provide forecasting of population events several months in advance, which greatly assists mobilization of targeted control efforts.

8 Conclusion

Governments all over the world and the United Nations have recognized the importance of coordinated monitoring and forecasting of grasshoppers, locusts, and katydids to control irruptions of their populations and migration from natal grounds onto cereals and other crops. Most locust outbreaks have declined in duration and severity as a result of international cooperation (Sword et al., 2010). However, regional strife can impede monitoring, and in the case of the recent desert locust outbreak, it may have prevented early suppression of gregarious populations (Meynard et al., 2020).

Although the successful reduction in outbreaks is largely due to surveys identifying hot spots at which the local outbreaks are suppressed with chemical insecticides, fungal insecticides have been adopted as alternatives that are becoming more widespread in their use (Lecoq, 2010). A long-term goal of research on integrated pest management of Orthoptera is to make agricultural systems more sustainable. Understanding the role that diet plays in the migration of grasshoppers, locusts, and katydids while also affecting their immunity to microbial control will improve efficacy of these alternatives, reducing costs and enhancing adoption.

Some Orthoptera have very long periods between outbreaks. The recent outbreak of the South American locust had followed 70 years of quiescence. Similarly, the outbreak of Mormon crickets in the first decade of this century was of a magnitude that had not been experienced since the historic outbreak of the 1930s. During these long periods of quiescence, expert knowledge of the natural history of these species faded in memory or was lost altogether. Because the response to insect pests is often reactionary rather than preventative (e.g. Branson et al., 2008), interest and funding to support research waxes and wanes with the periodicity of the pests' population cycles. For example, the outbreak of grasshoppers across the western USA in the 1980s resulted in the nearly simultaneous hiring of personnel as monitoring and suppression programs expanded. Forty years later, these same people are set to retire.

The boom and bust cycles have focused attention on how to transfer the knowledge gained. In some cases, new hires may overlap and be trained by retiring personnel. For example, locust forecasting is part science and part art (K. Cressman, personal communication), and new personnel will require experience to gain the scientific knowledge necessary for accurate forecasting (K. Cressman suggests 5 years of overlap with senior forecasters). Knowledge that is not published will be lost as funding wanes and researchers retire. If the current trends with global warming continue and cyclones in the Indian Ocean increase in frequency (Salih et al., 2020) while the pests of cereals grow faster and consume more forage (Deutsch et al., 2018), this knowledge will be needed to avert the impending crises affecting global food security.

9 Future trends in research

9.1 Tracking migratory insects

Automated tracking of swarming Orthoptera provides an early warning of infestation of new habitat. Doppler radar weather stations are capable of detecting aerial swarms of locusts and other pest insects, but weather stations are not available in many migratory areas. In regions where weather radar is available, quantity of individuals, direction of movement, and use of syntopic winds to optimize migration could be gleaned. Collective movement and mass settling of grasshoppers and locusts could be used to pinpoint locations for egg laying and forecast emergence of hatchlings.

Because individuals can be captured, radio-tagged, released back into the band, and tracked, Mormon crickets have become a model system for understanding collective motion (Bazazi et al., 2008, 2010; Vicsek and Zafeiris, 2012). Automated telemetry systems for migratory birds, such as Motus, can also be adapted for tracking radio-tagged individual insects migrating aerially or terrestrially. Mobile receiver stations can be set up to track individuals during band formation and directional changes to understand how group decisions are made. Combined with information on intake targets that optimize fitness components, radio-tracking can elucidate the role of dietary deficiencies in group movement and immunity to fungi and other microbial agents for controlling Orthoptera.

9.2 Parasitoid biology

There has been little effort to characterize the parasitoids of Orthoptera. Digger wasps that parasitize adults are extremely effective in controlling locusts and katydid populations because they preferentially attack and paralyze females (Srygley and Lorch, 2016) and even females in copula (Kemp, 2012). Egg parasitoids (Yoder et al., 2014) are also effective, but their populations tend to lag behind that of their host. Assisted migration of parasitoids from former outbreak sites to new outbreaks could prove an effective short-term solution.

9.3 Remote sensing and smart technologies

The wide use of cell phones has made it possible to distribute software to identify grasshoppers and other Orthoptera so that farmers and other land managers can collect data in regions that had been too remote in the past. The FAO's eLocust3 is one new management tool that integrates field surveys so that GPS coordinates can be integrated with control operations. More applications that combine satellite data of vegetation greening with ground observations are needed to be able to detect nymphal bands so that microbial and conventional control operations can be mobilized in the earliest stages to minimize loss of crops and other forage.

10 Where to look for further information

- Lecoq, M. and Zhang, L. eds. (2019). Encyclopedia of Pest Orthoptera of the World. China Agricultural University Press, Beijing. - most recent information on species identification, biological, cultural, and chemical control methods, and surveying.
- Klein, I., Oppelt, N. and Kuenzer, C. (2021). 'Application of remote sensing data for locust research and management- A review', Insects, 12, 233.
- Zhang, L., Lecoq, M., Latchininsky, A. and Hunter, D. (2019). 'Locust and grasshopper management', Annu. Rev. Entomol., 64, 15-34.
- Global Locust Initiative: https://sustainability-innovation.asu.edu/global-locust-initiative/.
- Australian Plague Locust Commission: https://www.awe.gov.au/biosecurity-trade/pests-diseases-weeds/locusts.
- United Nations FAO: https://www.fao.org/locusts/en/.

- Motus: www.motus.org- automated radiotracking of migratory animals.
- Orthopterists' Society: orthsoc.org.

11 Acknowledgements

The author would like to thank Stefan Jaronski for reading a draft of this manuscript. Mention of any trade names or commercial products in this publication is solely for the purpose of providing specific information and does not imply recommendation or endorsement by the USDA. USDA is an equal opportunity provider and employer.

12 References

Bazazi, S., Buhl, J., Hale, J. J., Anstey, M. L., Sword, G. A., Simpson, S. J. and Couzin, I. D. (2008). Collective motion and cannibalism in locust migratory bands, *Curr. Biol.* 18(10), 735-739.

Bazazi, S., Ioannou, C. C., Simpson, S. J., Sword, G. A., Torney, C. J., Lorch, P. D. and Couzin, I. D. (2010). The social context of cannibalism in migratory bands of the Mormon cricket, *PLoS ONE* 5(12), e15118.

Behmer, S. T. and Joern, A. (2008). Coexisting generalist herbivores occupy unique nutritional feeding niches, *Proc. Natl Acad. Sci. U. S. A.* 105, 1977-1982.

Branson, D. H., Joern, A. and Sword, G. A. (2008). Sustainable management of insect herbivores in grassland ecosystems: new perspectives in grasshopper control, *BioScience* 56(9), 743-755.

Branson, D. H. and Vermeire, L. T. (2016). Grasshopper responses to fire and postfire grazing in the Northern Great Plains vary among species, *Rangeland Ecol. Manag.* 69(2), 144-149.

Buhl, J., Sumpter, D. J. T., Couzin, I. D., Hale, J. J., Despland, E., Miller, E. R. and Simpson, S. J. (2006). From disorder to order in marching locusts, *Science* 312(5778), 1402-1406.

Campbell, A., Frazer, B. D., Gilbert, N., Gutierrez, A. P. and Mackauer, M. (1974). Temperature requirements of some aphids and their parasites, *J. Appl. Ecol.* 11(2), 431-438.

Cease, A. J., Elser, J. J., Fenichel, E. P., Hadrich, J. C., Harrison, J. F. and Robinson, B. E. (2015). Living with locusts: connecting soil nitrogen, locust outbreaks, livelihoods, and livestock markets, *BioScience* 65(6), 551-558. DOI: 10.1093/biosci/biv048.

Cease, A. J., Harrison, J. F., Hao, S., Niren, D. C., Zhang, G., Kang, L. and Elser, J. J. (2017). Nutritional imbalance suppresses migratory phenotypes of the Mongolian locust (*Oedaleus asiaticus*), *R. Soc. Open Sci.* 4(6), 161039. DOI: 10.1098/rsos.161039.

Cerritos, R. and Cano-Santana, Z. (2008). Harvesting grasshoppers *Sphenarium purpurascens* in Mexico for human consumption: a comparison with insecticidal control for managing pest outbreaks, *Crop Prot.* 27(3-5), 473-480. DOI: 10.1016/j.cropro.2007.08.001.

Cerenius, L. and Söderhäll, K. (2004). The prophenoloxidase-activating system in invertebrates. *Immunol. Rev.* 198(1), 116-126. DOI: 10.1111/j.0105-2896.2004.00116.x

Chambers, P. G., Simpson, S. J. and Raubenheimer, D. (1995). Behavioural mechanisms of nutrient balancing in *Locusta migratoria* nymphs, *Anim. Behav.* 50(6), 1513-1523.

Deutsch, C. A., Tewksbury, J. J., Tigchelaar, M., Battisti, D. S., Merrill, S. C., Huey, R. B. and Naylor, R. L. (2018). Increase in crop losses to insect pests in a warming climate, *Science* 361(6405), 916–919. DOI: 10.1126/science.aat3466.

Deveson, E. D. (2013). Satellite normalized difference vegetation index data used in managing Australian plague locusts, *J. Appl. Remote Sens.* 7(1), 075096. DOI: 10.1117/1.JRS.7.075096.

Dingle, H., Mousseau, T. A. and Scott, S. M. (1990). Altitudinal variation in life cycle syndromes of California populations of the grasshopper, *Melanoplus sanguinipes* (f.), *Oecologia* 84(2), 199–206.

Dossey, A. T., Morales-Ramos, J. A. and Rojas, M. G. (2016). Insects as Sustainable Food Ingredients: Production, Processing and Food Applications. Academic Press: New York.

FAO (2021). Available at: http://www.fao.org/news/story/en/item/1393635/icode/.

Fielding, D. J. (2004). Developmental time of *Melanoplus sanguinipes* (Orthoptera: Acrididae) at high latitudes, *Environ. Entomol.* 33(6), 1513–1522. DOI: 10.1603/0046-225X-33.6.1513.

Fielding, D. J. and DeFoliart, L. S. (2005). Density and temperature-dependent melanization of fifth-instar *Melaniplus sanguinipes*: interpopulation comparisons, *J. Orthop. Res.* 14(1), 107–113.

Gillespie, J. P., Burnett, C. and Charnley, A. K. (2000). The immune response of the desert locust *Schistocerca gregaria* during mycosis of the entomopathogenic fungus, *Metarhizium anisopliae* var acridum, *J. Insect Physiol.* 46(4), 429–437.

Gurulingappa, P., Sword, G. A., Murdoch, G. and McGee, P. A. (2010). Colonization of crop plants by fungal entomopathogens and their effects on two insect pests when in planta, *Biol. Control* 55(1), 34–41.

Hinks, C. F. and Olfert, O. (1992). Cultivar resistance to grasshoppers in temperate cereal crops and grasses: a review, *J. Orthop. Res.* 1(1), 1–9.

Humphreys, J. M., Srygley, R. B. and Branson, D. H. (2022a). Geographic variation in migratory grasshopper recruitment under projected climate change, *Geographies* 2(1), 12–30. DOI: 10.3390/geographies2010003.

Humphreys, J. M., Srygley, R. B., Lawton, D., Hudson, A. R. and Branson, D. H. (2022b). Grasshoppers exhibit asynchrony and spatial non-stationarity in response to the El Niño/Southern and Pacific Decadal Oscillations. *Ecol. Model.* 471: 110043. DOI: 10.1016/j.ecolmodel.2022.110043

Hunter, D. M. (2005). Mycopesticides as part of integrated pest management of locusts and grasshoppers, *J. Orthop. Res.* 14(2), 197–201.

Jaronski, S. T. (2010). Ecological factors in the inundative use of fungal entomopathogens, *BioControl* 55(1), 159–185. DOI: 10.1007/s10526-009-9248-3.

Jaronski, S. T. (2013). Mycosis inhibits cannibalism by *Melanoplus sanguinipes, M. differentialis, Schistocerca americana*, and *Anabrus simplex*, *J. Insect Sci.* 13, 122. DOI: 10.1673/031.013.12201.

Kearney, M. R. (2019). microclimUS: hourly estimates of historical microclimates for the United States of America with example applications, *Ecology* 100(11), e02829. DOI: 10.1002/ecy.2829.

Kearney, M. R., Simpson, S. J., Raubenheimer, D. and Helmuth, B. (2010). Modelling the ecological niche from functional traits, *Phil. Trans. R. Soc. Lond. B Biol. Sci.* 365(1557), 3469–3483. DOI: 10.1098/rstb.2010.0034.

Kemp, D. J. (2012). Costly copulation in the wild: mating increases the risk of parasitoid-mediated death in swarming locusts, *Behav. Ecol.* 23(1), 191-194. DOI: 10.1093/beheco/arr173.

Kistner-Thomas, E., Kumar, S., Jech, L. and Woller, D. A. (2021). Modeling rangeland grasshopper (Orthoptera: Acrididae) population density using a landscape-level predictive mapping approach, *J. Econ. Entomol.* 114(4), 1557-1567.

Klein, I., Oppelt, N. and Kuenzer, C. (2021). Application of remote sensing data for locust research and management - a review, *Insects* 12(3), 233.

Lactin, D. J. and Johnson, D. L. (1996). Effect of insolation and body orientation on internal thoracic temperature of nymphal *Melanoplus packardii* (Orthoptera: Acrididae), *Environ. Entomol.* 25(2), 423-429. DOI: 10.1093/ee/25.2.423.

Lawton, D., Scarth, P. Deveson, E., Piou, C., Spessa, A., Waters, C. and Cease, A. J. (2022). Seeing the locust in the swarm: Accounting for spatiotemporal hierarchy improves ecological models of insect populations. *Ecography* 2022(2), 05763. DOI: 10.111/ecog.05763

Le Gall, M., Overson, R. and Cease, A. (2019). *A global review on locusts (Orthoptera: Acrididae) and their interactions with livestock grazing practices*, *Front. Ecol. Evol.* 7, 263. DOI: 10.3389/fevo.2019.00263.

Lecoq, M. (2010). Integrated pest management for locusts and grasshoppers: are alternatives to chemical pesticides credible?, *J. Orthop. Res.* 19(1), 131-132. DOI: 10.1665/034.019.0107.

Lockwood, J. A., Anderson-Sprecher, R. and Schell, S. P. (2002). When less is more: optimization of reduced agent-area treatments (RAATs) for management of rangeland grassshoppers, *Crop Prot.* 21(7), 551-562. DOI: 10.1016/S0261-2194(01)00145-4.

Lockwood, J. A. and Schell, S. P. (1997). Decreasing economic and environmental costs through reduced area and agent insecticide treatments (RAATs) for the control of rangeland grasshopers. Empirical results and their implications for pest management, *J. Orthop. Res.* 6, 19-32.

Lomer, C. J., Bateman, R. P., Johnson, D. L., Langewald, J. and Thomas, M. (2001). Biological control of locusts and grasshoppers, *Annu. Rev. Entomol.* 46, 667-702. DOI: 10.1146/annurev.ento.46.1.667.

Lorch, P. D., Sword, G. A., Gwynne, D. T. and Anderson, G. L. (2005). Radiotelemetry reveals differences in individual movement patterns between outbreak and non-outbreak Mormon cricket populations, *Ecol. Entomol.* 30(5), 548-555.

Mascarin, G. M. and Jaronski, S. T. (2016). The production and uses of *Beauveria bassiana* as a microbial insecticide, *World J. Microbiol. Biotechnol.* 32(11), 177. DOI: 10.1007/s11274-016-2131-3.

Matojo, N. D. and Njau, M. A. (2010). Plasticity and biosystematics of swarming of the conehead *Ruspolia differens* Serville (Orthoptera: Conocephalidae), *Int. J. Integr. Biol.* 9, 97-103.

Meynard, C. N., Lecoq, M., Chapuis, M.-P. and Piou, C. (2020). On the relative role of climate change and management in the current desert locust outbreak in East Africa. *Global Change Biol.* 26(7), 3753-3755. DOI: 10.1111/gcb.15137

Mullen, L. M. and Goldsworthy, G. J. (2006). Immune responses of locusts to challenge with the pathogenic fungus *Metarhizium* or high doses of laminarin, *J. Insect Physiol.* 52(4), 389-398.

Murray, D. A. H., Clarke, M. B. and Ronning, D. A. (2013). Estimating invertebrate pest losses in six major Australian grain crops, *Aust. J. Entomol.* 52(3), 227-241. DOI: 10.1111/aen.12017.

Oulebsir-Mohandkaci, H., Khemili-Talbi, S., Benzina, F. and Halouane, F. (2015). Isolation and identification of entomopathogenic bacteria from Algerian desert soil and their effects against the migratory locust, *Locusta migratoria* (Linnaeus, 1758) (Orthoptera: Acrididae), *Egypt. J. Biol. Pest Control* 25, 739–746.

Ovadia, O. and Schmitz, O. J. (2004). Weather variation and tropic interaction strength: sorting the signal from the noise, *Oecologia* 140(3), 398–406.

Payne, C. L. R. and Van Itterbeeck, J. (2017). Ecosystem services and edible insects in agricultural systems: a review, *Insects* 8(1), 24. DOI: 10.3390/insects8010024.

Quesada-Moraga, E., Garcia-Tovar, E., Valverde-Garcia, P. and Santiago-Alvarez, C. (2004). Isolation, geographic diversity and insecticidal activity of *Bacillus thuringiensis* from soils in Spain, *Microbiol. Res.* 159(1), 59–71. DOI: 10.1016/j.micres.2004.01.011.

Rangel, D. E. N., Bighayan, H. G., Golez, H. G., Keyser, C. A., Evans, E. W. and Roberts, D. W. (2021). Virulence of the insect-pathogenic fungi *Metarhizium* spp. to Mormon crickets, *Anabrus simplex* (Orthoptera: Tettigoniidae), *Bull. Entomol. Res.* 112, 179–186. DOI: 10.1017/S0007485321000663.

Rangel, D. E. N., Braga, G. U. L., Fernandes, É. K. K., Keyser, C. A., Hallsworth, J. E. and Roberts, D. W. (2015). Stress tolerance and virulence of insect-pathogenic fungi are determined by environmental conditions during conidial formation, *Curr. Genet.* 61(3), 383–404. DOI: 10.1007/s00294-015-0477-y.

Rebaudo, F. and Rabhi, V.-B. (2018). Modeling temperature-dependent development rate and phenology in insects: review of major developments, challenges, and future directions, *Entomol. Exp. Appl.* 166(8), 607–617.

Rebaudo, F., Struelens, Q. and Dangles, O. (2018). Modelling temperature-dependent development rate and phenology in arthropods: the DEVRATE package for R, *Methods Ecol. Evol.* 9(4), 1144–1150. DOI: 10.1111/2041-210X.12935.

Régniere, J., Powell, J., Bentz, B. and Nealis, V. (2012). Effects of temperature on development, survival and reproduction of insects: experimental design, data analysis and modeling, *J. Insect Physiol.* 58(5), 634–647. DOI: 10.1016/j.jinsphys.2012.01.010.

Salih, A. A. M., Baraibar, M., Mwangi, K. K. and Artan, G. (2020). Climate change and locust outbreak in East Africa, *Nat. Clim. Change* 10(7), 584–585. DOI: 10.1038/s41558-020-0835-8.

Shrestha, S., Thakur, G., Gautam, J., Acharya, N., Pandey, M. and Shrestha, J. (2021). Desert locust and its management in Nepal: a review, *J. Agric. Nat. Resour.* 4(1), 1-28. DOI: 10.3126/janr.v4i1.33197.

Simpson, S. J., Raubenheimer, D., Behmer, S. T., Whitworth, A. and Wright, G. A. (2002). A comparison of nutritional regulation in solitarious- and gregarious-phase nymphs of the desert locust *Schistocerca gregaria*, *J. Exp. Biol.* 205(1), 121-129.

Simpson, S. J., Sword, G. A., Lorch, P. D. and Couzin, I. D. (2006). Cannibal crickets on a forced march for protein and salt, *Proc. Natl Acad. Sci. U. S. A.* 103(11), 4152–4156. DOI: 10.1073/pnas.0508915103.

Srygley, R. B. (2012). Ontogenetic changes in immunity and susceptibility to fungal infection in Mormon crickets *Anabrus simplex*, *J. Insect Physiol.* 58(3), 342–347. DOI: 10.1016/j.jinsphys.2011.12.005.

Srygley, R. B. (2014). Effects of temperature and moisture on Mormon cricket reproduction with implications for responses to climate change, *J. Insect Physiol.* 65, 57-62. DOI: 10.1016/j.jinsphys.2014.05.005.

Srygley, R. B. (2016). Diet drives the collective migrations and affects the immunity of Mormon crickets and locusts: a comparison of these potential superspreaders of disease, *Integr. Comp. Biol.* 56(2), 268-277. DOI: 10.1093/icb/icw035.

Srygley, R. B. (2017). Mormon crickets maximize nutrient intake at the expense of immunity, *Physiol. Entomol.* 42(1), 1-9. DOI: 10.1111/phen.12155.

Srygley, R. B. (2020). Coping with drought: diapause plasticity in katydid eggs at high, mid-, and low elevations, *Ecol. Entomol.* 45(3), 485-492.

Srygley, R. B. (2021a). Elevational changes in Mormon cricket life histories: minimum temperature for nymphal growth declines with elevation, *Environ. Entomol.* 50(1), 167-172. DOI: 10.1093/ee/nvaa151.

Srygley, R. B. (2021b). Elevational and latitudinal changes in cold tolerance of nymph and adult Mormon crickets *Anabrus simplex* (Orthoptera: Tettigoniidae), *Environ. Entomol.* 50(3), 699-705. DOI: 10.1093/ee/nvab009.

Srygley, R. B. (2022). Temperature effects on Mormon cricket *Anabrus simplex* embryo development, hatching and nymphal growth: Thermal performance curves change with ontogeny. *J. Thermal Biol.* 110, 103356. DOI: 10.1016/j.jtherbio.2022.103356

Srygley, R. B., Dudley, R., Oliveira, E. G., Aizprúa, R., Pelaez, N. Z. and Riveros, A. J. (2010). El Niño and dry season rainfall influence hostplant phenology and an annual butterfly migration from Neotropical wet to dry forests, *Glob. Change Biol.* 16(3), 936-945. DOI: 10.1111/j.1365-2486.2009.01986.x.

Srygley, R. B., Dudley, R., Oliveira, E. G. and Riveros, A. J. (2014). El Niño, hostplant growth, and migratory butterfly abundance in a changing climate, *Biotropica* 46(1), 90-97. DOI: 10.1111/btp.12081.

Srygley, R. B. and Jaronski, S. T. (2011). Immune response of Mormon crickets that survived infection by *Beauveria bassiana*, *Psyche* 2011, 1-5. DOI: 10.1155/2011/849038.

Srygley, R. B. and Jaronski, S. T. (2018). Protein deficiency lowers resistance of Mormon crickets to the pathogenic fungus *Beauveria bassiana*, *J. Insect Physiol.* 105, 40-45. DOI: 10.1016/j.jinsphys.2018.01.005.

Srygley, R. B. and Jaronski, S. T. (2022). Increasing temperature reduces cuticular melanism and immunity to fungal infection in a migratory insect, *Ecol. Entomol.* 47(1), 109-113. DOI: 10.1111/een.13088.

Srygley, R. B. and Lorch, P. D. (2011). Weakness in the band: Nutrient-mediated trade-offs between migration and immunity of Mormon crickets, *Anabrus simplex. Anim. Behav.* 81(2), 395-400. DOI: 10.1016/j.anbehav.2010.11.006

Srygley, R. B. and Lorch, P. D. (2013). Coping with uncertainty: Nutrient deficiencies motivate insect migration at a cost to immunity. *Integr. Comp. Biol.* 53(6), 1002-1013. DOI: 10.1093/icb/ict047

Srygley, R. B. and Lorch, P. D. (2016). Loss of safety in numbers and a novel driver of mass migration: radiotelemetry reveals heavy wasp predation on a band of Mormon crickets, *R. Soc. Open Sci.* 3(5), 160113. DOI: 10.1098/rsos.160113.

Srygley, R. B., Lorch, P. D., Simpson, S. J. and Sword, G. A. (2009). Immediate protein dietary effects on movement and the generalised immunocompetence of migrating Mormon crickets *Anabrus simplex* (Orthoptera: Tettigoniidae), *Ecol. Entomol.* 34(5), 663-668.

Stokstad, E. (2020). In Somalia, an unprecedented effort to kill massive locust swarms with biocontrol, *Science*. DOI: 10.1126/science.abb2759.

Sword, G. A., Lorch, P. D. and Gwynne, D. T. (2005). Insect behaviour: migratory bands give crickets protection, *Nature* 433(7027), 703. DOI: 10.1038/433703a.

Sword, G. A., Lecoq, M., and Simpson, S. J. (2010). Phase polyphenism and preventative locust management. *J. Ins. Physiol.* 56(8), 949-957. DOI: 10.1016/j.jinsphys.2010.05.005

Todd, M. C., Washington, R., Cheke, R. A. and Kniveton, D. (2002). Brown locust outbreaks and climate variability in southern Africa, *J. Appl. Ecol.* 39(1), 31-42. DOI: 10.1046/j.1365-2664.2002.00691.x.

Van Huis, A. (2013). Potential of insects as food and feed in assuring food security. *Annu. Rev. Entomol.* 58, 563-583. DOI: 10.1146/annurev-ento-120811-153704

Van Huis, A. (2021). Harvesting desert locusts for food and feed may contribute to crop protection but will not suppress upsurges and plagues, *J. Insects Food Feed* 7(3), 245-248.

Vega, F. E. (2018). The use of fungal entomopathogens as endophytes in biological control: a review, *Mycologia* 110(1), 4-30. DOI: 10.1080/00275514.2017.1418578.

Vicsek, T. and Zafeiris, A. (2012). Collective motion, *Phys. Rep.* 517(3-4), 71-140.

Wang, B., Deveson, E. D., Waters, C., Spessa, A., Lawton, D., Feng, P. and Liu, L. (2019). Future climate change likely to reduce the Australian plague locust (*Chortoicetes terminifera*) seasonal outbreaks, *Sci. Total Environ.* 668, 947-957. DOI: 10.1016/j.scitotenv.2019.02.439.

Yoder, M. J., Valerio, A. A., Polaszek, A., van Noort, S., Masner, L. and Johnson, N. F. (2014). Monograph of the Afrotropical species of *Scelio latreille* (Hymenoptera, Platygastridae), egg parasitoids of acridid grasshoppers (Orthoptera, Acrididae), *ZooKeys* (380), 1-188. DOI: 10.3897/zookeys.380.5755. Available at: https://www.waspweb.org/Platygastroidea/Keys/Index.htm.

Zanotto, F., Gouveia, S., Simpson, S. and Calder, D. (1997). Nutritional homeostasis in locusts: is there a mechanism for increased energy expenditure during carbohydrate overfeeding?, *J. Exp. Biol.* 200(18), 2437-2448.

Zhang, L., Lecoq, M., Latchininsky, A. and Hunter, D. (2019). Locust and grasshopper management, *Annu. Rev. Entomol.* 64, 15-34.

Published by Burleigh Dodds Science Publishing Limited, 2023.

Part 2

Gall midges and stem feeding pests

Chapter 3

The Hessian fly: a destructive pest of wheat and barley

Ming-Shun Chen, Hard Winter Wheat Genetics Research Unit, Center for Grain and Animal Health Research – USDA-ARS, USA; Nida Ghori, Kansas State University, USA; and Guihua Bai and Xuming Liu, Hard Winter Wheat Genetics Research Unit, Center for Grain and Animal Health Research – USDA-ARS, USA

1 Introduction

The Hessian fly, *Mayetiola destructor*, is a gall midge that belongs to the family Cecidomyiidae (Gagne et al., 1991). In addition to the Hessian fly, there are two other closely related gall midges, the barley midge *M. hordei* and the oat midge *M. avenae*, which are all important agricultural pests (Gagne et al., 1991; Barnes, 1956; Cherif and Jemaa 2018). Morphologically, these three species are very similar and hard to be distinguished from each other with the naked eye. They also produce very similar symptoms in attacked plants. The main differences among the three gall midge species are their host specificities. Barley midge attacks only barley and oat midge attacks only oat. The Hessian

http://dx.doi.org/10.19103/AS.2022.0114.09

fly, however, has a broader host range in addition to its primary host, wheat, including barley, triticale, rye, and some wild grass species (Jones, 1936, 1938, 1939; Li, 2012). The Hessian fly can cause significant damage to barley near the east coast and southern plains in the United States in some years even though Hessian fly larvae grow less favorably on barley than wheat in the laboratory (Chen et al., 2009b).

2 Distribution and biology

The Hessian fly is a destructive pest of wheat in North Africa, North America, southern Europe, and West Asia (Harris et al., 2003, 2015). The insect is believed to have originated from West Asia in the Fertile Crescent along with its host wheat. The insect became well known as a pest after its first serious outbreak on the East Coast of the United States in the 1780s, which attracted the attention of prominent Americans including President George Washington, many of whom were farmers. The common name of the Hessian fly is based on the belief that the insect was brought to the United States via straw shoes and beddings used by soldiers from Hesse, Germany, who were mercenaries during the American Revolutionary period between 1765 and 1791 (Osborn, 1898; Pauly, 2002).

Like other dipterans, the Hessian fly has four distinct stages: eggs, larvae, pupae, and adults (Gagné and Hatchett, 1989). Hessian fly adults do not feed, living for only 3–5 days. Females deposit eggs on the adaxial surface of young leaves of the host plant. After hatching, neonates migrate down the leaf surface to the node and enter the wheat plant at the whorl. Neonate larvae can move freely between leaf sheaths and actively puncture wheat stems near the base of seedlings or close to the node of adult plants with their small, but sharp mandibles to establish a permanent feeding site (Hatchett et al., 1990; Stuart et al., 2012). As larvae grow bigger, they no longer migrate within wheat plants and rely on the established feeding sites for nutrients to support their development. First and second instars, each lasting for ~5 days under room temperatures, are the actively feeding stages. At the end of the second instar, the cuticle of the larva becomes brown and hardened, forming a puparium within which is the third non-feeding instar larva (Hatchett et al., 1990). Hessian fly puparia are called 'flaxseeds' based on their morphology, size, and color, which resemble flaxseeds (*Linum usitatissimum*). The puparium is an important stage for the insect to adapt to changes in living environments since it can enter diapause, which can be triggered by either high or low temperatures (Zhukovskii, 1950; Stuart et al., 2012). In the United States, Hessian fly puparia enter diapause during hot summer or at temperatures below 5°C during winter. Once the larva enters diapause, the dormancy lasts for 2–3 months before it can be broken. After that interval, diapause can be broken if the puparia are exposed to room temperatures (≥ 21°C) for ~10 consecutive days.

3 Feeding mechanisms and damage

Hessian fly larvae are insects with so-called piecing-sucking mouthparts. A pair of small, but very sharp mandibles insert into wheat tissues to facilitate saliva injection and nutrient ingestion. Because of the small size of the mandibles, which can hardly penetrate a single cell physically, Hessian fly larvae attack host plants mainly by injecting saliva into host plant tissues during feeding (Byers and Gallun, 1972; Hatchett et al., 1990; Chen et al., 2008; Wang et al., 2018; Aljbory et al., 2018; Aljbory et al., 2020). The substances injected into plants by the insect suppress host defense, reprogram host metabolism, induce the formation of nutritive cells at the feeding site, inhibit host plant growth, and protect attacked plants from secondary infection (Byers and Gallun, 1972; Zhu et al., 2008; Harris et al., 2006; Giron et al., 2016). Upon Hessian fly attack, defense-related genes are downregulated, whereas nutritive accumulation pathways are upregulated in susceptible plants (Liu et al., 2007; Zhu et al., 2008; Zhu et al., 2012; Khajuria et al., 2013). Nutritive cells formed at the feeding site are morphologically different from normal cells. The cell wall of nutritive cells is weakened and very thin and might be easily broken by the negative pressure resulting from insect sucking. The nuclei of nutritive cells are enlarged and cellular substructures are fragmented. Broken cell wall, cytoplasm, and disintegrated cellular organelles comprise a nutritional fluid that may be readily ingested by the larva (Harris et al., 2006). The concentration of chlorophyll increases in the center leaves of Hessian fly-attacked plants, giving them a darker green appearance, which is likely due to the inhibition of expansion of these leaves caused by larval feeding (Byers and Gallun, 1972). The elevated levels of chlorophyll may result in higher rates of photosynthesis per unit area in the infested compared to non-infested plants. Alternatively, source-sink perturbations due to Hessian fly attack may result in altered photosynthesis rates in plant tissues under different conditions (Gramig and Harris, 2015). The combination of weakened plant defense, shift of metabolic pathways toward the accumulation of soluble nutrients, and enhanced photosynthesis benefit Hessian fly parasitism.

Wheat yield losses due to Hessian fly attacks vary from year to year and from region to region. Some studies suggest an average 5% loss of wheat production due to Hessian fly damage (Buntin et al., 1992; Smiley et al., 2004). At the seedling stage, attacked plants eventually die after Hessian fly larvae cease feeding unless new tillers grow out from the attacked main tillers. In the East Coast and southern plains, seedling mortality is as high as 80% after Hessian fly attack and farmers simply abandon the wheat fields if this occurs (Comis, 2007; Smith, 2007). At mature plant stages, Hessian fly attack can cause plant lodging, inhibit full heading, reduce the number of grains per head, and/or reduce grain quality (Buntin, 1999; Schwarting et al., 2016).

4 Hessian fly monitoring and population dynamics

Because Hessian fly larvae live between the leaf sheath and stem of wheat plants creating no visible physical damage at early infestation stages, it is hard to monitor the presence of Hessian fly in a field. When plant death or lodging occurs in the field, it is too late to deploy any control measures. Pheromone traps have been developed to monitor the population dynamics of adult flies in a field (Schwarting et al., 2015). However, historical data of Hessian fly adult population dynamics are not very useful for predicting larval occurrence and density because many other factors affect larval populations in the field. One of the most important factors affecting larval density in the field is humidity because Hessian fly larvae will not hatch if the humidity within the micro-environment close to the leaf surface (the leaf boundary layer) is too low during egg development. Low ambient humidity plus strong wind can produce a dry boundary layer. Temperature also affects Hessian fly populations. First, temperature affects Hessian fly development at all stages. In the Northern Great Plains such as in Kansas, the Hessian fly has only two generations annually: a fall generation and a spring generation (Whitworth et al., 2009). In the coastal regions and Southern Great Plains such as in Georgia, the Hessian fly can have four generations annually, which can greatly increase fly population densities. Temperature also affects Hessian fly survival and the interactions between the Hessian fly and its host plants (Buntin and Chapin, 1990). Higher temperatures accelerate Hessian fly development and weaken host basal defense, leading to an increase in Hessian fly populations and plant damage (Cherif et al., 2021). On the other hand low humidity in combination with low temperatures and strong wind results in low larval hatching rates and therefore, low larval density even when egg density is high.

5 Advances in integrated management

In the first few decades when the Hessian fly was brought to the United States, the pest was mainly controlled by cultural practices including crop rotation, deep tilling to bury puparia, destruction of volunteer wheat, and delayed planting (planting during so-called fly-free dates) (Parks, 1917; Drake et al., 1924). The purpose of crop rotation, deep tilling, and destruction of volunteer wheat is to reduce the source of flies for initial infestations in the new season. Hessian fly puparia diapause during the late summer in wheat stubble from the previous year and can emerge as adults to infest new seedlings in the subsequent season. Therefore, planting continuous wheat should be avoided to break the chain of Hessian fly infestation. Deep tilling can bury puparia so that emerging adults are prevented from reaching the surface to mate and search for hosts. Since major hosts of the Hessian fly are only from the tribe Triticeae, non-Triticeae tribe crops such as corn can be rotated with wheat to break the chain infestation from one year to another. Cultural measures are effective, but hard to execute area-wide among wheat growers with different practices (till vs no-till, for example) and purposes (grazing vs grain production) (Derpsch et al., 2010).

Wheat planted after fly-free dates can reduce fly damage. Hessian fly adults and larvae die before fly-free dates because wheat seedlings are not available for egg deposition and feeding. However, this strategy is not effective in the southern United States where Hessian fly adults emerge throughout the entire winter season. In the Pacific Northwest, the Hessian fly has been a problem in spring wheat, where fly-free planting date has not been feasible. Climate change and annual temperature fluctuation affect Hessian fly emergence and cause the fly-free dates to differ from year to year. Delayed planting may also result in smaller plants, resulting in forage losses that are especially important for dual purposes. Delayed planting of winter wheat may also result in lower germination, higher winter cold injury, and lower grain yields (Buntin et al., 1992). In addition, current fly-free dates were established in the 1960s. Recent pheromone trap data have shown the presence of Hessian fly adults even in late November in Kansas (Schwarting et al., 2015). Therefore, the current fly-free dates are outdated and need to be re-evaluated due to temperature increases.

Seed treatment with systemic insecticides can kill fly larvae and is effective for up to three weeks after the emergence of seedlings (Wilde et al., 2001). Field application of foliar insecticides for Hessian fly control based on thresholds needs accurate monitoring of fly populations, which is difficult to achieve as mentioned previously. Another factor to diminish the effectiveness of insecticides for Hessian fly control is that once the nutritive cells are formed, the damage to the plants is irreversible even though the larvae are killed (Brown, 1960; Buntin and Hudson, 1991; Harris et al., 2006).

Currently, the main emphasis on controlling Hessian fly damage is to develop and deploy resistant wheat cultivars (Carver et al., 2001; Berzonsky et al., 2003). Resistance in wheat to Hessian fly larvae is antibiotic, as the larvae die on resistant wheat without further development. Breeding wheat with resistance to Hessian fly is one of the focuses in major wheat-growing regions of the United States. Currently, ~15% of commercial wheat cultivars released by public and private breeders in the United States have resistance to Hessian flies (Ehmke, 2022). There is also an international effort for breeding wheat with resistance to Hessian fly, which is coordinated by the International Center for Agricultural Research in Dry Areas (ICARDA) and the International Maize and Wheat Improvement Center (CIMMYT) (https://www.icarda.org/media/news/strengthening-wheat-resistance-insect-pests).

In addition to host plant resistance, biological control for Hessian fly has also been explored. A total of 41 species of Hessian fly natural enemies have been reported in North America and Europe (REF). Most are parasitoid Hymenoptera belonging to the superfamily Chalcidoidea. The majority of these parasitoids parasitize puparia of the spring fly generation, while only a few of them parasitize eggs of fall generations (Gahan, 1933). The parasitoids that attack Hessian fly puparia are not very effective since the damage of Hessian fly to wheat is caused by larvae. Three parasitoids including *Platygaster hiemalis*, *Homoporus destructor,* and *Eupelmus allynii* are widespread, but their effectiveness to suppress Hessian fly populations in fields remains to be determined (Hill, 1933; Schuster and Lidell, 1990; Chen et al., 1991; Bullock et al., 2004). Among the three widespread parasitoids, *P. hiemalis* is more promising in suppressing Hessian fly damage since it parasites eggs whereas *H. destructor* and *E. allynii* both parasite puparia (Pachard, 1916; Schuster and Lidell, 1990; Chen et al., 1991). Overall variation in parasitism rates from year to year and from location to location also makes biological control difficult (Wise, 2007).

6 Advances in identification and mapping of resistance genes

Ever since resistance to Hessian fly in wheat was observed in the field (Painter, 1951), identification and characterization of new Hessian fly resistance genes in wheat and related species have been a continuous effort. To date, 37 Hessian fly resistance genes, named *H1-H36* and *Hdic*, have been identified (Zhang et al., 2021). As shown in Table 1, most resistance genes are from common, durum, and discoccum wheat, and are clustered on chromosome 1A. Among the named genes, *H1*, *H2*, *H3*, *h4*, *H5*, *H7*, *H8*, *H12*, *H32*, and *H34* are from common wheat. *H6*, *H9*, *H10*, *H11*, *H14*, *H15*, *H16*, *H17*, *H18*, *H19*, *H20*, *H28*, *H29*, *H31*, and *H33* are from durum wheat (Liu, et al., 2005b), *H13*, *H22, H23*, *H24*, and *H26* are from *Aegilops tauschii* (Gill et al., 1987; Raupp et al., 1993; Wang et al., 2006), *H21* and *H25* are from Secale cereal (rye) (Friebe et al., 1999), *H27* is from *Aegilops ventricose* (Delibes et al., 1997; Ohm et al., 1997), *H30* is from *Aegilops triuncialis* (Martín-Sánchez et al., 2003), and *Hdic* is from Emmer wheat *Triticum* discoccum (Liu et al., 2005a). Although these genes are distributed on different chromosomes of all 3 genomes, 19 (about half) of them are from the A genome, with 16 of A genome genes, including *H3*, *h4*, *H5*, *H6*, *H9*, *H10*, *H11*, *H12*, *H14*, *H15*, *H16*, *H17*, *H19*, *H28*, *H29,* and *Hdic*, clustered within an ~1 cM region at the distal end of chromosome 1A (Liu et al., 2005b; Niu et al., 2020). Resistance genes on other chromosomes of the A genome include *H7* (6AL), *H33* (3AS), and *H36* (7AS) (Liu et al., 2020; Niu et al., 2020). Only seven genes, including*, H13*, *H22*, *H23*, *H24*, *H26*, *H30,* and *H32,* were

Table 1 Hessian fly resistance genes: sources, cultivars, and chromosome locations

Name	Original source (genome)	Germplasm or Cultivar	Chromosome location	References
H1	*Triticum aestivum*	Dawson, Seneca	/	Noble, 1943
H2	*T. aestivum*	Clawson, Dawson, Seneca	/	Noble, 1943
H3	*T. aestivum*	Arthur, Carol, Larned, Monon, Oasis, W38	1AS^	Stebbins et al., 1982; Liu et al., 2005b
*h4**	*T. aestivum*	Dixon, Java (Cltr 12763)	1AS	Suneson and Noble, 1950; Niu et al., 2020
H5	*T. aestivum*	Arthur71, Erin, Magnum	1AS	Roberts and Gallun, 1984; Liu et al., 2005b
H6	*T. turgidum subsp. durum*	Caldwell, Flynn, Knox62, Lathrop	1AS^	Stebbins et al., 1982; Liu et al., 2005b
H7H8#	*T. aestivum*	Adena, Seneca	6AL(H7) 2B(H8)	Liu et al., 2020
H9	*T. turgidum subsp. durum*	Ella, Elva, Iris	1AS	Liu et al., 2005b; Kong et al., 2005
H10	*T. turgidum subsp. durum*	Elva, Joy	1AS	Liu et al., 2005b
H11	*T. turgidum subsp. durum*	Karen, Kay	1AS	Liu et al., 2005b
H12	*T. turgidum subsp. durum*	Lola, Luso	1AS^	Liu et al., 2005b
H13	*T. tauschii (DD)*	Molly	6DS	Gill et al., 1987; Liu et al., 2005c
H14	*T. turgidum subsp. durum*	ELS 6404-160-5, D6647 (921676A3-5)	1AS^	Mass et al., 1989; Liu et al., 2005b
H15	*T. turgidum subsp. durum*	ELS 6404-160-5, 81602C5-3-3-8-1	1AS^	Mass et al., 1989; Liu et al., 2005b
H16	*T. turgidum subsp. durum*	PI 94587, D6647 (921680A4-6)	1AS	Liu et al., 2005b; Kong et al., 2008
H17	*T. turgidum subsp. durum*	DW28, D6647 (921680D1-7)	1AS	Liu et al., 2005b; Kong et al., 2008
H18	*T. turgidum subsp. durum*	Marquillo, Redland	/	Maas et al., 1987; Obanni et al., 1988
H19	*T. turgidum subsp. durum*	PI 422297, Sincape 90, 84702B14-1-3-4-3	1AS^	Ohm et al., 1997; Liu et al., 2005b

(Continued)

Table 1 (*Continued*)

Name	Original source (genome)	Germplasm or Cultivar	Chromosome location	References
H20	*T. turgidum subsp. durum*	Jori	2B	Amri et al., 1990
H21	*Secale cereale* (Rye, RR)	Hamlet, KS89WGRC8, KS09WGGRC51	T2BS.2RL	Friebe et al., 1990; Hatchett et al., 1993
H22	*T. tauschii*	TA 1644, KS85WGRC1	1DS	Raupp et al., 1993; Zhao et al., 2006
H23	*T. tauschii*	TA1642, KS89WGRC3	6DS	Raupp et al., 1993; Ma et al., 1993
H24	*T. tauschii*	TA1645, KS89WGRC6	3DL	Raupp et al., 1993
H25	*S. cereale* (Rye, RR)	KS92WGRC17, 18, 19, 20; KS98WGRC41	6BL, 4BL, 4BL, 4AL, 4AL	Friebe et al., 1991
H26	*T. tauschii*	TA2473, KS93WGRC26	3DL	Wang et al., 2006; Yu et al., 2009
H27†	*Aegilops ventricosa (DDMvMv)*	H-93-33, PI422297	4D/4Mv	Delibes et al., 1997; Ohm et al., 1997
H28	*T. turgidum subsp. durum*	PI 59190	1AS^	Liu et al., 2005b
H29†	*T. turgidum subsp. durum*	Sincape 90	1AS^	Liu et al., 2005b
H30	*Ae. triuncialis*	TR-3531	/	Martín-Sánchez et al., 2003
H31	*T. turgidum subsp. durum*	P921696 (PI 633876), CI3984	5BS	Williams et al., 2003
H32	Synthetic wheat	W-7984	3DL	Sardesai et al., 2005
H33	*T. turgidum subsp. durum*	PI 134942	3AS	McDonald et al., 2014
H34	*T. aestivum*	Clark	6B	Li et al., 2013
H35	*T. aestivum*	SD06165, Overland	3BS	Zhao et al., 2020
H36	*T. aestivum*	SD06165, Overland	7AS	Zhao et al., 2020
Hdic	*T. turgidum subsp. dicoccum*	PI 94641, KS99WGRC42	1AS	Brown-Guedira et al., 2005; Liu et al., 2005a

*Recessive; # Effective only when both genes are together as a combination; ^Reassigned based on new data; †*H27* and *H29* are Likely the same gene (Ohm et al., 1997); / No information available.

mapped to the D genome (Liu et al., 2015c; Wang et al., 2006), and eight genes, including *H8*, *H18*, *H20*, *H21*, *H25*, *H31*, and *H34*, were mapped to the B genome (Wang et al., 2006; Li et al., 2013). In addition to the named genes, two new major quantitative trait loci (QTLs) conferring HF resistance, *QH.icd -2A* and *QHara.icd-6B*, were identified by association mapping and transferred to wheat from *T. dicoccum* and *T. araraticum*, respectively (Bassi et al., 2019). The allelic relationships between the two QTLs and other known Hessian fly resistance genes in these chromosome regions remain to be determined. Several minor QTLs may also contribute additive effects to major genes to enhance the levels of Hessian fly resistance in wheat cultivars (Bassi et al., 2019; Tan et al., 2013; Li et al., 2015). Many of the identified resistance genes are being used in various breeding programs to combat Hessian fly infestations in the field. Significant efforts have been made in the fine mapping and cloning of several major Hessian fly resistance genes. Recently, a candidate gene for *H13* has been identified (Joshi, 2018). However, no cloned Hessian fly resistance genes have been officially published.

In addition to resistance genes, a Hessian fly susceptibility gene, named *Mayetiola destructor susceptibility gene-1* (*Mds-1*), has been identified from wheat. Disruption of the susceptibility gene conferred wheat resistance, and this resistance appears to be biotype independent. In addition to susceptibility to Hessian fly, *Mds-1* encodes a small heat shock protein that is essential to wheat development. Because of this, further research is needed before *Mds-1* can be employed to breed Hessian fly resistant wheat (Liu et al., 2013).

7 Advances towards understanding the mechanism of wheat resistance to Hessian fly

In resistant plants, Hessian fly larvae die without any apparent development. Based on transcriptomic and metabolic evidence, two types of defensive reactions, rapid induction of toxicity and swift fortification of cell wall, may be crucial for wheat resistance to Hessian fly (Wu et al., 2008; Liu et al., 2010; Khajuria et al., 2013). Upon Hessian fly attack, genes encoding toxic substances including protease inhibitors, enzymes for synthesizing toxic chemicals such as reactive oxygen species and secondary metabolites, and defensive proteases, are upregulated swiftly (Liu et al., 2007; Wu et al., 2008; Zhu et al., 2012; Khajuria et al., 2013; Subramanyam et al., 2019). Toxic chemicals in resistant plants are evidently not lethal since the Hessian fly larvae from resistant plants can survive if they are transferred to susceptible plants (Khajuria et al., 2013). However, toxic chemicals from host plants can slow down the speed of Hessian fly larval attack on wheat plants, giving more time for wheat plants to launch additional defenses. Rapid fortification of cell walls may play an essential role in wheat resistance to Hessian fly larvae (Liu et al., 2007; Williams et al., 2011; Khajuria et al., 2013). After Hessian fly attack, genes encoding proteins that participate in the synthesis of cell wall components and cell fortification enzymes are rapidly upregulated. Primary metabolism is almost shut down temporarily to save metabolic intermediates that can be used for cell wall fortification (Zhu et al., 2008). Membrane lipids are rapidly mobilized for producing components such as wax for cell wall strengthening (Kosma et al., 2010; Zhu et al., 2012; Khajuria et al., 2013). As a result of cell wall fortification, Hessian fly effectors cannot penetrate host tissues (Aljbory et al., 2020). Cell wall permeability in resistant plants is greatly reduced in comparison with susceptible plants attacked by Hessian fly larvae (Williams et al., 2011; Khajuria et al., 2013).

8 Advances towards understanding the mechanisms of Hessian fly biotype differentiation

Like many pathogen-plant interactions, the Hessian fly interacts with wheat in a gene-for-gene manner (Flor, 1971). The product of a resistance gene in wheat recognizes the product of the corresponding avirulence gene in Hessian fly to initiate plant defense, resulting in host plant resistance phenotype. Many different Hessian fly biotypes have been identified in the field that can overcome resistance conferred by certain resistance genes (Kudagamage et al., 1990; Ratcliffe et al., 2000; Naber et al., 2003; Chen et al., 2009a; Garcés-Carrera et al., 2014). The rapid shift in prevalent biotypes after the deployment of resistant wheat cultivars poses a major challenge to the host plant resistance strategy for the Hessian fly. Typically, a cultivar containing a specific resistance gene loses effectiveness within 5-7 years after its deployment (Gould, 1986; Shukle et al., 2016). It is essential to understand the molecular mechanisms for rapid biotype shifts in order to improve the durability of resistant cultivars.

Based on the gene-for-gene interactions, major efforts have been implemented to map and clone avirulence genes from the Hessian fly. To date, *vH6*, *vH9*, *vH13*, and *vHdic* genes have been mapped to specific chromosome regions (Schulte et al., 1999; Aggarwal et al., 2014; Zhao et al., 2016; Navarro-Escalante et al., 2020). Candidate genes for these avirulence genes have been identified and confirmed in some cases. All avirulence genes identified so far encode proteins with a secretion signal peptide and are expressed in salivary glands, the major tissue to produce effector proteins for host plant injection in insects. The genes encoding secreted proteins and expressed in salivary glands are called Secreted Salivary Gland Proteins (SSGPs), which are likely responsible for rapid Hessian fly biotype shift (Chen et al., 2010; Johnson et al., 2015).

The genes encoding SSGPs are systematically analyzed via transcriptomic analyses with dissected salivary glands (Chen et al., 2008; Aljbory et al.,

2018). Over 1500 genes were found encoding SSGPs. The SSGP-encoding genes belong to several families (Chen et al., 2004; Chen et al., 2016). Genes within a family are clustered within short chromosome regions and exhibit unconventional conservation patterns among family members (Chen et al., 2004; Chen et al., 2010). Among the family members of SSGP-encoding genes, the 5′-untranslated regions (UTRs), introns, and 3′-UTRs are highly conserved, whereas the regions encoding mature proteins are highly diversified. The clustered organization and unconventional conservation among SSGP-encoding genes may elevate mutation rates specifically in SSGP-coding regions via unequal crossings during meiosis (Chen et al., 2010). Indeed, SSGP proteins are highly diversified even among family members. The presence of many SSGPs (putative effectors) and a mechanism to produce high rates of mutations within SSGP-encoding genes may explain virulence diversification and rapid shifts of Hessian fly biotypes in the field.

9 Climate change and Hessian fly management

Global climate change poses a serious challenge to agriculture in general. The increase in temperatures due to climate change has a significant impact on wheat-Hessian fly interactions and the management of Hessian fly damage in fields (Sosa and Foster, 1976; Chen et al., 2014). The impact of higher temperatures on the interactions between wheat and the Hessian fly is complex. On the one hand, higher temperatures accelerate Hessian fly development, which could result in more generations of the insect within a year and could also spread to new wheat-growing regions where Hessian fly was not a problem due to colder temperatures. Higher temperatures can also speed up Hessian fly feeding on wheat plants, resulting in more severe damage to wheat plants and enhanced wheat susceptibility. On other hand, higher temperatures could change or even disrupt certain metabolic pathways of host plants. The impact of higher temperatures on wheat-Hessian fly interactions is also dependent on specific wheat cultivars and Hessian fly biotypes (Chen et al., 2014)

Nearly all identified Hessian fly resistance genes are temperature sensitive, losing effectiveness above a certain temperature. However, the temperature limits above which resistance genes lose effectiveness may vary with different resistance genes (Tyler and Hatchett, 1983; Chen et al., 2014; Tang et al., 2018). For example, cultivars containing resistance genes *H17*, *H18*, and *H19* lose resistance above 18°C, cultivars containing *H3*, *H29*, *H5*, *H20*, and *H28* lose resistance above temperature 24°C, cultivars containing *H10*, *Hdic*, *H11*, *H6*, *H22*, and *H25* lose resistance above temperature 28°C, whereas cultivars containing *H13*, *H9*, *H32*, *H15*, and *H26* are more temperature tolerant, and they become susceptible above 32°C (Tang et al., 2018). With elevated temperatures due to climate changes, those resistance genes that are more sensitive to temperatures, such as *H17*, *H18*, and *H19*, may not provide any protection against Hessian fly damage. Those resistance genes that are more tolerant to temperature changes including *H3*, *H29*, *H5*, *H20*, and *H28* are likely still effective to protect wheat seedlings from Hessian fly attack during fall and winter seasons, but may not offer good protection of wheat adult plants against Hessian fly attack during late spring and early summer. Resistance genes such as *H13*, *H9*, *H32*, *H15*, and *H26* are more heat resistant and should offer good protection for wheat at both seedling and adult plant stages (Tang et al., 2018).

10 Conclusion and future trends

The Hessian fly has been a major challenge to wheat production for a long time and will likely remain as one of the major problems for agriculture in the future. Therefore, we need to improve our existing technologies and develop new means to combat this destructive pest. We need to improve our monitoring and prediction systems to more accurately predict where and when Hessian fly outbreaks will occur. Pheromone traps can monitor population dynamics of adult flies in the field, but adult data alone cannot predict the actual levels of larval infestation. That's because egg development and larval hatch rates are profoundly affected by temperatures, humidity, and wind. During the past decades, supercomputing and artificial intelligence are advancing rapidly to solve complex practical problems such as pest-host-environment interactions and predictions. Accumulated data on temperatures, humidity, wheat cultivars, fly dynamics, applied controlling measures, and actual wheat damage levels across years and regions may allow us to establish more reliable models and monitoring systems for Hessian fly outbreak prediction.

With higher temperatures due to climate change, no-till practice and dual (grazing and harvest) purpose wheat in some regions, fly-free dates become not feasible in some areas or need to be reevaluated in other regions. To offer early protection of young wheat seedlings from Hessian fly attack, seed treatments with systemic insecticides shall be continued and improved based on local practices. Since seed treatment is a common practice to manage several types of plant diseases, it may be economic to combine seed treatments for controlling the Hessian fly and other pests at the same time (Hunger et al., 2017).

Developing and deploying resistant cultivars are likely to remain the main strategy to control Hessian fly damage in the future. At present, only 13–25%, depending on different regions, released cultivars by public and private breeders carry Hessian fly resistance genes. The percentage of Hessian fly resistant cultivars will have to increase, especially in the Southern states such as Oklahoma and Texas, in order to control Hessian fly damage effectively. Markers and marker-assisted breeding will accelerate breeding for Hessian fly-resistant cultivars and also facilitate the pyramiding of multiple resistance genes in the same cultivars to improve durability of host resistance. Unfortunately, only a few functional markers for Hessian fly resistance genes are currently available for marker-assisted breeding. More markers useful for breeding are critical to the success of genetic improvements in Hessian fly resistance in wheat. Continuing and constant monitoring of changes in Hessian fly biotypes in the field is also needed to select effective resistance genes in different regions.

New advances in biotechnologies such as gene editing and RNA interference provide new opportunities for Hessian fly management. In order to use these and other emerging new technologies to combat the Hessian fly, a better understanding of wheat and Hessian fly interactions at the molecular level is needed. More fundamental research on wheat defense mechanisms and Hessian fly attacking mechanisms is needed to provide a solid foundation for utilizing new biotechnologies in Hessian fly management.

11 Acknowledgements

Mention of trade names or commercial products in this publication is solely for the purpose of providing specific information and does not imply recommendation or endorsement by the US Department of Agriculture (USDA). The USDA is an equal opportunity provider and employer.

12 References

Aggarwal, R., Subramanyam, S., Zhao, C., Chen, M. S., Harris, M. O. and Stuart, J. J. (2014). Avirulence effector discovery in a plant galling and plant parasitic arthropod, the Hessian fly (*Mayetiola destructor*). *PLoS ONE* 9(6), e100958.

Aljbory, Z., Aikins, M. J., Park, Y., Reeck, G. R. and Chen, M. S. (2020). Differential localization of Hessian fly candidate effectors in resistant and susceptible wheat plants. *Plant Direct* 4(8), e00246.

Aljbory, Z., Anderson, K. M., Harris, M. O., Mittapalli, O., Whitworth, R. J. and Chen, M. S. (2018). Transcriptomic analyses of secreted proteins from the salivary glands of wheat midge larvae. *J. Insect Sci.* 18(1), 17.

Amri, A., Cox, T. S., Gill, B. S. and Hatchett, J. H. (1990). Chromosomal location of the Hessian fly resistance gene *H20*. *J. Hered.* 81(1), 71–72.

Barnes, H. F. (1956). Gall midges of economic importance. *Gall Midges Cereal Crops* VII, 95–141.

Bassi, F. M., Brahmi, H., Sabraoui, A., Amri, A., Nsarellah, N., Nachit, M. M., Al-Abdallat, A., Chen, M. S., Lazraq, A. and El Bouhssini, M. (2019). Genetic identification of loci for Hessian fly resistance in durum wheat. *Mol. Breed.* 39(2), 24.

Berzonsky, W. A., Ding, H., Haley, S. D., Harris, M. O., Lamb, R. J., McKenzie, R. I. H., Ohm, H. W., Patterson, F. L., Peairs, F., Porter, D. R., Ratcliffe, R. H. and Shanower, T. (2003). Breeding wheat for resistance to insects. *Plant Breed. Rev.* 22, 221–296.

Brown, H. E. (1960). Insecticidal control of the Hessian fly. *J. Econ. Entomol.* 53(4), 501–503.

Brown-Guedira, G. L., Hatchett, J. H., Liu, X. M., Fritz, A. K., Owuoche, J. O., Gill, B. S., Sears, R. G., Cox, T. S. and Chen, M. S. (2005). Registration of KS99WGRC42 Hessian fly-resistant hard red winter wheat germplasm. *Crop Sci.* 45(2), 804-805.

Bullock, D. G., Bosque-Pérez, N. A., Johnson, J. B. and Merickel, F. W. (2004). Species composition and distribution of Hessian fly (Diptera: Cecidomyiidae) parasitoids.

Buntin, G. D. (1999). Hessian fly (Diptera: Cecidomyiidae) injury and loss of winter wheat grain yield and quality. *J. Econ. Entomol.* 92(5), 1190-1197.

Buntin, D. G. and Chapin, J. W. (1990). Biology of Hessian fly (Diptera: Cecidomyiidae) in the Southeastern United States: Geographic variation and temperature-dependent phenology. *J. Econ. Entomol.* 83(3), 1015-1024.

Buntin, D. G. and Hudson, R. D. (1991). Spring control of the Hessian fly (Diptera: Cecidomyiidae) in winter wheat using insecticides. *J. Econ. Entomol.* 84(6), 1913-1919.

Buntin, G. D., Ott, S. L. and Johnson, J. W. (1992). Integration of plant resistance, insecticides, and planting date for management of the Hessian fly (Diptera: Cecidomyiidae) in Winter wheat. *J. Econ. Entomol.* 85(2), 530-538.

Byers, R. A. and Gallun, R. L. (1972). Ability of the Hessian fly to stunt winter wheat: Effect of larval feeding on elongation of leaves. *J. Econ. Entomol.* 65(4), 955-958.

Carver, B. F., Khalil, I., Krenzer, E. G. and MacKown, C. T. (2001). Breeding winter wheat for a dual-purpose management system. *Euphytica* 119(1/2), 231-234.

Chen, B. H., Foster, J. E., Araya, J. E. and Taylor, P. L. (1991). Parasitism of *Mayetiola destructor* (Diptera: Cecidomyiidae). *J. Entomol. Sci.* 26(2), 237-243.

Chen, M. S., Echegaray, E., Whitworth, R. J., Wang, H., Sloderbeck, P. E., Knutson, A., Giles, K. L. and Royer, T. A. (2009a). Virulence analysis of Hessian fly populations from Texas, Oklahoma, and Kansas. *J. Econ. Entomol.* 102(2), 774-780.

Chen, M. S., Liu, X., Wang, H. and El-Bouhssini, M. (2009b). Hessian fly (Diptera: Cecidomyiidae) interactions with barley, rice, and wheat seedlings. *J. Econ. Entomol.* 102(4), 1663-1672.

Chen, M. S., Fellers, J. P., Stuart, J. J., Reese, J. C. and Liu, X. M. (2004). A group of related cDNAs encoding secreted proteins from Hessian fly [Mayetiola destructor (Say)] salivary glands. *Insect Mol. Biol.* 13(1), 101-108.

Chen, M. S., Liu, S., Wang, H., Cheng, X., El Bouhssini, M. and Whitworth, R. J. (2016). Genes expressed differentially in Hessian fly larvae feeding in resistant and susceptible plants. *Int. J. Mol. Sci.* 17(8), 1324.

Chen, M. S., Liu, X., Yang, Z., Zhao, H., Shukle, R. H., Stuart, J. J. and Hulbert, S. (2010). Unusual conservation among genes encoding small secreted salivary gland proteins from a gall midge. *BMC Evol. Biol.* 10(1), 296.

Chen, M. S., Wheeler, S., Wang, H. and Whitworth, R. J. (2014). Impact of temperatures on Hessian fly (Diptera: Cecidomyiidae). *J. Econ. Entomol.* 107(3), 1266-1273.

Chen, M. S., Zhao, H. X., Zhu, Y. C., Scheffler, B., Liu, X. M., Liu, X., Hulbert, S. and Stuart, J. J. (2008). Analysis of transcripts and proteins expressed in the salivary glands of Hessian fly (*Mayetiola destructor*) larvae. *J. Insect Physiol.* 54(1), 1-16.

Cherif, A. and Jemaa, J. M. B. (2018). Distribution, population dynamics and damage potential of barley stem gall midge, *Mayetiola hordei* (Diptera: Cecidomyiidae) on cultivated barley in two semi-arid areas of North Tunisia. *Crop Prot.* 112, 295-303.

Cherif, A., Rezgui, M., Titouhi, F., Youssfi, S., Soltani, A., Barg, S., El Bouhssini, M. and Mediouni Ben Jemâa, J. (2021). Distribution, population dynamics and damage of Hessian fly, *Mayetiola destructor* (Diptera: Cecidomyiidae) in North Tunisia. *J. Appl. Entomol.* 145(3), 223-238.

Comis, D. (2007). Virulent Hessian flies renew attack on U.S. wheat. *United States Department of Agriculture Agricultural Research - News and Events* (January 24). Available at: http://www.ars.usda.gov/is/pr/2007/070124.htm.

Delibes, A., Moral, J. D. Del, Martín-Sánchez, J. A., Mejías, A., Gallego, M., Casado, D., Sin, E. and López-Braña, I. (1997). Hessian fly resistance gene transferred from chromosome *4MV* of *Aegilops ventricosa* to *Triticum aestivum*. *Theor. Appl. Genet.* 94(6-7), 858-864.

Derpsch, R., Friedrich, T., Kassam, A. and Li, H. (2010). Current status of adoption of no-till farming in the world and some of its main benefits. *Int. J. Agric. Biol. Eng.* 3, 1-25.

Drake, C. J., Fenton, F. A. and Butcher, F. D. (1924). The importance of the flaxseed count in predicting the actual fly-free date. *J. Econ. Entomol.* 17(4), 480-486.

Ehmke, L. (2022). *Wheat Varieties for Kansas and the Great Plains (Yearly Book)*. Healy, KS.

Flor, H. H. (1971). Current status of the gene-for-gene concept. *Annu. Rev. Phytopathol.* 9(1), 275-296.

Friebe, B., Hatchett, J. H., Gill, B. S., Mukai, Y. and Sebesta, E. E. (1991). Transfer of Hessian fly resistance from rye to wheat via radiation-induced terminal and intercalary chromosomal translocations. *Theor. Appl. Genet.* 83(1), 33-40.

Friebe, B., Hatchett, J. H., Sears, R. G. and Gill, B. S. (1990). Transfer of Hessian fly resistance from 'Chaupon' rye to hexaploid wheat via a 2BS/2RL wheat-rye chromosome translocation. *Theor. Appl. Genet.* 79(3), 385-389.

Friebe, B., Kynast, R. G., Hatchett, J. H., Sears, R. G., Wilson, D. L. and Gill, B. S. (1999). Transfer of wheat-rye translocation chromosomes conferring resistance to Hessian fly from bread wheat into durum wheat. *Crop Sci.* 39(6), 1692-1696.

Gagné, R. J. and Hatchett, J. H. (1989). Instars of the Hessian fly (Diptera: Cecidomyiidae). *Ann. Entomol. Soc. Am.* 82(1), 73-79.

Gagne, R. J., Hatchett, J. H., Lhaloui, S. and El Bouhssini, M. (1991). Hessian fly and barley stem gall midge, two different species of *Mayetiola* (Diptera: Cecidomyiidae) in Morocco. *Ann. Entomol. Soc. Am.* 84(4), 436-443.

Gahan, A. B. (1933). *The Serphoid and Chalcidoid Parasites of the Hessian Fly*. United States Department of Agriculture, 174-176.

Garcés-Carrera, S., Knutson, A., Wang, H., Giles, K. L., Huang, F., Whitworth, R. J., Smith, C. M. and Chen, M. S. (2014). Virulence and biotype analyses of Hessian fly (*Mayetiola destructor*) populations from Texas, Louisiana, and Oklahoma. *J. Econ. Entomol.* 107(1), 417-423.

Gill, B. S., Hatchett, J. H. and Raupp, W. J. (1987). Chromosomal mapping of Hessian fly-resistance gene *H13* in the D genome of wheat. *J. Hered.* 78(2), 97-100.

Giron, D., Huguet, E., Stone, G. N. and Body, M. (2016). Insect-induced effects on plants and possible effectors used by galling and leaf-mining insects to manipulate their host-plant. *J. Insect Physiol.* 84, 70-89.

Gould, F. (1986). Simulation models for predicting durability of insect-resistant germ plasm: Hessian fly (Diptera: Cecidomyiidae)-resistant winter wheat. *Environ. Entomol.* 15(1), 11-23.

Gramig, G. G. and Harris, M. O. (2015). Plant photosynthetic responses during insect effector-triggered plant susceptibility and immunity. *Environ. Entomol.* 44(3), 601-609.

Harris, M. O., Freeman, T. P., Rohfritsch, O., Anderson, K. G., Payne, S. A. and Moore, J. A. (2006). Virulent Hessian fly (Diptera: Cecidomyiidae) larvae induce a nutritive tissue during compatible interactions with wheat. *Ann. Entomol. Soc. Am.* 99(2), 305-316.

Harris, M. O., Friesen, T. L., Xu, S. S., Chen, M. S., Giron, D. and Stuart, J. J. (2015). Pivoting from *Arabidopsis* to wheat to understand how agricultural plants integrate responses to biotic stress. *J. Exp. Bot.* 66(2), 513-531.

Harris, M. O., Stuart, J. J., Mohan, M., Nair, S., Lamb, R. J. and Rohfritsch, O. (2003). Grasses and gall midges: Plant defense and insect adaptation. *Annu. Rev. Entomol.* 48(1), 549-577.

Hatchett, J. H., Kreitner, G. L. and Elzinga, R. J. (1990). Larval mouthparts and feeding mechanism of the Hessian fly (Diptera: Cecidomyiidae). *Ann. Entomol. Soc. Am.* 83(6), 1137-1147.

Hatchett, J. H., Sears, R. G. and Cox, T. S. (1993). Inheritance of resistance to Hessian fly in rye and wheat-rye translocation lines. *Crop Sci.* 33(4), 730-734.

Hill, C. C. (1933). Parasites of the Hessian fly in the North Central States. *USDA Circular*, 1-15.

Hunger, B., de Oliveira Silva, A., Royer, T. and Giles, K. (2017). Effect of planting date and seed treatment on diseases and insect pests of wheat. *Oklahoma Cooperative Extension Service Current Report CR-7088 Rev. 0812*. Available at: https://extension.okstate.edu/fact-sheets/print-publications/cr/effect-of-planting-date-and-seed-treatment-on-diseases-and-insect-pests-of-wheat-cr-7088.pdf.

Johnson, A. J., Shukle, R. H., Chen, M. S., Srivastava, S., Subramanyam, S., Schemerhorn, B. J., Weintraub, P. G., Abdel Moniem, H. E. M., Flanders, K. L., Buntin, G. D. and Williams, C. E. (2015). Differential expression of candidate salivary effector proteins in field collections of Hessian fly, *Mayetiola destructor*. *Insect Mol. Biol.* 24(2), 191-202.

Jones, E. T. (1936). hordeum grasses as hosts of the Hessian fly. *J. Econ. Entomol.* 29(4), 704-710.

Jones, E. T. (1938). Infestation of grasses of the genus *Aegilops* by the Hessian fly. *J. Econ. Entomol.* 31(3), 333-337.

Jones, E. T. (1939). Grasses of the tribe hordeae as hosts of the Hessian fly. *J. Econ. Entomol.* 32(4), 505-510.

Joshi, A. (2018). Map-based cloning of the Hessian fly resistance gene. In: *Wheat* (Doctoral dissertation, Kansas State University). Available at: https://www.proquest.com/docview/2118039884?pq-origsite=gscholar&fromopenview=trueH13.

Khajuria, C., Wang, H., Liu, X., Wheeler, S., Reese, J. C., El Bouhssini, M., Whitworth, R. J. and Chen, M. S. (2013). Mobilization of lipids and fortification of cell wall and cuticle are important in host defense against Hessian fly. *BMC Genom.* 14(1), 423.

Kong, L., Cambron, S. E. and Ohm, H. W. (2008). Hessian fly resistance genes *H16* and *H17* are mapped to a resistance gene cluster in the distal region of chromosome 1AS in wheat. *Mol. Breed.* 21(2), 183-194.

Kong, L., Ohm, H. W., Cambron, S. E. and Williams, C. E. (2005). Molecular mapping determines that Hessian fly resistance gene *H9* is located on chromosome 1a of wheat. *Plant Breed.* 124(6), 525-531.

Kosma, D. K., Nemacheck, J. A., Jenks, M. A. and Williams, C. E. (2010). Changes in properties of wheat leaf cuticle during interactions with Hessian fly. *Plant J.* 63(1), 31-43.

Kudagamage, C., Foster, J. E., Taylor, P. L. and Chen, B. H. (1990). Biotypes of the Hessian fly (Diptera: Cecidomyiidae) identified in the Southeastern United States. *J. Entomol. Sci.* 25(4), 575-580.

Li, C., Chen, M. S., Chao, S., Yu, J. and Bai, G. (2013). Identification of a novel gene, *H34*, in wheat using recombinant inbred lines and single nucleotide polymorphism markers. *Theor. Appl. Genet.* 126(8), 2065-2071.

Li, G., Wang, Y., Chen, M. S., Edae, E., Poland, J., Akhunov, E., Chao, S., Bai, G., Carver, B. F. and Yan, L. (2015). Precisely mapping a major gene conferring resistance to Hessian fly in bread wheat using genotyping-by-sequencing. *BMC Genom.* 16(1), 108.

Li, Y. (2012). *Crop and prairie grasses serving as hosts for the Hessian fly Mayetiola destructor (say) (Diptera: Cecidomyiidae).* Master Thesis, North Dakota State University, Fargo, ND.

Liu, G., Liu, X., Xu, Y., Bernardo, A., Chen, M., Li, Y., Niu, F., Zhao, L. and Bai, G. (2020). Reassigning Hessian fly resistance genes *H7* and *H8* to chromosomes 6A and 2B of the wheat cultivar 'Seneca' using genotyping-by-sequencing. *Crop Sci.* 60(3), 1488–1498.

Liu, X., Bai, J., Huang, L., Zhu, L., Liu, X., Weng, N., Reese, J. C., Harris, M., Stuart, J. J. and Chen, M. S. (2007). Gene expression of different wheat genotypes during attack by virulent and avirulent Hessian fly (*Mayetiola destructor*) larvae. *J. Chem. Ecol.* 33(12), 2171–2194.

Liu, X. M., Brown-Guedira, G. L., Hatchett, J. H., Owuoche, J. O. and Chen, M. S. (2005a). Genetic characterization and molecular mapping of a Hessian fly-resistance gene transferred from *T. turgidum* ssp. *dicoccum* to common wheat. *Theor. Appl. Genet.* 111(7), 1308–1315.

Liu, X. M., Fritz, A. K., Reese, J. C., Wilde, G. E., Gill, B. S. and Chen, M. S. (2005b). *H9*, *H10*, and *H11* compose a cluster of Hessian fly-resistance genes in the distal gene-rich region of wheat chromosome 1AS. *Theor. Appl. Genet.* 110(8), 1473–1480.

Liu, X. M., Gill, B. S. and Chen, M. S. (2005c). Hessian fly resistance gene *H13* is mapped to a distal cluster of resistance genes in chromosome 6DS of wheat. *Theor. Appl. Genet.* 111(2), 243–249.

Liu, X., Khajuria, C., Li, J., Trick, H. N., Huang, L., Gill, B. S., Reeck, G. R., Antony, G., White, F. F. and Chen, M. S. (2013). Wheat *Mds-1* encodes a heat-shock protein and governs susceptibility towards the Hessian fly gall midge. *Nat. Commun.* 4(1), 2070.

Liu, X., Williams, C. E., Nemacheck, J. A., Wang, H., Subramanyam, S., Zheng, C. and Chen, M. S. (2010). Reactive oxygen species are involved in plant defense against a gall midge. *Plant Physiol.* 152(2), 985–999.

Ma, Z. Q., Gill, B. S., Sorrells, M. E. and Tanksley, S. D. (1993). RELP markers linked to two Hessian fly resistance genes in wheat (*Triticum aestivum* L.) from *Triticum tauschii* (Coss.) Schmal. *Theor. Appl. Genet.* 85(6–7), 750–754.

Maas, F. B. III, Patterson, F. L., Foster, J. E. and Hatchett, J. H. (1987). Expression and inheritance of resistance of 'Marquillo' wheat to Hessian fly biotype D. *Crop Sci.* 27(1), 49–52.

Martín-Sánchez, J. A., Gómez-Colmenarejo, M., Del Moral, J., Sin, E., Montes, M. J., González-Belinchón, C., López-Braña, I. and Delibes, A. (2003). A new Hessian fly resistance gene (*H30*) transferred from the wild grass *Aegilops triuncialis* to hexaploid wheat. *Theor. Appl. Genet.* 106(7), 1248–1255.

McDonald, M. J., Ohm, H. W., Rinehart, K. D., Giovanini, M. P. and Williams, C. E. (2014). *H33* 334.amsi "http://example.com?id=journal"of 'Marquillo' wheat to Hessian fly biotype. *Crop Sci.* 54(5), 2045–2053.

Naber, N., El Bouhssini, M. and Lhaloui, S. (2003). Biotypes of Hessian fly (Dipt., Cecidomyiidae) in Morocco. *J. Appl. Entomol.* 127(3), 174–176.

Navarro-Escalante, L., Zhao, C., Shukle, R. and Stuart, J. J. (2020). BSA-Seq discovery and functional analysis of candidate Hessian fly (*Mayetiola destructor*) avirulence genes. *Front. Plant Sci.* 11, 956.

Niu, F., Xu, Y., Liu, X., Zhao, L., Bernardo, A., Li, Y., Liu, G., Chen, M. S., Cao, L., Hu, Z., Xu, X. and Bai, G. (2020). The Hessian fly recessive resistance gene *h4* mapped to chromosome 1a of the wheat cultivar 'Java' using genotyping-by-sequencing. *Theor. Appl. Genet.* 133(10), 2927-2935.

Noble, W. B. (1943). Differentiation of the two genetic factors for resistance to the Hessian fly in Dawson wheat. *J. Agric. Res.* 67, 27-32.

Obanni, M., Patterson, F. L., Foster, J. E. and Ohm, H. W. (1988). Genetic analyses of resistance of durum wheat PI 428435 to the Hessian fly. *Crop Sci.* 28(2), 223-226.

Ohm, H. W., Ratcliffe, R. H., Patterson, F. L. and Cambron, S. E. (1997). Resistance to Hessian fly conditioned by genes *H19* and proposed gene *H27* of durum wheat line PI 422297. *Crop Sci.* 37(1), 113-115.

Pachard, C. M. (1916). Life histories and method of rearing Hessian-fly parasites. *J. Agric. Res.* VI(10), 367-381.

Painter, R. H. (1951). *Insect Resistance in Crop Plants*. New York: The Macmillan Publishing Co., Inc., 520.

Parks, T. H. (1917). A county-wide survey to determine the effect of time of seeding and presence of volunteer wheat upon the extent of damage by the Hessian fly. *J. Econ. Entomol.* 10(2), 249-253.

Pauly, P. J. (2002). Fighting the Hessian fly: American and British responses to insect invasion; 1776-1789. *Environ Hist.* 7(3), 485-507.

Ratcliffe, R. H., Cambron, S. E., Flanders, K. L., Bosque-Perez, N. A., Clement, S. L. and Ohm, H. W. (2000). Biotype composition of Hessian fly (Diptera: Cecidomyiidae) populations from the Southeastern, Midwestern, and Northwestern United States and virulence to resistance genes in wheat. *J. Econ. Entomol.* 93(4), 1319-1328.

Raupp, W. J., Amry, A., Hatchett, J. H., Gill, B. S., Wilson, D. L. and Cox, T. S. (1993). Chromosomal location of Hessian fly resistance genes *H22, H23*, and *H24* derived from *Triticum tauschii* in the D genome of wheat. *J. Hered.* 84(2), 142-145.

Roberts, J. J. and Gallun, R. L. (1984). Chromosome location of the *H5* gene for resistance to the Hessian fly in wheat. *J. Hered.* 75(2), 147-148.

Sardesai, N., Nemacheck, J. A., Subramanyam, S. and Williams, C. E. (2005). Identification and mapping of *H32*, a new wheat gene conferring resistance to Hessian fly. *Theor. Appl. Genet.* 111(6), 1167-1173.

Schulte, S. J., Rider, S. D., Hatchett, J. H. and Stuart, J. J. (1999). Molecular genetic mapping of three X-linked avirulence genes, vH6, vH9 and vH13, in the Hessian fly. *Genome* 42(5), 821-828.

Schuster, M. F. and Lidell, M. C. (1990). Distribution and seasonal abundance of Hessian fly (Diptera, Cecidomyiidae) parasitoids in Texas. *J. Econ. Entomol.* 83(6), 2269-2273.

Schwarting, H. N., Whitworth, R. J., Chen, M. S., Cramer, G. and Maxwell, T. (2016). Impact of Hessian fly, *Mayetiola destructor*, on developmental aspects of hard red Winter wheat in Kansas. *Southwest. Entomol.* 41(2), 321-330.

Schwarting, H. N., Whitworth, R. J., Cramer, G. and Chen, M. S. (2015). Pheromone trapping to determine Hessian fly (Diptera: Cecidomyiidae) activity in Kansas. *J. Kans. Entomol. Soc.* 88(4), 411-417.

Shukle, R. H., Cambron, S. E., Moniem, H. A., Schemerhorn, B. J., Redding, J., David-Buntin, G., Flanders, K. L., Reisig, D. D. and Mohammadi, M. (2016). Effectiveness of genes for Hessian fly (Diptera: Cecidomyiidae) resistance in the Southeastern United States. *J. Econ. Entomol.* 109(1), 399-405.

Smiley, R. W., Gourlie, J. A., Whittaker, R. G., Easley, S. A. and Kidwell, K. K. (2004). Economic impact of Hessian fly (Diptera: Cecidomyiidae) on spring wheat in Oregon and additive yield losses with Fusarium crown rot and lesion nematode. *J. Econ. Entomol.* 97(2), 397–408.

Smith, R. (2007). Widespread outbreaks of Hessian fly reported in wheat. *Southwest Farm Report*. Available at: https://www.farmprogress.com/widespread-outbreaks-hessian-fly-reported-wheat.

Sosa, O. and Foster, J. E. (1976). Temperature and the expression of resistance in wheat to the Hessian fly. *Environ. Entomol.* 5(2), 333–336.

Stebbins, N. B., Patterson, F. L. and Gallun, R. L. (1982). Interrelationships among wheat genes *H3, H6, H9*, and *H10* for Hessian fly resistance. *Crop Sci.* 22(5), 1029–1032.

Stuart, J. J., Chen, M. S., Shukle, R. and Harris, M. O. (2012). Gall midges (Hessian flies) as plant pathogens. *Annu. Rev. Phytopathol.* 50(1), 339–357.

Subramanyam, S., Nemacheck, J. A., Hargarten, A. M., Sardesai, N., Schemerhorn, B. J. and Williams, C. E. (2019). Multiple molecular defense strategies in *Brachypodium distachyon* surmount Hessian fly (*Mayetiola destructor*) larvae-induced susceptibility for plant survival. *Sci. Rep.* 9(1), 2596.

Suneson, C. A. and Noble, W. B. (1950). Further differentiation of genetic factors in wheat for resistance to the Hessian fly. *USDA Bull.* 1004.

Tan, C. T., Carver, B. F., Chen, M. S., Gu, Y. Q. and Yan, L. (2013). Genetic association of OPR genes with resistance to Hessian fly in hexaploid wheat. *BMC Genom.* 14, 369.

Tang, G., Liu, X., Chen, G. H., Whitworth, R. J. and Chen, M. S. (2018). Increasing temperature reduces wheat resistance mediated by major resistance genes to *Mayetiola destructor* (Diptera: Cecidomyiidae). *J. Econ. Entomol.* 111(3), 1433–1438.

Tyler, J. M. and Hatchett, J. H. (1983). Temperature influence on expression of resistance to Hessian fly (Diptera: Cecidomyiidae) in wheat derived from *Triticum tauschii*. *J. Econ. Entomol.* 76(2), 323–326.

Wang, T., Xu, S. S., Harris, M. O., Hu, J., Liu, L. and Cai, X. (2006). Genetic characterization and molecular mapping of Hessian fly resistance genes derived from *Aegilops tauschii* in synthetic wheat. *Theor. Appl. Genet.* 113(4), 611–618.

Wang, Z., Ge, J. Q., Chen, H., Cheng, X., Yang, Y., Li, J., Whitworth, R. J. and Chen, M. S. (2018). An insect nucleoside disphosphate kinase (NDK) functions as an effector protein in wheat-Hessian fly interactions. *Insect Biochem. Mol. Biol.* 100, 30–38.

Whitworth, R. J., Sloderbeck, P. E., Davis, H. and Cramer, G. (2009). Kansas crop pests: Hessian fly. *Kans. State Univ. Agric. Exp. Stn Coop. Extension Serv. MF* 2866.

Wilde, G. E., Whitworth, R. J., Claassen, M. and Shufran, R. A. (2001). Seed treatment for control of wheat insects and its effect on yield. *J. Agric. Urban Entomol.* 18(1), 1–11.

Williams, C. E., Collier, C. C., Sardesai, N., Ohm, H. W. and Cambron, S. E. (2003). Phenotypic assessment and mapped markers for *H31*, a new wheat gene conferring resistance to Hessian fly (Diptera: Cecidomyiidae). *Theor. Appl. Genet.* 107(8), 1516–1523.

Williams, C. E., Nemacheck, J. A., Shukle, J. T., Subramanyam, S., Saltzmann, K. D. and Shukle, R. H. (2011). Induced epidermal permeability modulates resistance and susceptibility of wheat seedlings to herbivory by Hessian fly larvae. *J. Exp. Bot.* 62(13), 4521–4531.

Wise, I. L. (2007). Parasitism of the Hessian fly, *Mayetiola destructor* (Say) (Diptera: Cecidomyiidae), on spring wheat (Poaceae) in southern Manitoba. *Proc. Entomol. Soc. Manitoba* 63, 23–32.

Wu, J. X., Liu, X., Zhang, S., Zhu, Y. C., Whitworth, R. J. and Chen, M. S. (2008). Differential responses of wheat inhibitor-like genes to Hessian fly, *Mayetiola destructor*, attacks during compatible and incompatible interactions. *J. Chem. Ecol.* 34(8), 1005–1012.

Yu, G. T., Cai, X., Harris, M. O., Gu, Y. Q., Luo, M. C. and Xu, S. S. (2009). Saturation and comparative mapping of the genomic region harboring Hessian fly resistance gene *H26* in wheat. *Theor. Appl. Genet.* 118(8), 1589–1599.

Zhang, L., Xu, Y., Chen, M.-S., Su, Z., Liu, Y., Xu, Y., La, G. and Bai, G. (2022). Identification of a major QTL for Hessian fly resistance in wheat cultivar 'Chokwang.' *Crop J.* 10(3), 775-782.

Zhao, C., Escalante, L. N., Chen, H., Benatti, T. R., Qu, J., Chellapilla, S., Waterhouse, R. M., Wheeler, D., Andersson, M. N., Bao, R., Batterton, M., Behura, S. K., Blankenburg, K. P., Caragea, D., Carolan, J. C., Coyle, M., El-Bouhssini, M., Francisco, L., Friedrich, M., Gill, N., Grace, T., Grimmelikhuijzen, C. J., Han, Y., Hauser, F., Herndon, N., Holder, M., Ioannidis, P., Jackson, L., Javaid, M., Jhangiani, S. N., Johnson, A. J., Kalra, D., Korchina, V., Kovar, C. L., Lara, F., Lee, S. L., Liu, X., Löfstedt, C., Mata, R., Mathew, T., Muzny, D. M., Nagar, S., Nazareth, L. V., Okwuonu, G., Ongeri, F., Perales, L., Peterson, B. F., Pu, L. L., Robertson, H. M., Schemerhorn, B. J., Scherer, S. E., Shreve, J. T., Simmons, D., Subramanyam, S., Thornton, R. L., Xue, K., Weissenberger, G. M., Williams, C. E., Worley, K. C., Zhu, D., Zhu, Y., Harris, M. O., Shukle, R. H., Werren, J. H., Zdobnov, E. M., Chen, M. S., Brown, S. J., Stuart, J. J. and Richards, S. (2015). A massive expansion of effector genes underlies gall-formation in the wheat pest *Mayetiola destructor*. *Curr. Biol.* 25(5), 613-620.

Zhao, C., Shukle, R. H., Escalante, L. N., Chen, M. S., Richards, S. and Stuart, J. J. (2016). Avirulence gene mapping in the Hessian fly (Mayetiola destructor) reveals a protein phosphatase 2C effector gene family. *J. Insect Physiol.* 84, 22–31.

Zhao, H. X., Liu, X. M. and Chen, M. S. (2006). *H22*, a major resistance gene to the Hessian fly (*Mayetiola destructor*), is mapped to the distal region of wheat chromosome 1DS. *Theor. Appl. Genet.* 113(8), 1491–1496.

Zhao, L., Abdelsalam, N. R., Xu, Y., Chen, M. S., Feng, Y., Kong, L. and Bai, G. (2020). Identification of two novel Hessian fly resistance genes *H35* and *H36* in a hard winter wheat line SD06165. *Theor. Appl. Genet.* 133(8), 2343-2353.

Zhu, L., Liu, X., Liu, X., Jeannotte, R., Reese, J. C., Harris, M., Stuart, J. J. and Chen, M. S. (2008). Hessian fly (*Mayetiola destructor*) attack causes a dramatic shift in carbon and nitrogen metabolism in wheat. *Mol. Plant Microbe Interact.* 21(1), 70–78. https://doi.org/10.1094/MPMI-21-1-0070.

Zhu, L., Liu, X., Wang, H., Khajuria, C., Reese, J. C., Whitworth, R. J., Welti, R. and Chen, M. S. (2012). Rapid mobilization of membrane lipids in wheat leaf sheaths during incompatible interactions with Hessian fly. *Mol. Plant Microbe Interact.* 25(7), 920–930.

Zhukovskii, A. V. (1950). Diapause of the larvae of the Hessian fly. *Vsesoiuznsia Akademiia Sel'skokhoziaistvennykh Nauk* 15, 26–29.

Chapter 4

Wheat midge (*Sitodiplosis mosellana*): management in the Northern Great Plains of the United States and Canada

Govinda Shrestha, Oregon State University, USA; and Gadi V. P. Reddy, Southern Insect Management Research Unit, USA

1 Introduction

The wheat midge, also known as orange wheat blossom midge, *Sitodiplosis mosellana* (Géhin) (Diptera: Cecidomyiidae), is one of the most destructive pests of wheat, *Triticum aestivum* L. worldwide. *S. mosellana* is an oligophagous pest of the Poaceae family. It feeds on several grass host plants, including rye (*Secale cereal* L.) (Basedow, 1972), barley (*Hordeum vulgare* L.) (Wise et al., 2002), couch grasses (*Elymus repens* L.), and slender meadow foxtail (*Alopecurus myosuroides* Huds.) (Barnes, 1955). However, it is widely known as a wheat pest (Elliott and Mann, 1996; Jacquemin et al., 2014). This pest is native to Europe and was accidentally introduced to North America in the 1800s (Fitch, 1856; Elliott and Mann, 1996; Knodel and Ganehiarachchi, 2016) and Asia in the 1900s (Ding and Guo, 1992). It is a serious pest of different wheat crops depending on the geographical region. For example, in North America, wheat midge is a major pest of spring wheat and durum wheat (Elliott et al., 2009; Stougaard et al., 2014; Knodel and Ganehiarachchi, 2016), while it is a major pest of winter wheat in Europe (Flodrops and Taupin, 2010; Gaafar et al., 2011; Jacquemin et al., 2014) and Asia (Duan et al., 2013).

http://dx.doi.org/10.19103/AS.2022.0114.10
Published by Burleigh Dodds Science Publishing Limited, 2023.

The larva is the only damaging stage of wheat midge. It feeds on developing wheat kernels, causing kernels to shrivel, reducing yield and lowering grain grade quality (Stougaard et al., 2014). Without *S. mosellana* pest control measures and under heavy infestation, crop losses can reach 100% (Blodgett, 2007). The purpose of this chapter is to: (a) review the general morphology, life cycle, crop damage, and the current wheat midge integrated pest management (IPM) strategies and (b) identify needed research and approaches to managing the pest in the future in the Northern Great Plains of the United States and Canada.

2 Morphology of wheat midge

2.1 Adult

The adult wheat midge is an orange-colored, delicate nematoceran fly, about half the size of a mosquito (i.e. 1.5-2 mm long) (Reeher, 1945) (Fig. 1). Males are smaller than females. Legs are as long as to their body length. Adults have a pair of wings that are oval shaped, transparent, and with a fringe of fine hairs. Eyes are large, black colored, and cover the majority of the brown head. The antennae are long, slender, and dark brown colored. Male antennae are longer and more plumose than female antennae (Reeher, 1945; Ganehiarachchi, 2008). Because of small size, the wheat midge adults can be confused with lauxanid fly adults, *Camptoprosopella borealis* Shewell (Diptera: Lauxaniidae), a non-pest, which is also common in wheat in the Northern Great Plains of the United States and Canada (Knodel and Ganehiarachchi, 2016). Lauxanid flies are larger than the wheat midge and have relatively shorter legs and smaller eyes. They are active during the day, while wheat midge adults are active at dusk (Knodel and Ganehiarachchi, 2016).

2.2 Egg

Eggs are tiny (0.5 mm long) and are barely visible to the human eye. They are pale yellow when first laid (Fig. 1), orange color in about 2 days, and elongate in shape (Reeher, 1945).

2.3 Larva

The newly hatched larvae are pale yellow (Fig. 1). The matured larvae (final instar) appearance resembles wheat anthers in size (3-5 mm), shape, and color.

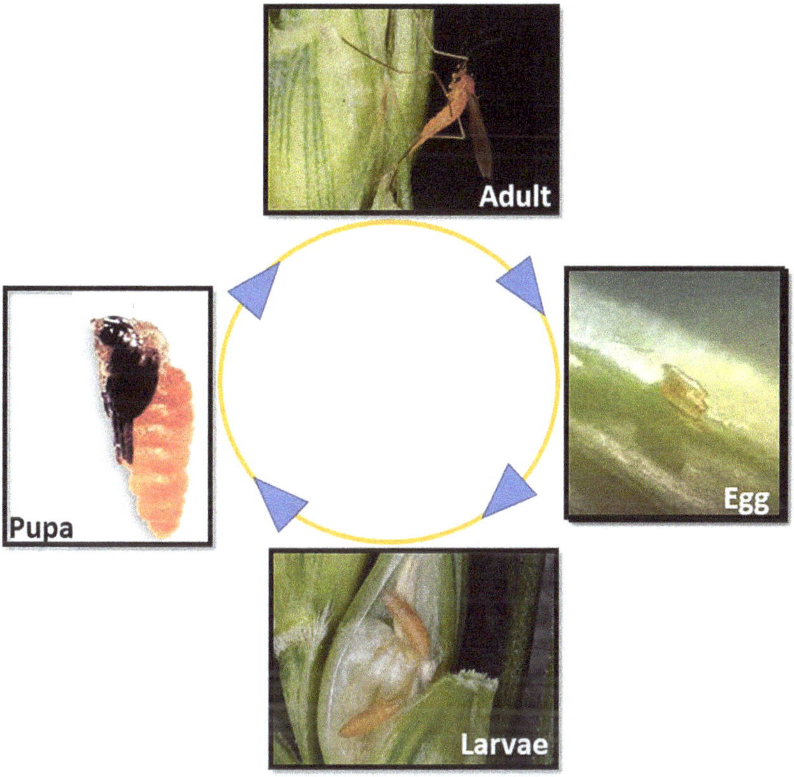

Figure 1 Life cycle of wheat midge. Photo sources: Adult and larvae - Mike Dolinski, Agriculture and Agri-Food Canada; egg and pupa - Saskatoon Research Centre, Canada.

2.4 Pupa

The pupae are orange colored, conical shaped, and similar in size to adults (Fig. 1).

3 Wheat midge life cycle

The wheat midge is a univoltine species wherever it occurs as a pest. The mature larvae overwinter within cocoons (about half the size of a canola seed) in the soil (Doane et al., 1987). In the spring, when soil temperatures start to rise (>10°C) and soil is under moist conditions, the overwintered larvae leave their cocoons and move to the soil surface to pupate. In some instances, larvae can pupate without leaving their cocoons (Doane et al., 1987). Depending on the temperature, the pupal duration is 2-4 weeks. With sufficient rainfall and soil temperature, pupation ends and adults emerge (Jacquemin et al., 2014). But, if there is insufficient warmth and moisture, larvae or pupae are capable of a long dormant period, up to 12 years (Barnes, 1956).

In North America, for example in Montana, adults emerge over a 5-week period, starting in mid-June, coinciding with heading of the crop (Stougaard et al., 2014; Shrestha and Reddy, 2019). Adults emerge in late afternoon and early evening (Pivnick and Labbé, 1992). After emergence, females walk several centimeters; then stop briefly before releasing sex pheromone ((2S, 7R)-2, 7-nonadiyl dibutyrate) to attract males for mating during scotophase (Gries et al., 2000). After mating, females fly up to the wheat head at dusk and when the air is still (Reeher, 1945). Adults are comparatively poor fliers, but they can move long distances on thermal updrafts and wind (Miao et al., 2013; Stougaard et al., 2014).

Females lay eggs on the surface of newly emerged wheat heads, mainly just before anthesis (Mukerji et al., 1988; Elliott and Mann, 1996). Eggs are deposited either singly or in groups of two to five on all parts of the head, especially between the glume (middle of wheat heads) and palea (tip of the wheat heads) (Ding and Lamb, 1999). The life span of wheat midge adults is about 1 week during which each adult female lays an average of 84 eggs (Pivnick and Labbé, 1993). Depending on temperature, eggs hatch within 4–7 days. After hatching, larvae crawl from the surface of the wheat heads to the developing kernels where they begin to feed (Dexter et al., 1987). Usually, 3–4 larvae can be found in each kernel, but during heavy infestations, as many as 26 larvae can feed in a single kernel (Alberta Agriculture Center, 2014). The larvae develop through three instars and when mature, they reside in the wheat head. Mainly, when rainfall events occur, the larvae move or drop to the ground where they excavate a tunnel in the soil as deep as 10 cm (Affolter, 1990). When buried beneath soil, the larvae spin a cocoon and go to an overwintering stage.

4 Effect of wheat midge on crops

Only the wheat midge larvae can cause crop damage. Among the several hosts of the midge, wheat is the most preferred. The crop is susceptible through the heading and flowering stages, with susceptibility usually dropping after anthesis, depending on the wheat variety (Elliott and Mann, 1996; Gaafar et al., 2011). Since larvae feed inside kernels, the larval damage to wheat kernels is barely apparent; externally, there are no changes in color, size, or shape of the affected kernels (Alberta Agriculture Center, 2014). Therefore, kernels should be opened to inspect for midge damage.

Damage to kernels within the wheat head can vary considerably. In some instances, a few kernels will be completely aborted while others are uninjured. In other instances, all kernels can be affected, reduced in size and underdeveloped (Alberta Agriculture Center, 2014). Regardless of the distribution of damage, it eventually reduces grain yield and quality. Wheat midge damage to kernels also often facilitates fungal pathogen attacks that lead to fungal diseases such as fusarium head blight (*Fusarium graminearium* Schwabe) (Miao et al., 2021) and glume blotch (*Septoria nodorum* Berk.) (Wellso and Freed, 1982).

5 Pest management of wheat midge

5.1 Monitoring and economic threshold

The wheat midge population monitoring tools include soil core sampling for larval cocoons and a variety of traps (e.g. plate traps, emergence traps, yellow sticky traps, and sex pheromone traps) for adult sampling (Knodel and Ganehiarachchi, 2016). Among these, soil core sampling (to assess potential infestations in the crop) and sex pheromone traps (to assess need for control measures within growing season) are most commonly used tools. Soil core sampling should be conducted in the fall/spring after the wheat is harvested using a soil core sampler (2.54 cm diameter by 15 cm deep) (Doane et al., 1987). A single core can represent a sampling unit.

Soil samples should be collected along a transect starting 30 m from the edge of the field, taking 10 samples per transect with samples spaced ≥10 m apart. Cores can be kept in individual plastic bags and stored at 2°C until processed (Olfert, 2013). Wet sieving can be used to process the samples to detect larval cocoons (Doane et al., 1987). Densities of 600 larval cocoons/m^2 within a field are considered indicative of damaging pest levels. This method has been used by researchers, extension agents, and crop consultants to inform wheat growers about the probability of wheat midge pest pressure in the wheat crop in the following year.

Sex pheromone traps, delta traps baited with female sex pheromone attractive to males, are effective for monitoring wheat midge populations in the field. Wheat midge pheromone traps are commercially available in North America, Europe, and Asia. For effective monitoring, traps should be placed in the field about a week before wheat plants reach heading. Traps should be installed 15-25 m from the field edge. Trap heights should be positioned at crop height. The traps should be examined at 1-2 days intervals. Cumulative trap counts exceeding 9-10 adults/trap over 3-day periods after heading indicate that the wheat crop can be at economic risk, and insecticide application may be warranted (Knodel and Ganehiarachchi, 2016).

Although monitoring tools provide pest risk assessment information, it is recommended to inspect/scout individual wheat heads for midges in the fields to determine if densities of adult midges have exceeded established economic thresholds. The common recommended method for wheat midge scouting is walking out into the field, sitting down, and watching adult activities in 10 heads; then turn 90° in second and third directions; and in each direction watch adult activities in 10 heads. Follow the same process in three to four other places of the field to determine adult density/head. Since wheat midge adults are sensitive to weather conditions, scouting should be done at dusk when there is no wind. Economic thresholds in North America differ with wheat class. For example, the economic threshold level is one or more midge adults for every four or five heads for hard red spring wheat and one or more adults for every seven to eight wheat heads for durum wheat (Elliott et al., 2011a; Knodel and Ganehiarachchi, 2016). After 50% of the heads have flowered, it is not recommended to apply insecticides due to reduced efficacy levels, and to protect parasitoids that attack the wheat midge larvae (Knodel and Ganehiarachchi, 2016).

5.2 Cultural control

Although cultural practices may not provide economic control of wheat midge, this approach helps to minimize pest populations in the agroecosystem. Planting date, tillage system, and crop rotation have been reported to influence wheat midge populations in spring wheat fields in North America (Elliott et al., 2002, 2011a). Elliott et al. (2002) showed that late-seeded and high tillage systems had one to two times lower wheat midge larval and adult populations than early-seeded and zero tillage systems. Elliott et al. (2011a) recommended that early planting coupled with high seeding rates can delay wheat heading helping to avoid it coinciding with the adult wheat midge population peak in mid- to late July in Canadian Prairies. Elliott et al. (2002) concluded that wheat production in the same field for 2 consecutive years should be avoided to break wheat midge population cycle. Rotational crops such as canola, flax, and legumes are effective.

5.3 Host-plant resistance

Host-plant resistance has a great potential to minimize wheat midge populations (Lamb et al., 2002; Blake et al., 2014). Several wheat midge-resistant wheat varieties have been developed in Canada, Europe, and the United States. Antibiosis, based on the single gene *Sm1*, is the most common type of resistance investigated for wheat midge. The effect of *Sm1* is inducibility of two phenolic acids (*p*-coumaric and ferulic acid) in plants in response to wheat midge attack of developing kernels (Ding et al., 2000). Two acids cause larval mortality or cause the larvae to move from kernels where they are feeding, resulting in larval death due to starvation (Ding et al., 2000). The *Sm1* has been bred into plants to develop wheat midge-resistant cultivars by selecting plants that affect midge biology or using molecular markers associated to this gene followed by bioassay to verify the resistance (Vera et al., 2013). In North America, the first wheat midge-resistant variety was based on *Sm1* and commercially released in 2010 (Fox et al., 2010) in Canada and in 2014 in Montana, USA (Blake et al., 2014). There are currently several midge-resistant varieties commercially available in North America, but mainly in Canada.

Also, wheat varietal blends have been released with a ratio of 90% midge-resistant wheat and 10% susceptible wheat to preserve the effectiveness of the *Sm1* (Smith et al., 2007; Vera et al., 2013). The concept behind varietal blends is that the maintenance of 10% susceptible seeds can maintain sufficient avirulent adult populations to ensure that nearly all virulent adults will likely mate with avirulent adults. Thus, it reduces or discourages the selection of virulent wheat midge population (Tabashnik et al., 2008; Vera et al., 2013).

In addition to antibiosis, wheat lines/cultivars that demonstrate wheat midge oviposition deterrence have been identified (Lamb et al., 2002, 2016; Gharalari, 2008; Blake et al., 2011). For example, Lamb et al. (2002) found that antixenosis in six spring wheat lines resulted in fewer eggs per spike relative to susceptible checks under field conditions. Lamb et al. (2016) found that two North American winter wheat cultivars (Goens and Rawhide) showed high levels of oviposition deterrence. These cultivars could be valuable sources in breeding programs seeking to incorporate wheat midge resistance into winter wheat or spring wheat.

5.4 Biological control

Natural enemies including parasitoids, predators, and entomopathogens can reduce wheat midge populations and the severity of pest outbreaks. Lists of these natural enemies have been compiled in Basedow (1973), Affolter (1990), Floate et al. (1990), Olfert et al. (2003), Elliott et al. (2011b), and Chavalle et al. (2018).

Wheat midge parasitoids include Pteromalidae and Platygastridae (Chavelle et al., 2018). Based on our knowledge, all identified parasitoids, including eight Platygastridae species (*Amblypasis tritici* Walker, *Euxestonotus error* Fitch, *Euxestonotus* sp. Fouts, *Platygaster gracilipes* Huggert, *Platygaster nisus* Walker, *Platygaster tuberosula* Kieffer, *Inostemma mosellanae* Vlug, and *Leptacis* sp. Foerster) and one Pteromalidae species (*Macroglenes penetrans* Kirby), are of European origin (Affolter, 1990; Vlug, 1991; Olfert et al., 2003; Chavalle et al., 2018).

Out of the nine species, *M. penetrans*, a wheat midge-specific parasitoid, plays an important role in reducing wheat midge infestations in Europe and North America (Olfert et al., 2003). *Platygaster tuberosula* and *E. error* parasitoids complement to *M. penetrans* in controlling wheat midge populations, though their individual roles in reducing infestation levels is often low compared with *M. penetrans* (Olfert et al., 2003; Echegaray et al., 2016). These species are complementary to *M. penetrans* because they emerge throughout the wheat midge emergence period, while *M. penetrans* emerges 3-5 days after the emergence of the midge adults (Olfert et al., 2003; Thompson and Reddy, 2016). *M. penetrans* probably arrived in North America along with the pest (Beirne, 1971) since wheat midge populations were found to be parasitized by this parasitoid in 1984 in Canada without its introduction. The other two species were mainly introduced as part of a biological control program in North America from Europe in 1993 and 1994 (Olfert et al., 2003).

All three species are egg-larval parasitoids, which oviposit into eggs of the host after which larvae develop within and eventually kill the host larvae (Doane et al., 1989). Like the midge, they have one generation per year. Olfert et al. (2003) estimated a combined average parasitism by *M. penetrans* and *P. tuberosula* of 64% in 2001 in Saskatchewan, Canada. Shanower (2005) found an average of 35% parasitism by *M. penetrans* in Montana in 2004. In addition, the cropping system was found to play an important role in *M. penetrans* abundance, with higher populations in irrigated fields over dryland fields in Montana (Thompson and Reddy, 2016).

Ground beetles (Coleoptera: Carabidae) are considered the most important predator of wheat midge, particularly for larvae overwintering in the soil. Fourteen species of ground beetles have shown to predate on wheat midge larvae under laboratory conditions (Floate et al., 1990). In the absence of alternate prey, *Pterostichus corvus* LeConte was the most effective predator of wheat midge larvae (killing 48 larvae per adult per day) followed by *Agonum placidum* Say (43 larvae/adult/day), *Bembidion obscurellum* Motschulsky (43 larvae/adult/day), and *Bembidion quadrimaculatum* L. (43 larvae/adult/day) (Floate et al., 1990). Under field conditions, each ground beetle adult can predate from <1 to 86 wheat midge larvae per square meter per day depending upon the species (Floate et al., 1990). Although rates of predation are unknown, spiders may prey upon eggs, larvae, and adults of wheat midge, potentially reducing the severity of wheat midge outbreak in North America (Barnes, 1955; Basedow, 1973).

Entomopathogenic fungi, bacteria, and viruses on reducing wheat midge populations are mostly unknown. In Switzerland, Keller and Wilding (1985) reported that the naturally occurring fungal pathogen *Entomophthora brevinucleate* was pathogenic to adults in winter wheat fields. In Montana, Shrestha and Reddy (2019) reported two commercially available entomopathogens (fungus *Beauveria bassiana* and nematode *Steinernema feltiae*) that can reduce midge larval populations under field conditions.

5.5 Chemical control

Foliar insecticides such as organophosphates and pyrethroids can be used to control wheat midge populations. Insecticides are usually applied when wheat midge adult populations exceed economic thresholds (Elliot, 1988; Elliott and Mann, 1997; Chavalle et al., 2015). In spring wheat, the insecticide application timing is also based on the percentage of heading in the field in conjunction with economic thresholds. Elliot (1988) showed that insecticide efficacy will be enhanced when application is waited for 3-4 days if 30% of the wheat is heading in the field, but apply immediately if 70% of the wheat is between heading and flowering stage. Insecticide efficacy will be reduced when 30-60% of wheat heads are at flowering stage (i.e. anther visible) in the field (Elliot, 1988). To provide effective control of wheat midge eggs and early larval instars, insecticide sprays should uniformly cover the wheat heads. Also, when spraying insecticides, managers should be encouraged to follow the label instructions and only apply if wheat midge densities exceed the economic threshold. If it is necessary to apply multiple sprays, active ingredients should be rotated among insecticide groups to prevent development of insecticide resistance.

6 Conclusion and future trends in research

Wheat is an important staple crop for global food security, and it provides about 20% of the total dietary calories and proteins worldwide (Shiferaw et al., 2013). Nevertheless, wheat crop production is continually challenged by a variety of pests, including kernel-feeding insect pests, that need comprehensive IPM practices. Overall, the biology, pest monitoring tools, and economic thresholds of wheat midge have been well-developed in wheat crops. Parasitoids (*M. penetrans*, *P. tuberosula*, and *E. error*) and antibiosis resistance (wheat midge-resistant varieties) have been identified as major pillars to suppress the wheat midge populations. Therefore, future research should focus on strategies that aim to conserve parasitoids, preserve host-plant resistance, and find the ecological balance between resistant varieties and parasitoids in agricultural environment. In addition, cultural control practices (e.g. crop rotation, tillage methods, planting methods, and dates) and additional sources of host-plant resistance need to be further explored for these practices to remain economical, environmentally sound, and sustainable for wheat production. Since insecticide control is the primary wheat midge control method, future research priorities must emphasize more sustainable IPM strategies in wheat crops.

7 References

Affolter, F. (1990). *Structure and Dynamics of the Parasitoid Complex of the Wheat Midges Sitodiplosis Mosellana (Géhin) and Contarinia Tritici (Kirby) Cecidomyiidae*, International Institute of Biological Control, Delemont, Switzerland.

Alberta Agriculture Center (2014). Wheat midge fact sheet. Available at: https://agresearch.montana.edu/wtarc/producerinfo/entomology-insect-ecology/OrangeWheatBlossomMidge/AlbertaFactsheet.pdf. (Accessed September 26, 2021).

Barnes, H. F. (1955). Avoiding wheat blossom midge attacks. *Plant Pathology* 4(4), 147–147.

Barnes, H. F. (1956). *Gall Midges of Economic Importance: Gall Midges of Cereal Crops*. Crosby Lockwood and Son Ltd., London.

Basedow, T. (1972). Relations between phenology and food-plants in the wheat gall-midges Contarinia tritici (Kirby) and *Sitodiplosis mosellana* (Gehin). *Zeitschrift fur Angewandte Entomologie* 71, 359–367.

Basedow, T. (1973). The influence of predatory arthropods of the soil surface on the abundance of phytophagous insects in agriculture. *Pedobiologia* 13, 410–422.

Beirne, B. P. (1971). Pest Insects of annual crop plants in Canada. I. Lepidoptera (vol. II) Diptera, III. Coleoptera. *Memoirs of the Entomological Society of Canada* 78, 124.

Blake, N. K., Stougaard, R. N., Weaver, D. K., Sherman, J. D., Lanning, S. P., Naruoka, Y., Xue, Q., Martin, J. M. and Talbert, L. E. (2011). Identification of a quantitative trait locus for resistance to *Sitodiplosis mosellana* (Géhin), the orange wheat blossom midge, in spring wheat. *Plant Breeding* 130(1), 25–30.

Blake, N. K., Stougaard, R. N., Bohannon, B., Weaver, D. K., Heo, H.-Y., Lamb, P. F., Nash, D., Wichman, D. M., Kephart, K. D., Miller, J. H., Reddy, G. V. P., Eckhoff, J. L., Grey, W. E., Lanning, S. P., Sherman, J. D. and Talbert, L. E. (2014). Registration of 'Egan' wheat with resistance to orange wheat blossom midge. *Journal of Plant Registrations* 8(3), 298–302.

Blodgett, S. (2007). Orange wheat blossom midge, high plains integrated pest management. *org/Uploads/OrangeWheatBlossomMidge-SmallGrains.Pdf.* Available at: http://wiki.bugwood. (Accessed 18 November 2021).

Chavalle, S., Buhl, P. N., San Martin, G. and De Proft, M. (2018). Parasitism rates and parasitoid complexes of the wheat midges, *Sitodiplosis mosellana*, *Contarinia tritici* and *Haplodiplosis marginata*. *BioControl*, 63(5), 641–653.

Chavalle, S., Censier, F., San Martin Y Gomez, G. and De Proft, M. (2015) Protection of winter wheat against orange wheat blossom midge, *Sitodiplosis mosellana* (Géhin) (Diptera: Cecidomyiidae): efficacy of insecticides and cultivar resistance. *Pest Management Science* 71(5), 783–790.

Dexter, J. E., Preston, K. R., Cooke, L. A., Morgan, B. C., Kruger, J. E., Kilborn, R. H. and Elliott, R. H. (1987). The influence of orange wheat blossom midge (*Sitodiplosis mosellana* Géhin) damage on hard red spring wheat quality and the effectiveness of insecticide treatments. *Canadian Journal of Plant Science* 67(3), 697–712.

Ding, H. J. and Guo, Y. Y. (1992). Research dynamic on wheat midge. *Agriculture World* 5, 31–32.

Ding, H. and Lamb, R. J. (1999). Oviposition and larval establishment of *Sitodiplosis mosellana* (Diptera: Cecidomyiidae) on wheat (Gramineae) at different growth stages. *The Canadian Entomologist* 131(4), 475–481.

Ding, H., Lamb, R. J. and Ames, N. (2000). Inducible production of phenolic acids in wheat and antibiotic resistance to *Sitodiplosis mosellana*. *Journal of Chemical Ecology* 26(4), 969–985.

Doane, J. F., Olfert, O. O. and Mukerji, M. K. (1987). Extraction precision of sieving and brine flotation for removal of wheat midge, *Sitodiplosis mosellana* (Diptera: Cecidomyiidae), cocoons and larvae from soil. *Journal of Economic Entomology* 80(1), 268–271.

Doane, J. F., DeClerck-Floate, R., Arthur, A. P. and Affolter, F. (1989). Description of the life stages of *Macroglenes penetrans* (Kirby) (Hymenoptera: Chalcidoidea, Pteromalidae), a parasitoid of the wheat midge, *Sitodiplosis mosellana* (Géhin) (Diptera: Cecidomyiidae). *The Canadian Entomologist* 121(12), 1041-1048.

Duan, Y., Wu, Y. Q., Luo, L. Z., Miao, J., Gong, Z. J., Jiang, Y. L. and Li, T. (2013). Genetic diversity and population structure of *Sitodiplosis mosellana* in Northern China. *PLoS ONE* 8(11), e78415.

Echegaray, E. R., Stougaard, R. N. and Bohannon, B. (2016). First record of *Euxestonotus error* (Hymenoptera: Platygastridae) in the Pacific Northwest, United States of America. *The Canadian Entomologist* 148(5), 616-618.

Elliot, R. H. (1988). Evaluation of insecticides for protection of wheat against damage by the wheat midge, *Sitodiplosis mosellana* (Géhin) (Diptera: Cecidomyiidae). *The Canadian Entomologist* 120(7), 615-626.

Elliott, R. H. and Mann, L. W. (1996). Susceptibility of red spring wheat, *Triticum aestivum* L. cv. katepwa, during heading and anthesis to damage by wheat midge, *Sitodiplosis mosellana* (Géhin)(Diptera: Cecidomyiidae). *The Canadian Entomologist* 128(3), 367-375.

Elliott, R. H. and Mann, L. W. (1997). Control of wheat midge, *Sitodiplosis mosellana* (Gehin), at lower chemical rates with small-capacity sprayer nozzles. *Crop Protection* 16(3), 235-242.

Elliott, R. H., Mann, L. W., Braun, M. and Olfert, O. O. (2002). Integrated management of crop pests - abundance of wheat midge and its parasite in different management systems. *Proceedings of the Soils & Crops*, 21-22.

Elliott, R. H., Mann, L. and Olfert, O. (2009). Calendar and degree-day requirements for emergence of adult wheat midge, *Sitodiplosis mosellana* (Géhin) (Diptera: Cecidomyiidae) in Saskatchewan, Canada. *Crop Protection* 28(7), 588-594.

Elliott, R. H., Olfert, O. O. and Hartley, S. (2011a). Management practices for wheat midge, *Sitodiplosis mosellana* (Géhin). *Prairie Soil & Crops* 4, 8-13.

Elliott, R. H., Mann, L. W. and Olfert, O. (2011b). Calendar and degree-day requirements for emergence of adult *Macroglenes penetrans* (Kirby), an egg-larval parasitoid of wheat midge, *Sitodiplosis mosellana* (Géhin). *Crop Protection* 30(4), 405-411.

Fitch, A. (1856). *Insects Infesting Grain Crops*. New York.

Floate, K. D., Doane, J. F. and Gillott, C. (1990). Carabid predators of the wheat midge (Diptera: Cecidomyiidae) in Saskatchewan. *Environmental Entomology* 19(5), 1503-1511.

Flodrops, Y. and Taupin, P. (2010). Soft wheat: combating the orange blossom midge. *Perspective Agriculture* 365, 46-50.

Fox, S. L., McKenzie, R. I. H., Lamb, R. J., Wise, I. L., Smith, M. A. H., Humphreys, D. G., Brown, P. D., Townley-Smith, T. F., McCallum, B. D., Fetch, T. G., Menzies, J. G., Gilbert, J. A., Fernandez, M. R., Despins, T., Lukow, O. and Niziol, D. (2010). Unity hard red spring wheat. *Canadian Journal of Plant Science* 90(1), 71-78.

Gaafar, N., El-Wakeil, N. and Volkmar, C. (2011) Assessment of wheat ear insects in winter wheat varieties in central Germany. *Journal of Pest Science* 84(1), 49-59.

Ganehiarachchi, G. A. S. M. (2008). Host finding behavior of two gall midges specializing on wheat. PhD. Thesis, North Dakota State University.

Gharalari, H. (2008). Oviposition behavior of wheat midge *Sitodiplosis mosellana* (Géhin) (Diptera: Cecidomyiidae) and inheritance of deterrence resistance in spring wheat. PhD Thesis, University of Manitoba, Canada.

Gries, R., Gries, G., Khaskin, G., King, S., Olfert, O., Kaminski, L. A., Lamb, R. and Bennett, R. (2000). Sex pheromone of orange wheat blossom midge, *Sitodiplosis mosellana*. *Naturwissenschaften* 87(10), 450-454.

Jacquemin, G., Chavalle, S. and De Proft, M. (2014). Forecasting the emergence of the adult orange wheat blossom midge, *Sitodiplosis mosellana* (Géhin) (Diptera: Cecidomyiidae) in Belgium. *Crop Protection* 58, 6-13.

Keller, S. and Wilding, N. (1985). *Entomophthora brevinucleata* sp. nov. [Zygomycetes, Entomophthoraceae], a pathogen of gall midges [Dip.: Cecidomyiidae]. *Entomophaga* 30(1), 55-63.

Knodel, J. J. and Ganehiarachchi, M. (2016). Integrated pest management of the wheat midge in North Dakota. NDSU Extension Service, North Dakota State University, E1330.

Lamb, R. J., Wise, I. L., Smith, M. A. H., McKenzie, R. I. H., Thomas, J. and Olfert, O. O. (2002). Oviposition deterrence against *Sitodiplosis mosellana* (Diptera: Cecidomyiidae) in spring wheat (Gramineae). *The Canadian Entomologist* 134(1), 85-96.

Lamb, R. J., Smith, M. A. H., Wise, I. L. and McKenzie, R. I. H. (2016). Resistance to wheat midge (Diptera: Cecidomyiidae) in winter wheat and the origins of resistance in spring wheat (Poaceae). *The Canadian Entomologist* 148(2), 229-238.

Miao, J., Wu, Y. Q., Gong, Z. J., He, Y. Z., Duan, Y. and Jiang, Y. L. (2013). Long-distance wind-borne dispersal of *Sitodiplosis mosellana* Géhin (Diptera: Cecidomyiidae) in Northern China. *Journal of Insect Behavior* 26(1), 120-129.

Miao, J., Zhang, G. P., Zhang, S. J., Ma, J. Q. and Wu, Y. Q. (2021). The orange wheat blossom midge promotes fusarium head blight disease, posing a risk to wheat production in northern China. *Acta Ecologica Sinica*. https://doi.org/10.1016/j.chnaes.2021.10.010.

Mukerji, M. K., Olfert, O. O. and Doane, J. F. (1988). Development of sampling designs for egg and larval populations of the wheat midge, *Sitodiplosis mosellana* (Géhin) (Diptera: Cecidomyiidae), in wheat. *The Canadian Entomologist* 120(6), 497-505.

Olfert, O. (2013). Wheat Midge: *Sitodiplosis mosellana* monitoring protocol, Paraire Pest Monitoring Network. Available at: https://westernforum.org/Documents/IPMN%20Protocols/2013_Wheat%20midge%20protocol.pdf. (Accessed September 26, 2021)

Olfert, O., Doane, J. F. and Braun, M. P. (2003). Establishment of *Platygaster tuberosula*, an introduced parasitoid of the wheat midge, *Sitodiplosis mosellana*. *The Canadian Entomologist* 135(2), 303-308.

Pivnick, K. A. and Labbé, E. (1992). Emergence and calling rhythms, and mating behaviour of the orange wheat blossom midge, *Sitodiplosis mosellana* (Géhin)(Diptera: Cecidomyiidae). *The Canadian Entomologist* 124(3), 501-507.

Pivnick, K. A. and Labbé, E. (1993). Daily patterns of activity of females of the orange wheat blossom midge, *Sitodiplosis mosellana* (Géhin)(Diptera: Cecidomyiidae). *The Canadian Entomologist* 125(4), 725-736.

Reeher, M. M. (1945). *The Wheat Midge in the Pacific Northwest* (vol. 732), United States Department of Agriculture.

Shanower, T. G. (2005). Occurrence of *Sitodiplosis mosellana* (Diptera: Cecidomyiidae) and its parasitoid, *Macroglenes penetrans* (Hymenoptera: Platygastridae), in northeastern Montana. *The Canadian Entomologist* 137(6), 753-755.

Shiferaw, B., Smale, M., Braun, H. J., Duveiller, E., Reynolds, M. and Muricho, G. (2013). Crops that feed the world 10. Past successes and future challenges to the role played by wheat in global food security. *Food Security* 5(3), 291-317.

Shrestha, G. and Reddy, G. V. P. (2019). Field efficacy of insect pathogen, botanical, and jasmonic acid for the management of wheat midge *Sitodiplosis mosellana* and the impact on adult parasitoid *Macroglenes penetrans* populations in spring wheat. *Insect Science* 26(3), 523–535.

Smith, M. A. H., Wise, I. L. and Lamb, R. J. (2007). Survival of *Sitodiplosis mosellana* (Diptera: Cecidomyiidae) on wheat (Poaceae) with antibiosis resistance: implication for the evolution of virulence. *The Canadian Entomologist* 139(1), 133–140.

Stougaard, B., Bohannon, B., Picard, D., Reddy, G. V. P., Talbert, L. and Wanner, K. (2014). Orange wheat blossom midge. MontGuide, Montana State University.

Tabashnik, B. E., Gassmann, A. J., Crowder, D. W. and Carrière, Y. (2008). Insect resistance to Bt crops: evidence versus theory. *Nature Biotechnology* 26(2), 199–202.

Thompson, B. M. and Reddy, G. V. P. (2016). Status of *Sitodiplosis mosellana* (Diptera: Cecidomyiidae) and its parasitoid, *Macroglenes penetrans* (Hymenoptera: Pteromalidae), in Montana. *Crop Protection* 84, 125–131.

Vera, C. L., Fox, S. L., DePauw, R. M., Smith, M. A. H., Wise, I. L., Clarke, F. R., Procunier, J. D. and Lukow, O. M. (2013). Relative performance of resistant wheat varietal blends and susceptible wheat cultivars exposed to wheat midge, *Sitodiplosis mosellana* (Géhin). *Canadian Journal of Plant Science* 93(1), 59–66.

Vlug, H. J. (1991). Two new species of Platygastridae (Hymenoptera), reared from wheat gall midges (Diptera: Cecidomyiidae). *Entomologische Berichten* 51, 23–26.

Wellso, S. G. and Freed, R. D. (1982). Positive association of the wheat midge (Diptera: Cecidomyiidae) with glume blotch. *Journal of Economic Entomology* 75(5), 885–887.

Wise, I. L., Lamb, R. J. and Smith, M. A. H. (2002). Susceptibility of hulled and hulless barley (Gramineae) to *Sitodiplosis mosellana* (Diptera: Cecidomyiidae). *The Canadian Entomologist* 134(2), 193–203.

Chapter 5

Wheat stem sawfly (*Cephus cinctus* Norton)

David Weaver, Montana State University, USA

1 Introduction

1.1 Historical distribution

The wheat stem sawfly (WSS), *Cephus cinctus* Norton, has been a major pest of cereal grain production in the Northern Great Plains of North America for well over a century (Beres et al., 2011c; Weiss and Morrill, 1992; Edde, 2022). It was first described from a Colorado grass (Norton, 1872), and Beres et al. (2011c) summarize species characters that may be used to identify this cephid. Currently, damaging populations exist in the contiguous region spanning the states Washington, Idaho, Montana, Wyoming, North Dakota, South Dakota, Utah, Minnesota, Nebraska, Kansas and Colorado and provinces Alberta, Saskatchewan and Manitoba (Fig. 1). These populations are centered on the area of greatest historic damage encompassing Montana, North Dakota, Alberta and Saskatchewan (Beres et al., 2011b).

Wheat stem sawflies are in the family Cephidae (Hymenoptera), which includes all grass-feeding stem sawflies, as well as others that develop in woody plant stems, twigs and branches. Globally, pest members of the Cephidae attacking cereal grains are found exclusively in the northern hemisphere. Three congeners responsible for most of the losses occur in distinct geographic areas. In Eastern Eurasia, the dominant species associated with crop loss is

http://dx.doi.org/10.19103/AS.2022.0114.13

Figure 1 Areas of historical (dark shading) and recent (light shading) damaging wheat stem sawfly populations in the USA. Source: Adapted from: Bekkerman and Weaver (2018).

Cephus fumipennis Eversmann (Chen et al., 2004), while in Western Eurasia and North Africa, the dominant species is *C. pygmaeus* (L.) (Wallace and McNeal, 1966).

The most damaging cephid species in North America is the native WSS, although small populations of introduced *C. pygmaeus* are found in eastern North America (Shanower and Hoelmer, 2004). A fourth species, *Trachelus tabidus* (F.), causes crop losses in North Africa, as well as in the south central and Mediterranean regions of Western Eurasia (Wallace and McNeal, 1966). As for *C. pygmaeus*, small populations of introduced *T. tabidus* have been found damaging wheat grown in the eastern USA (Shanower and Hoelmer, 2004; Filipy et al., 1985). Populations of *C. pygmaeus* and *T. tabidus* in western Eurasia have a longer history of damaging wheat, *Triticum aestivum* L. (Shanower and Hoelmer, 2004). Where wheat was first domesticated, a greater diversity of cephid species is also locally associated with crops (Korkmaz et al., 2010; Çalmaşur and Özbek, 2010; Mutlu et al., 2019). Damaging populations of *C. fumipennis* have been reported in China for approximately 50 years (Chen et al., 2004) and likely have existed for longer in central Asia, although available records are few (Shanower and Hoelmer, 2004).

In North America, the intensification of cereal crop production in the areas where WSS damage occurs is much more recent than where other Palearctic cephids cause losses (Beres et al., 2011a). The recorded distribution of WSS from known host feral and native grasses (Cockrell et al., 2017) also encompasses a much greater area of North America than where damaging populations in cereal crops are currently found (Ivie, 2001; Olfert et al., 2019). The species was first described from a specimen in native grass collected in Colorado (Norton, 1872) and was subsequently reared from larvae that were collected in grass stems near Alameda, California (Riley and Marlatt, 1891). Ainslie (1920) contains the following prophetic quote from Albert Koebele, who collected the specimens in California: 'The economic importance of this species arises from the fact that it may be expected at any time to abandon its natural food-plant in favor of small grains, on which it can doubtless successfully develop'.

The first record of a member of the Cephidae damaging wheat in North America was in New York in 1887 (Comstock, 1889) and was likely actually introduced *C. pygmaeus*, which is now endemic and rarely occurs in damaging numbers in eastern North America (Ivie, 2001). For WSS, the first record of damage in western wheat crops was in 1895 at locations in southeastern Saskatchewan and southwestern Manitoba (Fletcher, 1896). A historical progression of damaging populations followed the initial westward spread of wheat production first in spring wheat crops in Manitoba, Saskatchewan, Alberta, North Dakota and Montana (Ainslie, 1929). Subsequently, damaging populations emerged in winter wheat, first in Montana in the 1970s (Morrill and Kushnak, 1996) and likely afterward in southern North Dakota, South Dakota and Nebraska, although precise dates are unknown.

1.2 Future range of damaging populations

More recently, damaging populations arose in Colorado winter wheat, with the first observations in 2010 (Cockrell et al., 2021). It is important to note that new, larger populations were found in wheat in southeast Colorado, subsequent to the first records from more northerly counties. This is most likely due to the local adaptation of existing populations from feral grasses (Lesieur et al., 2016) rather than a progression of range expansion by an invasive population shifting southward. Genetic variation within the WSS populations suggests three main haplogroups who are somewhat geographically separated into clusters - with a notable northern cluster of haplogroups spanning the border between the USA and Canada (Lesieur et al., 2016). In Canada, all populations are from the northern haplogroup, while just south of the border in Montana this cluster contains small populations of the two other haplogroups: southern and mountain. Both the mountain and southern clusters contain populations from this northern haplogroup, while the southern cluster in Colorado contains several populations from the northern haplogroup and none from the mountain haplogroup. Grass- and wheat-associated populations exist in all haplogroups (Lesieur et al., 2016), while the southern and northern haplogroups have the greatest uniformity at the more northern and more southern extremes. Local adaptation to winter wheat progressed southward in Colorado from 2016 to 2021 (Cockrell et al., 2021).

There is a very large area of wheat production where WSS inhabits adjacent grasses only (Ivie, 2001). It is reasonable to assume that the assertion made by Painter (1953) that the life cycles of northwest Kansas populations in grass were too asynchronous from the crop cycle to successfully use large areas of winter wheat is no longer applicable. From the first use of winter wheat by WSS in Montana representing a local change of host use of more than a month sooner (Morrill and Kushnak, 1996) through the southward progression of adaptation in Colorado, this is an ominous scenario. This perspective is further justified by the recent expansion of large damaging populations of WSS into barley (*Hordeum vulgare* L.) in Montana (Varella et al., 2018; Achhami et al., 2020b) representing a further increase in the dryland cropping landscape. Among cereals, hexaploid (*T. aestivum* L.) and tetraploid (*T. durum* Desfontaines) wheat are commonly used as hosts, while another seven species of the genus *Triticum* are reported to be infested to some degree, as well as rye (*Secale cereale* L.) and barley (Wallace and McNeal, 1966). A significant list of host grasses have been recently described by Cockrell et al. (2017) and Peirce et al. (2022) along with Cárcamo et al. (2022), who reported that triticale (× *Triticosecale*), a hybrid of bread wheat and rye, is a suitable host that could be used as a trap crop or host for parasitoids. Recently, WSS life cycles in grasses are becoming more synchronous with wheat under climate warming coupled with frequent longer-term drought, leading to an ongoing expansion of southern adaptation. It is inevitable that WSS will damage wheat in Oklahoma and New Mexico, followed by Texas and more eastern states. In 2015, damage was reported in Sherman and Wallace counties in Kansas (Colorado Wheat Stem Sawfly Resources, 2020 - https://coloradowheat.org/2020/08/wheat-stem-sawlfy-resources/).

Equally important is the potential for damaging populations to occur more northerly in Canada under climate warming, with suitable habitat for expanding larger populations also spreading east and west (Olfert et al., 2019). These authors focused on potential impacts in western provinces under specific warming and drying scenarios, with alarming findings of greater suitability for populations over a much larger northerly area than currently impacted. Although not discussed in detail, the potential distribution of WSS suitable habitat illustrated by Olfert et al. (2019) includes parts of North Central Mexico and southeastern USA (Fig. 2). It is imperative that the long-standing, locally adapted presence of the insect in grass, coupled with a capacity to locally adapt to damage wheat quite readily, be recognized as a large and current challenge to wheat production.

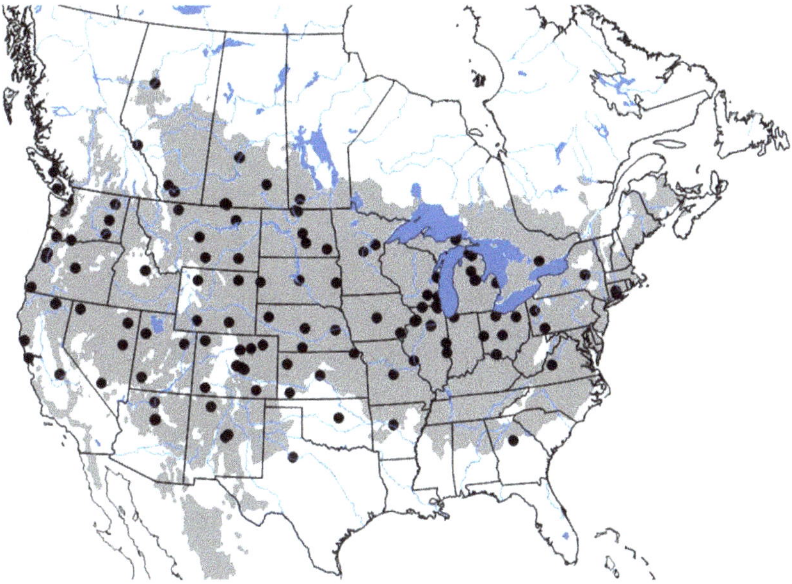

Figure 2 Potential distribution of wheat stem sawfly in North America. Dark points are actual records and dark shading indicates the modeled area where damaging populations might occur. Source: Adapted from: Olfert et al. (2018).

2 Damage

2.1 Unique life cycle-driven losses

The life cycle of WSS is shown in Fig. 3. Unlike most herbivorous pests, the biology of WSS ensures not only decreased yield due to feeding injury, in this case while fully protected inside a stem but also includes obligate cutting of the ripening stem by the larva to provide a secure overwintering chamber for hibernaculum. This is undoubtedly due to an evolved need for protection from exposure to desiccation, excess moisture and pathogens that would occur over up to 8 months of temperate and winter exposure without the frass plug constructed at the time of stem cutting. The frass plugs and diaphanous hibernacula both protect the overwintering larvae and facilitate easy egress of new adults the following year. The stem cutting also allows for a lengthy obligate larval diapause to be accomplished readily within protective hibernacula (Ainslie, 1929), and diapause can be reentered or maintained under unfavorable conditions for an additional year (Holmes, 1982).

In ripening crops, this stem cutting typically leads to lodging with the head of the infested plant on the ground, where it is difficult to recover even with specialized equipment. In practice, it is not possible to recover all lodged

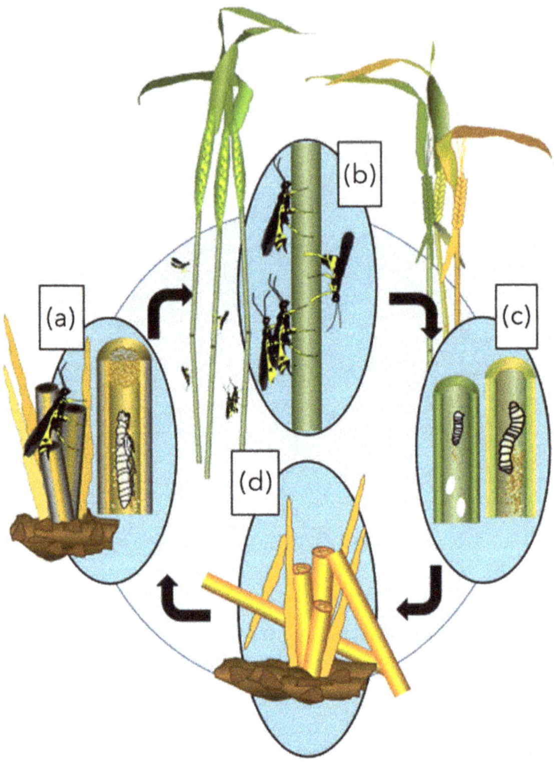

Figure 3 Life cycle of the wheat stem sawfly. (a) Overwintering chambers with pupae and newly emerged adults; mid-May-early July; (b) mating and oviposition; late May-mid-July; (c) eggs, feeding neonate and late-instar larvae with frass and early senescence of injured hosts; June-late July; (d) stems newly cut by late-instar larvae, with larvae in hibernacula inside until metamorphosis; mid-July-late May of the subsequent year. Source: Adapted from: Robertson et al. (2018).

heads that are due to stem cutting by WSS, and the degree of severity of this stem cutting influences the probability that a cut stem will be recovered. Prior to stem cutting, dissection of stems will yield clear evidence of extensive dry frass that is diagnostic of the species. The species is haplodiploid, meaning unmated females can lay eggs that produce male progeny (Holmes, 1982), with new females tending to emerge from larger stems (Morrill et al., 2000) while males are more likely to be found across smaller stems (Morrill and Weaver, 2000). Although each female typically lays a single egg per stem (Holmes, 1982; Buteler et al., 2009), females cannot detect the presence of conspecific eggs or feeding larvae (Buteler et al., 2009). This greatly increases the chances for larval cannibalism within stems (Criddle, 1923; Buteler et al., 2015; Achhami et al., 2020a).

2.2 Physiological yield loss

Impaired photosynthetic capacity from feeding damage in infested stems combines with abiotic stress to limit head weight by as much as 30% (Macedo et al., 2005, 2006, 2007; Delaney et al., 2010). This is corroborated by findings of 10–20% yield loss in spring wheat (Seamans et al., 1944; Holmes, 1977) and winter wheat (Morrill et al., 1992). Morrill et al. (1992, 1994) noted that larger stems were more likely to be infested, which can challenge simple estimation of loss because of innate differences in potential yield. Holmes (1977) reported that among infested stems those that were cut had head weights decreased by 17%, while those that were not cut yielded 11% less. Using an extensive dataset, the model of Bekkerman and Weaver (2018) similarly reported a further 9% decrease in yield for stems infested with larger larvae that proceeded through to stem cutting. Early larval feeding is confined to parenchymal tissues, but when these resources are exhausted in an internode the larva bores through the node impacting the vascular architecture. Late infestation is characterized by extensive feeding on parenchyma in all internodes, along with boring injury at most, if not all, nodes. Frass can be densely packed throughout the entire stem by the time a stem is cut, and it is not known whether loss of parenchyma or loss of vascular integrity contributes more to yield loss, since both occur. Morrill et al. (1994) attribute a darkened region below the node of ripening stems to accumulation of assimilate after vascular transport is disrupted.

2.3 Yield loss from unrecovered heads

Growers may not have any easy in-season estimate of potential head weight loss due to WSS feeding, which complicates management, although sampling guidance (Cockrell et al., 2021; McCullough et al., 2020; Nansen et al., 2005a, 2005c; Weaver et al., 2005) and sampling plans exist for characteristic 'edge-effect' populations (Nansen et al., 2005b; Cárcamo et al., 2007; Pesho et al., 1971). In contrast, at harvest, growers readily detect newly cut stems lodging due to the clear disruption of the standing row structure. In heavy infestations, fallen stems are difficult to harvest, especially after a wind event (Ainslie, 1929; Criddle,1921, 1922, 1923). Early estimates of crop loss due to stem cutting range from 25% of a crop (Criddle, 1922) to 30% of potential yield (Ainslie, 1920), while Beres et al. (2007) estimated losses across cultivars ranging from 6% to 20% due to unrecovered heads. These numbers approximate those derived from the loss of 11 heads per square meter suggested by Holmes (1979) which he determined to be 67 kg/ha and subsequently 53 kg/ha (Beres et al., 2007), due to varietal differences.

2.4 Overall losses

Historical overall crop loss estimates range from over 500 000 tons across the prairies of Canada (Platt and Farstad, 1946) to over 150 000 tons in North Dakota (Davis, 1955), while Beres et al. (2011c) estimate aggregate losses of US$350 million over the entire impacted area of wheat production in North America. The price of wheat currently would make these economic losses 60% greater, before considering that the area associated with damaging populations has increased in Nebraska, Colorado and Kansas. Calculation of current overall losses due to WSS remains challenged by increasing costs of on-farm inputs, inefficiency and additional operations at harvest associated with fallen stems, plus associated equipment costs and repairs (Beres et al., 2007; Bekkerman and Weaver, 2018). It is also difficult to quantify impacts such as increased soil erosion due to stem lodging that perturbs the wind boundary layer in stubble fields, which are also likely to have lower overall snow retention (Beres et al., 2011a; 2011c). Such environmental costs may decrease the available soil moisture, having a direct impact on crop yield, but estimates of current losses due to WSS considering all interactions with environmental factors are beyond the scope of this review.

3 Management

3.1 Insecticides

Adult WSS are likely to emerge over a period ranging from several weeks to longer than a month (Perez-Mendoza and Weaver, 2006), with variation likely to arise from location and wheat variety, suggesting a role of landscape-level population structure and host suitability (Lesieur et al., 2016). A lengthy adult flight period is not conducive to management using contact insecticides because it requires frequent repeat applications and is not efficacious or cost-effective (Knodel et al., 2009). Larvae feeding inside the stems are fully protected from contact insecticides, and only high rates of the systemic insecticide heptachlor (Wallace, 1962; Holmes and Peterson, 1963; Wallace and Butler, 1967) yielded results worthy of further study. More recently, the Montana Department of Agriculture issued a special local need pesticide registration [Section 24(c)] for phorate (Thimet®20-G), an organophosphate soil and systemic granular insecticide to control WSS larvae (Montana Department of Agriculture, 2015). This was due to a 367 kg/ha average yield increase and an 82% decrease in average stem cutting in winter and spring wheat (Wanner and Tharp, 2015). However, post-crop emergence soil incorporation of this insecticide poses potentially unacceptable health and environmental risks and has a long preharvest interval of 85 days in wheat (Portman et al., 2018). For these reasons, the use of phorate was limited, and the 'special local need' registration against WSS larvae in wheat in Montana was not renewed after expiration on 31 December 2019.

Research on alternative or low-impact insecticides, such as neem and Azadirachtin®, the plant defense elicitor Actigard® and signaling molecules had variable and limited, inconsistent benefits against WSS (Tangtrakulwanich et al., 2014; Shrestha et al., 2018) except for WSS adults preferring not to settle on Azadirachtin- or Actigard-treated wheat plants when there were untreated control plants available. In the absence of control wheat plants, and under no-choice conditions, adult WSS settled on treated wheat (Shrestha et al., 2018). Preliminary results for two applications of Actigard® in winter wheat during the growing season had greater WSS larval mortality, reduced larval body weight and wheat stem lodging (Shrestha et al., 2018). Current management of WSS relies entirely on integrated pest management without the planned inclusion of insecticides.

3.2 Cultural tactics

Several approaches have been used to potentially destroy WSS larvae. Among these is burning crop residue, suggested in Fletcher (1904), but Ainslie (1929) reported that there was limited impact on larvae due to their location inside cut stems below the soil surface. There was insufficient heating due to the rapid transient nature of the fire, even when excess straw was layered over residue to increase the intensity and longevity of the fire.

It is now generally considered that any process that removes soil residue has a detrimental impact on desirable soil attributes and moisture (Lafond et al., 1996; Lindwall and Anderson, 1981), which would include burning or moldboard ploughing (Mathews, 1945). The heavy erosion associated with the latter practice led to the use of limited disturbance tillage, such as the Noble blade, that was asserted to have increased WSS damage on the landscape (Holmes, 1982). However, it is difficult to find corroborating data; Ainslie (1920) reported that adult emergence was 90% when stubs containing larvae were buried at depths of 10 cm and 65% when buried at a depth of 15 cm. Farstad et al. (1945) stated that good control could be achieved by deep ploughing residue to a depth of 15 cm or more or by deep tillage of summer fallow that buried the developing WSS. The same publication also suggested shallow tillage that exposed the stubs on the soil surface was also effective in causing mortality before adult emergence, which was confirmed by Morrill et al. (1993). These authors also suggest that the high levels of mortality observed were due to the desiccation of the larvae in the arid environment of dryland wheat fields. In a later study, Runyon et al. (2002) observed that the process of burying residue had greater detrimental effects on beneficial parasitoids than on WSS, leading to a surprising observation that fields adjacent to herbicide-managed no-till cropping had less WSS damage - due to suppression by parasitoids - than heavily tilled fields. More recent WSS publications are replete with declarations that no-till practices have worsened WSS problems due to the loss of mortality caused by tillage, although Weiss et al. (1987) reported no effect of tillage on WSS survival. There is likely some mortality to WSS due to modern tillage operations, especially when stubble is sufficiently exposed or even by direct impact with an implement, but those considering such practices should look for evidence of a reduction of damage in the following crop before undertaking these efforts on a large scale (Morrill et al., 1993).

Historically, considerable attention was also given to trap crops that would intercept and retain immigrating WSS females, thus concentrating infestation on the perimeter of fields (Criddle, 1922). Most recommendations focused on the destruction of infested stems of annual crops such as rye, *S. cereale* L. (Criddle, 1921) or wheat by mowing or tilling (Ainslie, 1929; Farstad et al., 1945). Another observation was that smooth brome, *Bromus inermis* L., was suitable as a permanent trap because it elongated earlier than wheat and was highly attractive while also having limited survival and elevated parasitism (Criddle, 1922; Seamans, 1928; Farstad et al., 1945). Historically, growers may not want to give up productive land to sowing a crop that must be destroyed before harvest - or be willing to concede crop area to a permanent trap (Beres et al., 2011c).

Recently, trap crops such as non-host or resistant traps such as oats or solid-stem wheat have been revisited (Beres et al., 2009; Morrill et al., 2001a). However, Beres et al. (2009) found that this practice was ineffective due to large WSS populations overwhelming any benefits of trapping. Morrill et al. (2001b) used the phenological advantage of planting more favored winter wheat (a resistant solid-stem variety) as a border around hollow-stem spring wheat to demonstrate the efficacy of the concept. One disadvantage was that two separate seeding operations were required, and the benefits may not justify this costly approach. There may be a methodological flaw in how trap cropping is approached, where a potential trap crop is selected without considering variability among cultivars in attractiveness to the sawfly. Several volatile compounds from wheat were found to have behavioral activity for WSS females (Piesik et al., 2008), and these can explain field preferences between cultivars (Weaver et al., 2009) These differences were evident in winter (Buteler et al., 2010; Buteler and Weaver, 2012) and spring wheat (Weaver et al., 2009). Peirce et al. (2022b) found that triticale was a potential trap crop for WSS but was not more attractive than winter wheat to ovipositing WSS. This crop may also be quite suitable as a reservoir for parasitoids of WSS (Cárcamo et al., 2022). Thus, the key to developing effective trap crops remains a characterization of behavioral and oviposition preferences to identify the most attractive cultivars within any germplasm considered for deployment (Weaver et al., 2009; Sharma et al., 2019), as well as consideration of any impacts on parasitoids. Without more careful screening to assess the suitability of the trap crop and target crop, it will continue to be reported that trap cropping is of limited suitability or only applicable for light infestations (Weiss et al., 1990; Beres et al., 2009; Peirce et al., 2022a).

Swathing is a late-season operation that may be required to windrow heavily infested crops to prevent stem losses due to lodging (McCullough et al., 2020) and can be considered for either field margins or entire fields, depending on the severity of infestation (Nansen et al., 2005b). The practice was evaluated to determine if there is potential to use a swather to manage larval WSS populations by cutting the stem below where they are feeding (Holmes and Peterson, 1965). The authors found that the process was only partially effective when stems were swathed at a moisture content that was too high to maintain quality at harvest and did not recommend it if the WSS population was large. In addition, the idea of recoverable yield only strictly applies to stems that would be cut, overlooking the smaller heads due to physiological yield loss; simulations suggest the practice is costly and less efficient than managing large populations by repeated planting of WSS-resistant solid-stem wheat (Bekkerman and Weaver, 2018).

Wheat is an optimal crop for dryland agriculture and monocultures, whether or not interspersed with fallow fields, which can cause potentially severe WSS damage across large landscapes (Beres et al., 2011a; Weiss and Morrill, 1992; Adhikari et al., 2019a). Early in the last century, crop diversity was recognized as an important component in regulating outbreaks of WSS (Criddle, 1922, 1923). Logically, planting of non-hosts is a key part of the effective management of long-term outbreaks as are numerous other aspects of cropping systems (Beres et al., 2011c; Zentner et al., 2006).

Other considerations, including production type and cropping system (Adhikari et al., 2018; Malone et al., 2020, 2022), impacts of weed populations (Weaver et al., 2004; Perez-Mendoza et al., 2006; Keren et al., 2015; Cockrell et al., 2017), impacts of fertilizer use (Luginbill and McNeal, 1954; Delaney et al., 2010; Beres et al., 2011a; Beres et al., 2012), increased row spacing and seeding rates (Luginbill and McNeal, 1958; Miller et al., 1993; Beres et al., 2011b; Beres et al., 2012), later planting dates (Jacobson and Farstad, 1952; McNeal et al., 1955; Weiss and Morrill, 1992; Morrill and Kushnak, 1999), residue management (Weiss et al., 1987; Beres et al., 2011a; Beres et al., 2012) and harvest operations (Beres et al., 2011a; Bekkerman and Weaver, 2018), must be included when planning wheat cropping strategies in WSS-impacted areas. A thorough review of these is found in Beres et al. (2011a, 2011c), but in general, less disturbance, moderation in the use of inputs and conservation of ecosystem services all are favored in a proactive plan for economically optimal wheat production in the presence of damaging WSS populations.

3.3 Biological control

The management of WSS depends significantly on biological controls, primarily by parasitoids but also by insect predators, pathogens and nematodes.

3.3.1 Parasitoids

Two species of native parasitoids, *Bracon cephi* (Gahan), first described in 1918 (Gahan, 1918), and *B. lissogaster* Muesebeck, first described in 1953 (Muesebeck, 1953), can provide significant biological control of WSS (Buteler et al., 2015; Rand et al., 2017; Rand et al., 2020b). Their biology has been reviewed in detail [*B. cephi* - Nelson and Farstad (1953); *B. lissogaster* - Somsen and Luginbill (1956)]. WSS (late instars) is the only known host of either species, other than for the infrequent occurrence of hyperparasitism in both species (Davis et al., 1955). Although *B. cephi* and *B. lissogaster* were responsible for more than 80% mortality of larval WSS in grasses in Manitoba (Criddle, 1922), WSS in nearby heavily damaged wheat was not parasitized. This was more than 25 years after damaging WSS were first observed in wheat (Ainslie, 1920). Criddle (1923) made the first report of *B. cephi* attacking WSS larvae in wheat and was hopeful that *B. cephi* would eventually move into wheat and provide useful levels of biological control. These first observations (Criddle, 1923) were of first-generation presence in the wheat, possibly only occasional early-season foragers. WSS population suppression by *B. cephi* was recorded in wheat over a period of favorable years in North Dakota (Munro et al., 1949) and the Canadian Prairies (Nelson and Farstad, 1953; Davis et al., 1955; Holmes et al., 1963).

Currently, WSS populations are successfully suppressed by these parasitoids in many areas - with possible parasitism rates of greater than 90% (Holmes et al., 1963; Morrill et al., 1998; Runyon et al., 2002; Weaver et al., 2004; Wu et al., 2011, 2013; Cárcamo et al., 2016). Their bivoltine life cycles on a univoltine host are a major part of the frequently reported successful suppression of WSS. The life cycle of both *B. cephi* and *B. lissogaster* both feature adult emergence at approximately the same time as the host species (Nelson and Farstad, 1953; Somsen and Luginbill, 1956), with females requiring a pre-oviposition period in both species, although it is more protracted in *B. cephi*. This is likely to allow for WSS hosts to feed and develop to late-instar larvae suitable for detection (Somsen and Luginbill, 1956; Mankin et al., 2000, 2004) and subsequent parasitism (Nelson and Farstad, 1953). On the other hand, as first noted by Criddle (1923), the second generation may not be fully successful in wheat crops due to asynchrony, presumably because of early crop ripening and advanced harvest date that occurs under drought conditions that are frequent in these dryland cropping systems (Holmes et al.,

1963; Holmes, 1979; Rand et al., 2014; Peirce et al., 2021). This is difficult to corroborate experimentally. Rand et al. (2020b) conducted experiments with supplemented and constrained annual water limitations in caged wheat systems, but the treatments do not represent landscape-level drought, where WSS larval host availability in standing wheat stems is truncated by very early cutting of prematurely ripening wheat, which ultimately greatly constrains the success of second-generation parasitoids. Thus, the potential for suppression of WSS by these parasitoids is high, with the caveat of population declines as a result of drought, and especially multi-year drought (Holmes et al., 1963).

Changes in WSS life history associated with its attacking wheat necessitate subsequent adaptation for *B. cephi* and *B. lissogaster* (Criddle, 1923; Ainslie, 1929; Nelson and Farstad, 1953; Cárcamo et al., 2012; Peirce et al., 2021). This is especially true for more southerly regions of southern Montana, Colorado, Nebraska and Kansas where both host and natural enemies must utilize earlier maturing crops. For example, the transition to successful parasitism of WSS larvae in winter wheat in Montana was governed by a shift to earlier WSS adult emergence of approximately a month - probably in the 1970s (Morrill and Kushnak, 1996). *B. cephi* and *B. lissogaster* rapidly followed from nearby fields of spring wheat, a long-standing host of WSS and its parasitoids (Morrill et al., 1994). In winter wheat in Nebraska, Colorado and Kansas, where spring wheat is not grown, an equally rapid shift from grasses to winter wheat by the parasitoids is unlikely (Lesieur et al., 2016; Cárcamo et al., 2012) because this requires a shift in host plant species as well as a shift in phenology (Morrill et al., 1994, 1998; Cockrell et al., 2017; Peirce et al., 2021). Similar to the observations of Criddle (1923) and almost a century later, both parasitoids are present in grasses surrounding Colorado wheat fields, but there is as yet no parasitism of WSS in the wheat (Peirce et al., 2021) despite WSS first appearing in Colorado wheat in 2010 (Cockrell et al., 2017). Nonetheless, eventual biological control of WSS by these parasitoids will likely be important for its management in more southern regions where varietal choices on agricultural landscapes can trigger local intensification of WSS populations (Rand et al., 2012, 2014; Peirce et al., 2021), likely due to a massive increase in the density of highly suitable host stems per unit area. The cryptic and secure WSS developing inside wheat stems (Ainslie, 1920; Holmes, 1982; Weiss and Morrill, 1992; Weaver et al., 2005) are protected from natural enemies other than the bivoltine parasitoids (Holmes and Peterson, 1963). Although more than 66% of the stems cut by WSS were reported to be cut below the soil surface (Runyon et al., 2002), reports of elevated, unexpected levels of parasitism were reported in WSS cut stems (Rand et al., 2011; Cárcamo et al., 2011). It is not known whether these parasitized, cut stems were cut above the soil surface, but it does illustrate adaptation by both WSS and parasitoids to the agricultural system. Rand et al. (2017) suggested that 68% parasitism of

WSS larvae was required to suppress population growth, whether larvae were actively feeding or were in chambers created after stem cutting (called winter or summer larvae).

3.3.2 Enhancement of native parasitoids

Success of parasitism could be improved with habitat management. Pre-oviposition stage parasitoids have greater longevity, egg load and egg size when supplemental carbohydrates and amino acids are available (Reis et al., 2019; Rand et al., 2020a; Cavallini et al., 2022). This includes both nectar and honeydew (Rand et al., 2020a), as well as artificially supplemented resources (Reis et al., 2019; Cavallini et al., 2022). Management to provide these resources could improve parasitism. The overwintering strategy of *B. cephi* involves the accumulation of glycerol to enable supercooling and even freeze tolerance for the survival of prolonged exposure above the surface of the frozen substrate during the winter (Salt, 1956, 1959). Nelson and Farstad (1953) report that synovigenic *B. cephi* also foraged yellow flowers for nectar, while Somsen and Luginbill (1956) included honey-water as a nutritional supplement to increase longevity while collecting early data on synovigenic *B. lissogaster*.

Possibly consistent with a reliance on floral resources by these parasitoids, where WSS has had large impacts on intensified production of wheat crops, there has often been very limited cultivation of flowering species (Adhikari et al., 2019a) except for limited acreage of organic wheat production (Adhikari et al., 2018). The trend in crop production in dryland agriculture to favor increased floral diversity, with benefits for native and introduced pollinators (Adhikari et al., 2019b, 2019c), potentially can benefit WSS parasitoids. Although WSS can also benefit slightly from floral resources (Rand et al., 2019), this has no other impact on reproductive parameters. WSS and parasitoid preference for yellow flowers (Wallace and McNeal, 1966) could reflect a need for water [also corroborated in the experiments of Rand et al. (2019)] for greater longevity in the dryland habitat. It has been suggested that efforts to conserve and even mass produce the native braconid parasitoid species have merits (Portman et al., 2018).

3.3.3 Biological control and host plant resistance

The importance of parasitism in solid-stem WSS-resistant wheat (Weaver et al., 2004; Cárcamo et al., 2011) was explored using models to estimate the combined effect of solid-stem wheat-based resistance and parasitism on WSS larvae (Rand et al., 2020a). The findings showed that parasitism at 22% will attenuate WSS population growth in solid-stem wheat, which has great value for wheat growers. There are concerns about comparison of parasitoid efficiency (Weaver et al., 2004) and numerical responses (Cárcamo et al., 2011; Rand et al., 2012) across resistant and susceptible wheat that warrant further exploration, considering the somewhat concerning findings of Rand et al. (2012). In a recent review, Peirce et al. (2022) tabulate the benefits of host plant resistance, including contradictory reports on the interactions of solid stems with the specialist braconid parasitoids; nonetheless, efforts to increase host plant resistance have great merits.

Parasitism is complementary to host plant resistance because it delivers mortality to late-stage larvae, while mortality to host plant resistance is greater against younger and especially neonate larvae (Nelson and Farstad, 1953; Somsen and Luginbill, 1956; Peterson et al., 2011; Buteler et al., 2015). Mortality by parasitism is irreplaceable because later instars are protected within stems and protected from most other sources of mortality (Rand et al., 2017, 2020a). Although host plant resistance impedes establishment by early instars, the parasitoids have immediate economic value because they are idiobionts that paralyze hosts, preventing further injury to the stem, as shown by Buteler et al. (2008) and corroborated in the model of Bekkerman and Weaver (2018). Losses due to parasitizable late-stage larvae proceed a further 8% if injury continues to stem cutting, while there is a relative increase of 11% for such stems containing parasitized larvae. Thus, to incentivize wheat producers to conserve and enhance parasitoids, both decreased yield losses in the current crop and decreased risks to subsequent crops must be emphasized.

3.3.4 Other parasitoids

Efforts to identify additional parasitoid species for use against WSS have been well summarized (Shanower and Hoelmer, 2004; Hoelmer and Shanower, 2004). Much of this was based on the idea that WSS relatively recently colonized North America (Ivie, 2001). Foreign exploration in its presumed native range in Asia identified *Collyria catoptron* Wahl, which parasitizes *Cephus fumipennis* Eversmann (Wahl et al., 2007) as a candidate for further evaluation against WSS. The results showed that the parasitoid could successfully detect WSS larvae but could not complete development (Rand et al., 2016). Subsequently, it was shown that WSS was likely native to North America rather than a recent colonist (Lesieur et al., 2016) helping to explain why *C. catoptron* failed to develop in WSS. Multiple native parasitoid species attack WSS larvae in grasses, but they are very infrequently found in WSS-infested wheat plants (Davis et al., 1955; Morrill et al., 1998). The reasons for this are unclear, but several of the species are generalists that use multiple stem inhabiting hosts that are abundant in grass, including WSS.

3.3.5 Predatory species

A predatory clerid, *Phyllobaenus dubius* (Wolcott), occasionally causes low levels of mortality of WSS larvae, most often detected in overwintering stubs, as was the case for the specimens used for initial identification (Morrill et al., 2001b). Although there have been reports by others confirming this (Beres et al., 2009; Cárcamo et al., 2011), and alluding to other, related clerid species that could be doing the same thing (Beres et al., 2011c), very little is known of the life cycle of this predatory species or other predators nor are there any estimates of potential impacts on WSS.

3.3.6 Pathogens and nematodes

There has been work on entomopathogenic nematodes (Tangtrakulwanich et al., 2014; Portman et al., 2016, 2018), including efforts to identify native species in WSS habitat (Sandhi et al., 2020). In laboratory experiments, Portman et al. (2016) showed that nematodes were successful in killing WSS larvae in stubs especially with certain adjuvants, but these results did not translate into reliable efficacy in the field (Portman et al., 2016). Further exploration is needed (Portman et al., 2018), perhaps using species isolated from dryland wheat production (Sandhi et al., 2020).

Some fungal pathogens can impact WSS, including a complex of facultative *Fusarium* species isolated from larval cadavers (Wenda-Piesik et al., 2009), and commercially marketed strains of *Beauveria* and *Metarhizium* species (Tangtrakulwanich et al., 2014; Portman et al., 2018). Limited efficacy occurs for commercial pathogens that are applied externally to infested plants, perhaps because persistence is limited in these arid settings, but there may be potential for future development of these as endophytes for commercial use in wheat (Portman et al., 2018). The isolates of the *Fusarium* spp. are facultative entomopathogens that also cause plant disease in wheat and have limited utility unless a strain is found that is asymptomatic in wheat while maintaining the ability to kill WSS larvae (Wenda-Piesik et al., 2006, 2009; Portman et al., 2018).

4 Host plant resistance

The foundation of WSS management is planting rotational crops that are not hosts or are poor hosts of WSS (Weiss and Morrill, 1992). Dryland crops such as legumes and oilseeds may help to curtail persistent large populations of WSS as part of a broad production practice, even in a wheat-focused cropping scenario (Miller et al., 2002). None of the broadleaves are hosts except for one early report of oviposition in flax without successful development (Farstad, 1944). Among grasses, frequently cultivated cereals range in susceptibility

(Cockrell et al., 2017). Hexaploid bread wheat is more susceptible than durum wheat, which is followed by barley (Varella et al., 2018; Achhami et al., 2020c) and finally by oat (Beres et al., 2011c). Oat species are readily accepted for oviposition but are lethal to WSS eggs and larvae (Criddle, 1923; Ainslie, 1929; Weaver et al., 2004). Criddle (1923) attributed mortality to observations of excessive sap production in oat and a possible nutritional deficiency creating an unfavorable environment for larvae. The moisture content of oat is not different from that of either resistant or susceptible wheat, and WSS mortality occurs at an early stage, suggesting that a nutritional deficiency in oat is an unlikely cause (McGinnis and Kasting, 1961).

4.1 Pith containing wheat stems

Kemp (1934) showed that wheat stems containing pith were cut less by WSS larvae. Pith was an early focus on heritability of traits in wheat (Engledow and Hutchinson, 1925), even before potential benefits in WSS suppression were known. This observation and the ability to breed wheat cultivars targeting complete expression in all internodes (Dhariwal et al., 2022) soon led screening and breeding for solid stems rather than tests of resistance to WSS (Sherman et al., 2015; Talbert et al., 2014) because it was possible to readily assay pith in the absence of consistent and vigorous WSS populations. While solid-stem wheat varieties are the foundation of WSS management (Beres et al., 2011a, 2013a, 2013b), it is likely incorrect to strive solely for maximal expression of stem solidness throughout the stem. This is because WSS mortality is caused by several aspects of pith solidness. Simple stem phenotyping does not detect temporal variations such as 'early solidness' (Talbert et al., 2014; Varella et al., 2016) that can influence WSS behavior and survival. Understanding of the effects of pith on WSS oviposition and increased larval mortality has lagged (Varella et al., 2017b). Indeed, according to recent reviews, mortality in solid-stem wheat is reported to be due to physical factors impeding immatures or due to early drying of pith in solid stems (Peirce at al., 2022) or 'crushing' eggs (Beres et al., 2011c). The former is now less common, with only a small proportion of larger larvae dying in pith that impedes their ability to cut the stem for overwintering - and there is limited evidence in the scientific literature that truly supports that early drying of pith causes increased larval mortality. Rather, desiccated pith itself, arising under early, normal, or late drying, becomes a lethal barrier to some larger larvae. The inability to move and consume nutrients in desiccated pith of solid-stem wheat causes mortality and the cadaver dehydrates rapidly, but there is no evidence of larval desiccation as a direct mortality factor (Kasting et al., 1958; McGinnis and Kasting, 1961). In solid-stem pith, unhatched eggs containing embryos that have died appear deflated due to loss of liquid, but the outer surface of such eggs are intact

- there is no evidence that the integrity of the egg is lost via 'crushing' that breaches the chorion. Loss of water in egg or larval cadavers is expected, so the occurrence of dehydrated cadavers does not support desiccation-causing mortality. In a survey of the literature, 'crushing' of the larvae by the stem is only reported for overwintered hibernacula within cut stubs of timothy, *Phleum pratense* L., in which larvae are crushed by early-season 'bulb-like' growth of new plants (Farstad, 1940). This effect is limited, and otherwise timothy is susceptible to stem cutting by WSS (Criddle, 1922; Ainslie, 1929) and should be considered a host.

The key cause of mortality across all resistance to WSS is antibiosis, whether it is 'absolute' as in oat, which has hollow stems (Criddle, 1923; Ainslie, 1929), 'relative' in barley, which also has hollow stems (Varella et al., 2018; Achhami et al., 2020b, 2020c), or 'relative' in solid-stem wheat, which kills more larvae than hollow-stem wheat (Buteler et al., 2015; Varella et al., 2017b). Partial or 'relative' antibiosis, like in solid-stem wheat, has other fitness impacts on surviving larvae that inhibit population dynamics (Cárcamo et al., 2005). The cumulative advantage of the sublethal effects, combined with moderate mortality, has led to research with repeated planting (Holmes and Peterson, 1957) to evaluate longer-term impacts and contributions to population suppression. Data-driven models show that repeated planting of solid-stem wheat to counter large infestations is better than planting agronomically desirable hollow-stem wheat and swathing (Bekkerman and Weaver, 2019). Solid-stem wheat has better yields than hollow-stem wheat under large, damaging WSS infestations (Beres et al., 2007, 2009), despite its perceived agronomic liabilities in the absence of WSS. Yield under moderate-to-heavy sawfly infestation is often a key component of the registration of new solid-stem germplasm (Bruckner et al., 2020) and complements yield data obtained across multiple environments that encompass all levels of WSS infestation.

To further enhance mortality by isolating toxic compounds in the solid-stem pith of wheat, or from other host plant resistance, which could subsequently be targeted in breeding efforts is difficult. It is not possible to rear WSS larvae on an artificial diet (Kasting and McGinnis, 1958; McGinnis and Kasting, 1962; Macedo et al., 2005) long enough to quantify the mortality of candidate toxic compounds in bioassays. The potential for solid-stem pith components to cause antibiosis in WSS was inferred using a surrogate pest cutworm species (McGinnis and Kasting, 1959; Kasting and McGinnis, 1961), which showed a clear inhibition of growth when solid-stem pith was incorporated into an artificial diet, relative to other tissues. Molecular methods are being used to identify other traits that could confer WSS resistance (Biyiklioglu et al., 2018). Of interest are miRNAs as well as short- and long-noncoding RNAs that may differentially regulate the expression of host plant responses within infested plants (Akpinar et al., 2023; Cagirici et al., 2017; Biyiklioglu et al., 2018; Brant and Budak, 2018; Budak et al., 2020; Muslu et al., 2021; Robertson et al., 2018). The goal of hollow-stem wheat lethal to immature WSS (as in oat) based on classical breeding with non-wheat sources of resistance seems remote. The ability to scrutinize mechanisms causing mortality by comparing the relative expression of genes and metabolites in infested versus uninfested stems may provide insights (Biyiklioglu et al., 2018; Peirce et al., 2022).

4.2 Loci conferring stem solidness in wheat

Breeding for solid-stem wheat can be enabled by determining quantitative trait loci contributing to stem solidness (Dhariwal et al., 2022). These include loci on the 3B chromosome in hexaploid and tetraploid wheat (Cook et al., 2004; Lanning et al., 2006; Houshmand et al., 2007) and the 3D chromosome in hexaploid wheat (Lanning et al., 2006). Allelic variations at the 3B locus confer early solidness or lack of pith (Cook et al., 2017; Wong et al., 2022). Exploration of resistance in targeted populations or broad screening of landraces (Varella et al., 2015, 2017a, 2019a, 2019b; Cook et al., 2017; Peirce et al., 2022) in hexaploid and tetraploid wheat identified several other loci conferring useful stem solidness on chromosomes 1B, 2A, 3A, 4A, 5A, 5B, 5D and 6B (Sherman et al., 2010; Varella et al., 2017a; Nilsen et al., 2017; Cook et al., 2017; Varella et al., 2019a, 2019b), most of which were corroborated using WSS infestation and mortality data. Recently, genes on different chromosomes have been identified that are associated with stem solidness, including those expressed primarily in the basal or lower internodes, which may govern lignification processes that result in greater stem strength (Kebrom et al., 2012; Hyles et al., 2017; Oiestad et al., 2017). Nilsen et al. (2020) identified that copy number of *TdDof*, encoding a Dof zinc finger protein, controls stem solidness at the 3B locus in tetraploid and hexaploid wheat. Greater copy number was associated with increased retention of solid-stem characteristics. There is a great need for a comprehensive review of stem solidness and WSS resistance in wheat, focusing on consolidating information across publications focusing on genetics and molecular mechanisms.

4.3 Maximizing stem solidness in crops in the field

Early solid-stem varieties expressed pith inconsistently under environmental conditions that reduce light levels (Beres et al., 2017; Subedi et al., 2021). Thus, a major goal for breeding remains complete solidness across all internodes that is maintained even under low light levels (Dhariwal et al., 2022; Liang et al., 2022). Meanwhile, current recommendations to growers using existing germplasm may include cultivation practices such as avoiding high seeding and fertilizer rates to reduce undesirable autoshading within the crop that reduces stem solidness (Luginbill and McNeal, 1954, 1958; Beres et al., 2011a, 2011b; De Pauw and Read, 1982). Equally important is the comparative field assessment of solid-stem commercial cultivars (Beres et al., 2013a) and all other germplasm that has potential utility (Beres et al., 2013b) under the conditions imposed by large and damaging WSS populations. Often, it is not possible to combine agronomic and host plant resistance studies over a large geographic area, which makes those studies that do this all the more important (Subedi et al., 2021). Ideally, it will be possible to someday deploy wheat varieties with resistance levels more like those found in barley or even oat.

4.4 Other considerations for host plant resistance

Modern comparative techniques also show great promise in understanding underlying mechanisms for resistance. These include comparative metabolomics of leaf surface compounds (Lavergne et al., 2018), which may play a role in antixenosis, or non-preference, for ovipositing WSS. These, along with volatile attractants, may play a critical role in oviposition choices made by females (Weaver et al., 2009; Lavergne et al., 2018; Varella et al., 2017b). Equally important are those studies that have compared transcriptomic responses to WSS infestation (Biyiklioglu et al., 2018) as well as proteomic and metabolomic responses to infestation (Lavergne et al., 2020). These studies provide important insights into candidate germplasm for mechanistic exploration of relative responses to infestation across cultivars (Lavergne et al., 2020). For example, the relatively susceptible cultivar 'Hatcher' had the greatest number of induced changes in expression, while more resistant 'Conan' germplasm had fewer changes (Lavergne et al., 2020), suggesting underlying useful genetic traits are available even in more susceptible wheat germplasm. Multitrophic interactions could also be exploited. A *Spiroplasma* symbiont was recently characterized from larval and adult wheat stem sawflies from grass and crop populations (Yeoman et al., 2019). Wheat germplasm could be evaluated for its effects on this symbiont as an avenue for a novel type of host plant resistance.

5 Future trends in research

5.1 Distributions

There is evidence that WSS will continue to spread by local adaptation of populations in grasses to wheat. Climate warming could influence ripening synchronization between endemic grass populations and wheat, facilitating this colonization. Although there are extensive records for WSS at the level of states or provinces (Ivie, 2001), there have been no strategic surveys of local populations since the extensive Northern Great Plains surveys led by Davis (1948, 1952, 1953, 1955). Given recent reports from Washington, Minnesota and Kansas (Bekkerman and Weaver, 2018; Peirce et al., 2022) as well as recent outbreaks in Colorado and Nebraska, a combination of efforts from state-level extension personnel and even citizen scientists (Maistrello et al., 2016) could be employed to document where populations are occurring in grass adjacent to wheat fields and in the wheat crop. Sampling locations could be guided or supplemented by remote-sensed distributions, and the models evaluated in Lestina et al. (2016) using Moderate Resolution Imaging Spectroradiometer (MODIS) imagery may be informative. Sweep net sampling the distinctive adults in late spring to early summer, as well as dissection of wheat and grass stems on the interface of crop edges in mid-summer, would provide valuable information. Sampling near the edges of the current range of WSS could look for smaller larvae in wheat stems on the field edge that may not be viable but indicate the potential for further range expansion.

5.2 Basic research needs

The last major review on WSS (Beres et al., 2011c) and a recent concise review (Peirce et al., 2022) mention the need for work on the chemical ecology of plant-insect interactions across tritrophic levels made feasible by technological advances since early work on the chemical basis of WSS oviposition choice (Holmes and Peterson, 1960). Understanding of the sensory mechanisms used by WSS during the selection of wheat or barley, including responses to host odors (Gress et al., 2013; Robertson et al., 2018) and contact cues (Varella et al., 2017b; Robertson et al., 2018), and knowledge of variability in key volatile attractants that facilitate the use of multiple hosts (Weaver et al., 2009; Achhami et al., 2021) are needed. This should be augmented by efforts to determine whether plant-plant communication (Piesik et al., 2013) occurs when WSS feeds within the wheat stem. Inducible compounds and chemical messengers are not yet known for the type of plant injury caused by WSS larval feeding.

Sensory ecophysiology could also help improve biological control. It is not known how *B. cephi* and *B. lissogaster* detect WSS larvae that are hidden inside stems. The size of the infested stem relative to the much smaller and hidden WSS larva and the diminutive female parasitoid makes it likely that acoustic cues (Mankin et al., 2000, 2004) are important for localized search while induced volatile compounds, and possibly reflected wavelength differences (Nansen et al., 2009), provide cues to which stem is infested.

In many regions, *B. cephi* and *B. lissogaster* are spatially and temporally sympatric (Morrill et al., 1998; Peirce et al., 2021; Cárcamo et al., 2012), and it is not known what maintains species integrity and how these species might partition the host resource (Runyon et al., 2001). Mechanisms facilitating species isolation in these congeneric braconids are not known, including any potential pheromones - nor is the pheromonal system of the host WSS itself fully understood (Bartelt et al., 2002; Cossé et al., 2002). The mechanisms of incapacitation of hosts by the braconid species have not been elucidated, although differences in Dufour's gland hydrocarbon content in the sting complexes of the two species have been reported (Baker et al., 2005). Care should be taken to avoid the convenience of only reporting on one species in areas where both are present (Rand et al., 2012; Cárcamo et al., 2012), especially when working with adults because the key to separate these based on external characters (Runyon et al., 2001) is not difficult (Peirce et al., 2021).

There has been a large amount of research conducted on host plant resistance mechanisms, with a major focus on stabilizing and maximizing the expression of stem solidness in WSS-resistant wheat (Dhariwal et al., 2022; Liang et al., 2022) and equally on better understanding stem solidness (Nilsen et al., 2020). However, other sources of resistance are possible, given the consistent inability of WSS to complete development in oat that has been known for over a century. Recently, an increase in the use of barley by WSS allowed for the identification of relatively resistant and susceptible cultivars (Achhami et al., 2020a, 2020b), but mortality, and in particular neonate mortality, in barley still exceeds that in solid-stem wheat (Achhami et al., 2020c). This could enable crossing experiments to identify genes and loci that confer resistance. The ability to produce putative toxic compounds by barley may be latent and conserved in more distantly related wheat. It may also be possible to modify an oat population to be more susceptible to WSS so that a similar approach to scrutinizing the cause of mortality can begin.

Better understanding of what makes solid-stem wheat pith moderately toxic to WSS may allow for greater expression of resistance in commercial wheat varieties (Peirce et al., 2022). To achieve this, a careful study of mortality is key (Buteler et al., 2015; Varella et al., 2017b; Achhami et al., 2020c). Historically, reports suggested that pith obstruction of large larvae attempting to descend to cut stems was the dominant form of mortality (Holmes, 1954; Holmes and Peterson, 1962; Platt and Farstad, 1946; Morrill et al., 1992, 1994), but there was a debate on the relative importance of this and neonate mortality (Roberts, 1960; Holmes and Peterson, 1961; Kasting and McGinnis, 1961). Roberts (1960) indicated that the cause of mortality varied with plant stage and when infestation occurred (Fig. 4). Neonate mortality and plant response (Buteler et al., 2015) evidenced by the 'oviposition scar type' mortality of Roberts (1960) seem more important in modern datasets. Further exploration of how critical the newly identified symbiont (Yeoman et al., 2019) is to the survival of WSS, especially when feeding interactions include resistant germplasm, is needed.

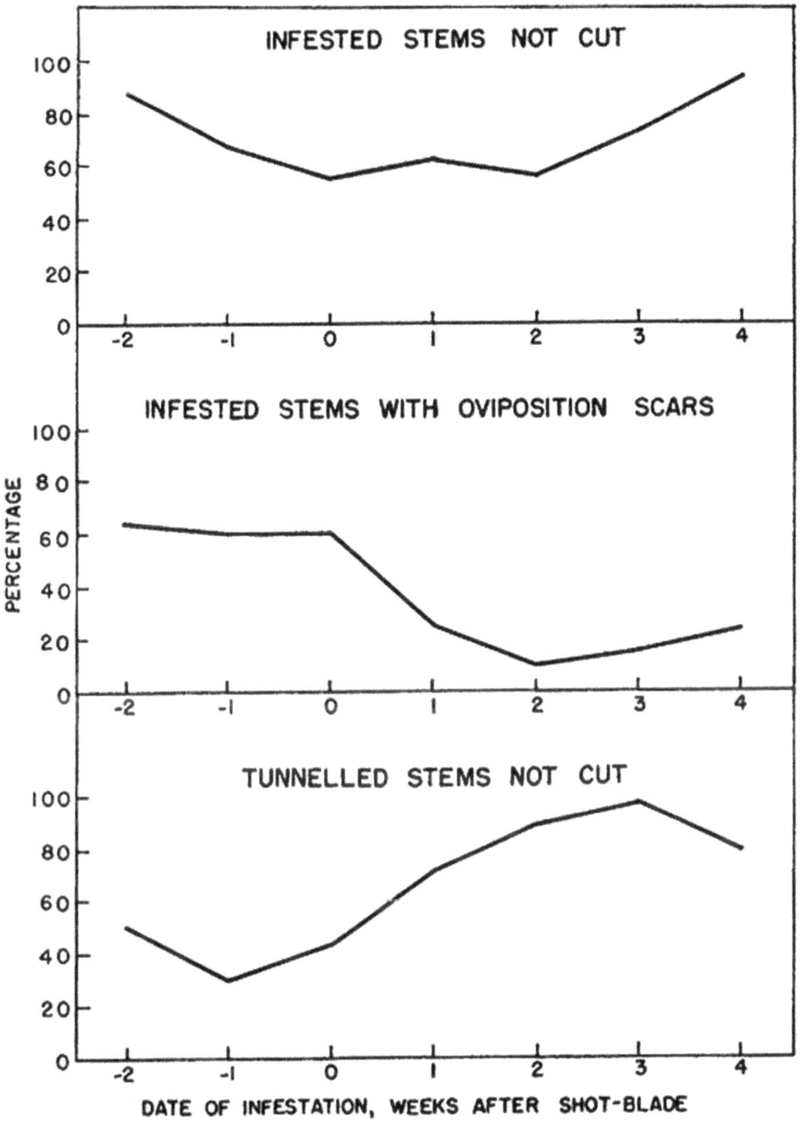

Figure 4 Mortality of wheat stem sawfly larvae from specific infestation over weeks after full emergence of the flag leaf (called 'shot-blade' below) of 'Rescue' solid-stem wheat. In this image, -2 weeks would correspond to the first possible infestation at the beginning of stem elongation. Top graph: the percentage of stems infested that were not cut for each infestation interval. Middle graph: stems with oviposition scars over the same interval. These scars occur for dead eggs and newly hatched larvae. Lower graph: stems tunneled by late-instar larvae from the same infestation interval showing the effect of desiccated pith. There is greater mortality of neonates in early infestations and greater mortality of late-instar larvae in later infestations. Source: Adapted from: Roberts (1960).

5.3 Applied research needs

The only two systemic insecticides effective against WSS larvae feeding in wheat stems, phorate and heptachlor, are no longer available for use in wheat. All new systemic insecticides should be evaluated for potential activity against WSS larvae, especially if the systemic expression extends to the interior lining of the wheat stem. In contrast, targeting WSS adults with repeated applications of contact insecticides is not viable, because of the protracted period of adult emergence and flight driving up costs of repeated applications (Knodel et al., 2009).

Cultural tactics that could target WSS include revisiting the use of cultivation implements to expose WSS larvae on the soil surface in their overwintering chambers inside stubs causing desiccation and mortality. Tillage to achieve this can cause undesirable disturbance of the soil surface, leading to loss of soil water and erosion and additional production costs which must be a consideration. Effectiveness must be assessed in terms of not only WSS destruction by tillage activity but whether the WSS populations are significantly smaller the following year. Use of tillage should also consider detrimental negative outcomes for parasitoids.

Harvest technology in dryland cropping has changed, where header stripper-type harvesters are used more frequently. These are desirable because they maximize the residue structure after harvest, ensuring better moisture retention and protecting the soil surface from erosion. It is unlikely that parasitoids can survive passing through the implement in straw processed at harvest by a conventional combine, while the header stripper may leave more parasitoids undisturbed to overwinter in tall standing stubble. A researchable question is whether these parasitoids survive the gentle harvest process. In addition, there may be ancillary value in the use of solid-stem wheat to manage WSS, which provides a standing stubble structure to retain moisture and limit erosion. It was recently shown that there is no advantage to this tactic when comparing solid- and hollow-stem wheat (Simão et al., 2021) using typical harvest implements under unknown (or perhaps limited) WSS numbers. More study is needed to make more meaningful comparisons when sufficient populations of WSS occur. Greatly increased stem cutting by abundant WSS will reduce the standing stubble structure considerably in susceptible cultivars relative to solid-stem cultivars.

More research is needed on conservation biological control for *B. cephi* and *B. lissogaster*. Beres et al. (2011a) outlined a plausible cropping system to minimize losses due to WSS, but additional effort is required to provide adjacent nectar-providing species that will benefit these parasitoids. This could be accomplished by cropping systems or rotations that have spatially staggered arrays of wheat and flowering crop species. Provisioning must consider the

Published by Burleigh Dodds Science Publishing Limited, 2023.

foraging ecology of these parasitoids (Rand and Waters, 2020), which may access nectar more readily from more open and shallow floral structures, from extrafloral droplets or from honeydew, when available. Research is needed on late-ripening grass species like smooth brome (*Bromus inermis* L.) that harbor WSS larvae as reservoir hosts for the parasitoids (Criddle, 1922a; Seamans, 1928). These reservoirs will be especially important when second-generation crashes occur during historical and frequent drought (Holmes et al., 1963; Holmes, 1982).

Pathogens are likely to be greatly challenged in the arid conditions that allow WSS populations to thrive, but more investigation of the possible endophytic establishment of candidate pathogens early in the growing season might identify a vehicle to deliver these organisms in drylands (Portman et al., 2018). However, caution needs to be exercised when considering endophytic organisms because of the feeding habits of WSS larvae. The principal diet of the larvae is the layer of tissue lining the stem lumen as well as vascular conduits consumed when boring through nodes. Thus, the establishment of an endophyte in host plant leaves will not necessarily provide a mode of action against WSS larvae, just as most systemic insecticides fail to reach feeding larvae. In the case of WSS, it will help if the organism colonizes interstitial spaces such as the stem interior by localizing on the surface of the inner stem lining.

Trap crops merit further consideration, especially if research can identify both highly attractive and less attractive cultivars (within and across species) to be deployed in a push-pull scenario (Khan et al., 2016) where immigrating female WSS are retained on a preferred cultivar on the crop periphery. This was previously demonstrated by using a phenologically more desirable winter wheat crop planted around spring wheat (Morrill et al., 2001a), but growers would rather not have two temporally separated seeding operations for the same field. There may also be an opportunity to use a WSS-preferred solid-stem cultivar to concentrate the pest and cause mortality as a 'dead-end' trap crop. Current technology should allow for the rapid collection of volatile compounds as part of this effort under controlled conditions, but field evaluations will be more challenging.

Finally, the single greatest challenge is providing growers information that allows them to maximize use and benefits from planting solid-stem WSS-resistant wheat varieties, which remains the single best management strategy for WSS (Beres et al., 2011a, 2011c). Fundamental to this is the recommendation for plant solid-stem wheat varieties, which growers are reluctant to do. The elite solid-stem germplasm currently available is typically slightly lower yielding than comparable hollow-stem cultivars, but the gap between these lessens with the continual efforts made by solid-stem variety development programs. There is an urgent need for nurseries specifically targeting yield characterization under large and damaging WSS populations. The reason for this is twofold. First, solid-stem wheat is intended for use when WSS problems occur, but not when the populations of the pest are small or non-existent. The yield of solid-stem wheat in the absence of WSS is unimportant but appears as tantamount in the way data are presented. Second, and equally critical, accurate and detailed data are urgently needed to demonstrate that susceptible cultivars have poor agronomic performance when damaging WSS populations exist, even when very successful swathing of ripened crops recovers nearly all of the smaller heads caused by WSS feeding (Bekkerman and Weaver, 2018).

6 Conclusion

It is very likely that WSS will become an even greater focus for wheat and barley growers in western North America. Local adaptation of populations in grassy areas near wheat fields will continue, resulting in spread of damaging populations. The natural regulation of WSS in such areas by specialist parasitoids appears to lag behind the host but can provide significant suppression of damaging populations when conditions are right. The foundation of management remains host plant resistance, and major efforts are underway to incorporate existing germplasm into crop production areas where outbreaks are expected, or where they may worsen. Ongoing production practices can be continually adapted to minimize losses as much as possible - and need to be supported by careful research. Exploratory efforts targeting new sources of host plant resistance and alternative biological control approaches will get additional attention. Outreach efforts should continue to remind growers of proactive management benefits and should also include resources to increase familiarity with the pest and the damage it causes.

7 Where to look for further information

There are a number of good resources listed in the following References. Additional on-line resources would include web pages of institutions leading research on this insect. Here are a few examples to get started:

- Edde, P. A. (2022). Chapter 9: Arthropod pests of small grains: Wheat (Triticum aestivum L.) and barley (Hordeum vulgare L.). In: Edde, P. A. and Crop, F. (Eds.), Arthropod Pests of Economic Importance, pp. 536-611. doi: 10.1016/B978-0-12-818621-3.00001-X.
- Beres, B. L., Dosdall, L. M., Weaver, D. K., Cárcamo, H. A. and Spaner, D. M. (2011). Biology and integrated management of wheat stem sawfly and the need for continuing research. The Canadian Entomologist 143(2): 105-125. doi: 10.4039/n10-056
- Peirce, E. S., Cockrell, D. M., Mason, E., Haley, S., Peairs, F. B. and Nachappa, P. (2022). Solid stems and beyond: Challenges and future directions of

resistance to wheat stem sawfly (Hymenoptera: Cephidae). Journal of Integrated Pest Management 13(1). doi: 10.1093/jipm/pmac023.

- Wallace, L. E. and McNeal, F. H. (1966). Stem Sawflies of Economic Importance in Grain Crops in the United States. Washington, DC: Agricultural Research Service, United States Department of Agriculture.
- E-1479. Integrated Pest Management of Wheat Stem Sawfly in North Dakota. www.ag.ndsu.edu/extensionentomology/recent-publications-main/publications/e-1479-ipm-of-wheat-stem-sawfly-in-north-dakota-2016/view.
- https://coloradowheat.org/wp-content/uploads/2022/07/WSS-Economic-Impact-Study_06212022.pdf.
- https://cropwatch.unl.edu/tags/wheat-stem-sawfly.

8 References

Achhami, B. B., Reddy, G. V. P., Hofland, M. L., Sherman, J. D., Peterson, R. K. D. and Weaver, D. K. (2021). Plant volatiles and oviposition behavior in the selection of barley cultivars by wheat stem sawfly (Hymenoptera: Cephidae). *Environmental Entomology* 50(4): 940–947. doi: 10.1093/ee/nvab035.

Achhami, B. B., Peterson, R. K. D., Sherman, J. D., Reddy, G. V. P., Weaver, D. K. and Weaver, D. K. (2020a). Multiple decrement life tables of *Cephus cinctus* Norton (Hymenoptera: Cephidae) across a set of barley cultivars: The importance of plant defense versus cannibalism. *PLoS ONE* 15(9): e0238527. doi: 10.1371/journal.pone.0238527.

Achhami, B. B., Reddy, G. V. P., Sherman, J. D., Peterson, R. K. D. and Weaver, D. K. (2020b). Effect of precipitation and temperature on larval survival of *Cephus cinctus* (Hymenoptera: Cephidae) in barley cultivars. *Journal of Economic Entomology* 113(4): 1982–1989. doi: 10.1093/jee/toaa097.

Achhami, B. B., Reddy, G. V. P., Sherman, J. D., Peterson, R. K. D. and Weaver, D. K. (2020c). Antixenosis, antibiosis, and potential yield compensatory response in barley cultivars exposed to wheat stem sawfly (Hymenoptera: Cephidae) under field conditions. *Journal of Insect Science* 20(5): 1–14. doi: 10.1093/jisesa/ieaa091.

Adhikari, S., Adhikari, A., Weaver, D. K., Bekkerman, A. and Menalled, F. D. (2019a). Impacts of agricultural management systems on biodiversity and ecosystem services in highly simplified dryland landscapes. *Sustainability* 11(11). doi: 10.3390/su11113223.

Adhikari, S., Burkle, L. A., O'Neill, K. M., Weaver, D. K., Delphia, C. M. and Menalled, F. D. (2019b). Dryland organic farming partially offsets negative effects of highly simplified agricultural landscapes on forbs, bees, and bee-flower networks. *Environmental Entomology* 48(4): 826–835. doi: 10.1093/ee/nvz056.

Adhikari, S., Burkle, L. A., O'Neill, K. M., Weaver, D. K. and Menalled, F. D. (2019c). Dryland organic farming increases floral resources and bee colony success in highly simplified agricultural landscapes. *Agriculture, Ecosystems and Environment* 270-271: 9-18. doi: 10.1016/j.agee.2018.10.010.

Adhikari, S., Seipel, T. F., Menalled, F. D. and Weaver, D. K. (2018). Farming system and wheat cultivar affect infestation of, and parasitism on, *Cephus cinctus* in the

Northern Great Plains. *Pest Management Science* 74(11): 2480–2487. doi: 10.1002/ps.4925.

Ainslie, C. N. (1920). The western grass-stem sawfly. *U.S. Department of Agriculture Bulletin* 841: 1–28.

Ainslie, C. N. (1929). The western grass-stem sawfly (*Cephus cinctus* Norton): A pest of small grains. *U.S Department of Agriculture Technical Bulletin* 157: 1–23.

Akpinar, B. A., Muslu, T., Reddy, G. V. P., Dogramaci, M. and Budak, H. (2023). Wheat long noncoding RNAs from organelle and nuclear genomes carry conserved microRNA Precursors which may together comprise intricate networks in insect responses. *International Journal of Molecular Sciences* 24: 1–18. doi: 10.3390/ijms24032226.

Baker, J. E., Howard, R. W., Morrill, W. L., Meers, S. B. and Weaver, D. K. (2005). Acetate esters of saturated and unsaturated alcohols (C12–C20) are major components in Dufour glands of *Bracon cephi* and *Bracon lissogaster* (Hymenoptera: Braconidae), parasitoids of the wheat stem sawfly, *Cephus cinctus* (Hymenoptera: Cephidae). *Biochemical Systematics and Ecology* 33(8): 757–769. doi: 10.1016/j.bse.2004.12.025.

Bartelt, R. J., Cossé, A. A., Petroski, R. J. and Weaver, D. K. (2002). Cuticular hydrocarbons and novel alkenediol diacetates from wheat stem sawfly *Cephus cinctus*: Natural oxidation to pheromone components. *Journal of Chemical Ecology* 28(2): 385–405. doi: 10.1023/A:1017994410538.

Bekkerman, A. and Weaver, D. K. (2018). Modeling joint dependence of managed ecosystems pests: The case of the wheat stem sawfly. *Journal of Agricultural and Resource Economics* 43: 172–194.

Beres, B. L., Cárcamo, H. A. and Bremer, E. (2009). Evaluation of alternative planting strategies to reduce wheat stem sawfly (Hymenoptera: Cephidae) damage to spring wheat in the Northern Great Plains. *Journal of Economic Entomology* 102(6): 2137-2145. doi: 10.1603/029.102.0617.

Beres, B. L., Cárcamo, H. A. and Byers, J. R. (2007). Effect of wheat stem sawfly damage on yield and quality of selected Canadian spring wheat. *Journal of Economic Entomology* 100(1): 79–87. doi: 10.1603/0022-0493(2007)100[79:eowssd]2.0.co;2.

Beres, B. L., Cárcamo, H. A., Byers, J. R., Clarke, F. R., Pozniak, C. J., Basu, S. K. and DePauw, R. M. (2013a). Host plant interactions between wheat germplasm source and wheat stem sawfly Cephus cinctus Norton (Hymenoptera: Cephidae) I. Commercial cultivars. *Canadian Journal of Plant Science* 93(4): 607–617. doi: 10.4141/cjps2012-088.

Beres, B. L., Cárcamo, H. A., Byers, J. R., Clarke, F. R., Ruan, Y., Pozniak, C. J., Basu, S. K. and DePauw, R. M. (2013b). Host plant interactions between wheat germplasm source and wheat stem sawfly Cephus cinctus Norton (Hymenoptera: Cephidae). II. Other germplasm. *Canadian Journal of Plant Science* 93(6): 1169-1177. doi: 10.4141/cjps2013-035.

Beres, B. L., Cárcamo, H. A., Weaver, D. K., Dosdall, L. M., Cárcamo, H. A., Evenden, M. L., Hill, B. D., McKenzie, R. H., Yang, R.-C. and Spaner, D. M. (2011a). Integrating the building blocks of agronomy and biocontrol into an IPM strategy for wheat stem sawfly. *Prairie Soils & Crops Journal* 4: 54–65.

Beres, B. L., Cárcamo, H. A., Yang, R. C. and Spaner, D. M. (2011b). Integrating spring wheat sowing density with variety selection to manage wheat stem sawfly. *Agronomy Journal* 103(6): 1755-1764. doi: 10.2134/agronj2011.0187.

Beres, B. L., Dosdall, L. M., Weaver, D. K., Cárcamo, H. A. and Spaner, D. M. (2011c). Biology and integrated management of wheat stem sawfly and the need for continuing research. *The Canadian Entomologist* 143(2): 105–125. doi: 10.4039/n10-056.

Beres, B. L., Hill, B. D., Cárcamo, H. A., Knodel, J. J., Weaver, D. K. and Cuthbert, R. D. (2017). An artificial neural network model to predict wheat stem sawfly cutting in solid-stemmed wheat cultivars. *Canadian Journal of Plant Science* 97: 329–336. doi: 10.1139/CJPS-2016-0364.

Beres, B. L., McKenzie, R. H., Cárcamo, H. A., Dosdall, L. M., Evenden, M. L., Yang, R.-C. and Spaner, D. M. (2012). Influence of seeding rate, nitrogen management, and micronutrient blend applications on pith expression in solid-stemmed spring wheat. *Crop Science* 52(3): 1316–1329. doi: 10.2135/cropsci2011.05.0239.

Biyiklioglu, S., Alptekin, B., Akpinar, B. A., Varella, A. C., Hofland, M. L., Weaver, D. K., Bothner, B. and Budak, H. (2018). A large-scale multiomics analysis of wheat stem solidness and the wheat stem sawfly feeding response, and syntenic associations in barley, *Brachypodium*, and rice. *Functional and Integrative Genomics* 18(3): 241–259. doi: 10.1007/s10142-017-0585-5.

Brant, E. J. and Budak, H. (2018). Plant small non-coding RNAs and their roles in biotic stresses. *Frontiers in Plant Science* 9: 1038. doi: 10.3389/fpls.2018.01038.

Bruckner, P. L., Berg, J. E., Lamb, P. F., Kephart, K. D., Eberly, J. O., Miller, J. H., Chen, C., Torrion, J. A., Pradhan, G. P., Ramsfield, R., Nash, D. L., Holen, D. L., Cook, J. P., Gale, S., Jin, Y., Kolmer, J., Chen, X. and Bai, G. (2020). Registration of 'bobcat' hard red winter wheat. *Journal of Plant Registrations* 14(3): 371–376. doi: 10.1002/plr2.20057.

Budak, H., Kaya, S. B. and Cagirici, H. B. (2020). Long non-coding RNA in plants in the era of reference sequences. *Frontiers in Plant Science* 11: 276. doi: 10.3389/fpls.2020.00276.

Buteler, M., Peterson, R. K. D., Hofland, M. L. and Weaver, D. K. (2015). A multiple decrement life table reveals that host plant resistance and parasitism are major causes of mortality for the wheat stem sawfly. *Environmental Entomology* 44(6): 1571–1580. doi: 10.1093/ee/nvv128.

Buteler, M. and Weaver, D. K. (2012). Host selection by the wheat stem sawfly in winter wheat and the role of semiochemicals mediating oviposition preference. *Entomologia Experimentalis et Applicata* 143(2): 138–147. doi: 10.1111/j.1570-7458.2012.01237.x.

Buteler, M., Weaver, D. K., Bruckner, P. L., Carlson, G. R., Berg, J. E. and Lamb, P. F. (2010). Identification of winter wheat cultivars suitable as trap crops for the wheat stem sawfly based on screening using agronomic traits and semiochemical production. *The Canadian Entomologist* 142(3): 222–233. doi: 10.4039/n09-072.

Buteler, M., Weaver, D. K. and Miller, P. R. (2008). Wheat stem sawfly-infested plants benefit from parasitism of the herbivorous larvae. *Agricultural and Forest Entomology* 10(4): 347–354. doi: 10.1111/j.1461-9563.2008.00396.x.

Buteler, M., Weaver, D. K. and Peterson, R. K. D. (2009). Oviposition behavior of the wheat stem sawfly when encountering plants infested with cryptic conspecifics. *Environmental Entomology* 38(6): 1707–1715. doi: 10.1603/022.038.0624.

Cagirici, H. B., Biyiklioglu, S. and Budak, H. (2017). Assembly and annotation of transcriptome provided evidence of miRNA mobility between wheat and wheat stem sawfly. *Frontiers in Plant Science* 8: 1653. doi: 10.3389/fpls.2017.01653.

Çalmaşur, Ö. and Özbek, H. (2010). Distribution data on the Cephidae (Hymenoptera: Symphyta) fauna of Turkey. *Zoology in the Middle East* 50(1): 144–146. doi: 10.1080/09397140.2010.10638429.

Cárcamo, H. A., Beres, B. L., Clarke, F. R., Byers, R. J., Mundel, H.-h, May, K. and DePauw, R. M. (2005). Influence of plant host quality on fitness and sex ratio of the wheat stem sawfly (Hymenoptera: Cephidae). *Environmental Entomology* 34(6): 1579–1592. doi: 10.1603/0046-225X-34.6.1579.

Cárcamo, H. A., Beres, B. L., Herle, C. E., McLean, H. and McGinn, S. (2011). Solid-stemmed wheat does not affect overwintering mortality of the wheat stem sawfly, *Cephus cinctus*. *Journal of Insect Science* 11: 129. doi: 10.1673/031.011.12901.

Cárcamo, H. A., Beres, B. L., Larson, T. R., Klima, C. L. and Wu, X. H. (2016). Effect of wheat cultivars and blends on the oviposition and larval mortality of *Cephus cinctus* (Hymenoptera: Cephidae) and parasitism by *Bracon cephi* (Hymenoptera: Braconidae). *Environmental Entomology* 45(2): 397–403. doi: 10.1093/ee/nvv231.

Cárcamo, H. A., Beres, B. L., Wijerathna, A. N. and Schwinghamer, T. D. (2022). Triticale cultivars and seeding rates affect wheat stem sawfly survivorship and parasitism by *Bracon cephi*. *Canadian Journal of Plant Science* 102(5): 1001–1006. doi: 10.1139/cjps-2021-0269.

Cárcamo, H. A., Entz, T. and Beres, B. L. (2007). Estimating *Cephus cinctus* wheat stem cutting damage- can we cut stem counts? *Journal of Agricultural and Urban Entomology* 24(3): 117–124. doi: 10.3954/1523-5475-24.3.117.

Cárcamo, H. A., Weaver, D. K., Meers, S. B., Beres, B. L. and Mauduit, A. L. (2012). First record of *Bracon lissogaster* (Hymenoptera: Braconidae) in Canada – A potentially important parasitoid of *Cephus cinctus* (Hymenoptera: Cephidae) in the prairies. *Biocontrol Science and Technology* 22(3): 367–369. doi: 10.1080/09583157.2012.658355.

Cavallini, L., Peterson, R. K. D. and Weaver, D. K. (2022). Dietary sugars and amino acids increase longevity and enhance reproductive parameters of *Bracon cephi* and *B. lissogaster*, two parasitoids that specialise on wheat stem sawfly. *Physiological Entomology* 47: 1–11. doi: 10.1111/phen.12399.

Chen, S., Hoelmer, K. A., Chen, H., Liu, A. and Shanower, T. G. (2004). A review of wheat stem sawfly (Hymenoptera: Cephidae) research in China. *Journal of Agricultural and Urban Entomology* 21: 249–256.

Cockrell, D. M., Griffin-Nolan, R. J., Rand, T. A., Altilmisani, N., Ode, P. J. and Peairs, F. (2017). Host plants of the wheat stem sawfly (Hymenoptera: Cephidae). *Environmental Entomology* 46(4): 847–854. doi: 10.1093/ee/nvx104.

Cockrell, D. M., Randolph, T. L., Peirce, E. S. and Peairs, F. B. (2021). Survey of wheat stem sawfly (Hymenoptera: Cephidae) infesting wheat in eastern Colorado. *Journal of Economic Entomology* 114(2): 998–1004. doi: 10.1093/jee/toab015.

Colorado wheat stem sawfly resources. (2020). Available at: https://coloradowheat.org/2020/08/wheat-stem-sawlfy-resources/.

Comstock, J. H. (1889). On a sawfly borer in wheat. N.Y. (Cornell) agricultural experiment station. *Bulletin* 11: 127–142.

Cook, J. P., Blake, N. K., Heo, H. Y., Martin, J. M., Weaver, D. K. and Talbert, L. E. (2017). Phenotypic and haplotype diversity among tetraploid and hexaploid wheat accessions with potentially novel insect resistance genes for wheat stem sawfly. *Plant Genome* 10(1): 1–10. doi: 10.3835/plantgenome2016.03.0026.

Cook, J. P., Wichman, D. M., Martin, J. M., Bruckner, P. L. and Talbert, L. E. (2004). Identification of microsatellite markers associated with a stem solidness locus in wheat. *Crop Science* 44(4): 1397–1402. doi: 10.2135/cropsci2004.1397.

Cossé, A. A., Bartelt, R. J., Weaver, D. K. and Zilkowski, B. W. (2002). Pheromone components of the wheat stem sawfly: Identification, electrophysiology, and field bioassay. *Journal of Chemical Ecology* 28(2): 407–423. doi: 10.1023/A:1017946527376.

Criddle, N. (1921). The western wheat-stem sawfly in Canada. *Ontario Entomological Society Annual Report* 52: 18–22.

Criddle, N. (1922). The western wheat stem sawfly and its control. *Dominion of Canada Department of Agriculture Pamphlet No. 6 - New Series*: 1-8.

Criddle, N. (1923). The life habits of *Cephus cinctus* Norton in Manitoba. *The Canadian Entomologist* 55(1): 1-4. doi: 10.4039/Ent551-1.

Davis, E. G. (1948). Status of the wheat stem sawfly in 1947. *United States Agricultural Research Service, Plant Pest Control Division, Insect Pest Survey, Special Supplement* 2: 1-4.

Davis, E. G. (1952). Status of the wheat stem sawfly in the Northern Great Plains in 1951. *United States Agricultural Research Service, Plant Pest Control Division, Cooperative Economic Insect Report* 2: 27-30.

Davis, E. G. (1953). Status of the wheat stem sawfly in the Northern Great Plains in 1952. *United States Agricultural Research Service, Plant Pest Control Division, Cooperative Economic Insect Report* 3: 140-144.

Davis, E. G. (1955). Status of the wheat stem sawfly in the United States in 1954. *North Dakota Agri-Cultural Experiment Station Bimonthly Bulletin* 17: 171-175.

Davis, E. G., Benton, C. and Somsen, H. W. (1955). Natural enemies of the wheat stem sawfly in North Dakota and Montana. *North Dakota Agricultural Experimental Station Bimonthly Bulletin* 18: 63-65.

De Pauw, R. M. and Read, D. W. L. (1982). The effect of nitrogen and phosphorus on the expression of stem solidness in Canuck wheat in four locations in southwestern Saskatchewan. *Canadian Journal of Plant Science* 62: 593-598. doi: 10.4141/cjps82-089.

Delaney, K. J., Weaver, D. K. and Peterson, R. K. D. (2010). Photosynthesis and yield reductions from wheat stem sawfly (Hymenoptera: Cephidae): Interactions with wheat solidness, water stress, and phosphorus deficiency. *Journal of Economic Entomology* 103(2): 516-524. doi: 10.1603/ec09229.

Dhariwal, R., Hiebert, C. W. and Randhawa, H. S. (2022). QTL analysis identified two major all-internodes solidness loci from a completely solid-stemmed spring wheat line. *Frontiers in Plant Science* 13: 1035620. doi: 10.3389/fpls.2022.1035620.

Edde, P. A. (2022). Chapter 9: Arthropod pests of small grains: Wheat (*Triticum aestivum* L.) and barley (*Hordeum vulgare* L.). In: Edde, P. A. and Crop, F. (Eds.), *Arthropod Pests of Economic Importance*, pp. 536-611. doi: 10.1016/B978-0-12-818621-3.00001-X.

Engledow, F. L. and Hutchinson, J. B. (1925). Inheritance in wheat. II. *Triticum turgidum* X *T. durum* crosses, with notes on the inheritance of solidness of straw. *Journal of Genetics* 16(1): 19-32.

Farstad, C. W. (1940). The development of western wheat stem sawfly (*Cephus cinctus* Norton) in various host plants as an index of resistance. *Iowa State College Journal of Science* 15: 67-69.

Farstad, C. W. (1944). Wheat stem sawfly in flax. *Scientific Agriculture* 24: 383-386. doi: 10.4141/sa-1944-0025.

Farstad, C. W., King, K. M., Glen, R. and Jacobson, L. A. (1945). Control of the wheat stem sawfly in the Prairie provinces. *Wartime Production Series - Agricultural Supplies Board Special Publication* 59: 1-7.

Filipy, F. L., Burbutis, P. P. and Fuester, R. W. (1985). Biological control of the European wheat stem sawfly in Delaware (Hymenoptera: Cephidae). *Environmental Entomology* 14(6): 665-668. doi: 10.1093/ee/14.6.665.

Fletcher, J. (1896). The western wheat stem sawfly (*Cephus pygmaeus* L.). *Dominion of Canada, Department of Agriculture, Report of the Dominion Entomologist* 1896: 147–149.

Fletcher, J. (1904). Experimental farms reports: Report of the entomologist and botanist. *Appendix to the report of the Minister of Agriculture, Sessional Paper* 16: 172–173.

Gahan, A. B. (1918). Description of new hymenopterous parasite (Braconidae). *Proceedings of the Entomological Society of Washington* 20: 18–19.

Gress, J. C., Robertson, H. M., Weaver, D. K., Dlakić, M. and Wanner, K. W. (2013). Odorant receptors of a primitive hymenopteran pest, the wheat stem sawfly. *Insect Molecular Biology* 22(6): 659–667. doi: 10.1111/imb.12053.

Hoelmer, K. A. and Shanower, T. G. (2004). Foreign exploration for natural enemies of cephid sawflies. *Journal of Agricultural and Urban Entomology* 21: 223-238.

Holmes, N. D. (1954). Food relations of the wheat stem sawfly, *Cephus cinctus* Norton (Hymenoptera: Cephidae). *The Canadian Entomologist* 86(4): 159–167. doi: 10.4039/Ent86159-4.

Holmes, N. D. (1977). The effect of the wheat stem sawfly, *Cephus cinctus* (Hymenoptera: Cephidae), on the yield and quality of wheat. *The Canadian Entomologist* 109(12): 1591–1598. doi: 10.4039/Ent1091591-12.

Holmes, N. D. (1979). The wheat stem sawfly. *Proceedings of the Twenty-Sixth Annual Meeting of the Entomological Society of Alberta* 26: 2–13.

Holmes, N. D. (1982). Population dynamics of the wheat stem sawfly, *Cephus cinctus* (Hymenoptera: Cephidae), in wheat. *The Canadian Entomologist* 114(9): 775–788. doi: 10.4039/Ent114775-9.

Holmes, N. D., Nelson, W. A., Peterson, L. K. and Farstad, C. W. (1963). Causes of variations in effectiveness of *Bracon cephi* (Gahan) (Hymenoptera: Braconidae) as a parasite of the wheat stem sawfly. *The Canadian Entomologist* 95(2): 113–126. doi: 10.4039/Ent95113-2.

Holmes, N. D. and Peterson, L. K. (1957). Effect of continuous rearing in Rescue wheat on the survival of the wheat stem sawfly *Cephus cinctus* Norton (Hymenoptera: Cephidae). *The Canadian Entomologist* 89(8): 363–364. doi: 10.4039/Ent89363-8.

Holmes, N. D. and Peterson, L. K. (1960). The influence of the host on oviposition by the wheat stem sawfly, *Cephus cinctus* Norton (Hymenoptera: Cephidae). *Canadian Journal of Plant Science* 40(1): 29–46. doi: 10.4141/cjps60-004.

Holmes, N. D. and Peterson, L. K. (1961). Resistance of spring wheats to the wheat stem sawfly, *Cephus cinctus* Norton (Hymenoptera: Cephidae) I. Resistance to the egg. *The Canadian Entomologist* 93(4): 250–260. doi: 10.4039/Ent93250-4.

Holmes, N. D. and Peterson, L. K. (1962). Resistance of spring wheats to the wheat stem sawfly, *Cephus cinctus* Norton (Hymenoptera: Cephidae) II. Resistance to the larva. *The Canadian Entomologist* 94(4): 348–365. doi: 10.4039/Ent94348-4.

Holmes, N. D. and Peterson, L. K. (1963). Heptachlor as a systemic insecticide against the wheat stem sawfly, *Cephus cinctus* Norton. *The Canadian Entomologist* 95(8): 792–796. doi: 10.4039/Ent95792-8.

Holmes, N. D. and Peterson, L. K. (1965). Swathing wheat and survival of wheat stem sawfly. *Canadian Journal of Plant Science* 45(6): 579–581. doi: 10.4141/cjps65-109.

Houshmand, S., Knox, R. E., Clarke, F. R. and Clarke, J. M. (2007). Microsatellite markers flanking a stem solidness gene on chromosome 3BL in durum wheat. *Molecular Breeding* 20(3): 261–270. doi: 10.1007/s11032-007-9088-8.

Hyles, J., Vautrin, S., Pettolino, F., Macmillan, C., Stachurski, Z., Breen, J., Berges, H., Wicker, T. and Spielmeyer, W. (2017). Repeat-length variation in a wheat cellulose synthase-like gene is associated with altered tiller number and stem cell wall composition. *Journal of Experimental Botany* 68(7): 1519–1529. doi: 10.1093/jxb/erx051.

Ivie, M. A. (2001). On the geographic origin of the wheat stem sawfly (Hymenoptera: Cephidae): A new hypothesis of introduction from northeastern Asia. *American Entomologist* 47(2): 84–97. doi: 10.1093/ae/47.2.84.

Jacobson, L. A. and Farstad, C. W. (1952). Effect of time of seeding Apex wheat on infestation and sex ratio of the wheat stem sawfly, *Cephus cinctus* Norton (Hymenoptera: Cephidae). *The Canadian Entomologist* 84(3): 90–92. doi: 10.4039/Ent8490-3.

Kasting, R., Longair, E. L. and McGinnis, A. J. (1958). Moisture and nitrogen contents of stem tissues from solid- and hollow-stemmed varieties of spring wheat and their relation to sawfly resistance. *Canadian Journal of Plant Science* 38(3): 287–291. doi: 10.4141/cjps58-047.

Kasting, R. and McGinnis, A. J. (1958). Note on a method of artificially sustaining larvae of the wheat stem sawfly *Cephus cinctus* Norton (Hymenoptera: Cephidae). *The Canadian Entomologist* 20: 63–64. doi: 10.4039/Ent9063-1.

Kasting, R. and McGinnis, A. J. (1961). Comparison of tissues from solid- and hollow-stemmed spring wheats during growth II. Food values determined with the pale western cutworm, *Agrotis orthogonia* Morrison (Lepidoptera: Noctuidae). *Canadian Journal of Zoology* 39(3): 273–280. doi: 10.1139/z61-030.

Kebrom, T. H., Chandler, P. M., Swain, S. M., King, R. W., Richards, R. A. and Spielmeyer, W. (2012). Inhibition of tiller bud outgrowth in the tin mutant of wheat is associated with precocious internode development. *Plant Physiology* 160(1): 308–318. doi: 10.1104/pp.112.197954.

Kemp, H. J. (1934). Studies of solid stem wheat varieties in relation to wheat stem sawfly control. *Scientific Agriculture* 15: 30–38. doi: 10.4141/sa-1934-0076.

Keren, I. N., Menalled, F. D., Weaver, D. K. and Robison-Cox, J. F. (2015). Interacting agricultural pests and their effect on crop yield: Application of a Bayesian decision theory approach to the joint management of *Bromus tectorum* and *Cephus cinctus*. *PLoS ONE* 10(2): e0118111. doi: 10.1371/journal.pone.0118111.

Khan, Z. R., Midega, C. A. O., Hooper, A. M. and Pickett, J. A. (2016). Push-pull: Chemical ecology-based integrated pest management technology. *Journal of Chemical Ecology* 42(7): 689–697. doi: 10.1007/s10886-016-0730-y.

Knodel, J. J., Beauzay, P. B., Eriksmoen, E. D. and Pederson, J. D. (2009). Pest management of wheat stem maggot (Diptera: Chloropidae) and wheat stem sawfly (Hymenoptera: Cephidae) using insecticides in spring wheat. *Journal of Agricultural and Urban Entomology* 26(4): 183–197. doi: 10.3954/1523-5475-26.4.183.

Korkmaz, E. M., Budak, M., Örgen, S. H., Bağda, E., Gençer, L., Ülgentürk, S. and Başibüyük, H. H. (2010). New records and a checklist of Cephidae (Hymenoptera: Insecta) of Turkey with a short biogeographical consideration. *Turkish Journal of Zoology* 34. 203–211. doi: 10.3906/zoo-0812-19.

Lafond, G. P., Boyetchko, S. M., Brandt, S. A., Clayton, G. W. and Entz, M. H. (1996). Influence of changing tillage practices on crop production. *Canadian Journal of Plant Science* 76(4): 641–649. doi: 10.4141/cjps96-114.

Lanning, S. P., Fox, P., Elser, J., Martin, J. M., Blake, N. K. and Talbert, L. E. (2006). Microsatellite markers associated with a secondary stem solidness locus in wheat. *Crop Science* 46(4): 1701–1703. doi: 10.2135/cropsci2005.10-0379.

Lavergne, F. D., Broeckling, C. D., Brown, K. J., Cockrell, D. M., Haley, S. D., Peairs, F. B., Pearce, S., Wolfe, L. M., Jahn, C. E. and Heuberger, A. L. (2020). Differential stem proteomics and metabolomics profiles for four wheat cultivars in response to the insect pest wheat stem sawfly. *Journal of Proteome Research* 19(3): 1037-1051. doi: 10.1021/acs.jproteome.9b00561.

Lavergne, F. D., Broeckling, C. D., Cockrell, D. M., Haley, S. D., Peairs, F. B., Jahn, C. E. and Heuberger, A. L. (2018). GC-MS metabolomics to evaluate the composition of plant cuticular waxes for four *Triticum aestivum* cultivars. *International Journal of Molecular Sciences* 19(2): 1-21. doi: 10.3390/ijms19020249.

Liang, D., Zhang, M., Liu, X., Li, H., Jia, Z., Wang, D., Peng, T., Hao, M., Liu, D., Jiang, B., Huang, L., Ning, S., Yuan, Z., Chen, X. and Zhang, L. (2022). Development and identification of four new synthetic hexaploid wheat lines with solid stems. *Scientific Reports* 12: 4898. doi: 10.1038/s41598-022-08866-x.

Lindwall, C. W. and Anderson, D. T. (1981). Agronomic evaluation of minimum tillage systems for summer fallow in southern Alberta. *Canadian Journal of Plant Science* 61(2): 247–253. doi: 10.4141/cjps81-037.

Lesieur, V., Martin, J. F., Weaver, D. K., Hoelmer, K. A., Smith, D. R., Morrill, W. L., Kadiri, N., Peairs, F. B., Cockrell, D. M., Randolph, T. L., Waters, D. K. and Bon, M. C. (2016). Phylogeography of the wheat stem sawfly, *Cephus cinctus* Norton (Hymenoptera: Cephidae): Implications for pest management. *PLoS ONE* 11(12): e0168370. doi: 10.1371/journal.pone.0168370.

Lestina, J., Cook, M., Kumar, S., Morisette, J., Ode, P. J. and Peairs, F. (2016). MODIS imagery improves pest risk assessment: A case study of wheat stem sawfly (*Cephus cinctus*, Hymenoptera: Cephidae) in Colorado, USA in Colorado. *Environmental Entomology* 45(6): 1343–1351. doi: 10.1093/ee/nvw095.

Luginbill, P., Jr. and McNeal, F. H. (1954). Effect of fertilizers on the resistance of certain winter and spring wheat varieties to the wheat stem sawfly. *Agronomy Journal* 46(12): 570–573. doi: 10.2134/agronj1954.00021962004600120010x.

Luginbill, P., Jr. and McNeal, F. H. (1958). Influence of seeding density and row spacings on the resistance of spring wheats to the wheat stem sawfly. *Journal of Economic Entomology* 51(6): 804–808. doi: 10.1093/jee/51.6.804.

Macedo, T. B., Macedo, P. A., Peterson, R. K. D. , Weaver, D. K. and Morrill, W. L. (2005). Rearing the wheat stem sawfly on an artificial diet. *The Canadian Entomologist* 137(4): 497–500. doi: 10.4039/n04-100.

Macedo, T. B., Peterson, R. K. D., Weaver, D. K. and Morrill, W. L. (2005). Wheat stem sawfly, *Cephus cinctus* Norton, impact on wheat primary metabolism: An ecophysiological approach. *Environmental Entomology* 34(3): 719–726. doi: 10.1603/0046-225X-34.3.719.

Macedo, T. B., Weaver, D. K. and Peterson, R. K. D. (2006). Characterization of the impact of wheat stem sawfly, *Cephus cinctus* Norton, on pigment composition and photosystem II photochemistry of wheat heads. *Environmental Entomology* 35(4): 1115–1120. doi: 10.1603/0046-225X-35.4.1115.

Macedo, T. B., Weaver, D. K. and Peterson, R. K. D. (2007). Photosynthesis in wheat at the grain filling stage is altered by larval wheat stem sawfly (Hymenoptera: Cephidae) injury and reduced water availability. *Journal of Entomological Science* 42(2): 228-238. doi: 10.18474/0749-8004-42.2.228.

Maistrello, L., Dioli, P., Bariselli, M., Mazzoli, G. L. and Giacalone-Forini, I. (2016). Citizen science and early detection of invasive species: Phenology of first occurrences of *Halyomorpha halys* in Southern Europe. *Biological Invasions* 18(11): 3109–3116. doi: 10.1007/s10530-016-1217-z.

Malone, S. C., Menalled, F. D., Weaver, D. K., Seipel, T. F., Hofland, M. L., Runyon, J. B., Bourgault, M., Boss, D. L. and Trowbridge, A. M. (2022). Cropping systems alter plant volatile emissions in the field through soil legacy effects. *Renewable Agriculture and Food Systems* 37(5): 375–381. doi: 10.1017/S174217052200014X.

Malone, S. C., Weaver, D. K., Seipel, T. F., Menalled, F. D., Hofland, M. L., Runyon, J. B. and Trowbridge, A. M. (2020). Herbivore-induced volatile emissions are altered by soil legacy effects in cereal cropping systems. *Plant and Soil* 455(1–2): 171–186. doi: 10.1007/s11104-020-04674-2.

Mankin, R. W., Brandhorst-Hubbard, J., Flanders, K. L., Zhang, M., Crocker, R. L., Lapointe, S. L., McCoy, C. W., Fisher, J. R. and Weaver, D. K. (2000). Eavesdropping on insects hidden in soil and interior structures of plants. *Journal of Economic Entomology* 93(4): 1173–1182. doi: 10.1603/0022-0493-93.4.1173.

Mankin, R. W., Weaver, D. K., Grieshop, M. J. D., Larson, B. and Morrill, W. M. (2004). Acoustic system for insect detection in plant stems: Comparisons of *Cephus cinctus* in wheat and *Metamasius callizona* in bromeliads. *Journal of Agricultural and Urban Entomology* 21: 239 –248.

Mathews, O. R. (1945). Crop residue management in dry-land crop production. *Agronomy Journal* 37(4): 297–306. doi: 10.2134/agronj1945.00021962003700040007x.

McCullough, C. T., Hein, G. L. and Bradshaw, J. D. (2020). Phenology and dispersal of the wheat stem sawfly (Hymenoptera: Cephidae) into winter wheat fields in Nebraska. *Journal of Economic Entomology* 113(4): 1831–1838. doi: 10.1093/jee/toaa093.

McGinnis, A. J. and Kasting, R. (1959). Note on a method of comparing diets for the pale western cutworm, *Agrotis orthogonia* Morrison (Lepidoptera: Noctuidae). *The Canadian Entomologist* 91(11): 742–743. doi: 10.4039/Ent91742-11.

McGinnis, A. J. and Kasting, R. (1961) Comparison of tissues from solid- and hollow-stemmed spring wheats during growth I. Dry matter and nitrogen contents of pith and wall and their relation to sawfly resistance. *Canadian Journal of Plant Science* 41(3): 469–478. doi: 10.4141/cjps61-071.

McGinnis, A. J. and Kasting, R. (1962). A method of rearing larvae of the wheat stem sawfly (Hymenoptera: Cephidae) under artificial conditions. *The Canadian Entomologist* 94(6): 573–574. doi: 10.4039/Ent94573-6.

McNeal, F. H., Berg, M. A. and Luginbill, P.,Jr. (1955). Wheat stem sawfly damage in four spring wheat varieties as influenced by date of seeding. *Agronomy Journal* 47(11): 522-525. doi: 10.2134/agronj1955.00021962004700110008x.

Miller, P. R., Waddington, J., McDonald, C. L. and Derksen, D. A. (2002). Cropping sequence affects wheat productivity on the semiarid northern Great Plains. *Canadian Journal of Plant Science* 82(2): 307–318. doi: 10.4141/P01-116.

Miller, R. H., Masri, S. E. and Jundi, K. A. (1993). Plant density and wheat stem sawfly (Hymenoptera: Cephidae) resistance in Syrian wheats. *Bulletin of Entomological Research* 83(1): 95–102. doi: 10.1017/S0007485300041821.

Montana Department of Agriculture (MDA). (2015). *Thimet® 20-G. EPA SLN No. MT-150001.*

Morrill, W. L., Gabor, J. W., Hockett, E. A. and Kushnak, G. D. (1992). Wheat stem sawfly (Hymenoptera: Cephidae) resistance in winter wheat. *Journal of Economic Entomology* 85(5): 2008–2011. doi: 10.1093/jee/85.5.2008.

Morrill, W. L., Gabor, J. W., Weaver, D. K., Kushnak, G. D. and Irish, N. J. (2000). Effect of host plant quality on the sex ratio and fitness of female wheat stem sawflies (Hymenoptera: Cephidae). *Environmental Entomology* 29(2): 195–199. doi: 10.1093/ee/29.2.195.

Morrill, W. L., Gabor, J. W. and Wichman, D. (1993). Mortality of the wheat stem sawfly (Hymenoptera: Cephidae) at low temperatures. *Environmental Entomology* 22(6): 1358–1361. doi: 10.1093/ee/22.6.1358.

Morrill, W. L. and Kushnak, G. D. (1996). Wheat stem sawfly (Hymenoptera: Cephidae) adaptation to winter wheat. *Environmental Entomology* 25(5): 1128–1132. doi: 10.1093/ee/25.5.1128.

Morrill, W. L. and Kushnak, G. D. (1999). Planting date influence on the wheat stem sawfly (Hymenoptera: Cephidae) in spring wheat. *Journal of Agricultural and Urban Entomology* 16: 123–128.

Morrill, W. L., Kushnak, G. D., Bruckner, P. L. and Gabor, J. W. (1994). Wheat stem sawfly (Hymenoptera: Cephidae) damage, rates of parasitism, and overwinter survival in resistant wheat lines. *Journal of Economic Entomology* 87(5): 1373–1376. doi: 10.1093/jee/87.5.1373.

Morrill, W. L., Kushnak, G. D. and Gabor, J. W. (1998). Parasitism of the wheat stem sawfly (Hymenoptera: Cephidae) in Montana. *Biological Control* 12(3): 159–163. doi: 10.1006/bcon.1998.0629.

Morrill, W. L. and Weaver, D. K. (2000). Host plant quality and male wheat stem sawfly (Hymenoptera: Cephidae) fitness. *Journal of Entomological Science* 35(4): 478–482. doi: 10.18474/0749-8004-35.4.478.

Morrill, W. L., Weaver, D. K. and Johnson, G. D. (2001a). Trap strip and field border modification for management of the wheat stem sawfly (Hymenoptera: Cephidae). *Journal of Entomological Science* 36(1): 34–45. doi: 10.18474/0749-8004-36.1.34.

Morrill, W. L., Weaver, D. K., Irish, N. J. and Barr, W. F. (2001b). *Phyllobaenus dubius* (Wolcott) (Coleoptera: Cleridae), a new record of a predator of the wheat stem sawfly (Hymenoptera: Cephidae). *Journal of the Kansas Entomological Society* 74: 181–183.

Muesebeck, C. F. W. (1953). Three new reared Braconidae (Hymenoptera). *Proceedings of the Entomological Society of Washington* 55: 149–151.

Munro, J. A., Fastdahl, W. and Post, R. L. (1949). Wheat stem sawfly. *Agricultural Experiment Station, ND Bimonthly Bulletin* 11: 85–91.

Muslu, T., Akpinar, B. A., Biyiklioglu-Kaya, S.,Yuce, M. and Budak, H. (2021). Comparative analysis of coding and non-coding features within insect tolerance loci in wheat with their homologs in cereal genomes. *International Journal of Molecular Sciences* 22: 12349. doi: 10.3390/ijms222212349.

Mutlu, Ç., Karaca, V., Tonğa, A., Erol, Ş. and Mamay, M. (2019). Infestation and damage caused by wheat stem sawflies (Hymenoptera: Cephidae) to some wheat cultivars in southeast Anatolia region, Turkey. *Journal of the Kansas Entomological Society* 92(1): 359–375. doi: 10.2317/0022-8567-92.1.359.

Nansen, C., Macedo, T. B., Swanson, R. and Weaver, D. K. (2009). Use of spatial structure analysis of hyperspectral data cubes for detection of insect-induced stress in wheat plants. *International Journal of Remote Sensing* 30(10): 2447–2464. doi: 10.1080/01431160802552710.

Nansen, C., Macedo, T. B., Weaver, D. K. and Peterson, R. K. D. (2005a). Spatiotemporal distributions of wheat stem sawfly eggs and larvae in dryland wheat fields. *The Canadian Entomologist* 137(4): 428–440. doi: 10.4039/N04-094.

Nansen, C., Payton, M. E., Runyon, J. B., Weaver, D. K., Morrill, W. L. and Sing, S. E. (2005b). Pre- harvest sampling plan for larvae of the wheat stem sawfly, *Cephus cinctus* (Hymenoptera: Cephidae), in winter wheat fields. *The Canadian Entomologist* 137(5): 602–614. doi: 10.4039/N04-087.

Nansen, C., Weaver, D. K., Sing, S. E., Runyon, J. B., Morrill, W. L., Grieshop, M. J. D., Shannon, C. L. and Johnson, M. L. (2005c). Within-field spatial distribution of *Cephus cinctus* (Hymenoptera: Cephidae) larvae in Montana wheat fields. *The Canadian Entomologist* 137(2): 202–214. doi: 10.4039/N04-036.

Nelson, W. A. and Farstad, C. W. (1953). Biology of *Bracon cephi* (Gahan) (Hymenoptera: Braconidae), an important native parasite of the wheat stem sawfly, *Cephus cinctus* Norton (Hymenoptera: Cephidae), in Western Canada. *The Canadian Entomologist* 85(3): 103–107. doi: 10.4039/Ent85103-3.

Nilsen, K. T., N'diaye, A., MacLachlan, P. R., Clarke, J. M., Ruan, Y., Cuthbert, R. D., Knox, R. E., Wiebe, K., Cory, A. T., Walkowiak, S., Beres, B. L., Graf, R. J., Clarke, F. R., Sharpe, A. G., Distelfeld, A. and Pozniak, C. J. (2017). High density mapping and haplotype analysis of the major stem-solidness locus SSt1 in durum and common wheat. *PLoS ONE* 12(4): e0175285. doi: 10.1371/journal.pone.0175285.

Nilsen, K. T., Walkowiak, S., Xiang, D., Gao, P., Quilichini, T. D., Willick, I. R., Byrns, B., N'Diaye, A., Ens, J., Wiebe, K., Ruan, Y., Cuthbert, R. D., Craze, M., Wallington, E. J., Simmonds, J., Uauy, C., Datla, R. and Pozniak, C. J. (2020). Copy number variation of *TdDof* controls solid-stemmed architecture in wheat. *Proceedings of the National Academy of Sciences of the United States of America* 117(46): 28708–28718. doi: 10.1073/pnas.2009418117.

Norton, E. (1872). Notes on North American Tenthredinidae with descriptions of new species. *Transactions of the American Entomological Society* 4: 77–86. doi: 10.2307/25076265.

Oiestad, A. J., Martin, J. M., Cook, J., Varella, A. C. and Giroux, M. J. (2017). Identification of candidate genes responsible for stem pith production using expression analysis in solid-stemmed wheat. *Plant Genome* 10(2): 1–8. doi: 10.3835/plantgenome2017.02.0008.

Olfert, O. O., Weiss, R. M., Catton, H., Cárcamo, H. and Meers, S. M. (2019). Bioclimatic assessment of abiotic factors affecting relative abundance and distribution of wheat stem sawfly (Hymenoptera: Cephidae) in western Canada. *The Canadian Entomologist* 151(1): 16–33. doi: 10.4039/tce.2018.46.

Painter, R. H. (1953). The wheat stem sawfly in Kansas. *Transactions of the Kansas Academy of Science* 56(4): 432–434 doi: 10.2307/3625625.

Peirce, E. S., Cockrell, D. M., Mason, E., Haley, S., Peairs, F. B. and Nachappa, P. (2022a). Solid stems and beyond: Challenges and future directions of resistance to wheat stem sawfly (Hymenoptera: Cephidae). *Journal of Integrated Pest Management* 13(1). doi: 10.1093/jipm/pmac023.

Peirce, E. S., Cockrell, D. M., Ode, P. J., Peairs, F. B. and Nachappa, P. (2022b). Triticale as a potential trap crop for the wheat stem sawfly (Hymenoptera: Cephidae) in winter wheat. *Frontiers in Agronomy* 4: 779013. doi: 10.3389/fagro.2022.779013.

Peirce, E. S., Rand, T. A., Cockrell, D. M., Ode, P. J. and Peairs, F. B. (2021). Effects of landscape composition on wheat stem sawfly (Hymenoptera: Cephidae) and its associated braconid parasitoids. *Journal of Economic Entomology* 114(1): 72-81. doi: 10.1093/jee/toaa287.

Pesho, G. R., McGuire, J. U. and McWilliams, J. M. (1971). Sampling methods for surveys of damage caused by the wheat stem sawfly. *FAO Plant Protection Bulletin* 19: 122–130.

Perez-Mendoza, J. and Weaver, D. K. (2006). Temperature and relative humidity effects on postdiapause larval development and adult emergence in three populations of wheat stem sawfly (Hymenoptera: Cephidae). *Environmental Entomology* 35(5): 1222–1231. doi: 10.1093/ee/35.5.1222.

Perez-Mendoza, J., Weaver, D. K. and Morrill, W. L. (2006). Infestation of wheat and downy brome grass by wheat stem sawfly and subsequent larval performance. *Environmental Entomology* 35(5): 1279–1285. doi: 10.1093/ee/35.5.1279.

Peterson, R. K. D., Buteler, M., Weaver, D. K., Macedo, T. B., Sun, Z., Perez, O. G. and Pallipparambil, G. R. (2011). Parasitism and the demography of wheat stem sawfly larvae, *Cephus cinctus*. *BioControl* 56(6): 831–839. doi: 10.1007/s10526-011-9357-7.

Piesik, D., Pańka, D., Jeske, M., Wenda-Piesik, A., Delaney, K. J. and Weaver, D. K. (2013). Volatile induction of infected and neighbouring uninfected plants potentially influence attraction/repellence of a cereal herbivore. *Journal of Applied Entomology* 137(4): 296–309. doi: 10.1111/j.1439-0418.2012.01742.x.

Piesik, D., Weaver, D. K., Runyon, J. B., Buteler, M., Peck, G. E. and Morill, W. L. (2008). Behavioral responses of wheat stem sawflies to wheat volatiles. *Agricultural and Forest Entomology* 10(3): 245–253. doi: 10.1111/j.1461-9563.2008.00380.x.

Platt, A. W. and Farstad, C. W. (1946). The reaction of wheat varieties to wheat stem sawfly attack. *Scientific Agriculture* 26: 231–247.

Portman, S. L., Jaronski, S. T., Weaver, D. K. and Reddy, G. V. P. (2018). Advancing biological control of the wheat stem sawfly: New strategies in a 100-yr struggle to manage a costly pest in the Northern Great Plains. *Annals of the Entomological Society of America* 111(3): 85-91. doi: 10.1093/aesa/say002.

Portman, S. L., Krishnankutty, S. M. and Reddy, G. V. P. (2016). Entomopathogenic nematodes combined with adjuvants presents a new potential biological control method for managing the wheat stem sawfly, *Cephus cinctus* (Hymenoptera: Cephidae). *PLoS ONE* 11(12): e0169022. doi: 10.1371/journal.pone.0169022.

Rand, T. A., Morrill, W. L., Runyon, J. B., Hoelmer, K. A., Shanower, T. G., Littlefield, J. L. and Weaver, D. K. (2016). Assessing phenological synchrony between the Chinese sawfly, *Cephus fumipennis* (Hymenoptera: Cephidae), its egg-larval parasitoid, *Collyria catoptron* (Hymenoptera: Ichneumonidae), and the North American sawfly, *Cephus cinctus*: Implications for biological control. *The Canadian Entomologist* 148(4): 482–492. doi: 10.4039/tce.2015.64.

Rand, T. A., Richmond, C. E. and Dougherty, E. T. (2020a). Modeling the combined impacts of host plant resistance and biological control on the population dynamics of a major pest of wheat. *Pest Management Science* 76(8): 2818–2828. doi: 10.1002/ps.5830.

Rand, T. A., Waters, D. K., Srygley, R. B. and Branson, D. H. (2020b). Tri-trophic interactions are resilient to large shifts in precipitation levels in a wheat agroecosystem. *Agriculture, Ecosystems and Environment* 301: 106981. doi: 10.1016/j.agee.2020.106981.

Rand, T. A., Richmond, C. E. and Dougherty, E. T. (2017). Using matrix population models to inform biological control management of the wheat stem sawfly, *Cephus cinctus*. *Biological Control* 109: 27–36. doi: 10.1016/j.biocontrol.2017.03.007.

Rand, T. A., Titus, E. F. and Waters, D. K. (2019). Do floral resources benefit the herbivorous sawfly, *Cephus cinctus* (Hymenoptera: Cephidae), a major pest of wheat in North America? *Journal of Economic Entomology* 112(2): 565–570. doi: 10.1093/jee/toy408.

Rand, T. A. and Waters, D. K. (2020). Aphid honeydew enhances parasitoid longevity to the same extent as a high-quality floral resource: Implications for conservation biological control of the wheat stem sawfly (Hymenoptera: Cephidae). *Journal of Economic Entomology* 113(4): 2022–2025. doi: 10.1093/jee/toaa076.

Rand, T. A., Waters, D. K., Blodgett, S. L., Knodel, J. J. and Harris, M. O. (2014). Increased area of a highly suitable host crop increases herbivore pressure in intensified agricultural landscapes. *Agriculture, Ecosystems and Environment* 186: 135–143. doi: 10.1016/j.agee.2014.01.022.

Rand, T. A., Waters, D. K. and Shanower, T. G. (2011). Unexpectedly high levels of parasitism of wheat stem sawfly larvae in postcutting diapause chambers. *The Canadian Entomologist* 143(5): 455–459. doi: 10.4039/n11-023.

Rand, T. A., Waters, D. K., Shanower, T. G. and Berzonsky, W. A. (2012). Effects of genotypic variation in stem solidity on parasitism of a grass-mining insect. *Basic and Applied Ecology* 13(3): 250–259. doi: 10.1016/j.baae.2012.03.005.

Reis, D. A., Hofland, M. L., Peterson, R . K. D. and Weaver, D. K. (2019). Effects of sucrose supplementation and generation on life-history traits of *Bracon cephi* and *Bracon lissogaster*, parasitoids of the wheat stem sawfly. *Physiological Entomology* 44(3–4): 266–274. doi: 10.1111/phen.12303.

Riley, C. V. and Marlatt, C. L. (1891). *Wheat and Grass Sawflies* (Vol. 4). United States Department of Agriculture, Division of Entomology, Insect Life, pp. 168–179.

Roberts, D. W. A. (1960). Sawfly resistance in wheat. III. Changes in resistance during the development of the wheat plant. *Canadian Journal of Plant Science* 40(1): 7–20. doi: 10.4141/cjps60-002.

Robertson, H. M., Waterhouse, R. M., Walden, K. K. O., Ruzzante, L., Reijnders, M. J. M. F., Coates, B. S., Legeai, F., Gress, J. C., Biyiklioglu, S., Weaver, D. K., Wanner, K. W. and Budak, H. (2018). Genome sequence of the wheat stem sawfly, *Cephus cinctus*, representing an early-branching lineage of the Hymenoptera, illuminates evolution of hymenopteran chemoreceptors. *Genome Biology and Evolution* 10(11): 2997-3011. doi: 10.1093/gbe/evy232.

Runyon, J. B., Hurley, R. L., Morrill, W. L. and Weaver, D. K. (2001). Distinguishing adults of *Bracon cephi* and *B. lissogaster* (Hymenoptera: Braconidae), parasitoids of the wheat stem sawfly (Hymenoptera: Cephidae). *The Canadian Entomologist* 133(2): 215–217. doi: 10.4039/Ent133215-2.

Runyon, J. B., Morrill, W. L., Weaver, D. K. and Miller, P. R. (2002). Parasitism of the wheat stem sawfly (Hymenoptera: Cephidae) by *Bracon cephi* and *B. lissogaster* (Hymenoptera: Braconidae) in wheat fields bordering tilled and untilled fallow in Montana. *Journal of Economic Entomology* 95(6): 1130–1134. doi: 10.1603/0022-0493-95.6.1130.

Salt, R. W. (1956). Influence of moisture content and temperature on cold-hardiness of hibernating insects. *Canadian Journal of Zoology* 34(4): 283-294. doi: 10.1139/z56-042.

Salt, R. W. (1959). Role of glycerol in the cold-hardening of *Bracon cephi* (Gahan). *Canadian Journal of Zoology* 37(1): 59-69. doi: 10.1139/z59-007.

Sandhi, R. K., Pothula, R., Pothula, S. K., Adams, B. J. and Reddy, G. V. P. (2020). First record of native entomopathogenic nematodes from Montana agroecosystems. *Journal of Nematology* 52(1): 1-11. doi: 10.21307/jofnem-2020-060.

Seamans, H. L. (1928). The value of trap crops in the control of the wheat stem sawfly in Alberta. *Annual Report of the Entomological Society of Ontario* 59: 59-64.

Seamans, H. L., Manson, G. F. and Farstad, C. W. (1944). The effect of wheat stem sawfly (*Cephus cinctus*) on the heads and grain of infested stems. *Annual Report of the Entomological Society of Ontario* 75: 10-15.

Shanower, T. G. and Hoelmer, K. A. (2004). Biological control of wheat stem sawflies: Past and future. *Journal of Agricultural and Urban Entomology* 21: 197-221.

Sharma, A., Shrestha, G. and Reddy, G. V. P. (2019). Trap crops: How far we are from using them in cereal crops? *Annals of the Entomological Society of America* 112(4): 330-339. doi: 10.1093/aesa/say047.

Sherman, J. D., Blake, N. K., Martin, J. M., Kephart, K. D., Smith, J., Clark, D. R., Hofland, M. L., Weaver, D. K., Lanning, S. P., Heo, H.-Y., Pumphrey, M., Chen, J. and Talbert, L. E. (2015). Agronomic impact of a stem solidness gene in near-isogenic lines of wheat. *Crop Science* 55(2): 514-520. doi: 10.2135/cropsci2014.05.0403.

Sherman, J. D., Weaver, D. K., Hofland, M. L., Sing, S. E., Buteler, M., Lanning, S. P., Naruoka, Y., Crutcher, F. K., Blake, N. K., Martin, J. M., Lamb, P. F., Carlson, G. R. and Talbert, L. E. (2010). Identification of novel QTL for sawfly resistance in wheat. *Crop Science* 50(1): 73-86. doi: 10.2135/cropsci2009.03.0145.

Shrestha, G., Briar, S. S. and Reddy, G. V. P. (2018). Plant defense elicitors: Plant fitness versus wheat stem sawfly. *PeerJ* 6: e5892. doi: 10.7717/peerj.5892.

Simão, L. M., Easterly, A. C., Kruger, G. R. and Creech, C. F. (2021). Winter wheat residue impact on soil water storage and subsequent corn yield. *Agronomy Journal* 113(1): 276-286. doi: 10.1002/agj2.20459.

Somsen, H. W. and Luginbill, P., Jr. (1956). *Bracon lissogaster* Muesebeck., a parasite of the wheat stem sawfly. *U.S Department of Agriculture Technical Bulletin* 1153: 1-7.

Subedi, M., Cárcamo, H. A., Knodel, J. J., Weaver, D. K., Cuthbert, R. D., Pozniak, C. J., Nilsen, K. T. and Beres, B. L. (2021). Stability analysis of stem solidness, grain yield, and grain protein concentration in spring wheat. *Canadian Journal of Plant Science* 101(4): 456-475. doi: 10.1139/cjps-2020-0089.

Talbert, L. E., Sherman, J. D., Hofland, M. L., Lanning, S. P., Blake, N. K., Grabbe, R., Lamb, P. F., Martin, J. M. and Weaver, D. K. (2014). Resistance to *Cephus cinctus* Norton, the wheat stem sawfly, in a recombinant inbred line population of wheat derived from two resistance sources. *Plant Breeding* 133(4): 427-432. doi: 10.1111/pbr.12184.

Tangtrakulwanich, K., Reddy, G. V. P., Wu, S., Miller, J. H., Ophus, V. L. and Prewett, J. (2014). Efficacy of entomopathogenic fungi and nematodes, and low risk insecticides against wheat stem sawfly, *Cephus cinctus* (Hymenoptera: Cephidae). *Journal of Agricultural Science* 6(5): 1-9. doi: 10.5539/jas.v6n5p1.

Varella, A. C., Talbert, L. E., Achhami, B. B., Blake, N. K., Hofland, M. L., Sherman, J. D., Lamb, P. F., Reddy, G. V. P. and Weaver, D. K. (2018). Characterization of resistance to *Cephus cinctus* (Hymenoptera: Cephidae) in Barley Germplasm. *Journal of Economic Entomology* 111(2): 923-930. doi: 10.1093/jee/toy025.

Varella, A. C., Talbert, L. E., Hofland, M. L., Buteler, M., Sherman, J. D., Blake, N. K., Heo, H.-Y., Martin, J. M. and Weaver, D. K. (2016). Alleles at a quantitative trait locus for stem solidness in wheat affect temporal patterns of pith expression and level of resistance to the wheat stem sawfly. *Plant Breeding* 135(5): 546–551. doi: 10.1111/pbr.12398.

Varella, A. C., Weaver, D. K., Blake, N. K., Hofland, M. L., Heo, H. Y., Cook, J. P., Lamb, P. F., Jordan, K. W., Akhunov, E., Chao, S. and Talbert, L. E. (2019a). Analysis of recombinant inbred line populations derived from wheat landraces to identify new genes for wheat stem sawfly resistance. *TAG. Theoretical and Applied Genetics. Theoretische und Angewandte Genetik* 132(8): 2195–2207. doi: 10.1007/s00122-019-03347-8.

Varella, A. C., Zhang, H., Weaver, D. K., Cook, J. P., Hofland, M. L., Lamb, P., Chao, S., Martin, J. M., Blake, N. K. and Talbert, L. E. (2019b). A novel QTL in durum wheat for resistance to the wheat stem sawfly associated with early expression of stem solidness. *G3 Genes|Genomes|Genetics* 9(6): 1999–2006. doi: 10.1534/g3.119.400240.

Varella, A. C., Weaver, D. K., Cook, J. P., Blake, N. K., Hofland, M. L., Lamb, P. F. and Talbert, L. E. (2017a). Characterization of resistance to the wheat stem sawfly in spring wheat landrace accessions from targeted geographic regions of the world. *Euphytica* 213(7). doi: 10.1007/s10681-017-1945-x.

Varella, A. C., Weaver, D. K., Peterson, R. K. D., Sherman, J. D., Hofland, M. L., Blake, N. K., Martin, J. M. and Talbert, L. E. (2017b). Host plant quantitative trait loci affect specific behavioral sequences in oviposition by a stem-mining insect. *TAG: Theoretical and Applied Genetics. Theoretische und Angewandte Genetik* 130(1): 187–197. doi: 10.1007/s00122-016-2805-0.

Varella, A. C., Weaver, D. K., Sherman, J. D., Blake, N. K., Heo, H. Y., Kalous, J. R., Chao, S., Hofland, M. L., Martin, J. M., Kephart, K. D. and Talbert, L. E. (2015). Association analysis of stem solidness and wheat stem sawfly resistance in a panel of North American spring wheat germplasm. *Crop Science* 55(5): 2046–2055. doi: 10.2135/cropsci2014.12.0852.

Wahl, D. B., Shanower, T. G. and Hoelmer, K. A. (2007). A new species of *collyria* Schiødte (Hymenoptera: Ichneumonidae: Collyriinae), a parasitoid of *Cephus fumipennis* (Hymenoptera: Cephidae) in China, and potential biological control agent for *Cephus cinctus* in North America. *Journal of the Kansas Entomological Society* 80(1): 43–50. doi: 10.2317/0022-8567(2007)80[43:ANSOCS]2.0.CO;2.

Wallace, L. E. (1962). Field-plot tests of chemicals for wheat stem sawfly control. *Journal of Economic Entomology* 55(6): 908–912. doi: 10.1093/jee/55.6.908.

Wallace, L. E. and Butler, L. I. (1967). Residues of heptachlor and heptachlor epoxide in wheat treated to control the wheat stem sawfly123. *Journal of Economic Entomology* 60(1): 254–256. doi: 10.1093/jee/60.1.254.

Wallace, L. E. and McNeal, F. H. (1966). *Stem Sawflies of Economic Importance in Grain Crops in the United States*. Washington, DC: Agricultural Research Service, United States Department of Agriculture.

Wanner, K. W. and Tharp, C. I. (2015). Thimet 20-G® for wheat stem sawfly management. Montana State University, Extension, *Pesticide News*. Available at: https://pesticides.montana.edu/news/msupestnews/20150506_PN_Thimet.html.

Weaver, D. K., Buteler, M., Hofland, M. L., Runyon, J. B., Nansen, C., Talbert, L. E., Lamb, P. F. and Carlson, G. R. (2009). Cultivar preferences of ovipositing wheat stem sawflies as influenced by the amount of volatile attractant. *Journal of Economic Entomology* 102(3): 1009–1017. doi: 10.1603/029.102.0320.

Weaver, D. K., Nansen, C., Runyon, J. B., Sing, S. E. and Morrill, W. L. (2005). Spatial distributions of *Cephus cinctus* Norton (Hymenoptera: Cephidae) and its braconid parasitoids in Montana wheat fields. *Biological Control* 34(1): 1–11. doi: 10.1016/j.biocontrol.2005.04.001.

Weaver, D. K., Sing, S. E., Runyon, J. B. and Morrill, W. L. (2004). Potential impact of cultural practices on wheat stem sawfly (Hymenoptera: Cephidae) and associated parasitoids. *Journal of Agricultural and Urban Entomology* 21: 271–287.

Weiss, M. J. and Morrill, W. L. (1992). Wheat stem sawfly (Hymenoptera: Cephidae) revisited. *American Entomologist* 38(4): 241–245. doi: 10.1093/ae/38.4.241.

Weiss, M. J., Morrill, W. L. and Reitz, L. L. (1987). Influence of planting date and spring tillage on the wheat stem sawfly. Montana State University, Montana Agricultural Experiment Station. *Montana AgResearch* 4: 2–5.

Weiss, M. J., Riveland, N. R., Reitz, L. L. and Olson, T. C. (1990). Influence of resistant and susceptible cultivar blends of hard red spring wheat on wheat stem sawfly (Hymenoptera: Cephidae) damage and wheat quality parameters. *Journal of Economic Entomology* 83(1): 255–259. doi: 10.1093/jee/83.1.255.

Wenda-Piesik, A., Morrill, W. L., Grey, W. E. and Weaver, D. K. (2006). Entomopathogenic capacity of *Fusarium* crown rot on wheat stem sawfly larvae. *Progress in Plant Protection* 46: 380–387.

Wenda-Piesik, A., Sun, Z. T., Grey, W. E., Weaver, D. K. and Morrill, W. L. (2009). Mycoses of wheat stem sawfly (Hymenoptera: Cephidae) larvae by *Fusarium* spp. isolates. *Environmental Entomology* 38(2): 387–394. doi: 10.1603/022.038.0211.

Wong, M. L. , Bruckner, P. L., Berg, J. E., Lamb, P. F., Hofland, M. L., Caron, C. G., Heo, H.-Y., Blake, N. K., Weaver, D. K. and Cook, J. P. (2022). Evaluation of wheat stem sawfly resistant solid stem Qss. *Crop Science* 63: msub-3BL. doi: 10.1002/csc2.20866.

Wu, X.-H., Cárcamo, H. A., Beres, B. L., Clarke, F. R., DePauw, R. M. and Pang, B.-P. (2013). Effects of novel solid-stemmed wheat genotype on *Cephus cinctus* Norton and its parasitoid *Bracon cephi*. *Cereal Research Communications* 41(4): 647–660. doi: 10.1556/CRC.2013.0024.

Wu, X.-H., Cárcamo, H. A., Beres, B. L. and Pang, B.-P. (2011). Parasitoid (*Bracon cephi*) effects on grain yield of selected genotypes of wheat infested by *Cephus cinctus*. *Australian Journal of Crop Science* 5: 1102–1107.

Yeoman, C. J., Brutscher, L. M., Esen, Ö. C., Ibaoglu, F., Fowler, C., Eren, A. M., Wanner, K. and Weaver, D. K. (2019). Genome-resolved insights into a novel *Spiroplasma* symbiont of the wheat stem sawfly (*Cephus cinctus*). *PeerJ* 7: e7548. doi: 10.7717/peerj.7548.

Zentner, R. P., Campbell, C. A., Selles, F., Jefferson, P. G. and Lemke, R. (2006). Effect of fallow frequency, flexible rotations, legume green manure, and wheat class on the economics of wheat production in the Brown soil zone. *Canadian Journal of Plant Science* 86(2): 413–423. doi: 10.4141/P05-081.

Part 3

Phloem feeding pests, mites and root feeding pests

Chapter 6

Russian wheat aphid (*Diuraphis noxia*): an overview

Vicki L. Tolmay, Agricultural Research Council – Small Grain Institute, South Africa

1 Introduction

The Russian wheat aphid (RWA), *Diuraphis noxia* (Kurdjumov), is one of a small number of aphids associated with small grains such as barley (*Hordeum vulgare* L.), wheat (*Triticum aestivum* L., *T. durum* Desf.), triticale (× Triticosecale Wittmack), oats (*Avena sativa* L.) and rye (*Secale cereale* L.). It occurs throughout the world and is primarily problematic on barley and wheat (bread and durum) crops, causing severe damage with yield losses of up to circa 75% possible (Halbert and Stoetzel, 1998; Tolmay and van Deventer, 2005). It is unlike the other small grain aphids due to some distinctive attributes. This species feeds on the adaxial surface of the newest leaves, causing longitudinal white stripes because of severe structural damage caused to cells affected during probing and feeding (Fouché et al., 1984; Saheed et al., 2007a; Luna et al., 2018). Feeding by this pest triggers the disintegration of chloroplasts and other cell organelles, which results in plasmolysis of stylet punctured cells and disruption of the vascular function in the infested leaf. In addition, the infested leaf rolls closed (or fails to unroll), exhibiting thickening of the epidermal and

http://dx.doi.org/10.19103/AS.2022.0114.01

other lignin-containing tissues (Fouché et al., 1984). This pseudo-gall, formed by the tightly rolled leaf, provides protection from predatory and parasitic insects, as well as contact insecticides. Furthermore, the rolled flag leaf often traps the awn during its emergence, causing a distinctive 'fish hook' appearance.

The aphids themselves are elongated and spindle shaped and seldom exceed 2 mm in length. The legs and antenna are short, and the cornicles (siphunculi) appear absent to the naked eye. They are in fact present but are very short. Together these features enable *D. noxia* to move around within the tightly rolled leaves they cause. *D. noxia* are pale green in colour, sometimes with a white/grey waxy covering, but the eyes are always dark and easily visible. It is also often possible to see the black eyespots of developing nymphs within the abdomens of mature adult viviparous females. In addition to these traits, *D. noxia* has a prominent supracaudal process, which looks like a 'double tail' when the aphid is viewed from the side. On susceptible hosts, feeding by *D. noxia* causes an increase in the levels of all ten essential amino acids in the phloem, resulting in a more nutritious diet and increased growth rate (Telang et al., 1999). Sandström et al. (2000) postulate that this may be due to *D. noxia*'s ability to maintain a sufficient supply of plant nutrients through the characteristic damage it causes. Luna et al. (2018) identified numerous bacterial genera associated with *D. noxia* feeding, pointing to a complex interaction involving several organisms. These aphid-associated bacteria facilitate aphid virulence by contributing to the severe damage symptomology associated with *D. noxia* infestation. In contrast to many other aphid species, *D. noxia* has been reported to be less negatively influenced by host plants experiencing drought (Oswald and Brewer, 1997) and is known for occurring on stressed and sparse stands in contrast to lush, dense crop stands. Baugh (1991) determined that the formation of alates in *D. noxia* was not related to aphid density on the host plant but was in fact mediated by the water pressure (turgor) of the leaf tissue.

2 Global occurrence and pest status

Historic literature documents the primary range of *D. noxia* as the mountainous regions of the Iranian-Turkestanian area (Kovalev et al., 1991). Halbert and Stoetzel (1998) describe this as the area between the Caucasus Mountains and the 'Mountains of Heaven' known as 'Tian Shan', which straddle the border between the Xinjiang Province of China and the Central Asian countries of Kyrgyzstan and Kazakhstan. Recent molecular work on the genetic variation and genealogical and evolutionary relationships of different populations of *D. noxia* placed this pest in Western China since the period in which wheat was domesticated, concluding that this area forms the most eastern part of the primary range of this species (Zhang, 2012; Zhang et al., 2012).

The first record of *D. noxia* as pest was from 1901 in the Crimea in response to an increase of host plants in the Ukraine (Mokrzhetsky, 1901 as quoted by Kovalev et al., 1991). During the course of the 1900s, *D. noxia* was reported as pest in various countries. It was found in:

- Algeria in 1938 (Mimeur, 1941 as quoted by Laamari and Benyahia, 2018);
- Iran in 1954 (Davachi, 1954 as quoted by Ahmadi and Sarafrazi, 1992); and
- Turkey in 1959 (Elmali, 1998).

By 1963, it was known to occur in the former USSR (Ukraine, Caucasus, Crimea and Kazakhstan), Afghanistan, Israel, Egypt, East Africa, Southern Rhodesia (now Zimbabwe), Libya, Morocco, Spain and Great Britain (ARS-USDA, 1963).

Outbreaks of a pestiferous magnitude have been recorded on a regular basis, as *D. noxia* has spread around the barley and wheat production areas of the world. It has occurred:

- on barley in Ethiopia since 1972 (Haile, 1981);
- on wheat in South Africa since 1978 (Walters, 1984);
- on barley and wheat in Mexico since 1980 (Gilchrist et al., 1984);
- on barley and/or wheat in 16 States of the USA since 1986 (Stoetzel, 1987);
- in Chile since 1987 (Zerenè et al. as quoted by Halbert and Stoetzel, 1998);
- on wheat in Canada since 1988 (Jones et al., 1989);
- on barley, bread wheat and durum wheat in Argentina since 1992 (Castro et al., 1996);
- on wheat and barley in Kenya since 1995 (Macharia et al., 1999); and
- on bread wheat in Pakistan since 2010 (Akhtar et al., 2010).

The most recent incursion into the only *D. noxia*-free continent of the world occurred when this pest was reported from Australia in 2016 (Yazdani et al., 2018). This aphid species now occurs on all the continents where barley and wheat are grown and in all three major production regions of Australia. A comprehensive review of *D. noxia* in Australia was recently published (Ward et al., 2020) covering the biology, ecology and management of this pest since its incursion in 2016. The three incursions that have prompted the collection of large volumes of research data have been those in South Africa (1978 to date); Mexico, USA and Canada (1980 to date) and Australia (2016 to date). It is prudent to remember, however, that valuable *D. noxia* research outputs are not limited to the information generated as a result of only these incursions.

A common outcome of the introduction of *D. noxia* to a new area appears to be the widespread occurrence of newly established *D. noxia* populations, which in later years shrink to smaller areas where *D. noxia* persists as a regular problem. This phenomenon was observed in Ethiopia (Haile, 1981) where *D. noxia* was first reported on barley in Tigrai in 1972-1973 and Western Welo in 1974 during a period of drought, from where it spread to all the barley production areas within 2 years. The same pattern was reported for the incursions in South Africa (Walters, 1984) as well as North America (Jones et al., 1989; Quisenberry and Peairs, 1998) and South America (Castro et al., 1996). The occurrence of *D. noxia* in Chile and Argentina was not associated with as severe damage as in the other three incursions listed above. Rapid spread of this pest in Chile and Argentina during the drought of 1996 and 1997 to a large area of common occurrence (Ortego and Delfino, 1994 and Bellone and Almaraz, 1995, as quoted by Castro et al., 1998) does nonetheless appear to have narrowed over time.

Factors which have been implicated in the distribution of pestiferous populations of *D. noxia* are the extent of the area planted to the susceptible crop, the altitude at which the crop is grown and the prevalence of conditions that promote entomopathogenic fungal infections.

In South America, *D. noxia* was reported to be uncommon in the Central Valley of Argentina and close to the Pacific Ocean. This pest was more abundant above an altitude of 1000 m above sea level (Castro et al., 1998). Clúa et al. (2004) collected *D. noxia* in a survey conducted in Chile between 1996 and 1999 at two locations with altitudes of 1000 m and 2400 m above sea level. In Ethiopia, Haile (1981) reported that the most problematic *D. noxia* infestations occurred on barley grown in the Highlands (>2500 m above sea level) as well as in areas where large tracts of land were planted to barley. Belay and Stauffer (2011) confirmed years later that *D. noxia* infestations were severe in the Ethiopian Highlands and particularly on late maturing types that are susceptible to RWA. Recently, Damte (2015) reported that Ethiopia has begun to grow irrigated wheat and to intensify the production of rainfed wheat. Although the area under irrigated wheat at that time was less than 10 000 ha, it was highly infested by *D. noxia* (Damte, 2015).

In South Africa, the Eastern Free State (1400-1800 m above sea level) is the area where *D. noxia* had a sizeable annual presence four decades after its incursion. It is also the region where new biotypes of the aphid have been found most often (Jankielsohn, 2011). *D. noxia* continues to occur sporadically in other cereal production regions of South Africa, in order of importance namely the Central and Western Free State, the Western Cape and the irrigation areas. Most often, when *D. noxia* is found on irrigated wheat, it is in flood-irrigated fields where the leaves of the plants remain dry. It seldom occurs on crops irrigated with pivots or other overhead irrigation systems. Furthermore, Hatting et al. (1999) mentioned a correlation of *D. noxia* pest status to the macroclimatic conditions encountered in the various wheat-producing areas. These authors noted that *D. noxia* is considered a secondary, sporadic pest of wheat in the Western Cape which is typified by moderate temperatures (averaging 15-20°C) and wet conditions. They further concluded that the Mediterranean climate that prevails during the wheat growth season, combined with the fact that plant surfaces can remain wet until 11:00 am or even noon, is conducive to entomopathogenic epizootics.

In North America, *D. noxia* was reported from central and northern Mexico, 16 western States and three Canadian provinces (Brewer et al., 2019). It caused widespread losses to both barley and wheat. In an elliptical area in the Western Great Plains of the USA (Webster and Kenkel, 1999), particularly where Wyoming, Colorado and New Mexico border on Nebraska, Kansas, Oklahoma and Texas, *D. noxia* was a regular pest for more than a decade before it decreased in intensity and extent to become sporadic in nature. Small populations of *D. noxia*, however, continue to occur throughout the widespread area that was initially colonized, but incidences of pestiferous population size with accompanying yield loss are sporadic. In Canada, *D. noxia* was initially problematic; however, according to McCallum et al. (2021), it is now considered to have been a migratory pest from 1988 to 1991.

In Australia, the initial spread of *D. noxia* was rapid. Populations spread from Tarlee, in the mid-north of South Australia to Victoria, New South Wales and Tasmania (January 2017; Yazdani et al., 2018), within a 12-month period. In August 2020, *D. noxia* was found in Western Australia near Esperance and by July 2021 had moved further west reaching Broomehill, Western Australia, where it was reported on barley and wheat (Pestfax, 2021). It now remains to be seen if and where this aphid will establish as a persistent pest.

3 Population growth and conducive circumstances for outbreaks

3.1 Production circumstances and Diuraphis noxia *outbreaks*

It appears that *D. noxia* is feared more for its potential ability to cause devastating damage than for the sum of all actual annual damage it currently causes, if one considers all regions in which it occurs. This is because, for the majority of places the aphid is known to occur, population densities of this pest remain low due to climatic, cultural or biological factors that suppress aphid numbers sufficiently to mask their presence within a crop. However, two characteristics of *D. noxia* seem to contribute to its ability to unexpectedly 'appear' where scouting has not been properly done or where it was not expected to be found.

The first, similar to most aphids, is the ability to attain explosive population growth under conducive circumstances. The duration of the instars of *D. noxia* shortens as temperatures reach the ideal range for aphid growth (see Ma and Bechinski, 2009a). This results in a shorter period from birth to reproduction and contributes to the high total fecundity of *D. noxia*. Different studies of *D. noxia* life history have produced slightly different data (Aalbersberg et al., 1987; Webster and Starkes, 1987; Kieckhefer and Elliott, 1989; Girma et al.,

1990; Nowierski et al., 1995; Ma and Bechinski, 2009b; Brewer et al., 2019). A general synopsis can be made that the life span can range from 40 to 80 days and that somewhere around 60–80 progeny are produced during this time. Under optimal conditions, a period of only 9 days from birth to reproduction translates to three generations becoming reproductive in the span of a single month. It is important to consider that the host plant on which *D. noxia* feeds has a significant effect on nymphal development time, total fecundity and daily reproduction (Veisi et al., 2012; also see sections 'Case study' and 'Host plant resistance' in this chapter).

The second factor is the almost imperceptibly low population levels at which *D. noxia* can survive non-crop periods. A wide host range within the grasses and volunteer small grains can be utilized by this aphid at very low densities in specific environmental niches, to survive the non-crop season from which the population then increases. Numerous studies (Butts et al., 1997; Smit et al., 2010; Mulatu et al., 2011; Brewer et al., 2019) have identified the specific circumstances within particular regions that promote the carry-over from one crop season to the next. Usually the green-bridge effect is important over the summer, but it can also occur over the winter in specific areas. *D. noxia* can cause damage to susceptible crops in any season, depending on the specific environmental niche created by the 'environment-host' combination in which it occurs. Table 1 lists some examples of the production circumstances where damage is caused and how the infestation plays out over the season.

Furthermore, overwintering of *D. noxia* is intimately related to the nature of this aphids' life cycle and whether it is holocyclic or anholocyclic. In most cases, it is the latter, with obligate parthenogenetic females overwintering on host plants in sheltered niches. Anholocycly is also known as cyclical parthenogenesis. Holocycly, with the presence of sexual reproduction and overwintering eggs, from which the new fundatrices hatch to continue with parthenogenic reproduction, is known to occur in extremely cold climates and at high elevations. *D. noxia* is reported as being holocyclic in China (Zhang, 2012) as well as in Russia (Kiriac et al., 1990; Kaplin et al., 2015). Oviparae and sexual reproduction was confirmed in some *D. noxia* populations of the USA (Kiriac et al. 1990; Puterka et al. 2012), although the majority of the US populations reproduce through parthenogenetic reproduction. Ricci et al. (2011) have determined that *D. noxia* populations can be holocyclic, while obligatory and facultative parthenogenesis can coexist with these sexual populations in Argentina.

What is noteworthy is that holocycly usually occurs where *D. noxia* is not a regular pest. Zhang (2012) postulated that the particular *D. noxia* that plague cultivated crops may possibly be a viviparous clone adapted to a monoculture of cultivated crops. This separate *D. noxia* lineage may have spread more successfully around the world due to the occurrence of suitable host crops in many countries and different circumstances. The influence of climate change on the distribution and pest status of *D. noxia* is something that will have to be observed carefully in future. Mondor et al. (2003) noted that altered abiotic conditions such as rainfall and temperature changes could influence species to shift to different geographical ranges should their current distribution become unsuitable for habitation. In future, this could potentially influence overwintering in addition to other aspects of *D. noxia* life history.

3.2 Climate change and Diuraphis noxia *response to elevated CO_2 levels*

While the changing climate is usually perceived to be a change in macro-circumstances, its influence on microcircumstances and niches is what makes its effects so profound. Over the past 60 years, atmospheric CO_2 has increased by more than 30% from 330 ppm in 1960 to above 450 ppm in 2021, as measured by the NOAA's Mauna Loa Observatory in Hawaii. (https://climate.nasa.gov/news/2915/the-atmosphere-getting-a-handle-on-carbon-dioxide/, accessed 22/10/2021). It has been shown that the net photosynthetic capacity of plants under elevated CO_2 results in an increase in plant biomass, and this is particularly concerning when considering aphid pests (Douglas, 2003). The reason for this is that aphids feed on phloem sap, forming local sinks that withdraw plant nutrients and water once their stylets have penetrated the functional phloem.

Douglas (2003) further reported that phloem sap feeders are very sensitive to changes in plant's nutritional status, including those resulting from climate change. This was corroborated by the study of Mondor et al. (2005) with pea aphids (*Acyrthosiphon pisum*) feeding on broad bean (*Vicia faba*) plants under elevated CO_2 and O_3 conditions. Matsiliza (2003) verified that *D. noxia* probes the leaf between epidermal cells or through the stomata and proceeds on an intercellular pathway through the mesophyll cells to the vascular tissue. Upon reaching the vascular bundle, the stylet pathway becomes intracellular, near and inside the bundle. *D. noxia* tap into the growing plants' most robust sinks of minerals, carbohydrates and amino acids, due to their preferred feeding position on the youngest leaves of their host plant. Electropenetration graph studies by Tolmay (2006) showed that the interaction between *D. noxia* and each host plant is unique. *D. noxia* feeding parameters for resistant genotypes, TugelaDn, 1684/Tugela and 2199/Tugela, however, all showed phloem-based feeding differences indicative of difficulties in either reaching sustained ingestion of phloem sap or sufficient ingestion of phloem sap without obtaining adequate nutrition. Studies by Saheed et al. (2007a, 2007b, 2009, 2010) in barley reported that *D. noxia* feeding causes severe damage to the phloem as well as to the xylem transport systems within its hosts, with dire consequences for the host plant.

Using two South African biotypes of *D. noxia* (RWASA1 and RWASA2), Jimoh et al. (2013) studied plant growth (plant biomass, C:N ratios and leaf nitrogen concentrations), feeding damage (chlorosis and leaf roll) and aphid population growth on susceptible and resistant barley under ambient and elevated CO_2 levels (450 ppm and 550 ppm, respectively). Both feeding damage and aphid population growth were noticeably increased at elevated CO_2 levels, with less damage being inflicted overall by the biotype RWASA1 than the biotype RWASA2. Resistant genotypes sustained less damage than susceptible genotypes under ambient CO_2 levels. However, *D. noxia* feeding under elevated CO_2 caused very high loss of biomass, particularly in the susceptible barley genotype. Botha et al. (2017) showed that a relatively small increase of CO_2 (i.e. from 400 ppm to 450 ppm) could possibly nullify inherent plant resistance to *D. noxia*. These studies demonstrated that both barley and wheat support larger *D. noxia* populations under elevated CO_2. These increased aphid populations lead to an increase in plant stress, an increase in vascular tissue damage in the host plants, an increase in cellular disruption and an increase in leaf chlorosis and rolling. This damage is also associated with a decrease in foliar N concentrations and a loss of plant biomass, resulting in potential plant death.

Besides CO_2 levels, climate change is expected to influence temperatures, rainfall and the location of suitable regions for crop production. A change in these factors will undoubtedly lead to the creation of new habitats for *D. noxia*. Niches within the agro-ecological landscape may facilitate the survival of the small populations that are required to make it through the non-crop season for carry-over to the next crop to occur. At moderate temperatures within the range of 5–30°C (Burd et al., 1998), *D. noxia* growth and development occurs easily. Within the 5–30°C range, *D. noxia* developmental time, lifespan and number of progeny produced varies, becoming less suitable towards the outer boundaries of this range. Optimal growth, used for colony maintenance in research, is between 18°C and 22°C.

4 Case study

The nature, extent and ensuing control measures for *D. noxia* in South Africa have advanced through various stages during the past four decades since this pest was first observed in the grassland plateau of the Eastern Free State. Here, RWA maintained a consistent pest status over many years (1978 to mid-2000s) before declining to a sporadic pest of lesser status. This production area was characterized by the planting of large tracts of facultative wheat, with a medium vernalization requirement, during the winter months at low seeding densities (Smit et al., 2010). The wheat crop used moisture stored in the soil from the preceding summer rainfall for early season growth and rainfall for later season growth. To access the moisture in the soil profile, cultivars with a long coleoptile and large tillering capacity were planted. At the end of winter, just before and during the tillering phase, the plants were at their most vulnerable to *D. noxia* infestation. One of the reasons for this was a short period of limited moisture between the depletion of the soil water profile and the first spring rainfall (see Fig. 1). This 'mini-drought' coincided with rising temperatures, which increased both plant and aphid growth.

Population studies from 1983 to 1985 by Aalbersberg et al. (1989) found that exponential increase of *D. noxia* numbers in terms of both the percentage infestation and the number of *D. noxia* per stem took place during the stem elongation stage of wheat. A mean maximum of 150–160 aphids was recorded per stem on two susceptible cultivars Betta and Flamink. A follow-up study was undertaken a decade later using two susceptible cultivars Betta and Tugela. These cultivars were compared to two resistant backcross (BC5) breeding lines of Tugela containing the genes *Dn1* (from PI 137739) and *Dn2* (from PI

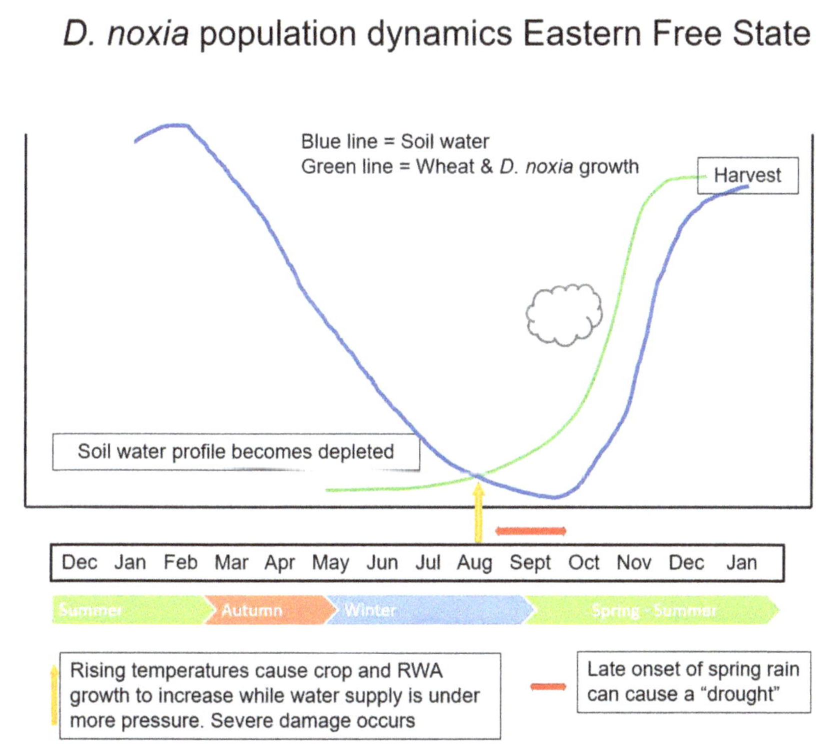

Figure 1 Infographic depicting conditions in the grassland plateau of the Eastern Free State, South Africa. Depletion of stored soil moisture from the profile (blue line) coincides with the growth of wheat and increase of *D. noxia* (green line) from August until September. From mid-August (yellow arrow) as temperatures begin to rise, the wheat withdraws more moisture from the profile. When plants are infested with *D. noxia* and a 'mini-drought' occurs due to the late onset of spring rain, severe damage occurs. This scenario was common during the 1980s to the mid-2000s when late spring rain was defined as after the 15th of October.

262660) in a randomized block design trial with six replications during 1993 and 1994. Trials were left to become infested naturally. Samples were collected weekly, and the percentage of tillers that became infested was determined. The number of RWA per infested tiller was also recorded.

The 1993 season was characterized by high yields and low *D. noxia* numbers in contrast to the 1994 season where severe *D. noxia* infestations and lower yields occurred. During both seasons, the percentage of infested tillers was significantly higher on susceptible cultivars Betta and Tugela. The number of *D. noxia* per infested tiller was also significantly higher on the susceptible cultivars during both seasons. *D. noxia* numbers were lower during 1993 than during 1994, with the maximum percentage of infested tillers on the susceptible cultivars reaching 34% and 90%, respectively. The percentage of infested tillers over the seasons is shown in Fig. 2 (1993) and Fig. 3 (1994). The maximum percentage of infested tillers on the resistant lines was lower than that on the susceptible cultivars, reaching a maximum of 12% and 50% during 1993 and 1994, respectively. The number of *D. noxia* per infested tiller was also much lower on the resistant lines during both seasons, as seen in Fig. 4 (1993) and Fig. 5 (1994). During both years, the infestation was lower on TugelaDn2 than on TugelaDn1. During the 2005 season, the same cultivars were replanted, but the trial did not become sufficiently infested with *D. noxia* to obtain useful data (Fig. 6), confirming the reported decrease in *D. noxia* pest status by producers.

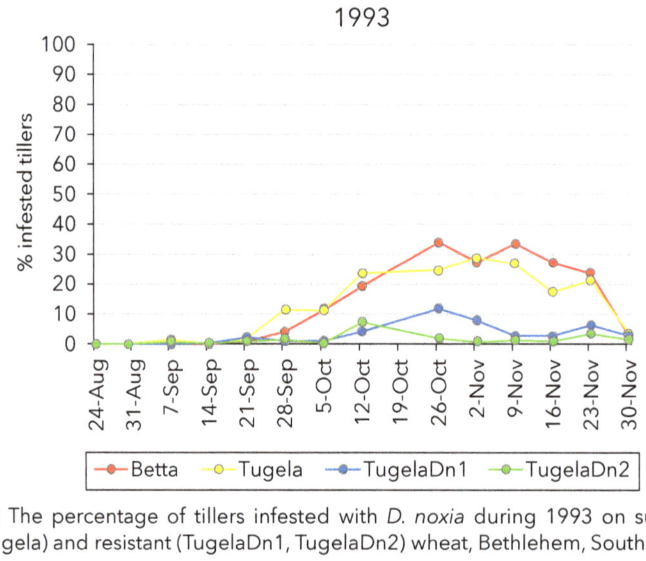

Figure 2 The percentage of tillers infested with *D. noxia* during 1993 on susceptible (Betta, Tugela) and resistant (TugelaDn1, TugelaDn2) wheat, Bethlehem, South Africa.

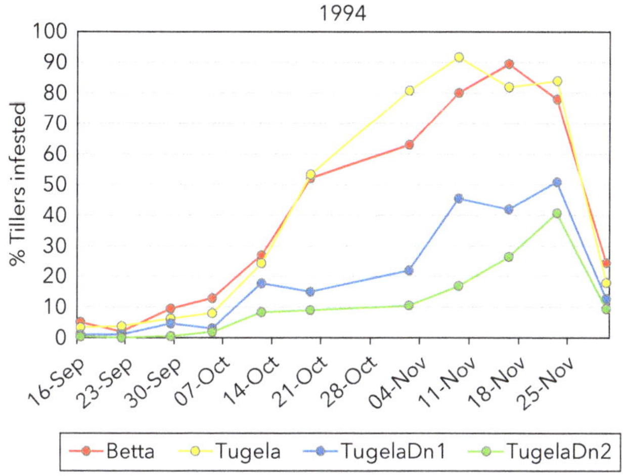

Figure 3 The percentage of tillers infested with *D. noxia* during 1994 on susceptible (Betta, Tugela) and resistant (TugelaDn1, TugelaDn2) wheat, Bethlehem, South Africa.

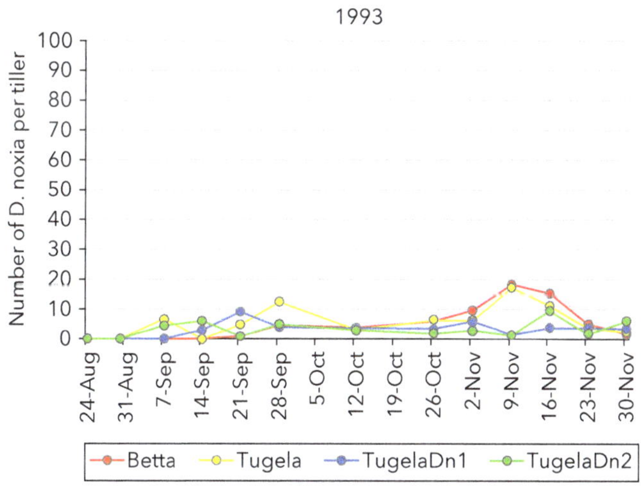

Figure 4 The number of *D. noxia* per infested tiller during 1993 on susceptible (Betta, Tugela) and resistant (TugelaDn1, TugelaDn2) wheat, Bethlehem, South Africa. Note difference in *Y*-axis values when compared to Figure 5.

In terms of the likelihood of infestation, a logistic analysis performed on the 1993 and 1994 data showed that all lines differ significantly from each other.

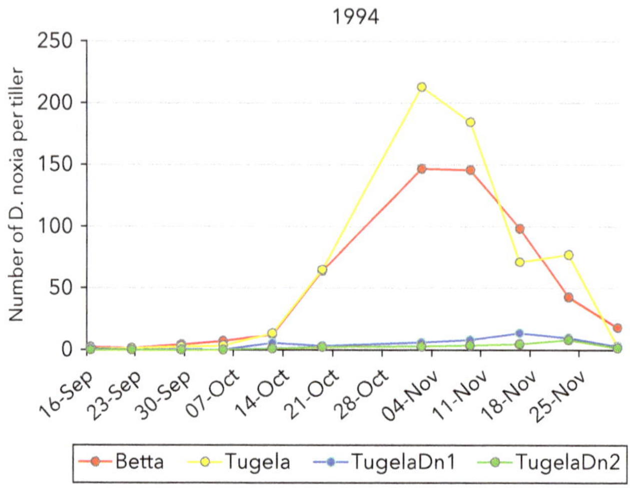

Figure 5 The number of *D. noxia* per infested tiller during 1994 on susceptible (Betta, Tugela) and resistant (TugelaDn1, TugelaDn2) wheat, Bethlehem, South Africa.

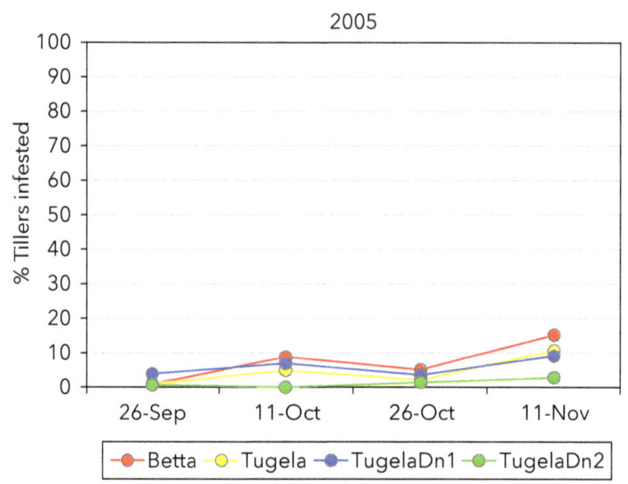

Figure 6 The percentage of tillers infested with *D. noxia* during 2005 on susceptible (Betta, Tugela) and resistant (TugelaDn1, TugelaDn2) wheat, Bethlehem, South Africa.

- Compared to Tugela, Betta ($_{exp}$0.1518) was 1.2x more likely to become infested; TugelaDn1 ($_{exp}$-1.1458) was 5x less likely to become infested and TugelaDn2 ($_{exp}$ -2.2390) was 10x less likely to be infested.
- Compared to Betta, TugelaDn1 ($_{exp}$1.5676) was 5x less likely to be infested; TugelaDn2 ($_{exp}$2.3908) was 11x less likely to be infested.
- Tugela/Dn2 was 2.2x less likely to become infested than TugelaDn1.

Mechanisms of resistance studies undertaken in the glasshouse for these four genotypes related well to the field data but not well to the mechanism of resistance data collected for the R-donor landraces PI 137739 and PI 262660.

The difference in yield between the resistant lines and susceptible cultivars was most pronounced during 1994, when higher RWA infestation occurred (Fig. 7) and susceptible Tugela sustained significant damage. The 2005 yields were in line with the long-term mean yields for the region and were not influenced by *D. noxia* damage. The more reliable yields of *D. noxia* resistant wheat cultivars over varying circumstances resulted in the area planted to these genotypes increasing rapidly (Marasas et al., 1997) as seeds of resistant cultivars became available.

To record the level of *D. noxia* infestation on resistant cultivars during the period that resistance was deployed, a field trial was performed at Bethlehem, Eastern Free State, from 1995 to 2000. Figure 8 illustrates the increase in area planted to *D. noxia*-resistant cultivars from 22% in 1995 (Marasas et al., 2005) to an estimated 85% in 2000. Modelled after the AUDPC (area under the disease progression curve) (Van der Plank, 1963) which is commonly used to characterize disease intensity over time, for comparison across years, locations or management tactics, the AUDnIPC (area under the *D. noxia* infestation progression curve) was used to visually represent the abundance of *D. noxia* each year. This was calculated as:

AUDnIC = [(1st *D. noxia* count + 2nd *D. noxia* count)/2 × (number of days between the 1st and 2nd count) + (2nd *D. noxia* count + 3rd *D. noxia* count)/2 ×

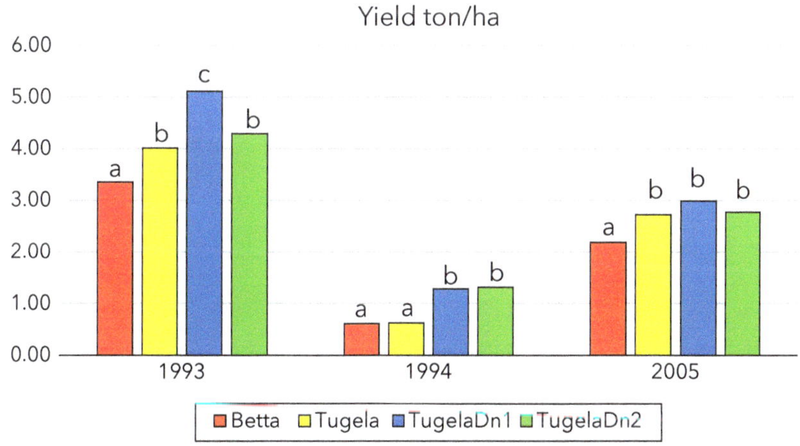

Figure 7 Yield (ton/ha) of susceptible (Betta, Tugela) and resistant (TugelaDn1, TugelaDn2) wheat during 1993, 1994 and 2005, Bethlehem, South Africa.

(number of days between the 2nd and 3rd count) + (3rd *D. noxia* count + 4th *D. noxia* count)/2 × (number of days between the 3rd and 4th count) + …].

Values were plotted to create the curve for each year. Total AUDnIC values from 1995 to 2000 were as follows: 3905.1, 1502.0, 4540.5, 3022.1, 1323.5 and 965.1, respectively. During the first 3 years of the trial, a distinct peak was observed annually in *D. noxia* population growth. In 1995 (Fig. 9a) and 1997 (Fig. 9e), peak *D. noxia* infestation occurred in the second half of October, while that of 1996 (Fig. 9c) occurred in the first half of November. From 1998 to 2000 (Fig. 9g, 9i and 9k), the AUDnIPC curve started flattening out, no longer representing the characteristic sigmoid shape associated with *D. noxia* population growth on susceptible cultivars. The trend in AUDnIPC graphs shows a definite decline from 1995 to 2000. This decline corresponds well with a steady increase in the use of RWA resistant cultivars in the Eastern Free State province (Fig. 8).

Weather conditions, in particular minimum and maximum temperatures and rainfall (Fig. 9b, 9d, 9f, 9h, 9j and 9l), were obviously an influencing factor on aphid abundance. In July of 1996, heavy snowfall occurred (data not shown), combined with cooler than usual temperatures in September. This contributed to the slower increase in *D. noxia* numbers, as well as the shift in the peak from October to November of the AUDnIC graph and an overall lower AUDnIC. By 2000, 8 years after the first *D. noxia*-resistant cultivar was released (Smit et al., 2010), the large-scale use of insecticides had reduced significantly and *D. noxia* was considered a less problematic pest.

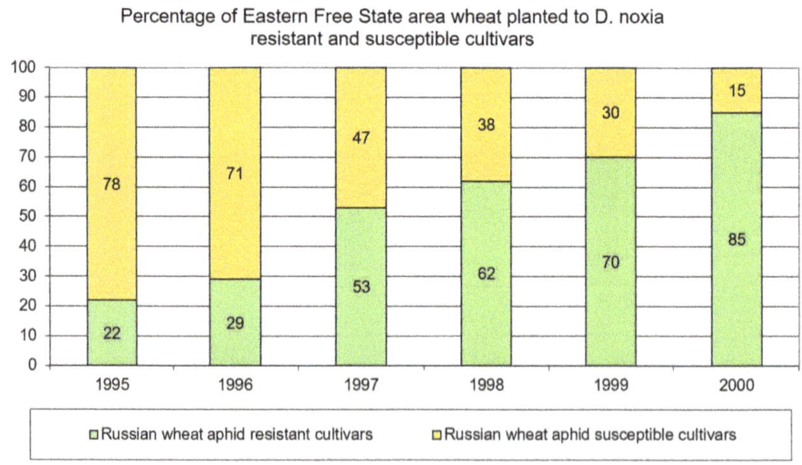

Figure 8 Proportional representation of the area of the Eastern Free State planted to *D. noxia*-resistant and susceptible cultivars.

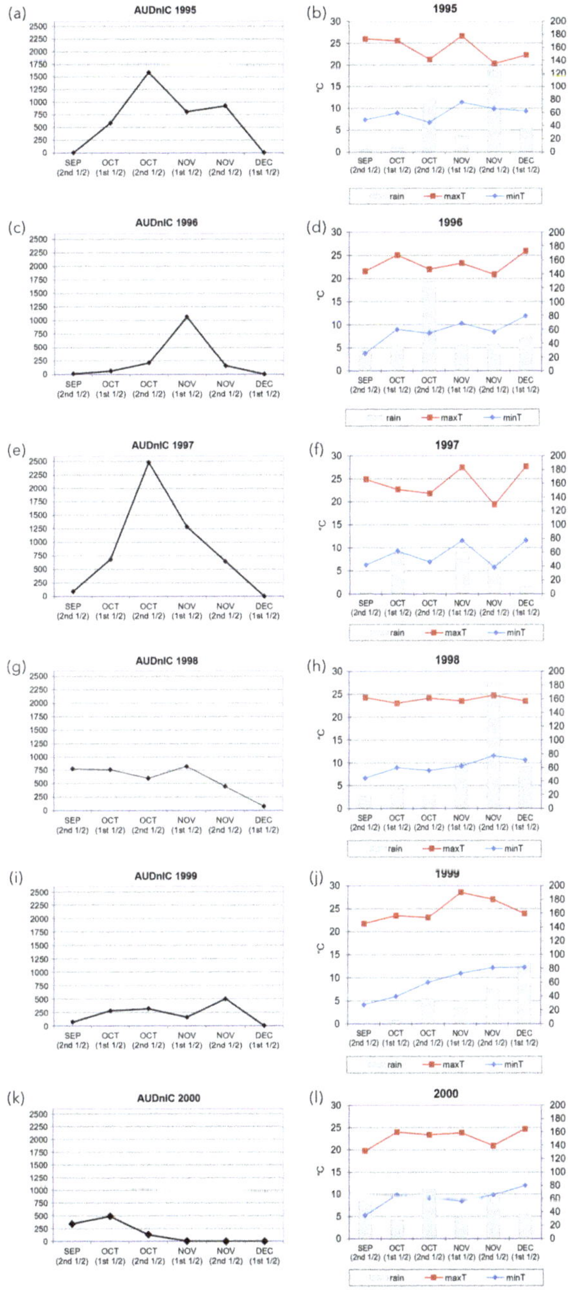

Figure 9 Area under the *D. noxia* infestation progression curve (AUDnIPC) as well as rainfall and minimum and maximum temperatures for the years 1995 (a, b), 1996 (c, d), 1997 (e, f), 1998 (g, h), 1999 (I, j) and 2000 (k, l).

5 Managing *Diuraphis noxia* populations

In the long term, *D. noxia* is best managed proactively using an integrated approach. Annual surveys of the occurrence and incidence of *D. noxia* throughout production regions of barley and wheat are conducted in South Africa (Jankielsohn, 2011, 2016), the USA (Puterka et al., 2007) and Australia (https://imappests.com.au/). These data give producers area-wide information on which they can base their choice of *D. noxia* control option and contribute significantly to an informed, more targeted pest management strategy. The most reliable monitoring of this aphid pest is however done on farm level using proper scouting techniques to determine the presence, exact location and quantity of *D. noxia* on the farm.

An economic impact analysis by Marasas et al. (1997) of the integrated *D. noxia* control program initiated in the Eastern Free State between 1992 and 2000 found that the benefit to society from the yield gains of resistant cultivars alone was between 35% and 39%. Including other cost savings due to the reduction in insecticide use and the efficacy of biological control interventions, the internal rate of return was as high as 49%. With a social opportunity cost of 6.73% at the time, and assuming only domestic benefits, the integrated control strategy of *D. noxia* was a profitable investment.

Multiple different options are available for the control of *D. noxia*, and the most appropriate will depend on the agro-ecological production area and the scenario of *D. noxia* damage which occurs in that area (see Table 1 for examples of different damage scenarios). Control can be broadly classified into pre-emptive or reactive options. Pre-emptive options include cultural and agronomic practices that minimize *D. noxia* presence in the vicinity of crops. They also specifically include activities or practices that promote biological control entities, be they predators, parasitoids or entomopathogens. The use of crops with inherent resistance to the aphid has successfully been used in both barley and wheat. Insecticides can be used proactively or curatively, and in the curative action, they are the most valuable tool for control of damaging levels of *D. noxia* populations.

5.1 Cultural control

Sound cultural practices based on the awareness of *D. noxia* presence and movements on and in the vicinity of the farm can contribute significantly to delaying or reducing the initial infestation in a crop. The prevention or reduction of *D. noxia* carry-over and build-up during the non-crop period can be done by eliminating the so-called green bridge as much as is feasible. This includes the pre-seasonal control of volunteer barley and wheat growing adjacent to fields as well as in the road reserves close to fields while actively supporting

Table 1 Various scenarios under which *D. noxia* reaches damaging population levels in different agro-ecological areas

Production circumstances where damage is caused	How the *D. noxia* infestation plays out
Winter wheat planted in autumn/fall	Idaho, North America, where *D. noxia* entered in 1988 and has been a sporadic pest since 1996: University of Idaho | College of Agriculture and Life Sciences (uidaho.edu) • Emerging winter wheat planted in autumn/fall can become infested from infested volunteer grain or, more often, a very late planting of spring barley or wheat. Damage can be restricted by controlling infested volunteer plants 2 weeks before planting or by planting later. • In most years, D. noxia do not overwinter in high numbers, but when more aphids overwinter (i.e. in mild dry winters), extensive damage can occur very quickly in early spring. Damage can be restricted by spraying to control D. noxia at 5% infestation. Canada, North America: Butts et al. (1997) reported infestation of winter wheat, planted in autumn resulted in wheat with less resilience to winterkill, causing stand reductions of 28% to 35%. *D. noxia* on these plants often do not survive the winter; however, plants that do survive fully recover from the *D. noxia* damage.
Facultative wheat planted in winter	Eastern Free State, South Africa: intermediate or facultative wheat planted in winter (May-July) does not die during the winter from infestation but rather experiences a rapid increase of *D. noxia* populations in spring to early summer. This increase often coincides with a dry period (see the section 'Case study' in this chapter) as the stored soil moisture on which the crop has relied since planting is depleted. Warmer temperatures promote both plant and *D. noxia* growth exacerbating the 'drought' the plants are experiencing, thus causing severe yield loss.
Spring wheat planted in spring or summer	Central Production Region, South Africa: spring wheat planted from August to October under dryland or flood irrigation can become infested from winter plantings, volunteer wheat or alternate hosts. Damage can be restricted by using a seed dressing; however, the cost of this option is often prohibitive due to higher seeding rates used for spring wheat in comparison to those used for facultative wheat. This is no longer seen as a viable practice and does not occur in significant volumes.
Spring barley planted in spring	Ethiopia, Northern Africa: Mulatu et al. (2011) reported that early planting (May) of barley in the Meher rainfall season (June-October) in the eastern parts of South Gondar, Wollo and North Shewa regions results in early infestation of the crop, which is exacerbated by a dry period in June, resulting in major losses if not total crop failure. Farmers now delay planting until late June, using shorter growth period barley genotypes which have a lower potential yield than those planted in May but yield better than damaged June plantings due to the avoidance of *D. noxia* infestation. Forest steppe of the Middle Volga region, Russia: Kaplin et al. (2015) report that monoecious *D. noxia* overwinter on winter wheat as eggs that hatch in spring and spread to additionally infect spring barley and wheat in May. Spring barley yield losses were estimated at 19%.
Facultative or spring wheat planted in summer	Lesotho, Southern Africa: Tolmay et al. (2000) reported that dryland production of facultative or spring wheat at high altitudes (2440 m above sea level) during the summer (planted in October-November) in Lesotho are sporadically infested with *D. noxia* during drought years. This pest is, however, not usually problematic due to summer rainfall and low temperatures prevalent in the Mokhotlong valley where wheat is produced. Infestation of this wheat occurs from *D. noxia* dispersal flights off winter wheat in the neighbouring Eastern Free State province of South Africa.

populations of natural enemies of *D. noxia*. Crop planting should be timed so that vulnerable young crops do not coincide with aphid flights. This can result in severe early infestation, with later re-infestations requiring an additional control intervention. Close observation of the particular conditions (rainfall and temperature) before the crop season and while the crop progresses through its growth stages can give producers a good indication of the potential pest problem for that season. Scouting is imperative for effective management of *D. noxia*.

5.2 Host plant resistance

The deployment of host plant resistance in the form of resistant cultivars is an exceptionally effective contribution to the integrated control of *D. noxia*. Well-adapted cultivars with resistance to *D. noxia* are inherently protected against the pest from crop emergence until physiological maturity. For the development of resistant cultivars, suitable sources of resistance are required to introgress into adapted cultivars.

Genetic resistance to *D. noxia* has been reported from a multitude of accessions encompassing numerous species in the family Gramineae. This includes *Aegilops* spp. (Ghannoum et al., 2016), *A. tauschii* (Lage et al., 2004; El-Bouhssini et al., 2011), *Hordeum bulbosum*, *H. brevisubulatum*, *H. vulgare* (Calhoun et al., 1991; Webster et al., 1991; Robinson and Skovmand, 1992; Araya et al., 2014, among others), *S. cereale*, *T. aestivum* (du Toit, 1987, among others), *T. dicoccon*, *T. monococcum*, *T. tauschii*, *T. turgidum* var. *durum* (Formusoh et al., 1992), × Triticosecale (Scott et al., 1991 among others) as well as wheat-rye translocation (Marais et al.1994) and substitution lines (Crespo-Herrera et al., 2019). Souza (1998) contains a list of studied *D. noxia*-resistant *Triticum* sp., *Hordeum* sp. and related species, which is very useful though not exhaustive.

Many more barley and wheat accessions with varying levels of *D. noxia* resistance are known following the systematic screening of >60 000 entries from the United States Department of Agriculture-Agriculture Research Service National Small Grain Collection in Aberdeen Idaho by the USDA - ARS in Stillwater, OK (Webster, 1991). This highly successful project, which initially utilized USA biotypes RWA1 and later RWA2, rapidly expanded the available R-donor options for both barley and wheat from the initial four wheat genotypes identified in South Africa by du Toit (1987; 1988; 1992). Screening was also conducted by CIMMYT. Two RWA nurseries, namely the 2007 RWASN (Russian wheat aphid screening nursery) ex Jacob Lage (El Batan, Mexico) and the 2011 FWWRWARN (facultative winter wheat Russian wheat aphid resistance nursery) ex Alexey Morgunov (TCI, Ankara), were made available worldwide.

Using the Focussed Identification of Germplasm Strategy, El-Bouhssini et al. (2011) identified 12 additional *D. noxia*-resistant sources.

These genotypes can be accessed by searching the databases of National and International germplasm collections where they are accessioned (see section 'Where to look for further information' in this chapter). The challenge, however, now remains to identify those accessions among the numerous available, with unique resistance genes. To date, in excess of 10 *Dn* genes have been named and markers for a number of these genes have been published (see review Tolmay et al., 2020). Regrettably, there is a measure of confusion regarding the naming and chromosomal location of, for example, *Dn5* and *DnUnknown* on chromosome 7D (Tolmay et al., 2020) as well as use of one name, i.e. *Dnx*, for two distinct loci (Khan et al., 2009). This is likely due to the complication that resistance genes are identified and characterized using *D. noxia* biotypes, while biotypes are characterized using resistant accessions. Furthermore, the fact that different biotypes 'recognize' different R-genes in the same germplasm (Khan et al., 2009; Tolmay et al., 2013; Xu et al., 2015; Selladurai, 2016) causes further confusion. A recent focus has been the phenotypic identification of R-donor accessions with multi-biotype resistance (Tolmay and Booyse, 2017; Xu et al., 2020; Bapela and Tolmay, 2021) to identify unique and effective resistance to utilize in breeding programmes.

There is growing evidence that *D. noxia* resistance as a trait is polygenic and that multiple loci of genetic control occur within single accessions. As mentioned above, breeding line KS94H871, containing *Dnx* from PI 220127, was thought to contain a single dominant gene. This line is resistant to both USA biotypes RWA1 and RWA2 (Khan et al., 2009); however, when examined closely, it contained two loci encoding resistance to RWA1, but only one locus encoding resistance to RWA2. The same phenomena was found by Tolmay et al. (2013) and Kisten et al. (2020) with respect to *Dn4* and *Dny* lines developed in the USA where they were unvaryingly resistant to USA biotype RWA1 but segregated when evaluated with South African biotypes RWASA1, RWASA2, RWASA3 and RWASA4. This aspect will be touched on again under the section 'Biotypes' in this chapter. A clearer, more detailed understanding of the genetic control of *D. noxia* genes is needed.

Notwithstanding the genetic control of resistance, resistant barley cultivars released in the USA, i.e. Burton (Bregitzer et al., 2005), RWA-1758 (Bregitzer et al., 2008), Sidney (Mornhinweg et al., 2009) and Stoneham (Mornhinweg et al., 2012), have revived feed barley production in arid areas of the Western Great Plains of the USA, which had been lost to *D. noxia* damage following the incursion of *D. noxia* there (Sotelo et al., 2014). In Ethiopia, Belay and Stauffer (2007) report the usefulness of resistant barley genotypes citing a similar recovery of production areas. In the period from 1992 to 2020, 43 *D. noxia*-resistant wheat cultivars were released in South Africa (Smit et al., 2010;

Tolmay and Booyse, 2020) and one named Puseletso released in Lesotho (Moremoholo and Purchase, 1999). In the USA, following the first resistant cultivar Halt released in 1994 (Quick et al., 1996), numerous cultivars including Ankor, Bond CL, Hatcher, Prowers99, Ripper, Stanton, Thunder CL and Yumar, among others, were commercialized.

Resistant cultivars dramatically reduce the number of aphids present in the field by inhibiting aphid growth and reproduction (Tolmay and Maré, 2000), leading to a reduction in both the percentage of infested tillers and the number of *D. noxia* per infested tiller. Furthermore, the leaves of the resistant cultivars do not roll closed, leaving *D. noxia* exposed on the leaf surface resulting in more effective predation and parasitism. In South Africa, to a large degree, the use of insecticides was discontinued once resistance became widely used. Field trials showed that not all cultivars were equally resistant (Tolmay and Van Deventer, 2005) but that it was only sometimes profitable to apply insecticide to resistant cultivars (Maré, 2003).

The downside of resistance is, however, the development of resistance breaking biotypes which have been reported from various countries. This is in contrast to the fact that no insecticide resistance has ever been reported in *D. noxia* despite the fact that a lot of chemical control of this pest occurs throughout the world. Cultivars containing natural resistance to this pest remain a very useful control option as these cultivars can negate the need for insecticide application when infested by resistance-sensitive biotypes. Furthermore, some resistant cultivars can still yield more than susceptible wheat lines when infested with virulent *D. noxia* biotypes, although this increased yield often still constitutes an economically significant loss.

5.3 Biological control

Classic biological control is a major investment in time, labour and funding when one considers the planning, collection, culturing, evaluation, releasing and monitoring of predators, parasitoids and entomopathogens of an invasive aphid species. Notwithstanding this input, classic biological control is one of the best 'insurance policies' to invest in regarding pest aphid control, with a potential long-term and adaptable impact. If the biocontrol agent establishes in the agro-ecosystem and surrounds, albeit at low densities there is an increase in mortality of the pest species and the prospect that these natural enemies will adapt to the new areas and circumstances, associating with the pest as it occupies its new territory or as circumstances change.

D. noxia occurs regularly at low levels in many localities around the world. Some examples of this phenomena can be found in Europe for example in the Czech Republic (Stary, 1996; Havelka et al., 2013) and Hungary (Basky, 1993); in the Middle East in countries like Yemen (Stary and Erdelen, 1982) and Israel (Tanigoshi et al., 1995); in Asia for example in China (references in Zhang, 2012); in Africa in specific areas in South Africa (Hatting et al., 2000; Jankielsohn, 2011, 2016) and Morocco (Tanigoshi et al., 1995); in the Americas in areas of Chile and Argentinia (Cluá et al., 2004); in areas outside of Colorado in the USA (Puterka et al., 2007) and even in Australia (van Helden et al., 2018).

While it is generally accepted that climate- and/or host plant-related factors are influential in these low populations of *D. noxia*, natural enemies of this pest undoubtedly contribute significantly to suppressing *D. noxia* populations. For the most part, this is significant where the aphids are either endemic or have been present for many years and biological control releases have been made (Hopper et al., 1989; Ward et al., 2020). Worldwide, a wide variety of natural enemies have been reported to attack *D. noxia*, including predators, parasitoids and entomopathogens (Hopper et al., 1989; Ward et al., 2020, among others).

A classical biological control program against *D. noxia* that was undertaken from 1987 to 1994 in the USA was one of the largest and most intense, including multiple agencies and in excess of 120 participants both in the USA and abroad (Prokrym et al., 1998). Multiple papers report on the successful establishment of biological agents. Some selected examples of successful establishments follow.

Prokrym et al. (1998) documents the release of some 15.5 million *D. noxia* natural enemies (individuals of various species) from 1987 to 1994 throughout the cereal-producing states of the USA where *D. noxia* occurred. A comprehensive list of species released is included in that paper. Recovery data (Brewer et al., 2001) of *Aphelinus albipodus* and *Diaeretiella rapae* showed that these species occurred in barley as well as in grassland sites adjacent to the crop a few years post release. Almost a decade later, Brewer et al. (2019) reported that many recovered natural enemies of *D. noxia* were indigenous species or long-established residents that had been introduced to control other aphid pests in the past. This phenomenon of expanding host ranges contributes to what is known as regional saturation, where a greater number of natural enemy species from more diverse taxa and wider seasonal occurrence are recorded for multiple aphid pests, in this case for *D. noxia*, greenbug (*Schizaphis graminum*) (Rondani) and sugarcane aphid (*Melanaphis sacchari*) (Zehntner).

It appears that while predators are regularly found in association with *D. noxia*, parasitoid wasps provide better control. In the Pacific North West, Tanigoshi et al. (1995) document the introduction and release of natural enemies from Morocco and the Middle East for *D. noxia* control in Washington State. In Idaho, Bosque-Pérez et al. (2002) documented numerous biocontrol agents, particularly parasitoids, occurring on both susceptible and resistant wheat genotypes. Among other biocontrol agents (Prinsloo, 1990), an introduction of *Aphelinus hordei* (initially named, *Aphelinus* sp. nr. *varipes*) collected in the Ukraine from *D. noxia* was made into South Africa (Prinsloo, 1998). This introduction was deemed successful when *A. hordei* were successfully recovered from sites between 100 km and 200 km from the release sites and their identity verified by polymerase chain reaction (Prinsloo et al., 2002). Reporting a study of parasitism of *D. noxia* in Australia, Heddle et al. (2020) comment on the success of the pre-emptive release of parasitoid wasps, which contributed to limiting the *D. noxia* invasion in Australia. The global move away from pesticides to options with better environmental impact opens new opportunities for augmentative and conservation biological control.

The contribution natural enemies make reaches further than simply that of suppressing the pest population. They play a critical role in the tri-trophic interaction that occurs between the host plant, the pest, in this case *D. noxia,* and the biological control agent. An important aspect of this interaction is that the different trophic levels may act in a mutualistic way, although this is not always the case (Prinsloo, 2006). Natural enemies of *D. noxia* support host plants by reducing *D. noxia* abundance, while the plants may benefit the natural enemies by making *D. noxia* in some way more vulnerable to the natural enemies. A good example is the flat leaves associated with resistant cultivars that make it possible for adults and larvae of larger predators like ladybirds to prey on *D. noxia* as they are easier to access. A field study by Prinsloo (2006), on the combination of *D. noxia*-susceptible and -resistant wheat cultivars and two parasitoids of this aphid, is another good example. On the susceptible cultivar Betta, *D. noxia* was highly parasitized while a positive interaction occurred on resistant Gariep, resulting in the enhancement of the resistance. Slightly lower control of *D. noxia* was recorded on resistant SST 333 in the presence of *A. hordei*.

Olfactometer studies (Prinsloo, 2006; Prinsloo et al., 2007) determined that the volatile profiles emitted by infested Betta plants and resistant Elands and SST 333 plants differed in terms of the concentration and composition of volatiles. These differences resulted in behavioural differences between parasitoid species; i.e. *A. hordei* could not distinguish infested from uninfested Elands, while *Diaeretiella rapae* did. *Diaeretiella rapae* could, however, not distinguish infested from uninfested SST 333 while *A. hordei* did. The difference was not only limited to infested and uninfested resistant wheat. *Aphelinus hordei* could not distinguish between infested Betta and Elands, while *D. rapae* significantly preferred infested susceptible Betta to infested resistant Elands. Parasitoids, therefore, responded to different volatiles from the same cultivars. This means that both *A. hordei* and *D. rapae* use different volatiles or volatile combinations from the same plant-herbivore combination in host habitat location.

Entomopathogens for *D. noxia* control occur naturally in many parts of the world and can decimate aphid populations when epizootics occur. The potential of entomopathogenic fungi for the management of *D. noxia* has been illustrated by various authors (Feng et al., 1990; Vandenberg, 1996; Hatting et al., 1999, 2000, 2004; Knudsen and Schotzko, 1999; Vandenberg et al., 2001), with some studies including the interaction of the entomopathogens and parasitoid wasps (Mesquita and Lacey, 2001). Quisenberry and Peairs (1998) contain further reading regarding fungal pathogens and ice-nucleating bacteria for the control of RWA.

Furthermore, when resistant plants are used, biocontrol agent-mediated pest mortality is likely indiscriminate regarding the pest biotype, thus assisting in reducing selection pressure for the development of new biotypes. Altogether, as reported by Wickremasinghe and Van Emden (1992), there is a potential for more beneficial interactions between biological control agents and partial plant resistance, suggesting that there is a disadvantage in seeking a level of plant resistance greater than necessary when other restraints (natural enemies) are present in a pest management system.

5.4 Insecticides

Preventative application of seed dressings is an option that is available in most areas where *D. noxia* occurs. Its use, however, is a cost-carrying decision that is made before the pest infests the crop, and these products are usually only effective for a limited part of the season. In South Africa's Free State wheat production area, where *D. noxia* was most problematic, studies from 1997 to 2002 (Maré, 2004) showed that the use of imidacloprid seed dressing on both susceptible and resistant cultivars did not necessarily result in a yield increase compared to a spray treatment. Seed-treated wheat did nonetheless successfully maintain hectolitre mass, resulting in a better grade and price for the wheat (Maré 2004). Seed treatments have been used widely as a pre-emptive action in the recent incursion of *D. noxia* into Australia (Ward et al., 2020). Imidacloprid is a good example of an insecticide option, which is currently being withdrawn or limited in use in various countries due to environmental concerns regarding non-target species.

Due to the quick reaction time that is possible with the use of insecticidal sprays, economic threshold values on susceptible crops were defined rapidly following incursion in South Africa (du Toit, 1986). These were later refined to account for the yield potential of the crop (Nienaber and Kilian, 2021). In the USA, thresholds that were initially used were later refined (Archer and Bynum, 1992) to better suit the dynamics of changing cost of insecticide and crop value (review by Legg and Archer, 1998) and much the same is currently happening in Australia (Ward et al., 2020; van Helden et al., 2022). Almost all threshold values for the chemical control of *D. noxia* are for use on susceptible cultivars and in most instances give separate values for infestation in young plants (before tillering) and a lower threshold for plants infested during tillering. Threshold values in barley are generally lower than those in wheat. Management of *D. noxia* with insecticides is the only remedial option available to large-scale wheat and barley farmers when severe infestation occurs and a large number of insecticides are registered throughout the world for this purpose.

In Table 2, countries about which registrations for chemical control of *D. noxia* could be found are listed alphabetically. It is of the utmost importance to confirm with the proper authority which pesticides hold current registration for the area of interest. The pesticide label should be consulted for detailed information before application of these remedies. Certain references are somewhat dated, and care should be taken to verify that the active ingredients are still registered for use in the applicable country before application.

Concerns have been raised about the potential for *D. noxia* to develop resistance to insecticides by various authors (Nicholson et al., 2015; Umina et al., 2020; Ward et al., 2020). To mitigate this potential problem, Umina et al. (2020) recommend that insecticides should be used taking cognisance of three important principles relating to their judicious use, namely:

- only applying chemicals when the pest infestation warrants it;
- avoiding the application of broad-spectrum formulations as much as is practical; and
- rotation of formulations whereby the same mode of action is not applied across consecutive generations of the target pest.

Despite widespread use of insecticides for the control of *D. noxia* in South Africa over a span of more than four decades, no insecticide resistance has been reported from the Eastern Free State production region of South Africa to date. Internationally, no reports of insecticide-resistant *D. noxia* could be found to include in this chapter. This is in contrast to the numerous host plant-resistant biotypes of *D. noxia* which have been reported from various countries.

6 *Diuraphis noxia* biotypes

D. noxia biotypes have until now mostly been defined by the resistant or susceptible reaction of entries in a set of wheat accessions purported to contain specific *Dn*-resistance gene(s). Using this methodology, but not always the same set of differential lines, resistance-breaking biotypes USA RWA2 (Haley et al., 2004); RWASA2 (Tolmay et al., 2007); USA RWA6, RWA7, RWA8 (Weiland et al., 2014); RWASA4 (Jankielsohn, 2014a) and RWASA5 (Jankielsohn, 2019), which have rendered commercial resistant cultivars susceptible in farmers' fields, have been defined. In addition to those biotypes characterized by their ability to inflict severe damage on resistant cultivars in commercial use, numerous other biotypes have also been identified using a differential set of lines. These aphids were collected from a variety of hosts during surveys to determine the distribution and host range of *D. noxia* (Belay et al., 2004; Burd et al., 2006; Jankielsohn, 2011; Puterka et al., 2007). Again different accessions were used in different studies. Harris-Shultz et al. (2019) provide a concise review of the

Table 2 Countries with registrations for the chemical control of *D. noxia* in small grains

Countries[a] where insecticides are registered for *D. noxia* control	Reference
Australia	Ward et al. (2020), APVIMA (2021)
Argentina	CASAFE (2022)
Canada	Health Canada (2021)
Ethiopia	Ethiopian Ministry of Agriculture (2021)
European and Mediterranean Plant Protection Organization	List of databases on registered plant protection products in the EPPO (2022) region
Kenya	Pest Control Products Board (2010)
South Africa	AVCASA (2018)
Turkey	Turkish Ministry of Agriculture (2021)
United States of America[b]	Crop Data Management System (2022)
• California	Godfrey and Barlow (2019)
• High Plains	Peairs et al. (2018)
• Nebraska	Hein and Thomas (2006)
• Pacific Northwest	Rashed and Tanner (2021)
• South Dakota	Varenhorst et al. (2020)
• Texas	Bynum et al. (2020)
• Utah	Hodgson and Karren (2008)

[a]Information could not be found for Chile, countries in the Middle East and Western Asia, Russia or China due to language constraints.
[b]In the USA, the CDMS Advanced Search database contains the official list of registered insecticides. State- or region-specific information can be found in the additional references.

designation of *D. noxia* biotypes in the USA, highlighting those of agricultural significance and those that occur on other hosts.

Several papers report on a difference in resistance reaction of accessions tested with *D. noxia* collected from different areas of the same county or in different countries, i.e. Hungary (Basky, 2003) and Chile (Smith et al., 2004), Kenya (Malinga et al., 2007), Argentina (Almaráz et al., 2011), USA (Lui et al., 2011) and Ethiopia (Araya et al., 2014). Other studies have utilized molecular methods to quantify genetic diversity of *D. noxia* (Malinga et al., 2007; Lui et al., 2010; Cui et al., 2012; Gong et al., 2012; Zhang et al., 2012). Additionally, the genome of *D. noxia* itself has been sequenced in two separate studies (Nicholson et al., 2015; Burger and Botha, 2017) and that of its *Buchnera* endosymbionts examined in various ways (Baumann et al., 1998; Thao et al., 1998; Swanevelder et al., 2010; Anathakrishnan et al., 2014). None of these findings led to clear-cut characterization of the biotypes tested.

Cooper et al. (2010), describing the method for the collection of *D. noxia* saliva, enabled the examination of saliva from two *D. noxia* biotypes that differ in virulence to wheat (Nicholson et al., 2012). Again, although large differences were identified between biotypes, the results could not be used to characterize them. Studies by Campillo et al. (2015) and Luna et al. (2018) have highlighted just how complex the interaction between the *D. noxia* and its host may be. These papers identified *Erwinia* and other bacterial genera associated with *D. noxia*. During feeding, the bacteria appear to contribute to the severe damage symptomology associated with *D. noxia*. Data however showed that there was little difference in the overall bacterial composition in the proteome between the biotypes used (RWA1 and RWA2) in this study. Epigenetics or the methylation status of *D. noxia* has received attention of late (Gong et al., 2012; Breeds et al., 2018; du Preez et al., 2020) and may possibly show promise as a method to characterize biotypes.

The various reproductive paths available to *D. noxia* result in multiple populations which are currently quite difficult to characterize as different biotypes. The particular concern is that a differential set of R-genes is a function of the biotype used to develop it. Additionally, most genotypes developed as differential lines are considered to contain the R-gene reported for the R-donor genotype used in their development. This assumption is not necessarily correct in all cases. Furthermore, the precise R-gene(s) in each line is seldom properly verified before using the differential set to define biotypes. Despite this, the evaluation of clonal colonies using differential sets (Jankielsohn, 2014b) has proven to be the most reliable method of biotype identification, at least within each specific differential set, to date. It is cumbersome and relatively expensive in terms of the culture of clone colonies and the phenotypic evaluation of plants, but currently, the more sophisticated methods that could be employed to differentiate biotypes are not yet sufficiently accurate. Until the genetic control of *D. noxia* resistance in wheat is more accurately defined, it may be worthwhile to refrain from associating *Dn*-gene designations with cultivars or lines in the differential set simply because they were developed from the same R-donor. Extending the current differential set to include diagnostic genotypes from various countries, irrespective of purported genes they contain, would enable the accurate comparison of biotypes from different parts of the world. The specific genes in each line should be researched and the differential set modified as required when the accurate genetic information becomes available.

7 Decision support for integrated *Diuraphis noxia* control

There is not a common decision support recommendation for controlling *D. noxia* in all cases. Rather there are a set of principles that should be followed to ensure that this pest is constrained throughout its annual activities. A clear understanding of the population dynamics of *D. noxia* in the specific region of interest is vital. Risk factors that encourage an increase in populations and how they may play out in each season should be known, monitored and then used to make informed decisions. Integrated control implies the simultaneous use of multiple, different control options and as many as are appropriate in the specific region of interest should be employed.

The basis of control lies with sustainable practices that impede *D. noxia* population growth, particularly in specific critical stages of the pest annual cycle. An area-wide, healthy natural enemy cadre of various mixes of parasitoids, predators and entomopathogens working in concert within a particular habitat type, together with practices like controlling the green bridge between crop seasons and ensuring that fields are clear of volunteer cereals and other grass hosts, gives a good foundation for integrated control.

In terms of the crop, it is essential to utilize practices and inputs to ensure optimum plant development and growth. appropriate planting date, good adaptation to the production region and cultivars with resistance should be planted taking cognisance of the prevalent biotype that occurs in the region. If it is not possible to adapt planting date to avoid early infestation, the use of insecticidal seed dressing can be considered. These seed dressings should, however, not be used indiscriminately. Should *D. noxia* populations reach threshold levels during the season, insecticide sprays should be applied promptly and in a manner that optimizes their efficacy.

8 Conclusion

Recorded reports tell us that *Diuraphis noxia* has periodically afflicted the production of small grain crops such as barley and wheat for more than a century. This may be an underestimate. During the past century, Russian wheat aphid has spread across the world to all continents that produce wheat and barley. Effective, albeit imperfect control methods have been developed and deployed against this pest in many different localities around the world. This has saved many millions of tons of food for human and animal consumption.

Though immense advances have been made in our understanding of Russian wheat aphid, there is still a good deal to be studied. The biggest challenge still lies ahead. It is to integrate knowledge in the form of cutting edge scientific methodologies and developments in artificial intelligence with the complex biological interactions that take place on a daily basis between this small green aphid, its various host plants, including crop species and other grasses, and the environments in which they occur.

As I see it, the goal for scientists in future will be to enable the sustainable control of this pest, amongst others, whilst preserving healthy ecosystems and biological diversity to sustain the kind of world we would like our grand-childrens' children to live in.

9 Future trends in research

In order to provide a rapid and effective reaction to outbreaks of *D. noxia* in future, there are a number of research questions that should be answered. The first and most important of these is to ensure that regions where *D. noxia* is problematic have access to all the information required to manage the pest. This includes:

- population dynamics of D. noxia over the entire year including alternate hosts and natural enemies. Weak links or opportunities for cultural control should be identified for exploitation;
- information of the distribution and occurrence of different biotypes of D. noxia; and
- valid threshold values for the region regarding chemical intervention.

In a more scientific context and to foster global and regional collaboration, particularly in the face of climate change, the following research areas will likely be important:

- the identification of a quick and reliable method to clearly differentiate D. noxia biotypes, particularly those from different countries;
- clarification of the genetic control of resistance in wheat to enable the development of virulence/avirulence formulae together with the biotype classification similar to those for the rust diseases. This will enable the formulation and management of deployment strategies for host plant resistance genes;
- studying the role of the plant pathogenic bacteria present in D. noxia saliva (Luna et al., 2018) and the potential disruption of this phenomenon as control method; and
- identifying robust molecular markers for specific D. noxia resistance genes in wheat.

10 Where to look for further information

Due to the widespread distribution of *D. noxia* throughout the world, research has been undertaken in a great many different locations. Unfortunately, some valuable studies are less accessible due to language barriers, difficulties in obtaining the papers because of inadequate indexing, limited journal availability or lack of computerized literature retrieval facilities for particularly pre-1981 literature. Where possible, this information has been included in this review. The resources listed below provide a good foundation:

- Poprawski, T. J., Underwood, N. L., Mercandier, G., et al. (1992), Diuraphis noxia (Kurdjumov) ~ A Bibliography on the Russian wheat aphid 1886-1992'.

United States Department of Agriculture - Agricultural Research Service. Plant, Soil and Nutrition Laboratory, Ithaca, NY 14853, USA.

- Quisenberry, S. S. and Peairs, F. B. (1998), Response model for an introduced pest - The Russian wheat aphid. Thomas Say Publications in Entomology: Proceedings. Entomological Society of America, Lanham, Maryland, USA. ISBN 0-938522-68-X.
- Data sheet report for Diuraphis noxia (Russian wheat aphid), Crop Protection Compendium. https://www.cabi.org/isc/datasheet/9887.

Databases of Germplasm Collections:

- Germplasm Resource Information Network, GRIN-Global http://www.grin-global.org/.
- Plant Genetic Resources for Food and Agriculture https://www.genesys-pgr.org/.
- USDA-ARS National Plant Germplasm System https://npgsweb.ars-grin.gov.

11 References

Aalbersberg, Y. K., Toit, F. D., Westhuizen, M. C. V. D. and Hewitt, P. H. (1987). Development rate, fecundity and lifespan of apterae of the Russian wheat aphid, *Diuraphis noxia* (Mordvilko) (Hemiptera: Aphididae), under controlled conditions. *Bulletin of Entomological Research* 77(4), 629–635. https://doi.org/10.1017/S000748530001213X.

Aalbersberg, Y. K., van der Westhuizen, M. C. and Hewitt, P. H. (1989). Characteristics of the population build-up of the Russian wheat aphid *Diuraphis noxia* and the effect on wheat yield in the eastern Orange Free State. *Annals of Applied Biology* 114(2), 231–242. https://doi.org/10.1111/j.1744-7348.1989.tb02100.x.

Ahmadi, A. A. and Sarafrazi, A. M. (1992). Distribution and natural enemies of Russian wheat aphid, *Diuraphis noxia* (Mordvilko) (Homoptera: Aphididae) in the Fars Province. *Iran Agricultural Research* 11, 105–118.

Akhtar, L. H., Hussain, R. M., Iqbal, A. M., Marghub, A. and Tariq, A. H. (2010). Losses in grain yield caused by Russian wheat aphid *Diuraphis noxia* (Mordvilko). *Sarhad Journal of Agriculture* 26(4), 625–628. Available at: http://aup.edu.pk/sj_pdf/LOSSES%20IN%20GRAIN%20YIELD%20CAUSED%20BY%20RUSSIAN.pdf (accessed 13/10 2021).

Almaráz, L. B., Bellone, B. M., Chidichimo, H. O., et al. (2011). Evaluation by principle component analysis of the behaviour of three Argentinean isolations of *Diuraphis noxia* (Homoptera: Aphidoidea) on eight wheat cultivars. *Interciencia* 36(4), 286–290. (In Spanish).

Anathakrishnan, R., Sinha, D. K., Murugan, M., Zhu, K. Y., Chen, M., Zhu, Y. C. and Smith, C. M. (2014). Comparative gut transcriptome analysis reveals differences between virulent and avirulent Russian wheat aphids, *Diuraphis noxia*. *Arthropod-Plant Interactions* 8(2), 79–88. https://doi.org/10.1007/s11829-014-9293-4.

APVIMA (2021). Public chemical registration information system database. Australian Pesticides and Veterinary Medicines Authority. Available at: https://portal.apvma.gov.au/pubcris search 'Russian wheat aphid'.

Araya, A., Belay, T. and Hussein, T. (2014). Variation between Ethiopian and North America barley varieties (*Hordeum vulgare*) in response to Russian wheat aphid (*Diuraphis noxia*) populations. *Journal of Insect Science* 14, 40. Available at: http://www.insectscience.org/14.40.

Archer, T. L. and Bynum, E. O. Jr. (1992). Economic injury level for the Russian wheat aphid (Homoptera: Aphididae) on dryland winter wheat. *Journal of Economic Entomology* 85(3), 987-992.

ARS-USDA (1963). *Cooperative Economic Insect Report: Insects Not Known to Occur in the United States, 13, 150-161 of Series*. Plant Pest Control Division, Agricultural Research Service, United States Department of Agriculture.

AVCASA (2018). *Insecticides*. Association of Veterinary and Crop Associations of South Africa. Available at: http://www.croplife.co.za/images/croplife/initiatives/Insecticides.pdf.

Bapela, T. M. and Tolmay, V. L. (2021). Evaluation of Russian wheat resistance sources with the spectrum of South African *Diuraphis noxia* biotypes. *Crop Science* 62(2), 564-574, http://doi.org/10.1002/csc2.20656.

Basky, Z. (1993). Incidence and population fluctuation of *Diuraphis noxia* in Hungary. *Crop Protection* 12(8), 605-609.

Basky, Z. (2003). Biotypic and pest status differences between Hungarian and South African populations of Russian wheat aphid, Diuraphis noxia (Kurdjumov) (Homoptera: Aphididae). *Pest Management Science*. https://doi.org/10.1002/ps.750.

Baugh, B. A. (1991). Influence if population density and plant water potential on Russian wheat aphid (Homoptera: Aphididae) alate production. MSc thesis, Science Faculty, Texas Tech University, Lubbock, TX.

Baumann, L., Baumann, P. and Moran, N. A. (1998). The endosymbiont (*Buchnera*) of the aphid *Diuraphis noxia* contains all the genes of the tryptophan biosynthetic pathway. *Current Microbiology* 37(1), 58-59.

Belay, T., Smith, C. M. and Stauffer, C. (2004). Biotypic status of the Russian wheat aphid, *Diuraphis noxia*, in Ethiopia. *Mitt. Dtsch. Ges. Allg. Angew. ENT* 14, 133-136.

Belay, T. and Stauffer, C. (2007). Low genetic variability among worldwide populations of the Russian wheat aphid (RWA), *Diuraphis noxia* (Mordvilko). *African Crop Science Conference Proceedings*, (vol. 8), pp. 621-626.

Belay, T. and Stauffer, C. (2011). Biotypic and genetic variability in the Russian wheat aphid (*Diuraphis noxia* (Mordvilko)) (Homoptera: Aphididae). In: Mulatu, B. and Grando, S. (Eds). *Barley Research and Development in Ethiopia. Proceedings of the 2nd National Barley Research and Development Review Workshop*. 28-30 November 2006, HARC, Holetta, Ethiopia. ICARDA. Aleppo, Syria, pp. xiv-391.

Bellone, B. and Almaraz, L. B. (1995). The Russian wheat aphid a new cereal pest in Argentina. *Proc. 3rd Latin Am. Cong. Entomol., Santiago*. 18-19.

Bosque-Pérez, N. A., Johnson, J. B., Schotzko, D. J. and Unger, L. (2002). Species diversity, abundance, and phenology of aphid natural enemies on spring wheats resistant and susceptible to Russian wheat aphid. *BioControl* 47(6), 667-684. Available at: https://doi.org/10.1023/A:1020579507817.

Botha, C. E. J., Sacranie, S., Gallagher, S. and Hill, J. M. (2017). Russian wheat aphids: breakfast, lunch, and supper: feasting on small grains in South Africa. *South African Journal of Botany* 109, 154-173.

Breeds, K., Burger, N. F. V. and Botha, A. M. (2018). New insights into the methylation status of virulent *Diuraphis noxia* (Hemiptera: Aphididae) biotypes. *Journal of

Economic Entomology 111(3), 1395–1403. Available at: https://doi.org/10.1093/jee/toy039.

Bregitzer, P., Mornhinweg, D. W., Hammon, R., Stack, M., Baltensperger, D. D., Hein, G. L., O'Neill, M. K., Whitmore, J. C. and Fiedler, D. J. (2005). Registration of 'Burton' barley. *Crop Science* 45(3), 1166–1167.

Bregitzer, P., Mornhinweg, D. W., Obert, D. E. and Windes, J. (2008). Registration of 'RWA 1758' Russian wheat aphid-resistant spring barley. *Journal of Plant Registrations* 2(1), 5–9.

Brewer, M. J., Nelson, D. J., Ahern, R. G., Donahue, J. D. and Prokrym, D. R. (2001). Recovery and range expansion of parasitoids (Hymenoptera: Aphelinidae and Braconidae) released for biological control of *Diuraphis noxia* (Homoptera: Aphididae) in Wyoming. *Environmental Entomology* 30(3), 578–588. Available at: https://doi.org/10.1603/0046-225X-30.3.578.

Brewer, M. J., Peairs, F. B. and Elliott, N. C. (2019). Invasive cereal aphids of North America: ecology and pest management. *Annual Review of Entomology* 64, 73–93. Available at: https://doi.org/10.1146/annurev-ento-011118-111838.

Burd, J. D., Butts, R. A., Elliot, N. C. and Shufran, K. A. (1998). Seasonal development, overwintering biology and host plant interactions of Russian wheat aphid (Homoptera: Aphididae) in North America. In: Quisenberry, S. S. and Peairs, F. B. (Eds). *Response Model for an Introduced Pest – the Russian Wheat Aphid. Thomas Say Publications in Entomology: Proceedings, Entomological Society of America*, Lanham, MD, pp. 12–30.

Burger, N., and Botha, A.-M. (2017). Genome of Russian wheat aphid an economically important cereal aphid. *Standards in Genomic Sciences* 12(1), 1–12, https://doi.org/10.1186/s40793-017-0307-6.

Bapela, T. M. and Tolmay, V. L. (2021). Evaluation of Russian wheat resistance sources with the spectrum of South African *Diuraphis noxia* biotypes. *Crop Science* 62(2), 564–574, http://doi.org/10.1002/csc2.20656.

Butts, R. A., Thomas, J. B., Lukow, O. and Hill, B. D. (1997). Effect of fall infestations of Russian wheat aphid (Homoptera: Aphididae) on winter wheat yield and quality on the Canadian prairies. *Journal of Economic Entomology* 90(4), 1005–1009. Available at: https://doi.org /10.1093/jee/90.4.1005.

Bynum, E., Knutson, A., Reed, B. and Gonzales, J. D. (2020). *Managing Insect and Mite Pests of Texas small grains*. Texas A&M Agrilife Extension, ENTO-084. Available at: http://texaswheat.org/wp-content/uploads/2020/03/Wheat-Pest-Guide-ENTO-084_final.pdf.

Calhoun, D. S., Burnett, P. A., Robinson, J. and Vivar, H. E. (1991). Field resistance to Russian wheat aphid in Barley. I. Symptom expression. *Crop Science* 31(6), 1464–1467.

Campillo, T., Luna, E., Portier, P., Fischer-Le Saux, M., Lapitan, N., Tisserat, N. A. and Leach, J. E. (2015). *Erwinia iniecta* sp. nov., isolated from Russian wheat aphid (*Diuraphis noxia*). *International Journal of Systematic and Evolutionary Microbiology* 65(10), 3625–3633. Available at: https://doi.org/10.1099/ijsem.0.000466.

CASAFE (2022). Câmara de Sanidad agropecuaria e fertilizantes. (In Spanish). Available at: https://guiaonline.casafe.org/index.php/ms_session_manager.

Castro, A. M., Vasicek, A., Ramos, S., Worland, A., Suárez, E., Muñoz, M., Giménez, D. and Clúa, A. A. (1996). Different types of resistance against greenbug, *Schizaphis graminum Rond, and the Russian Wheat Aphid, Diuraphis noxia Mordvilko, in Wheat. Plant Breeding* 118(2), 131–137. Available at: https://doi.org/10.1046/j.1439-0523.1999.118002131.x.

Castro, A. M., Vasicek, A., Ramos, S., Martin, A., Martin, L. M. and Dixon, A. F. G. (1998). Resistance against greenbug, Schizuphis graminum Rond., and Russian wheat aphid, Diuraphis noxia Mordvilko, in tritordeum amphiploids. *Plant Breeding* 117(6), 515–522. Available at: https://doi.org/10.1111/j.1439-0523.1998.tb02200.x.

CDMS (2022). Crop Data Management Systems, Inc. Las Vegas, NV. Available at: http://www.cdms.net/Label-Database/Advanced-Search.

Clúa, A., Castro, A. M., Ramos, S., Gimenez, D. O., Vasicek, A., Chidichimo, H. O. and Dixon, A. F. G. (2004). The biological characteristics and distribution of the greenbug, *Schizaphis graminum*, and Russian wheat aphid, *Diuraphis noxia* (Hemiptera: Aphididae), in Argentina and Chile. *European Journal of Entomology* 101(1), 193–198. Available at: https://doi.org/10.14411/eje.2004.024.

Cooper, W. R., Dillwith, J. W. and Puterka, G. J. (2010). Salivary proteins of Russian wheat aphid (Hemiptera: Aphididae). *Environmental Entomology* 39(1), 223–231. Available at: https://doi.org/10.1603/en09079.

Crespo-Herrera, L. A., Singh, R. P., Sabraoui, A. and El-Bouhssini, M. (2019). Resistance to insect pests in wheat–rye and *Aegilops speltoides* Tausch translocation and substitution lines. *Euphytica* 215, 123. Available at: https://doi.org/10.1007/s10681-019-2449-7.

Cui, F., Smith, C. M., Reese, J., Edwards, O. and Reeck, G. (2012). Polymorphisms in salivary-gland transcripts of Russian wheat aphid biotypes 1 and 2. *Insect Science* 19(4), 429–440. Available at: https://doi.org/10.1111/j.1744-7917.2011.01487.x.

Damte, T. (2015). Occurrence of, and yield response to, Russian wheat aphid, *Diuraphis noxia* (Hemiptera: Aphididae) in irrigated wheat in Ethiopia. *International Journal of Tropical Insect Science* 35(1), 3–10. Available at: https//doi.org/10.1017/S1742758414000587.

Davachi, A. (1954). Insect pests of cereals in Iran. Univ. Tehran, Public. No. 211, (in Persian).

Douglas, A. E. (2003). The nutritional physiology of aphids. *Advances in Insect Physiology* 31, 73–140.

du Preez, P. H., Breeds, K., Burger, N. F. V., Swiegers, H. W., Truter, J. C. and Botha, A. M. (2020). DNA methylation and demethylation are regulated by functional DNA methyltransferases and DnTET enzymes in *Diuraphis noxia*. *Frontiers in Genetics* 11, 452. Available at: https://doi.org/10.3389/fgene.2020.00452.

du Toit, F. (1986). Economic thresholds for *Diuraphis noxia* (Hemiptera: Aphididae) on winter wheat in the eastern Orange Free State. *Phytophylactica* 18, 107–109.

du Toit, F. (1987). Resistance in wheat (*Triticum aestivum*) to *Diuraphis noxia* (Hemiptera: Aphididae). *Cereal Research Communications* 15, 175–179.

du Toit, F. (1988). Another Source of Russian Wheat Aphid (Diuraphis noxia) Resistance in *Triticum estivum*. *Cereal Research Communications* 16, 105–106.

du Toit, F. (1992). Russian wheat aphid resistance in a wheat line from the Caspian Sea area. *Cereal Research Communications* 21, 55–61.

El-Bouhssini, M., Street, K., Amri, A., Mackay, M., Ogbonnaya, F. C., Omran, A., Abdalla, O., Baum, M., Dabbous, A. and Rihawi, F. (2011). Sources of resistance in bread wheat to Russian wheat aphid (*Diuraphis noxia*) in Syria identified using the Focused Identification of Germplasm Strategy (FIGS). *Plant Breeding* 130(1), 96–97. Available at: https://doi.org/10.1111/j.1439-0523.2010.01814.

Elmali, M. (1998). Russian wheat aphid in Konya Province. *Euphytica* 100(1/3), 69–76. Available at: https://doi.org/10.1023/a:1018351805644.

EPPO (2022). List of databases on registered plant protection products in the EPPO region, European and Mediterranean Plant Protection Organization. Available at: https://www.eppo.int/ACTIVITIES/plant_protection_products/registered_products.

Ethiopian Ministry of Agriculture (2021). List of registered pesticides. Available at: http://www.moa.gov.et/list-of-registered-pesticides.

Feng, M.-G., Johnson, J. B. and Kish, L. P. (1990). Survey of entomopathogenic fungi naturally infecting cereal aphids (Homoptera: Aphididae) of irrigated grain crops in Southwestern Idaho. *Environmental Entomology* 19(5), 1534-1542. Available at: https://doi.org/10.1093/ee/19.5.1534.

Formusoh, E. S., Wilde, G. E., Hatchett, J. H. and Collins, R. D. (1992). Resistance to Russian wheat aphid (Homoptera: Aphididae) in Tunisian wheats. *Journal of Economic Entomology* 85(6), 2505-2509. Available at: https://doi.org/10.1093/jee/85.6.2505.

Fouché, A., Verhoeven, R. L., Hewitt, P. H., Walters, M. C., Kriel, C. F. and DeJager, J. (1984). Russian wheat aphid (*Diuraphis noxia*) feeding damage on wheat, related cereals and a *Bromus* grass species. In: Walters, M. C. (Ed). *Progress in Russian wheat aphid (Diuraphis noxia Mordw.) Research in the Republic of South Africa*, Technical Communication 191, Republic of South Africa, Department of Agriculture, Republic of South Africa, RSA, Pretoria, 22-33.

Ghannoum, M. I., El-Bouhssini, M., Nachit, M. M. and Kornosor, S. (2016). Sources of resistance in durum wheat and its wild relatives to Russian wheat aphid (Hemiptera: Aphididae). *Revue Marocaine de Protection des Plantes* 9, 61-67.

Gilchrist, L. I., Rodriguez, R. and Burnett, P. A. (1984). The extent of Free State streak and *Diuraphis noxia* in Mexico. In: Burnett, P. A. and Cuellar, E. (Eds). *Barley Yellow Dwarf. Proceedings of the Workshop*, 6-8 December 1983, Mexico City, Mexico. CIMMYT, El Batan, Mexico, pp. 157-163.

Girma, M., Wilde, G. and Reese, J. C. (1990). Influence of temperature and plant growth stage on development, reproduction, life span and intrinsic rate of increase of the Russian wheat aphid (Homoptera: Aphididae). *Environmental Entomology* 19(5), 1438-1442.

Godfrey, L. D. and Barlow, V. M. (2019). Guidelines for Russian wheat aphid on Small Grains. University of California IPM Pest Management Guidelines: Small Grains, UC Cooperative Extension, Riverside County and UC IPM Program 'UC IPM: UC Management Guidelines for Russian Wheat Aphid on Small Grains (ucanr.edu)'.

Gong, L., Cui, F., Sheng, C., Lin, Z., Reeck, G., Xu, J. and Kang, L. (2012). Polymorphism and methylation of four genes expressed in salivary glands of Russian wheat aphid (Homoptera: Aphididae). *Journal of Economic Entomology* 105(1), 232-241. Available at: https://doi.org/10.1603/ec11289.

Haile, A. (1981). *Cereal Aphids: Their Distribution, Biology and Management on Highland Barley. MSc in Crop Protection*, School of Graduate Studies, Addis Ababa University, Addis Ababa, Ethiopia.

Halbert, S. E. and Stoetzel, M. B. (1998). Historical overview of the Russian wheat aphid (Homoptera: Aphididae). In: Quisenberry, S. S. and Peairs, F. B. (Eds). *Response Model for an Introduced Pest - the Russian Wheat Aphid. Thomas Say Publications in Entomology Proceedings*, Entomological Society of America, Lanham, MD, pp. 12-30.

Haley, S. D., Peairs, F. B., Walker, C. B., Rudolph, J. B. and Randolph, T. L. (2004). Occurrence of a new Russian wheat aphid biotype in Colorado. *Crop Science* 44(5), 1589-1592. Available at: https://doi.org/10.2135/cropsci2004.1589.

Harris-Shultz, K., Armstrong, S. and Jacobson, A. (2019). Invasive cereal aphids of North America: Biotypes, genetic variation, management, and lessons learned. *Trends Entomol* 15, 99–122.

Hatting, J. L., Humber, R. A., Poprawski, T. J. and Miller, R. M. (1999). A survey of fungal pathogens of aphids from South Africa, with special reference to cereal aphids. *Biological Control* 16(1), 1–12.

Hatting, J. L., Poprawski, T. J. and Miller, R. M. (2000). Prevalences of fungal pathogens and other natural enemies of cereal aphids (Homoptera: Aphididae) in wheat under dryland and irrigated conditions in South Africa. *BioControl* 45(2), 179–199.

Hatting, J. L., Wright, S. P. and Miller, R. M. (2004). Efficacy of *Beauveria bassiana* (Hyphomycetes) for control of Russian wheat aphid (Homoptera: Aphididae) on resistant wheat under field conditions. *Biocontrol Science and Technology* 14(5), 459-473.

Havelka, J., Žurovcová, M., Rychlý, S. and Starý, P. (2013). Russian wheat aphid, *Diuraphis noxia* in the Czech Republic – cause of the significant population decrease. *Journal of Applied Entomology* 138(4), 273–280. Available at: https://doi.org/10.1111/jen.12068.

Health Canada (2021). Consumer product safety, pesticides & pest management, product labels. Available at: https://pr-rp.hc-sc.gc.ca/ls-re/index-eng.php. Search 'Russian wheat aphid'.

Heddle, T., Van Helden, M., Nash, M. and Muirhead, K. (2020). Parasitoid communities and interactions with *Diuraphis noxia* in Australian cereal production systems. *BioControl* 65(5), 571–582. Available at: https://doi.org/10.1007/s10526-020-10030-1.

Hein, G. and Thomas, J. (2006). Insecticides for control of aphids in wheat. *Cropwatch*. Institute of Agriculture and Natural Resources, University of Nebraska, Extension. Available at: https://cropwatch.unl.edu/insect/wheataphids.

Hodgson, E. W. and Karren, J. B. (2008). Russian wheat aphid, Utah State University Extension and Utah Plant Pest Diagnostic Laboratory ENT-67-08. Available at: https://digitalcommons.usu.edu/cgi/viewcontent.cgi?article=1946&context=extension_curall.

Hopper, K. R., Coutinot, D., Chen, K., Kazmer, D. J., Mercadier, G., Halbert, S. E., Miller, R. H., Pike, K. S. and Tanigoshi, L. K. (1989). Exploration for natural enemies to control *Diuraphis noxia* (Homoptera: Aphididae) in the United States. In: Quisenberry, S. S. and Peairs, F. B. (Eds). *Response Model for an Introduced Pest – The Russian Wheat Aphid. Thomas Say Publications in Entomology: Proceedings*, Entomological Society of America, Lanham, MD, pp. 12–30.

Jankielsohn, A. (2011). Distribution and diversity of Russian wheat aphid (Hemiptera: Aphididae) biotypes in South Africa and Lesotho. *Journal of Economic Entomology* 104(5), 1736–1741.

Jankielsohn, A. (2014a). The Russian wheat aphid. *Farmer's Weekly*, 17 January 2014, 22–23.

Jankielsohn, A. (2014b). Guidelines for the sampling, identification and designation of Russian wheat aphid (*Diuraphis noxia*) biotypes in South Africa. *Journal of Dynamics in Agricultural Research* 1, 36–43.

Jankielsohn, A. (2016). Changes in the Russian wheat aphid (Hemiptera: Aphididae) biotype complex in South Africa. *Journal of Economic Entomology* 109(2), 907–912. Available at: https://doi.org/10.1093/jee/tov408.

Jankielsohn, A. (2019). New Russian wheat aphid biotype found in Free State. *SA Grain*, 19 July 2019, pp. 70–71. Available at: https://sagrainmag.co.za/2019/07/19/new-russian-wheat-aphid-biotype-found-in-free-state-2/.

Jones, J. W., Byers, J. R., Butts, R. A. and Harris, J. L. (1989). A new pest in Canada: Russian wheat aphid, *Diuraphis noxia* (Mordvilko) (Homoptera: Aphididae). *The Canadian Entomologist* 121(7), 623-624.

Jimoh, M. A., Kaehler, S. and Botha, C. E. J. (2013). Increased feeding damage under elevated CO2: the case of the Russian wheat aphid. *South African Journal of Botany* 84, 30-37.

Kaplin, V., Morozova, J. and Vikhrova, E. (2015). Influence of Russian wheat aphid *Diuraphis noxia* feeding on grain yield components of barley and wheat in the forest-steppe of the Middle Volga region. *Bulletin of Insectology* 68(1), 147–152.

Khan, S. A., Murugan, M., Starkey, S., Manley, A. and Smith, C. M. (2009). Inheritance and categories of resistance in wheat to Russian wheat aphid (Hemiptera: Aphididae) Biotype 1 and Biotype 2. *Journal of Economic Entomology* 102(4), 1654–1662. Available at: https://doi:10.1603/029.102.0433.

Kieckhefer, R. W. and Elliott, N. C. (1989). Effects of fluctuating temperatures on development of immature Russian wheat aphid and demographic statistics. *Journal of Economic Entomology* 82(1), 119–122.

Kiriac, I., Gruber, F., Poprawski, T., Halbert, S. and Elberson, L. (1990). Occurrence of sexual morphs of Russian wheat aphid, Diuraphis noxia (Homoptera: Aphididae), in several locations in the Soviet Union and the northwestern United States. *Proceedings of the Entomological Society of Washington* 92(3), 544–547.

Kisten, L., Tolmay, V. L., Mathew, I., Sydenham, S. L. and Venter, E. (2020). Genome-wide association analysis of Russian wheat aphid (*Diuraphis noxia*) resistance in Dn4 derived wheat lines evaluated in South Africa. *PLoS ONE* 15(12), e0244455. Available at: http://doi.org/10.1371/journal.pone.0244455.

Knudsen, G. R. and Schotzko, D. J. (1999). Spatial simulation of epizootics caused by *Beauveria bassiana* in Russian wheat aphid populations. *Biological Control* 16(3), 318–326.

Kovalev, O. V., Poprawski, T. J., Stekolshchikov, A. V., Vereshchagina, A. B. and Gandrabur, S. A. (1991). *Diuraphis* Aizenberg (Hom., Aphididae): key to apterous viviparous females, and review of Russian language literature on the natural history of Diuraphis noxia (Kurdjumov, 1913) .*Journal of Applied Entomology* 112(1-5), 425–436 .

Laamari, M. and Benyahia, L. (2018). Evaluation of the resistance of different barley accessions to the Russian wheat aphid *Diuraphis noxia*. *Tunisian Journal of Plant Protection* 13, 99–106.

Lage, J., Skovmand, B. and Andersen, S. B. (2004). Field evaluation of emmer wheat-derived synthetic hexaploid wheat for resistance to Russian wheat aphid (Homoptera: Aphididae). *Journal of Economic Entomology* 97(3), 1065–1070. Available at: https://doi.org/10.1093/jee/97.3.1065.

Legg, D. E. and Archer, T. L. (1998). Sampling methods, economic injury levels and economic thresholds for the Russian wheat aphid (Homoptera: Aphididae). In: Quisenberry, S. S. and Peairs, F. B. (Eds). *Response Model for an Introduced Pest - the Russian Wheat Aphid. Thomas Say Publications in Entomology: Proceedings*, Entomological Society of America, Lanham, MD, pp. 12–30.

Lui, X., Marshall, J. L., Stary, P., Edwards, O., Puterka, G., Dolatti, L., El Bouhssini, M., Malinga, J., Lage, J. and Smith, C. M. (2010). Global phylogenetics of *Diuraphis noxia* (Hemiptera: Aphididae), and invasive aphid species: evidence for multiple invasions into North America. *Journal of Economic Entomology* 103(3), 958–965.

Lui, X., Meng, J., Starkey, S. and Smith, C. M. (2011). Wheat gene expression is differentially affected by a virulent Russian wheat aphid biotype. *Journal of Chemical Ecology* 37(5), 472–482.

Luna, E., van Eck, L., Campillo, T., Weinroth, M., Metcalf, J., Perez-Quintero, A. L., Botha, A., Thannhauser, T. W., Pappin, D., Tisserat, N. A., Lapitan, N. L. V., Argueso, C. T., Ode, P. J., Heck, M. L. and Leach, J. E. (2018). Bacteria associated with Russian wheat aphid (*Diuraphis noxia*) enhance aphid virulence to wheat. *Phytobiomes Journal* 2(3), 151–164.

Ma, Z. and Bechinski, E. J. (2009a). Accelerated failure time (AFT) modeling for the development and survival of Russian wheat aphid, *Diuraphis noxia* (Mordvilko). *Population Ecology* 51(4), 543–548. Available at: https://doi.org/10.1007/s10144-009-0160-6.

Ma, Z. and Bechinski, E. J. (2009b). Life tables and demographic statistics of Russian wheat aphid (Hemiptera: Aphididae) reared at different temperatures and on different host plant growth stages. *European Journal of Entomology* 106(2), 205–210. Available at: https://doi.org/10.14411/eje.2009.026.

McCallum, B. D., Geddes, C. M., Chatterton, S., Peng, G., Carisse, O., Turkington, T. K., Olfert, O., Leeson, J., Sharpe, S., Stephens, E., Hervet, V., Aboukhaddour, R. and Vankosky, M. (2021). We stand on guard for thee: a brief history of pest surveillance on the Canadian prairies. *Crop Protection* 149, 105748. Available at: https://doi.org/10.1016/j.cropro.2021.105748.

Macharia, M., Muthangya, P. M. and Wanjama, J. K. (1999). Response to seed dressing aphicides in commercial varieties for preventing Russian wheat aphid (*Diuraphis noxia*) damage in Kenya. In: CIMMYT (Ed). *Proceedings of the Tenth Regional Workshop for Eastern, Central and Southern Africa*. 14-18 September 1998, Stellenbosch, South Africa.

Malinga, J. N., Kinuya, M. G., Kamau, A. W., Wanjama, J. K., Awalla, J. O. and Pathak, R. S. (2007). Biotypic and genetic variation within tropical populations of Russian wheat aphid, Diuraphis noxia (Kurdjumov) (Homoptera: Aphididae) in Kenya. *J. Entomol* 4, 350–361.

Marais, G. F., Horn, M. and du Toit, F. (1994). Intergeneric transfer (rye to wheat) of a gene(s) for Russian Wheat Aphid resistance. *Plant Breeding* 113(4), 265–271. Available at: https://doi.org/10.1111/j.1439-0523.1994.tb00735.x.

Marasas, C. N., Anandajayasekeram, P., Tolmay, V. L., Purchase, J., Van Rooyen, J., Martella, D. and Prinsloo, G. (1997). *The socio-economic Impact of the Russian Wheat Aphid Control Research Program. Southern African Centre for Cooperation in Agricultural and Natural Resources Research and Training*, Gaborone, Botswana.

Marasas, C. N., Anandajayasekeram, P., Millard, S. and van Rooyen, C. J. (2005). Farm-level adoption and impact of agricultural technology: the case of Russian wheat aphid resistant cultivars in South Africa. *South African Journal of Agricultural Extension* 34(2), 318–333.

Maré, R. (2003). Insecticide for RWA control: what you win on the swings, you lose on the roundabouts. *SA Grain* 4(9), 54–55.

Maré, R. (2004). Russian wheat aphid resistant cultivars: the end of the road for insecticides? *Wheat Focus* 22(1), 6–9.

Matsiliza, B. (2003). Effects of sustained Russian wheat aphid (Diuraphis noxia Mordvilko) feeding on leaf blades of wheat (*Triticum aestivum* L. cv Adamtas). PhD Thesis, Rhodes, University, Grahamstown, South Africa.

Mesquita, A. L. M. and Lacey, L. A. (2001). Interactions among the entomopathogenic fungus *Paecilomyces fumosoroseus* (Deuteromycotina: Hyphomycetes), the parasitoids, *Aphelinus asychis* (Hymenoptera: Aphelinidae), and their aphid host. *Biological Control* 22(1), 51-59.

Mimeur, J. M. (1941). Aphididae Nord Africains. *Bulletin de la Société des Sciences Naturelles du Maroc* 21, 67-70.

Mondor, E. B., Tremblay, M. N., Awmack, C. S. and Lindroth, R. L. (2005). Altered genotypic and phenotypic frequencies of aphid populations under enriched CO2 and O3 atmospheres. *Global Change Biology* 11(11), 051013014052002-051013014052

Moremoholo, L. and Purchase, J. L. (1999). The release of Puseletso, a Russian wheat aphid (*Diuraphis noxia*) resistant wheat cultivar, in Lesotho. In: CIMMYT (Ed). *Proceedings of the Tenth Regional Workshop for Eastern, Central and Southern Africa*. 14-18 September 1998, Stellenbosch, South Africa.

Mornhinweg, D. W., Bregitzer, P. P., Porter, D. R., Peairs, F. B., Baltensperger, D. D., Hein, G. L., Randolph, T. A., Koch, M. and Walker, T. (2009). Registration of 'Sidney' spring feed barley resistant to Russian wheat aphid. *Journal of Plant Registrations* 3(3), 214-218. Available at: https://doi.org/10.3198/jpr2009.04.0205crc.

Mornhinweg, D. W., Bregitzer, P. P., Porter, D. R., Peairs, F. B., Baltensperger, D. D., Hein, G. L., Randolph, T. A., Koch, M. and Walker, T. (2012). 'Registration of 'Stoneham' spring feed barley resistant to Russian wheat aphid. *Journal of Plant Registrations* 6(1), 1-5. Available at: https://doi.org/10.3198/jpr2010.09.0550crc.

Mokrzhetsky, K. A. (1901). Animal and PlantPests of Crimea in 1900. Simferopol (cited in Kovalev, OV, Poprawski, TJ, Stekolshchikov, AV, Vereshchagina, B.,. Gandrabur, SA, 1991). Diuraphis Aizenberg (Hom., Aphididae) key to apterous females, and review of Russian language literature on the natural history of Diuraphis noxia (Kurdjumov, 1913). *J. Appl. Entomol* 112, 425-436.

Mulatu, B., Belay, T., Dejen, A., et al. (2011). Research achievements for the Russian wheat aphid (*Diuraphis noxia* Mord.) on barley in Ethiopia. In: Mulatu, B. and Grando, S. (Eds). *Barley Research and Development in Ethiopia. Proceedings of the 2nd National Barley Research and Development Review Workshop*. 28-30 November 2006, HARC, Holetta, Ethiopia. ICARDA, Aleppo, Syria, pp. xiv-391.

Nicholson, S. J., Hartson, S. D. and Puterka, G. J. (2012). Proteomic analysis of secreted saliva from Russian wheat aphid (Diuraphis noxia Kurd.) biotypes that differ in virulence to wheat. *Journal of Proteomics* 75(7), 2252-2268.

Nicholson, S. J., Nickerson, M. L., Dean, M., Song, Y., Hoyt, P. R., Rhee, H., Kim, C. and Puterka, G. J. (2015). The genome of *Diuraphis noxia*, a global aphid pest of small grains. *BMC Genomics* 16, 429.

Nienaber, H. and Kilian, W. H. (2021). Guidelines for the production of small grains in the summer rainfall region 2021. *Arc-Small Grain Institute*, University of the Free State, and SAB Maltings (Pty) Ltd. Available at: http://www.arc.agric.za/arc-sgi/Documents/PRODUCTION%20GUIDELINES/2021/LNR%20SUMMER%20GUIDELINE%202021.pdf.

Nowierski, R. M., Zeng, Z. and Scharen, A. L. (1995). Age-specific life table modelling of the Russian wheat aphid (Homoptera: Aphididae) on barley grown in benzimidazole agar. *Environmental Entomology* 24(5), 1284-1290. Available at: https://doi.org/10.1093/ee/24.5.1284.

Ortego, J. and Delfino, M. A. (1994). Diuraphis noxia (Mordvilko) (Homoptera: Aphididae) in Argentina. *Rev. Fac. Agron* 70, 51-55.

Oswald, C. J. II and Brewer, M. J. (1997). Aphid-barley interactions mediated by water stress and barley resistance to Russian wheat aphid (Homoptera: Aphididae). *Environmental Entomology* 26(3), 591-602.

Peairs, F. B., Hein, G. L. and Brewer, M. J. (2018). Russian wheat aphid. High Plains Integrated Pest Management. Available at: https://wiki.bugwood.org/HPIPM:Russian_Wheat_Aphid.

Pest Control Products Board (2010). *Pest Control Products Registered for Use in Kenya* (6th edn.). Available at: http://projects.nri.org/adappt/docs/McKnight/6thPestControlProductsList.pdf.

Pestfax (2021). *Agriculture and Food Newsletter*. Department of Primary Industries and Regional Development and Government of Western Australia. Issue 9. Available at: https://www.agric.wa.gov.au/newsletters/pestfax/pestfax-issue-9-2-july-2021?page=0%2C3.

Prinsloo, G. J. (1990). Some aspects of the biology of the aphidophagous parasitoids *Aphidius colemani* (Viereck) and *Aphelinus asychis* Walker. MSc Thesis, University of the Orange Free State, Bloemfontein, South Africa.

Prinsloo, G. J. (1998). Aphelinus hordei (Kurdjumov) (Hymenoptera: Aphelinidae), a parasitoid released for the control of Russian wheat aphid, Diuraphis noxia (Kurdjumov) (Homoptera: Aphididae), in South Africa. *African Entomology* 6(1), 147-156.

Prinsloo, G. J. (2006). *Parasitoids and Aphid Resistant Plants: Prospects for Diuraphis noxia (Kurdjumov) Control. PhD in Entomology*, University of the Free State, Bloemfontein, South Africa.

Prinsloo, G. J., Chen, Y., Giles, K. L. and Grenstone, M. H. (2002). Release and recovery in South Africa of the exotic aphid parasitoid *Aphelinus hordei* verified by the polymerase chain reaction. *BioControl* 47(2), 127-136.

Prinsloo, G. J., Ninkovic, V., van der Linde, T. C., van der Westhuizen, A. J., Pettersson, J. and Glinwood, R. (2007). Test of semiochemicals and a resistant wheat variety for Russian wheat aphid management in South Africa. *Journal of Applied Entomology* 131(9-10), 637-644. Available at: https://doi.org/10.1111/j.1439-0418.2007.01213.x.

Prokrym, D. R., Pike, K. S. and Nelson, D. J. (1998). Biological control of *Diuraphis noxia* (Homoptera: Aphididae): implementation and evaluation of natural enemies. In: Quisenberry, S. S. and Peairs, F. B. (Eds). *Response Model for an Introduced Pest - the Russian Wheat Aphid. Thomas Say Publications in Entomology: Proceedings*, Entomological Society of America, Lanham, MD, pp. 12-30.

Puterka, G. J., Burd, J. D., Porter, D., Shufran, K., Baker, C., Bowling, B. and Patrick, C. (2007). Distribution and diversity of Russian wheat aphid (Hemiptera: Aphididae) biotypes in North America. *Journal of Economic Entomology* 100(5), 1679-1684. Available at: https://doi.org/10.1093/jee/100.5.1679.

Puterka, G. J., Hammon, R. W., Burd, J. D., Peairs, F. B., Randolph, T. L. and Cooper, W. R. (2012). Cyclical parthenogenetic reproduction in the Russian wheat aphid (Hemiptera: Aphididae) in the United States: sexual reproduction and its outcome on biotypic diversity. *Journal of Economic Entomology* 105(3), 1057-1068. Available at: https://doi.org/10.1603/ec11338.

Quick, J. S., Ellis, G. E., Normann, R. M., Stromberger, J., Shanahan, J., Peairs, F., Rudolph, J. B. and Lorenz, K. (1996). Registration of 'Halt' wheat. *Crop Science* 36(1), 210. Available at: https://doi.org/10.2135/cropsci1996.0011183X003600010046x.

Quisenberry, S. S. and Peairs, F. B. (Eds) (1998). Response model for an introduced pest - the Russian wheat aphid. *Thomas Say Publications in Entomology: Proceedings, Entomological Society of America*, Lanham, MD.

Rashed, A. and Tanner, C. (2021). Small grain pests, small grain - Russian wheat aphid. *Pacific Northwest Pest Management Handbooks*. Available at: https://pnwhandbooks.org/insect/agronomic/small-grain/small-grain-russian-wheat-aphid.

Ricci, M., Tocho, E., Dixon, A. F. G. and Castro, A. M. (2011). *Diuraphis noxia*: reproductive behaviour in Argentina. *Bulletin of Insectology* 64(2), 235-241.

Robinson, J. and Skovmand, B. (1992). Evaluation of emmer wheat and other Triticeae for resistance to Russian wheat aphid. *Genetic Resources and Crop Evolution*. Available at: https://doi.org/10.1007/BF00051929.

Saheed, S. A., Liu, L., Jonsson, L. and Botha, C. E. J. (2007a). Xylem - as well as phloem - sustains severe damage due to feeding by the Russian wheat aphid. *South African Journal of Botany* 73(4), 593-599.

Saheed, S. A., Botha, C. E. J., Liu, L. and Jonsson, L. (2007b). Comparison of structural damage caused by Russian wheat aphid (*Diuraphis noxia*) and Bird cherry-oat aphid (*Rhopalosiphum padi*) in a susceptible barley cultivar, *Hordeum vulgare* cv. Clipper. *Physiologia Plantarum* 129(2), 429-435.

Saheed, S. A., Cierlik, I., Larsson, K. A. E., Delp, G., Bradley, G., Jonsson, L. M. and Botha, C. E. (2009). Stronger induction of callose deposition in barley by Russian wheat aphid than Bird cherry-oat aphid is not associated with differences in callose synthase or β-1,3-glucanase transcript abundance. *Physiologia Plantarum* 135(2), 150-161.

Saheed, S. A., Jonsson, L. M. V. and Botha, C. E. J. (2010). Russian wheat aphid causes greater reduction in phloem transport capacity of barley leaves than bird cherry-oat aphid. *Acta Botanica Croatica* 69, 7-18.

Sandström, J., Telang, A. and Moran, N. A. (2000). Nutritional enhancement of host plants by aphids – a comparison of three aphid species on grasses. *Journal of Insect Physiology* 46(1), 33-40. Available at: https://doi.org/10.1016/S0022-1910(99)00098-0.

Scott, R. A., Worrall, W. D. and Frank, W. A. (1991). Screening for Resistance to Russian wheat aphid in Triticale. *Crop Science* 31(1), 32-36. Available at: https://doi.org/10.2135/cropsci1991.0011183x003100010007x.

Selladurai, S. (2016). Genetic mapping and molecular characterisation of Russian wheat aphid resistance loci in wheat. PhD Thesis, Murdoch University, Perth, WA, Australia.

Smit, H. A., Tolmay, V. L., Barnard, A., Jordaan, J. P., Koekemoer, F. P., Otto, W. M., Pretorius, Z. A., Purchase, J. L. and Tolmay, J. P. C. (2010). An overview of the context and scope of wheat (*Triticum aestivum*) research in South Africa from 1983 to 2008. *South African Journal of Plant and Soil* 27(1), 81-96. Available at: https://doi/10.1080/02571862.2010.10639973.

Smith, C. M., Havlickova, H., Starkey, S., Gill, B. S. and Holubec, V. (2004). Identification of Aegilops germplasm with multiple aphid resistance. *Euphytica* 135(3), 265-273.

Sotelo, P. A., Hein, G. L., Peairs, F. B., et al. (2014). Effects of planting date and barley variety on Russian wheat aphid (Hemiptera: Aphididae) populations in Colorado, Kansas, and Nebraska. *Journal of Economic Entomology* 107(5), 1969-1976. Available at: http://doi/10.1603/EC14055.

Souza, E. J. (1998). Host plant resistance to the Russian wheat aphid (Homoptera: Aphidiae) in wheat and barley. In: Quisenberry, S. S. and Peairs, F. B. (Eds). Response model for an introduced pest - The Russian wheat aphid. *Thomas Say Publications in Entomology: Proceedings*. Entomological Society of America, Lanham, MD, pp. 12-30.

Stary, P. (1996). The expansive Russian wheat aphid, *Diuraphis noxia* (Mordw.) detected in the Czech Republic. *Anzeiger für Schädlingskunde, Pflanzenschutz, Umweltschutz* 69(1), 19–20.

Stary, P. and Erdelen, C. (1982). Aphid parasitoids (Hymenomptera: Aphidiidae, Aphelinidae) from the Yemen Arab Republic. *Entomophaga* 27(1), 105–108.

Stoetzel, M. B. (1987). Information on and identification of *Diuraphis noxia* (Homoptera: Aphididae) and other aphid species colonizing leaves of wheat and barley in the United States. *Journal of Economic Entomology* 80(3), 696-704.

Swanevelder, Z. H., Surridge, A. K. J., Venter, E. and Botha, A. M. (2010). Limited endosymbiont variation in *Diuraphis noxia* (Hemiptera: Aphididae) biotypes from the United States and South Africa. *Journal of Economic Entomology* 103(3), 887-897. Available at: https://doi.org/10.1603/EC09257.

Tanigoshi, L. K., Pike, K. S. and Miller, R. H. (1995). Search for, and release of, parasitoids for the biological control of Russian wheat aphid on Washington State (USA). *Agriculture, Ecosystems and Environment* 52(1), 25–30.

Telang, A., Sandström, J., Dyreson, E. and Moran, N. A. (1999). Feeding damage by *Diuraphis noxia* results in a nutritionally enhanced phloem diet. *Entomologia Experimentalis et Applicata* 91(3), 403-412.

Thao, M. L., Baumann, L., Baumann, P. and Moran, N. A. (1998). Endosymbionts (Buchnera) from aphids *Schizaphis graminum* and *Diuraphis noxia* have different copy numbers of the plasmid containing the leucine biosynthetic genes. *Current Microbiology* 36(4), 238–240.

Tolmay, V. L. (2006). *Genetic Variability for Russian Wheat Aphid, Diuraphis noxia, Resistance in South African Wheat Genotypes. PhD in Plant Breeding*, University of the Free State, Bloemfontein, South Africa.

Tolmay, J. P. C., Rosenblum, M. L., Moletsane, M., Makula, M. and Pederson, T. (2000). The introduction of disease and pest resistant wheat to small-scale farming systems in the highlands of Lesotho. In: CIMMYT (Ed). *The Eleventh Regional Wheat Workshop for Eastern, Central and Southern Africa*, Addis Ababa, Ethiopia.

Tolmay, V. L. and Booyse, M. (2017). Valuable Russian wheat aphid-resistant bread wheat accessions identified using four South African *Diuraphis noxia* biotypes. *South African Journal of Plant and Soil* 34(1), 65–70. Available at: https://doi.org/10.1080/02571862.2016.1180716.

Tolmay, V. L. and Booyse, M. (2020). *Phenotyping South African Winter Wheat Cultivars for Russian Wheat Aphid Resistance to Five South African Biotypes of Diuraphis noxia (Kurdjumov). Poster Presented at 24th International Plant Resistance to Insects Workshop, 2–4 March 2020*, CIMMYT, Texcoco, Mexico.

Tolmay, V. L., Jankielsohn, A. and Sydenham, S. L. (2013). Resistance evaluation of wheat germplasm containing *Dn4* or *Dny* against Russian wheat aphid biotype RWASA3. *Journal of Applied Entomology* 137(6), 476–480.

Tolmay, V. L., Lindeque, R. C. and Prinsloo, G. J. (2007). Preliminary evidence of a resistance-breaking biotype of the Russian wheat aphid, *Diuraphis noxia* (Kurdjumov) (Homoptera: Aphididae), in South Africa. *African Entomology* 15(1), 228-230. Available at: https://doi.org/10.4001/1021-3589-15.1.228.

Tolmay, V. L. and Maré, R. (2000). Is it necessary to apply insecticides to Russian wheat aphid resistant cultivars? In: CIMMYT (Ed). *Proceedings of the Eleventh Regional*

Wheat Workshop for Eastern, Central and Southern Africa, 18-22 September 2000, Addis Ababa, Ethiopia.

Tolmay, V. L., Sydenham, S. L., Sikhakhane, T. N., Nhlapho, B. N. and Tsilo, T. J. (2020). Elusive diagnostic markers for Russian wheat aphid resistance in bread wheat: deliberating and reviewing the status quo. *International Journal of Molecular Sciences* 21(21), 8271. Available at: https://doi.org/10.3390/IJMS21218271.

Tolmay, V. L. and van Deventer, C. S. (2005). Yield retention of resistant wheat cultivars, severely infested with Russian wheat aphid, *Diuraphis noxia* (Kurdjumov), in South Africa. *South African Journal of Plant and Soil* 22(4), 246–250.

Turkish Ministry of Agriculture (2021). Database of registered products. Available at: https://bku.tarimorman.gov.tr/Kullanim/TavsiyeArama?csrt=14982784728530207561.

Umina, P. A., Reidy-Crofts, J., Babineau, M., Maino, J. L. and Edwards, O. R. (2020). Susceptibility of the bird cherry-oat aphid, Rhopalosiphum padi (Hemiptera: Aphididae), to four insecticides. *Austral Entomology* 59(4), 838–844.

Vandenberg, J. D. (1996). Standardised bioassay and screening of *Beauveria bassiana* and *Paecilomyces fumosoroseus* against the Russian wheat aphid (Homoptera: Aphididae). *Journal of Economic Entomology* 89(6), 1418–1423.

Vandenberg, J. D., Sandvol, L. E., Jaronski, S. T. and Jackson, M. A. (2001). Efficacy of fungi for control of Russian wheat aphid (Homoptera: Aphididae) in irrigated wheat. *Southwestern Entomologist* 26, 73–85.

Van der Plank, J. E. (1963). *Plant Diseases: Epidemics and Control*. Academic Press, New York.

van Helden, M., Heddle, T., Pirtle, E., et al. (2018). Russian wheat aphid – learnings from 2018 and how they can inform us in 2019. In A. Cook (Ed). *Section 10, Eyre Peninsula Farming Systems 2018 Summary*, SARDI, Minnipa Agricultural Centre, Australia.

van Helden, M., Heddle, T., Umina, P. A. and Maino, J. L. (2022). Economic injury levels and dynamic action thresholds for *Diuraphis noxia* (Hemiptera: Aphididae) in Australian cereal crops. *Journal of Economic Entomology*. Available at: https://doi.org/10.1093/jee/toab272, 115(2), 592–601.

Varenhorst, A., Wagner, P. and Bachmann, A. (2020). Scouting winter wheat for aphid pests. South Dakota State University Extension. Available at: https://extension.sdstate.edu/scouting-winter-wheat-aphid-pests.

Veisi, R., Safavi, S. and Karimpour, Y. (2012). Duration of life stages and fecundity of *Diuraphis noxia* (Hemiptera: Aphididae) on six wheat cultivars. *Urmia University Journal of Crop Protection* 1(3), 181–187.

Walters, M. C. (Eds) (1984). Progress in Russian wheat aphid (Diuraphis noxia Mordw.) Research in the Republic of South Africa. *Technical Communication* 191, Department of Agriculture, Republic of South Africa, RSA, Pretoria.

Ward, S., van Helden, M., Heddle, T., Ridland, P. M., Pirtle, E. and Umina, P. A. (2020). Biology, ecology and management of *Diuraphis noxia* (Hemiptera: Aphididae) in Australia. *Austral Entomology* 59(2), 238–252. Available at: https://doi.org/10.1111/aen.12453.

Webster, J. A. (1991). Host plant resistance: Germplasm Evaluation. The Russian Wheat Aphid Fourth Annual Report, United States Department of Agriculture-Agriculture Research Service Progress Report. In: Burton, R. L. (Ed). *PSWCL Progress Report 92-001*, pp. 10–11.

Webster, J. A., Baker, C. A. and Porter, D. R. (1991). Detection and mechanisms of Russian wheat aphid (Homoptera: Aphididae) resistance in barley. *Journal of Economic Entomology* 84(2), 669–673.

Webster, J. A. and Kenkel, P. (1999). Benefits of managing small grain pests with plant resistance. In: Wiseman, B. R. and Webster, J. A. (Eds). *Economic, Environmental, and Social Benefits of Resistance in Field Crops. Thomas Say Publications in Entomology: Proceedings*, Entomological Society of America, Lanham, MD, pp. 87–114.

Webster, J. A. and Starks, K. J. (1987). Fecundity of *Schizaphis graminum* and *Diuraphis noxia* (Aphididae) at three temperature regimes. *Journal of the Kansas Entomological Society* 60, 580–582.

Wickremasinghe, M. G. V. and van Emden, H. F. (1992). Reactions of adult female parasitoids, particularly *Aphidius rhopalosiphi*, to volatile chemical cues from the host plants of their prey. *Physiological Entomology* 17, 279–304.

Weiland, A. A., Peairs, F. B., Randolph, T. L., Rudolph, J. B., Haley, S. D. and Puterka, G. J. (2014). Biotypic diversity in Colorado Russian wheat aphid (Hemiptera: Aphididae) populations. *Journal of Economic Entomology* 101(2), 569–574.

Xu, X., Bai, G., Carver, B. F., Zhan, K., Huang, Y. and Mornhinweg, D. (2015). Evaluation and reselection of wheat resistance to Russian wheat aphid biotype 2. *Crop Science* 55(2), 695–701.

Xu, X., Li, G., Carver, B. F. and Puterka, G. (2020). Identification of wheat germplasm resistant to major Russian wheat aphid biotypes in the United States. *Crop Science* 60(3), 1428–1435. Available at: https://doi.org/10.1002/csc2.20041.

Yazdani, M., Baker, G., DeGraaf, H., Henry, K., Hill, K., Kimber, B., Malipatil, M., Perry, K., Valenzuela, I. and Nash, M. A. (2018). First detection of Russian wheat aphid *Diuraphis noxia* Kurdjumov (Hemiptera: Aphididae) in Australia: a major threat to cereal production. *Austral Entomology* 5(4), 410–417. Available at: https://doi.org /10.1111/aen.12292.

Zerenè, Z., Mireya: Caveglic , D., Milàn y Ramirez, A. and Ignacio. (1988). Un nuevo àfido de los cereales detectado en Chile. *Agricultura Tècnica (Chile)* 48, 60–61.

Zhang, B. (2012). Evolutionary genetics and human assisted movement of a globally invasive pest (Russian wheat aphid: *Diuraphis noxia*). PhD dissertation. Earth, Environment and Biological Sciences, Earth, Science and Engineering Faculty, Queensland University of Technology, Brisbane, Australia.

Zhang, B., Edwards, O. R., Kang, L. and Fuller, S. J. (2012). Russian wheat aphids (*Diuraphis noxia*) in China: native range expansion or recent introduction? *Molecular Ecology* 21(9), 2130–2144. Available at: https://doi.org/10.1111/j.1365-294X.2012.05517.x.

Chapter 7

Greenbug (*Schizaphis graminum*): an overview

Tom A. Royer, Oklahoma State University, USA

1 Introduction

Cereal grains (barley, maize, millet, oat, rice, rye, sorghum, and wheat) are a foundation of human diet worldwide. Comprising more than 50% of the world daily calorie intake (Awika, 2011), cereal grain crops are also a food host for more than 40 aphid species (Blackman and Eastop, 2000). The greenbug, *Schizaphis graminum* (Rondani) (Hemiptera: Aphididae), is distributed worldwide and is associated with all cereal crops except pearl millet (Blackman and Eastop, 2017). The greenbug's ubiquitous distribution coupled with its ability to feed on the major cereal food crops elevates its status as an important, worldwide pest of cereal grains (Blackman and Eastop, 2017).

1.1 Origin, description, and life history

First described by Camillo Róndani (1852), the greenbug is a non-host alternating aphid that feeds on 70 species of Poaceae (Table 1) (Michaels, 1986). Its specific geographic provenance is still in doubt but believed to be of Palearctic origin. However, Blackman and Eastop (2017) suggested that it might have originated from farther south in Europe or Asia because one form feeds on sorghum and reproduces sexually in warmer climates. Today, the greenbug occurs throughout Southern Europe, the Middle East, Central and Western Asia, Africa, and North and South America (CABI EPPO, 1999). It was first reported in the USA in oats in 1882 (Webster and Phillips, 1912).

http://dx.doi.org/10.19103/AS.2022.0114.04

Table 1 Chronology of greenbug biotype designations

Biotype	Host collected from	Year collected or reported
A	Wheat	1961
B	Wheat	1961
C	Sorghum	1968
D[a]	Sorghum	1975
E	Wheat	1979
F	Canada bluegrass	1986
G	Wheat	1987
H	Wheat	1987
I	Sorghum	1990
J[b]	Wheat	1995
K	Sorghum	1992
Florida isolate[c]	Seashore paspalum	2003

Source: Porter et al. (1997) and Nuessly et al. (2008).[a]Biotype D was originally classified as a biotype that was resistant to organophosphate insecticides (Teetes et al., 1975); however, it is no longer considered a biotype because the designation was not based on insect-plant relationships (Porter et al., 1997).

[b]Biotype J was described as a distinct biotype because it was avirulent to all known resistant wheat and barley varieties, including the susceptible wheat control 'Triumph 64' (Beregovoy and Peters, 1995).

[c]The Florida isolate was not given a letter designation but is considered to be a unique biotype because it attacks seashore paspalum, the greenbug-resistant wheat variety 'GRS1201', and all identified resistant sorghum varieties (Nuessly et al., 2008).

The greenbug has both winged (alate) and wingless (apterous) forms (Figs 1 and 2). Winged adults (male or female) have transparent wings, brownish yellow heads with black lobes on the thorax, and a yellowish-green to dark-green abdomen and measure about 1.6–2.1 mm (Blackman and Eastop, 2000). Males may be slightly smaller than females (Washburn, 1908). Wingless adults and nymphs are oval shaped, with a darker green stripe down the middle of their light green body (Fig. 3). Their cornicles have dark tips and their green legs include dark tarsi. Wingless adults are all females and reproduce by parthenogenesis.

Greenbugs are monoecious on Poaceae (do not alternate to a nongrass host for sexual reproduction). In southern climates, they are anholocyclic (reproduce exclusively by parthenogenesis and produce only females), which overwinter as nymphs or adults. The greenbug's lower and upper development thresholds are −3.5–6°C and 37°C, respectively (Walgenbach et al., 1988). The optimal temperature for their development is 24–30°C, but optimal development is slightly lower for wheat-adapted biotypes (Walgenbach et al., 1988) than sorghum-adapted biotypes (Kirkland et al., 1981). Arnold (1981) found that greenbug survival during winter was lower when nighttime temperatures dropped to −12.3°C for 3 days. Jones et al. (2008) reported

Figure 1 Winged (alate) *S. graminum*. Source: Photo by R. Grantham.

Figure 2 Wingless (apterous) *S. graminum*. Source: Photo by R. Grantham.

that adults were not able to survive below their supercooling temperature of −26°C.

In more northern climates (typically areas above the 35th parallel), winged females may be either parthenogenetic, or able to produce females and males that can mate and produce eggs which overwinter (Wadley, 1931). This suggests that day length and temperature are regulators of the asexual and sexual life cycles. Mayo and Starks (1974) demonstrated that temperature influences the production of winged forms, and Mittler and Gorder (1991) demonstrated that photoperiod also influenced the induction of females to produce sexual morphs. However, 'sorghum-adapted forms' of greenbugs do

Figure 3 *S. graminum* adult and nymphs. Source: Photo by R. Grantham.

reproduce sexually in warm latitudes (Blackman and Eastop, 2017), suggesting that there is variation in the life cycle among phenotypes.

As stated previously, the greenbug is associated with most of the cereals grown for human food. It can inflict direct injury to its host plant as well as transmit plant disease viruses (Webster, 1995). The greenbug, like other aphids, extracts plant sap from phloem tissues with its piercing/sucking mouthparts. As it feeds, it injects saliva into the plant to improve food uptake. The saliva contains compounds that enter the plant tissue (Burd, 2002). Sorghum and wheat plants react to those compounds by forming visible injury symptoms, such as yellow-orange spots (wheat) or reddish spots (sorghum), on the leaves. Continued feeding can result in seedling death in wheat (Dahms and Wood, 1957) and leaf death in sorghum (Teetes and Johnson, 1973). In older plants, greenbug feeding affects the general plant health and alters root and shoot development (Burton, 1985; Burton and Burd, 1993).

Greenbug feeding causes structural damage to sorghum cell organelles, including damage to chloroplast membranes, which collapses mesophyll tissues (Al-Mousawi et al., 1983). Morgham et al. (1994) found similar injury caused to wheat. Other effects include reduced chlorophyll content, reduced carbon assimilation, and reduced transpiration (Ryan et al., 1987; Girma et al., 1998). Besides the direct injury inflicted by greenbug feeding on its host, it can indirectly injure its host as well. It is a vector of the barley yellow dwarf virus in wheat (Johnson and Rachow, 1972; Gray et al., 1998) and potentially of maize dwarf mosaic virus in corn and sorghum (Berger et al., 1983). Greenbug feeding also predisposes sorghum to infection by charcoal rot (Teetes et al., 1973).

2 Management of greenbugs

Historically, the greenbug has been one of the most yield-limiting pests of wheat and grain sorghum (Royer, 2021). At least in the Southern Great Plains of the USA, the greenbug has become much less of an important pest of wheat and sorghum (Michaud, 2017). Insecticides are still used to manage greenbug infestations. However, combinations of other management tactics with chemical control, including biological, cultural, and host-plant resistance, have been shown to effectively manage greenbug infestations (Giles et al., 2017; Hesler and Berg, 2003; Hesler et al., 2005; Royer et al., 2015).

Farming practices affect greenbug establishment in cropping systems. The practice of reduced tillage resulted in lower greenbug numbers and associated crop injury to grain sorghum compared to clean tillage (Burton et al., 1987). Greenbug response to reduced tillage is variable in spring and winter wheat (Hesler and Berg, 2003; Burton and Krenzer, 1985). Planting date can also influence greenbug infestations. Grain sorghum, as a summer crop, can be grown either as a full-season or as a double crop (Crabtree et al., 1990). Burton et al. (1990) reported that greenbugs were less abundant in late-planted sorghum compared to early-planted sorghum and that planting date and tillage practices would be compatible with the deployment of greenbug-resistant sorghum hybrids.

In parts of the Southern Great Plains of the USA, farmers grow 'dual purpose' wheat that is used for winter forage for stocker cattle and for grain production. This system requires that the wheat is planted earlier to allow for forage accumulation (Hossain et al., 2003). Wheat that is planted earlier supports greater number of greenbugs compared to later plantings (Pike and Schaffner, 1985; Hesler et al., 2005; Royer et al., 2005). In addition, earlier planting is at a greater risk of infection with the viruses that cause barley yellow dwarf disease (Hesler et al., 2005; Royer et al., 2005). However, continued grazing during the winter by cattle reduces greenbug abundance (Arnold, 1981; Ismail et al., 2003) and lowers the incidence of barley yellow dwarf disease by up to 70% (Ismail et al., 2003).

Like most aphids, greenbugs are prey to numerous natural enemies, several of which are capable of suppressing greenbug infestations (Brewer and Elliott, 2004). They include predators, parasitoids, and insect pathogens (Elliott et al., 2006; Feng et al., 1990; Walker et al., 1973; Archer et al., 1974; Gilstrap et al., 1984; Michels et al., 2001). Generalist predators, such as lady beetles (Coleoptera: Coccinellidae), can reduce greenbug infestations in sorghum and wheat (Rice and Wilde, 1989; Kring and Gilstrap, 1984; Kring et al., 1986; Michels et al., 2001). One important component for lady beetle effectiveness as a greenbug predator in sorghum relies on the early appearance of corn leaf aphid *Rhopalosiphum maidis* (Fitch) (Hemiptera: Aphididae) to help maintain high numbers of *Hippodamia* spp. lady beetles that then feed on greenbugs as the sorghum panicle emerges (Kring and Gilstrap, 1986). In Brazil, a planting date study of grain sorghum revealed that aphid predators, the most abundant being *Scymnus* spp., *Cycloneda sanguinea* (L.) (Coleoptera: Coccinellidae), and *Chrysopa sancta* Schneider (Neuroptera: Chrysopidae), were associated with greenbugs. Ali et al. (2018) reported that the most important natural enemies of greenbugs in China are three coccinellids, *Coccinella septempunctata* (L.), *Harmonia axyridis* Pallas, and *Propylea japonica* (Thunberg), while Yunding et al. (1998) reported that a spider *Erigonidium graminicola* (Sundevall) (Araneae: Linyphiidae) was a predominate predator as well. More recent research conducted in the Southern and Central Great Plains is focused on between-crop movement of natural enemies of greenbugs and other aphids in an intercropping consisting of winter wheat, alfalfa, and cotton (Phoofolo et al., 2010).

The wasp parasitoid *Lysiphlebus testaceipes* L. (Hymenoptera: Braconidae) is a ubiquitous parasitoid of greenbugs in wheat and sorghum in the Southern Great Plains of the USA. It overwinters in southern Kansas, Oklahoma, and Texas, where it plays a major role in suppressing greenbug outbreaks in winter wheat (Jones, 2001; Giles et al., 2003) and sorghum (Jackson et al., 1970) and is active at low levels throughout the winter (Jones et al., 2007). However, it does not successfully overwinter northward of central Kansas, and geographic isolates have little genetic or phenotypic flexibility for winter survival (Royer et al., 2001; Shufran et al., 2004). Instead, it is believed to disperse northward through the migration of parasitized alate aphids (Jackson et al., 1970). Gravena (1979) reported that the hyperparasitoid *Aphidencyrtus aphidivorus* (Mayr) (Hymenoptera: Encyrtidae) had a significant impact on the two primary parasitoids of greenbug, *Aphidius colemani* Vier. (Hymenoptera: Braconidae) and *Diaeretiella rapae* (M'Intosh) (Hymenoptera: Braconidae), in Brazilian sorghum. The braconid parasitoid *A. gifuensis* (Ashmead) (Hymenoptera: Braconidae) is an important parasitoid of greenbug in northern China (Ali et al., 2018).

Greenbugs are host for a number of entomopathogenic fungi (Feng et al., 1992). However, epizootics of these fungi in greenbug populations are rare, typically associated with other aphid species in irrigated wheat (Feng et al., 1992). Environmental factors affect the persistence and initial infection process of entomopathogenic fungi (Jaronski, 2010). Precipitation, number of rainy days, relative humidity, solar radiation, temperature, and other factors such as leaf-surface chemistry can influence epizootics of these fungi (Jaronski, 2010). In addition, the competing microorganisms that live on the plant surface can limit their persistence (Ignoffo, 1992; Jaronski, 2010). Research is now focusing on factors that can extend fungal spore viability and persistence (Maina et al., 2018) and on the interactions of entomopathogenic fungi with other natural enemies (Furlong and Pell, 2005).

Current research on biological/natural control of greenbug has revolved around the interaction among different natural enemies with greenbug. Mullins et al. (2011) observed that field populations of coccinellids in winter wheat readily fed on aphids parasitized by *L. testaceipes*. Royer et al. (2008) found that two species of lady beetle larvae readily fed on aphids parasitized by *L. testaceipes* but that diet either delayed or prevented the completion of lady beetle larvae development, depending upon the mixture of live parasitized aphid and mummies consumed. Jessie et al. (2019) found that the green lacewing *Chrysoperla rufilabris* Burmeister (Neuroptera: Chrysopidae) readily fed on greenbugs at all stages of parasitism by *L. testaceipes* and that lady beetle development and survival were not affected by that diet, regardless of the mixture of live parasitized greenbugs or greenbug mummies consumed.

There is a rich literature addressing the history of host-plant resistance in greenbug, its deployment, and greenbug response to that resistance. The search for greenbug resistance in sorghum and wheat has been a long-term part of sorghum hybrid development (Teetes and Pendleton, 2000) and winter wheat variety development (Schuster and Starks, 1973; Martin, 1981; Tyler et al., 1987; Porter et al., 1997). More recently, molecular tools are being used to identify the specific genetic mechanisms for greenbug resistance in sorghum (Katsar et al., 2002) and locate them for more rapid deployment in breeding programs in wheat (Weng et al, 2005; Tan et al., 2017; Xu et al., 2020; Zhu et al., 2004) and sorghum (Radchenko et al., 2017). Resistance mechanisms in wheat or sorghum may include antibiosis and antixenosis (Dogramaci et al., 2007; Radchenko and Tyrishkin, 2004; Grover et al., 2019) and possibly tolerance (Dogramaci et al., 2007; Radchenko, 2010; Wu and Huang, 2008; Punnuri et al., 2012).

Greenbug-resistant wheat varieties were first deployed in the 1950s (Porter et al., 1997), and then additional sources of greenbug resistance were discovered and deployed (Martin, 1981; Tyler et al., 1985, 1987; Porter and Mornhinweg, 2004). In 1969, a greenbug biotype (Biotype C) was identified that infested sorghum (Harvey and Hackerott, 1969). Through the process of phenotypic screening of resistance developed by Painter and Peters (1956), 12 greenbug biotypes have been identified based on their ability to overcome a specific resistance gene or set of resistance genes in wheat or sorghum (Table 1).

Genes for greenbug resistance have been identified for wheat and sorghum (Table 2). Sources of greenbug resistance in winter wheat include genes *Gb2* and *Gb6* from chromosome 1R in rye (Purterka and Peters, 1989; Lu et al., 2010) and *Gb5* in *Aegilops speltoides* Tausch (Tyler et al., 1987; Dubcovsky et al., 1998; Crespo-Herrera et al., 2013). Other sources (*Gb3, Gb4*, Gb7, *Gba, Gbb, Gbc, Gbx*, and *Gbz*), have been found in synthetic wheat lines that were derived from chromosome 7Dl from *A. tauschii* (Puterka and Peters, 1989; Zhu et al.,

Table 2 Resistance genes for greenbug in sorghum (Radchenko, 2010) and wheat (Crespo-Herrera et al., 2019; Xu et al., 2020)

Sorghum resistance gene designation	Wheat resistance gene designation
Sgr1	*Gb1*
Sgr2	*Gb2*
Sgr3	*Gb3*
Sgr4	*Gb4*
Sgr5	*Gb5*
Sgr6	*Gb6*
Sgr7	*Gb7*
Sgr8	*Gb8*
Sgr9	*Gba*
Sgr10	*Gbx1*
Sgr11	*Gbb*
	Gbd

2005; Crespo-Herrera et al., 2014; Tan et al., 2017). In 1969, a greenbug biotype (Biotype C) was identified that infested sorghum (Harvey and Hackerott, 1969).

The deployment of greenbug resistance in winter wheat lines was met with new greenbug biotypes that could overcome the resistance genes (Porter et al., 1997). Originally, researchers believed that such biotypes were selected following widespread deployment of resistance (Eisenbach and Mittler, 1987; Lazar et al., 1995). Recent research suggests that greenbug populations diverge through genetic recombination of races that are already host adapted to various native grasses (Porter et al., 1997). Therefore, they are likely present independently of any particular deployment of resistant variety or grass species (Shufran et al., 1997, 2000; Anstead et al., 2003; Burd and Porter, 2006; Shufran, 2011).

The search and identification of greenbug resistance in wheat and sorghum has led to the discovery of host resistance to other aphid pests, including bird cherry-oat aphid *R. padi* in wheat (Dunn et al., 2011; Hesler, 2005) and sugarcane aphid *Melanaphis sacchari* in grain sorghum (Armstrong et al., 2015). The search for resistance has also led to complimentary research evaluating the compatibility of greenbug host resistance with biological control. Rice and Wilde (1989) demonstrated that greenbug-resistant sorghum hybrids had a negative effect on the development and survival of the convergent lady beetle (*Hippodamia convergens*). However, *L. testaceipes* activity was found to be compatible with resistant wheat (Starks et al., 1972; Fuentes-Granados et al., 2001) and resistant sorghum (Starks et al., 1974; Dogramaci et al., 2005) to the benefit of the crop.

Generally, insecticides play a major role in greenbug management. Until recently, insecticides registered for greenbug control in sorghum and wheat were restricted to four insecticide modes of action: carbamates, organophosphates, pyrethroids, and neonicotinoids. Unfortunately, these insecticides also presented some challenges. The carbamates, organophosphates, and pyrethroids are broad spectrum in their effects on nontargeted arthropods, fish, reptiles, birds, and mammals when applied. Some carbamates and organophosphates are highly toxic at low exposure levels to humans (Gupta, 2006). In addition, their continual use on wheat and sorghum contributed to the development of greenbug populations that are resistant to organophosphates and carbamates (Teetes et al., 1975; Shufran et al., 1997; Wilde et al., 2001). Pyrethroids are highly toxic to aquatic organisms (Hill, 1989). Neonicotinoids are often applied as seed treatments, but pollinators are exposed to them and they are known to have detrimental effects on honeybees and wild bee pollinators (Krupke et al., 2012; Feltham et al., 2014; Dively et al., 2015).

Effective and judicious use of insecticides for greenbug control depends on the accuracy of the tools used to make treatment decisions. These decision tools are based on several factors:

- the cost and effectiveness of the insecticide application;
- a research-based understanding of the greenbug's reproductive capacity;
- the effect of feeding on the crop; and
- the impact of natural enemies on the population.

Kindler et al. (2003) developed economic injury levels for greenbugs in winter wheat in the Southern Plains. Sampling plans for greenbugs in wheat are based on greenbug within-field distribution (Kring and Gilstrap, 1983; Elliott and Kieckhefer, 1986, Elliott et al., 1990; Feng and Nowierski, 1992a, 1992b; Giles et al., 2000). Giles et al. (2003) developed a presence–absence, sequential sampling system for greenbugs in the Southern Plains that considers seasonal changes in the within-field distribution of greenbugs in winter wheat and the presence, distribution, and impact of *L. testaceipes* on greenbug populations. The data were used to develop a binomial sequential sampling decision tool termed 'Glance-N-Go' (Elliott et al., 2004; Royer et al., 2022), which can be accessed in the myFields web platform at: https://www.myfields.info/ (McCornack and Johnson, 2016). It is also now available as a smartphone application at: www.dasnr.okstate.edu/apps.

In sorghum, economic injury levels have been developed (Teetes and Johnson, 1973, 1974), and greenbug distribution has been characterized (Archer and Bynum, 1986; Michels and Undersander, 1986). This research has been adapted into treatment thresholds throughout the Central Plains of the USA (Wright et al., 2006; Cronholm et al., 2007; Michaud et al., 2014; Royer,

2021). While some treatment thresholds encourage the documentation of natural enemies, (Kring and Gillstrap, 1984; Kring et al., 1986) most greenbug thresholds for sorghum do not formally include natural enemy incidence (Michaud et al., 2014; Royer et al., 2022). Sampling strategies have been developed for lady beetles (Michels and Behle, 1992), but most thresholds provide only a nominal recommendation to 'delay insecticide applications for greenbugs if 20% or more greenbugs are parasitized by *Lysiphlebus testaceipes*' (Knutson et al., 2018).

3 Future trends in research

Current research on greenbugs in winter wheat focuses on the development of remote-sensing tools to monitor greenbug infestations in wheat. Since greenbug feeding creates a visible symptom on its host plants, laboratory and greenhouse research has shown that reflected light and/or temperature from wheat in the green, red, and infrared spectra are altered and recognizable remotely when greenbugs feed (Michels et al., 1999; Riedell and Blackmer, 1999; Yang et al., 2005; Mirik et al., 2006; Elliott et al., 2009).

The research has shown correlations between greenbug infestations and hyperspectral imaging (i.e. hundreds of narrow spectral radiation bands) and multispectral imaging (i.e. 3–10 broad spectral radiation bands) in the field as well (Mirik et al., 2006; Elliott et al., 2009). Backoulou et al. (2016) demonstrated that this technology could differentiate injury in winter wheat fields caused by greenbugs from injury caused by Russian wheat aphid, based on the within-field distribution patterns and size of the infestations. In sorghum, work has been focused on the remote sensing of sugarcane aphid (Elliott et al., 2015), but presumably, the technology's capacity to distinguish between sugarcane aphid and greenbug injury is applicable as well. Future research will focus on the development and integration of more sophisticated remote-sensing capacity that can be linked to the identification of abiotic and biotic crop stressors and precision application of chemical controls based on their distribution within a field (Kelly and Guo, 2007; Iost Filho et al., 2020; Sishodia et al., 2020) and the integration of remote-sensing tools.

4 Where to look for further information

There are still important areas of research that need exploration for greenbug and other aphid pests of row crops. Future research areas include the continued development of new pest resistant cultivars (Luo et al., 2021), the circumvention of insecticide resistance coupled with the development of new insecticide modes of action (Jeschke, 2021; Onstad and Knolhoff, 2022), the incorporation of biological control into aphid pest management decision-making (Giles et al., 2017), the application nanotechnology to insect management (Rai and Ingle,

2012, Christiaens et al. 2020), the impacts of insect pests and their management on crops due to a warming climate (Aljaryian and Kumar, 2016; Deutsch et al., 2018; Skendzic et al., 2021), the impact of transgenic crops on insects (Christou et al., 2006), and the comprehensive impact of agricultural cropping systems on aphid establishment and movement (Simone and Peccoud, 2018).

A comprehensive list of research projects that involve remote sensing, plant breeding, or pest management that are funded by the National Institute of Food and Agriculture can be accessed through searching the USDA NIFA portal page at: https://portal.nifa.usda.gov/lmd4/recent_awards?report_title=Recent%20Awards&from_site=NIFA&search_label=Awards%20Listing.

5 References

Al-Mousawi, A. H., Richardson, P. E. and Burton, R. L. 1983. Ultrastructural studies of greenbug (Hemiptera: Aphididae) feeding damage to susceptible and resistant wheat cultivars. *Annals of the Entomological Society of America* 76(6): 964–971.

Ali, A., Desneux, N., Lu, Y. and Wu, K. 2018. Key aphid natural enemies showing positive effects on wheat yield through biocontrol services in northern China. *Agriculture, Ecosystems and Environment* 266: 1-9.

Aljaryian, R. and Kumar, L. 2016. Changing global risk of invading greenbug under climate change. *Crop Protection* 88: 137–148.

Anstead, J. A., Burd, J. D. and Shufran, K. A. 2003. Over-summering and biotypic diversity of *Schizaphis graminum* (Homoptera: Aphididae) population on noncultivated grass hosts. *Environmental Entomology* 32(3): 662–667.

Archer, T. L. and Bynum, E. D. Jr. 1986. Greenbug (Homoptera: Aphididae) and Banksgrass mite (Acari: Tetranychidae) distribution on sorghum plants. *Journal of Economic Entomology* 79(1): 116–121.

Archer, T. L., Cate, R. H., Eikenbary, R. D. and Starks, K. S. 1974. Parasitoids collected from greenbugs and corn leaf aphids in Oklahoma in 1972. *Annals of the Entomological Society of America* 67: 11–14.

Armstrong, J. S., Rooney, W. L., Peterson, G. C., Villenueva, R. T., Brewer, M. J. and Sekula-Ortiz, D. 2015. Sugarcane aphid (Hemiptera: Aphididae): Host range and sorghum resistance including cross-resistance from greenbug sources. *Journal of Economic Entomology* 108(2): 576–582.

Arnold, D. C. 1981. Effects of cold temperatures and grazing on greenbug populations in wheat in Noble County, Oklahoma 1975–76. *Journal of the Kansas Entomological Society* 54: 571–577.

Awika, J. M. 2011. Major cereal grains production and use around the world. In: Awika, J. M., Piironen, V. and Bean, S. (Eds.), *ACS Symposium Series*. Washington, DC: American Chemical Society, pp. 1–13.

Backoulou, G. F., Elliott, N. C. and Giles, K. L. 2016. Using multispectral imagery to compare the spatial pattern of injury to wheat caused by Russian wheat aphid and greenbug. *Southwestern Entomologist* 41(1): 1–8.

Beregovoy, V. H. and Peters, D. C. 1995. Biotype J, a unique greenbug (Homoptera: Aphididae) distinguished by plant damage characteristics. *Journal of the Kansas Entomological Society* 67: 248–252.

Berger, P. H., Tolar, R. W. and Harris, K. F. 1983. Maize dwarf mosaic virus transmission by greenbug biotypes. *Plant Disease* 67(5): 49–97.

Blackman, R. L. and Eastop, V. F. 2000. *Aphids on the World's Crops: An Identification and Information Guide*. West Sussex, England: John Wiley & Sons.

Blackman, R. L. and Eastop, V. F. 2017. *Taxonomic Issues: PP in Van Emden and Harrington* (9th edn.). West Sussex: Aphids as Crop Pests

Brewer, M. J. and Elliott, N. C. 2004. Biological control of cereal aphids in North America and mediating effects of host plant and habitat manipulations. *Annual Review of Entomology* 49: 219–242.

Burd, J. D. 2002. Physiological modification of host feeding site by cereal aphids (Homoptera: Aphididae). *Journal of Economic Entomology* 95(2): 463–468.

Burd, J. D. and Porter, D. R. 2006. Biotypic diversity in greenbug (Hemiptera: Aphididae): Characterizing new virulence and host associations. *Journal of Economic Entomology* 99(3): 959–965.

Burton, R. L., 1986. Effect of greenbug (Homoptera: Aphididae) damage on root and shoot biomass of wheat seedlings. *Journal of Economic Entomology* 79: 633–636.

Burton, R. L., Burd, J. D., Jones, O. R. and Wicks, G. A. 1990. Crop production strategies for managing greenbug (Homoptera: Aphididae) in sorghum. *Journal of Economic Entomology* 83(6): 2476–2479.

Burton, R. L., Jones, O. R., Burd, J. D., Wicks, G. A. and Krenzer, E. G., Jr. 1987. Damage by greenbug to grain sorghum as affected by tillage, surface residues, and canopy. *Journal of Economic Entomology* 80(4): 792–798.

Burton, R. L. and Krenzer, E. G. Jr. 1985. Reduction of greenbug (Homoptera: Aphididae) populations by surface residues in wheat tillage studies. *Journal of Economic Entomology* 78(2): 390–394.

Burton, R. W. and Burd, J. D. 1993. Relationship between number of greenbugs and damage to wheat seedlings. *Southwestern Entomologist* 18: 263–268.

CABI EPPO. (1999). Schizaphis graminum (Distribution map). Map 173. In: *Distribution Maps of Plant Pests*. Wallingford, UK: CAB International. DOI: 10.1079/DMPP/20066600173.

Christiaens, O., Petek, M., Smagghe, G. and Taning, C. N. T. 2020. The use of nanocarriers to improve the efficience of RNAi-based pesticides in agriculture. In: Fraceto, L. F., de Castro, V. L. S. S., Grillo, R., Ávila, D., Caixeta Oliveira, H. and Lima, R. (Eds) *Nanopesticides*. Springer, Cham. https://doi.org/10.1007/978-3-030-44873-8_3.

Christou, P., Capell, T., Kohli, A., Gatehouse, J. A. and Gatehouse, A. M. R. 2006. Recent developments and future prospects in insect pest control in transgenic crops. *Trends in Plant Science* 11(6): 302–308.

Crabtree, R. J., Prater, J. D. and Mbolda, P. 1990. Long-term wheat, soybean, and grain sorghum double-cropping under rainfed conditions. *Agronomy Journal* 82(4): 683–686.

Crespo-Herrera, L. A., Akhunov, E., Garkava-Sustavsson, L., Jordan, K. W., Smith, C. M., Sing, R. P. and Ahman, I. 2014. Mapping resistance to the bird cherry-oat aphid and the greenbug in wheat using sequence-based genotyping. *TAG. Theoretical and Applied Genetics. Theoretische und Angewandte Genetik* 127(9): 1963–1973.

Crespo-Herrera, L. A., Smith, C. M., Singh, R. P. and Ahman, I. 2013. Resistance to multiple cereal aphids in wheat-alien substitution and translocation lines. *Arthropod-Plant Interactions* 7(5): 535–545.

Crespo-Herrera, L. R. P. S., Reynolds, M. and Huerta-Espino, J. 2019. Genetics of greenbug resistance in synthetic hexaploid wheat derived germplasm. *Fronteirs of Plant Science* 13: 782. DOI: 10.3389/fpls.2019.00782.

Cronholm, G., Knutson, A., Parker, R. and Pendleton, B. 2007. *B-1220. Managing Insect and Mite Pests of Texas Sorghum*. Texas Agrilife Extension Service, College Station, TX.

Dahms, R. G. and Wood, E. A. Jr. 1957. Evaluation of greenbug damage to small grains. *Journal of Economic Entomology* 50(4): 443–446.

Deutsch, C. A., Tewksbury, J. J., Tigchelaar, M., Battisti, D. A., Merrill, S. C., Huey, R. B. and Naylor, R. L. 2018. Increase in crop losses to inset pests in a warming climate. *Science* 361: 916–919.

Dively, G. P., Embrey, M. S., Kamel, A., Hawthorne, D. J. and Pettis, J. S. 2015. Assessment of chronic sublethal effects of Imidacloprid on honeybee colony health. *PLoS ONE* 10(3): e0118748. DOI: 10.1371/journal.pone.0118748.

Dogramaci, M., Mayo, Z. B., Wright, R. J. and Reese, J. C. 2005. Tritrophic interaction of parasitoid *Lysiphlebus testaceipes* (Hymenoptera: Aphidiidae), greenbug *Schizaphis graminum* (Homoptera: Aphididae) and greenbug-resistant sorghum hybrids. *Journal of Economic Entomology* 98(1): 202–209.

Dogramaci, M., Mayo, Z. B., Wright, R. J. and Reese, J. C. 2007. Catetories ofresistance, antibiosis and tolerance, to biotype I greenbug (Schizaphis graminum (Rondani) Homoptera: Aphididae). *Journal of the Kansas Entomological Society* 80(3): 183–191.

Dubcovsky, J., Lukaszewski, A. J., Echaide, M., Antonelli, E. F. and Porter, D. R. 1998. Molecular characterization of two Triticum speltoides interstitial translocations carrying leaf rust and greenbug resistance genes. *Crop Science* 38(6): 1655–1660.

Dunn, B. L., Porter, D. R., Baker, C. A. and Carver, B. F. 2011. Screening USDA-ARS wheat germplasm for bird cherry-oat aphid tolerance. *Journal of Crop Improvement* 25(2): 176–182.

Eisenbach, J. and Mittler, T. E. 1987. Polymorphism of biotypes E and C of the Aphid Schizaphis graminum (Homoptera: Aphididae) in response to different scotophases. *Environmental Entomology* 16(2): 519–523.

Elliott, N. C., Backoulou, G. F., Brewer, M. J. and Giles, K. L. 2015. NDVI to detect sugarcane aphid injury to grain sorghum. *Journal of Economic Entomology* 108(3): 1452–1455.

Elliott, N. C. and Kieckhefer, R. W. 1986. Cereal aphid populations in winter wheat: Spatial distributions and sampling with fixed levels of precision. *Environmental Entomology* 15(4): 954–958.

Elliott, N. C., Kieckhefer, R. W. and Walgenbach, D. D. 1990. Binomial sequential sampling methods for cereal aphids in small grains. *Journal of Economic Entomology* 83(4): 1381–1387.

Elliott, N. C., Mirik, M., Yang, Z., Jones, D., Phoofolo, M., Catana, V., Giles, K. and Michels, G. J., Jr. 2009. Airborne remote sensing to detect greenbug stress to wheat. *Southwestern Entomologist* 34(3): 205–211.

Elliott, N. C., Royer, T. A., Giles, K. L., Kindler, S. D., Porter and D. R. 2004. A web-based decision support system for managing greenbugs in wheat. *Crop Management* 3(1). 1–6. DOI: 10.1094/CM-2004-1006-01-MG. Available at: Fenghttps://www.agronomy.org/publications/cm/tocs/3/1.

Elliott, N. C., Tao, F. L., Fuentes-Granados, R., Giles, K. L., Elliott, D. T., Greenstone, M. H., Shufran, K. A. and Royer, T. A. 2006. D-vack sampling for predatory arthropods in winter wheat. *Biological Control* 38(3): 325–330.

Feltham, H., Park, K. K. and Goulson, D. 2014. Field realistic doses of pesticide Imidacloprid reduce bumblebee pollen foraging efficiency. *Ecotoxicology* 23(3): 317-323.

Feng, M. G., Johnson, J. B. and Kish, L. P. 1990. Survey of entomopathogenic fungi naturally infecting cereal aphids (Homoptera: Aphididae) of irrigated grain crops in southwestern Idaho 1534–1542. *Environmental Entomology* 19: 1534–1542.

Feng, M. G. and Nowierski, R. M. 1992a. Spatial distribution and sampling plans for four species of cereal aphids (Homoptera: Aphididae) infesting spring wheat in southwestern Idaho. *Journal of Economic Entomology* 85(3): 830–837.

Feng, M. G. and Nowierski, R. M. 1992b. Spatial patterns and sampling plans for cereal aphids (Hom.: Aphididae) killed by entomophthoralean fungi and hymenopterous parasitoids in spring wheat. *Entomophaga* 37(2): 265–275.

Feng, M. G., Nowierski, R. M., Johnson, J. B. and Poprawski, T. I. 1992. Epizootics Caused by Entomophthoralean Fungi (Zygomycetes, Entomophthorales) in populations of cereal aphids (Homoptera: Aphididae) in irrigated small grains of southwestern Idaho, USA. *Journal of Applied Entomology* 113: 376–390.

Fuentes-Granados, R., Giles, K. and Elliott, N. C. 2001. Assessment of greenbug-resistant wheat germplasm on *Lysiphlebus testaceipes* Cresson (Hymenoptera: Aphididae) oviposition and development in greenbug over two generations. *Southwestern Entomologist* 26: 187–194.

Furlong, M. J. and Pell, J. K. 2005. Interactions between entomopathogenic fungi and arthropod natural enemies. In: Vega, F. E. and Blackwell, M. (Eds.), *Insect-Fungal Associations: Ecology and Evolution*. New York: Oxford University Press, pp. 51–73.

Giles, K. L., Jones, D. B., Royer, T. A., Elliott, N. C. and Kindler, S. D. 2003. Development of a sampling plan in winter wheat that estimates cereal aphid parasitism levels and predicts population suppression. *Journal of Economic Entomology* 96(3): 975–982.

Giles, K. L., McCornack, B. P., Royer, T. A. and Elliott, N. C. 2017. Incorporating biological control into IPM decision making. *Current Opinion in Insect Science* 20: 84–89.

Giles, K. L., Royer, T. A., Elliott, N. C. and Kindler, S. D. 2000. Development and validation of a binomial sequential sampling plan for greenbug (Homoptera: Aphididae) infesting winter wheat in the southern plains. *Journal of Economic Entomology* 93(5): 1522–1530.

Gilstrap, F. E., Kring, T. J. and Brooks, G. W. 1984. Parasitism of aphids (Homoptera: Aphididae) associated with Texas sorghum. *Environmental Entomology* 13(6): 1613–1617.

Girma, M., Kofoid, K. D. and Reese, J. C. 1998. Sorghum germplasm tolerant to greenbug (Homoptera: Aphididae) feeding damage as measured by reduced chlorophyll loss. *Journal of the Kansas Entomological Society* 71: 108–115.

Gravena, S. 1979. Population dynamics of the greenbug, Schizaphis graminum (Rondani, 1852) (Homoptera, Aphididae) and its natural enemies associated with grain sorghum in Jaboticabal, Sao Paulo, Brazil. *Anais da Sociedade Entomologica do Brasil* 8(2): 325–334.

Gray, S. M., Chapin, J. W., Smith, D. M., Banerjee, N. and Thomas, J. S. 1998. Barley yellow dwarf luteoviruses and their predominant aphid vectors in winter wheat grown in South Carolina. *Plant Disease* 82(12): 1328–1333.

Grover, S. B., Wojahn, B., Varsani, S. E., Sattler, S. E. and Louis, J. 2019 Resistance to greenbugs in the sorghum nested association mapping population. *Arthropod-Plant Interactions* 13(2): 261-269. Available at: https://digitalcommons.unl.edu/entomologyfacpub/774.

Gupta, R. C. 2006. *Toxicology of Organophosphate and Carbamate Compounds*. Burlington, MA: Elsevier Academic Press.

Harvey, T. A. and Hackerott, H. L. 1969. Recognition of a greenbug (*Schizaphis graminum*) biotype injurious to sorghum. *Journal of Economic Entomology* 62: 776–779.

Hesler, L. S. 2005. Resistance to Rhopalosiphum padi (Homoptera: Aphididae) in three triticale accessions. *Journal of Economic Entomology* 98(2): 603-610.

Hesler, L. S. and Berg, R. K. 2003. Tillage impacts cereal aphid (Homoptera: Aphididae) infestations in spring small grains. *Journal of Economic Entomology* 96(6): 1792–1797.

Hesler, L. S., Riedell, W. E., Langham, M. A. C. and Osborne, S. L. 2005. Insect infestations, incidence of viral plant diseases, and yield of winter wheat in relation to planting date in the Northern Great Plains. *Journal of Economic Entomology* 98(6): 2020–2027.

Hill, I. R. 1989. Aquatic organisms and pyrethroids. *Pesticide Science* 27(4): 429–457.

Hossain, I., Epplin, F. M. and Krenzer, E. G., Jr. 2003. Planting date influence on dual-purpose winter wheat forage yield, grain yield, and test weight. *Agronomy Journal* 95(5): 1179–1188.

Ignoffo, C. M. 1992. Environmental factors affecting persistence of entomopathogens. *Florida Entomologist* 75(4): 516–525.

Ismail, E. A., Giles, K. L., Coburn, L., Royer, T. A., Hunter, R. M., Verchot, J., Horn, G. W., Krenzer, E. G., Peeper, T. F., Payton, M. E., Michels, G. J., Bible, J. B. and Owings, D. A. 2003. Effects of aphids, barley yellow dwarf and grassy weeds on grazed winter wheat. *Southwestern Entomologist* 28: 121–130.

Jackson, H. B., Coles, L. W., Wood, E. A. Jr. and Eikenbary, R. D. 1970. Parasites reared from the greenbug and corn leaf aphid in Oklahoma in 1968 and 196912. *Journal of Economic Entomology* 63(3): 733–736.

Jaronski, S. T. 2010. Ecological factors in the inundative use of fungal entomopathogens. *BioControl* 55(1): 159–185.

Jeschke, P. 2021. Status and outlook for acaricide and insecticide discovery. *Pest Management Science* 77(1): 64–76.

Jessie, C. N., Giles, K. L., Royer, T. A., Payton, M. E., Elliott, N. C. and Jessie, W. P. 2019. Suitability of Schizaphis graminum parasitized by Lysiphlebus testaceipes as intraguild prey for Chrysoperla rufilabris. *Southwestern Entomologist* 44(1): 21–33.

Johnson, R. A. and Rochow, F. W. 1972. An isolate of barley yellow dwarf virus transmitted specifically by Schizaphis graminum. *Phytopathology* 62(8): 921–925.

Jones, D. B. 2001. *Natural enemy thresholds for greenbug, Schizaphis graminum Rondani, on winter wheat*. MS Thesis. Oklahoma State University, Stillwater.

Jones, D. B., Giles, K. L. and Elliott, N. C. 2008. Supercooling points of *Lysiphlebus testaceipes* and its host *Schizaphis graminum*. *Environmental Entomology* 37(5): 1063–1068.

Jones, D. B., Giles, K. L., Elliott, N. C. and Payton, M. E. 2007. Parasitism of greenbug, *Schizaphis graminum*, by the parasitoid *Lysiphlebus testaceipes* at winter temperatures. *Environmental Entomology* 36(1): 1–8.

Katsar, C. S., Paterson, R. H., Teetes, G. L. and Peterson, G. C. 2002. Molecular analysis of sorghum resistance to the greenbug (Homoptera: Aphididae). *Journal of Economic Entomology* 95(2): 448–457.

Kelly, M. and Guo, Q. 2007. Integrated agricultural pest management through remote sensing and spatial analyses. In: Ciancio, A. and Mukerji, K. G. (Eds) *General Concepts in Integrated Pest and Disease Management. Integrated Management of*

Plants Pests and Diseases (vol. 1). Springer, Dordrecht. https://doi.org/10.1007/978-1-4020-6061-8_7.

Kindler, S. D., Elliott, N. C., Giles, K. L. and Royer, T. A. 2003. Economic injury level for the greenbug, Schizaphis graminum in Oklahoma winter wheat. *Southwestern Entomologist* 28: 163–166.

Kirkland, R. L., Peries, K. D. and Hamilton, G. C. 1981. Differentiation and developmental rate of nymphal instars of green bug reared on sorghum. *Journal of the Kansas Entomological Society* 54: 743–747.

Knutson, A., Bynum, E., Kerns, D., Porter, P., Biles, S. and Reed, B. 2018. *Managing Insect and Mite Pests of Texas Sorghum. ENTO-085.* College Station: Texas A&M Agrilife Extension Service.

Kring, T. J. and Gilstrap, F. E. 1983. Within-field distribution of greenbug (Homoptera: Aphididae) and its parasitoids in Texas winter wheat. *Journal of Economic Entomology* 76(1): 57–62.

Kring, T. J. and Gilstrap, F. E. 1984. Efficacy of the natural enemies of grain sorghum aphids (Homoptera: Aphididae). *Journal of the Kansas Entomological Society* 57: 460–467.

Kring, T. J. and Gilstrap, F. E. 1986. Beneficial role of corn leaf aphid R*hopalosiphum maidis* (Fitch) (Homoptera: Aphididae) in maintaining *Hippodamia* spp. (Coleoptera: Coccinellidae) in grain sorghum. *Crop Protection* 5(2): 125–128.

Kring, T. J., Gilstrap, F. E. and Michels, G. J., Jr. 1986. Role of indigenous coccinellids in regulating greenbugs (Homoptera: Aphididae) on Texas grain sorghum. *Journal of Economic Entomology* 78(1): 269–273.

Krupke, C. H., Hunt, G. J., Eitzer, B. D., Andino, G. and Given, K. 2012. Multiple routes of pesticide exposure for honeybees living near agricultural fields. *PLoS ONE* 7(1): e29268.

Lazar, M. D., Michels, G. J. Jr. and Booker, J. D. 1995. Reproductive and developmental rates of two greenbug biotypes in relation to two wheat host resistance genes. *Southwestern Entomologist* 20: 467–482.

Lost Fihlo, F. H., Heldens, W. B., Kong, Z. and de Lange, E. S. 2020. Drones: innovative technology for use in precision pest management. *Journal of Economic Entomology* 113(1): 1-25.

Lu, H., Rudd, J. C., Burd, J. D. and Weng, Y. 2010. Molecular mapping of greenbug resistance genes Gb2 and Gb6 in T1AL.1RS wheat-rye translocations. *Plant Breeding* 129: 472–476.

Luo, K., Zhao, H., Wang, X. and Kang, Z. 2021. Prevalent pest management strategies for grain aphids: opportunities and challenges. *Frontiers in Plant Science* 12: 790919. https://doi.org/10.3389/fpls.2021.790919.

Maina, U. M., Galadima, I. B. Gambo, F. M., and Zakaria, D. 2018. A review on the use of entomopathobenic fungi in the management of insect pest of field crops. *Journal of Entomology and Zoology Studies* 6: 27-32.

Martin, W. 1981. Wild goatgrass gives greenbug resistance to wheat. *Agricultural Research USDA* 30: 13.

Mayo, Z. B. and Starks, K. J. 1974. Temperature influences on alary polymorphism in *Schizaphis graminum*1,2. *Annals of the Entomological Society of America* 67(3): 421–423.

McCornack, B. P. and Johnson, W. A. 2016. Getting growers to go digital: The power of a positive user experience. *Journal of Extension* 54: Article 19. Available at: https://tigerprints.clemson.edu/joe/vol54/iss4/19.

Michaud, J. P. 2017. IPM case studies: Sorghum. In: van Emden, H. F. and Harrington, R. (Eds.), *Aphids as Crop Pests*. London, UK: CABI Publishing, pp. 557–568.

Michaud, J. P., Whitworth, R. J. and Schwarting, H. 2014. *Sorghum Insect Management 2015. MF742, Kansas State University Agricultural Experiment Station and Cooperative Extension Service*. Manhattan, KS.

Michels, G. J. Jr. 1986. Graminaceous North American host plants of the greenbug with notes on biotypes. *Southwestern Entomologist* 11: 55–66.

Michels, G. J., Jr. and Behle, R. W. 1992. Evaluation of sampling methods for lady beetles (Coleoptera: Coccinellidae) in grain sorghum. *Journal of Economic Entomology* 85(6): 2251–2257.

Michels, G. J., Jr., Elliott, N. C., Romero, R. A., Owings, D. A. and Bible, J. B. 2001. Impact of indigenous coccinellids on Russian wheat aphids and greenbugs (Homoptera: Aphididae) infesting winter wheat in the Texas panhandle. *Southwestern Entomologist* 26: 97–114.

Michels, G. J., Jr., Fritts, D. A., Rush, C. M. and Piccinni, G. 1999. Using infrared transducers to sense greenbug (Homoptera: Aphididae) infestations in winter wheat. *Southwestern Entomologist* 24: 269–279.

Michels, G. J., Jr. and Undersander, D. J. 1986. Temporal and spatial distribution of the greenbug (Homoptera: Aphididae) on sorghum in relation to water stress 1221–1225. *Journal of Economic Entomology* 79: 122.

Mirik, M., Michels, G. J., Mirik, S. K., Elliott, N. C. and Catana, V. 2006. Spectral sensing of aphid (Hemiptera: Aphididae) density using field spectrometry and radiometry. *Turkish Journal of Agriculture and Forestry* 30: 421–428.

Mittler, T. E. and Gorder, N. K. N. 1991. Variation between clones of *Schizaphis graminum* (Homoptera: Aphididae) in the photoperiodic induction of sexual morphs. *Environmental Entomology* 20(2): 433–440.

Morgham, A. T., Richardson, P. E., Campbell, R. K., Burd, J. D., Eikenbary, R. D. and Sumner, L. C. 1994. Ultrastructural responses of resistant and susceptible wheat to infestation by greenbug biotype E (Homoptera: Aphididae). *Annals of the Entomological Society of America* 87(6): 908–917.

Mullins, C. B., Giles, K. L., Ye, C. M. and Phoofolo, M. W. 2011. Using PCR to detect intraguild predation of Lysiphlebus testaceipes by coccinellids. *Southwestern Entomologist* 36: 295–304.

Nuessly, G. S., Nagata, R. T., Burd, J. D., Hentz, M. G., Carroll, A. S. and Halbert, S. E. 2008. Biology and biotype determination of greenbug, *Schizaphis graminum* (Hemiptera: Aphididae), on seashore paspalum turfgrass (*Paspalum vaginatum*). *Environmental Entomology* 37(2): 586–591.

Onstad, D. W. and Knolhoff, L. 2022. Chapter 9: Arthropod resistance to crops. In: Onstad, D. W. (Ed.), *Insect Resistance Management: Biology, Economics and Prediction*. New York: Elsevier Ltd.

Painter, R. H. and Peters, D. C. 1956. Screening wheat varieties and hybrids for resistance to the greenbug. *Journal of Economic Entomology* 49(4): 546–548.

Phoofolo, M. W., Giles, K. L. and Elliott, N. C. 2010. Effects of relay-intercropping Sorghum with winter wheat, Alfalfa, and cotton on lady beetle (Coleoptera Coccinellidae) abundance and species composition. *Environmental Entomology* 39: 763–774.

Pike, K. S. and Schaffner, R. L. 1985. Development of autumn populations of cereal aphids, Rhopalosiphum padi (L.) and Schizaphis graminum (Rondani) (Homoptera:

Aphididae) and their effects on winter wheat in Washington State. *Journal of Economic Entomology* 78(3): 676–680.

Porter, D. R., Burd, J. D., Shufran, K. A., Webster, J. A. and Teetes, G. L. 1997. Greenbug (Homoptera: Aphididae) biotypes: Selected by resistant cultivars or preadapted opportunists? *Journal of Economic Entomology* 90(5): 1055–1065.

Porter, D. R. and Mornhinweg, D. W. 2004. Characterization of greenbug resistance in barley. *Plant Breeding* 123(5): 493–494.

Punnuri, S., Steets, J. and Wu, Y. 2012. Developing new markers and QTL mapping for greenbug resistance in sorghum [*Sorghum bicolor* (L.) Moench]. *Euphytica* 191: 191-203.

Puterka, G. J. and Peters, D. C. 1989. Inheritance of greenbug virulence to Gb2 and Gb3 resistance genes in wheat. *Genome* 32: 109-114.

Radchenko, E. E. 2010. Identification of sorghum genes responsible for resistance to greenbug. *Genetika* 36(4): 510-519.

Radchenko, E. E., Berim, M. N. and Zubov, A. A. 2017. Genetic control of different types of aphid resistance in cereal crops. *Entomological Review* 87: 253–262.

Radchenko, E. E. and Tyryshkin, L. G. 2004. Components of the greenbug (*Schizaphis graminum* Rond.) resistance and wheat and barley somaclonal variants. *Cereal Research Communications* 32(2): 255–258.

Rai, M. and Ingle, A. 2012. Role of nanotechnology in agriculture with special reference to management of insect pests. *Applied Microbiology and Biotechnology* 94(2): 287-293.

Rice, M. E. and Wilde, G. E. 1989. Antibiosis effect of sorghum on the convergent lady beetle (Coleoptera; Coccinellidae), a third-trophic level predator of the greenbug (Homoptera: Apididae). *Journal of Economic Entomology* 82(2): 570–573.

Riedell, W. E. and Blackmer, T. M. 1999. Leaf reflectance spectra of cereal aphid-damaged wheat. *Crop Science* 39(6): 1835–1840.

Rondani, C. 1852. *Aphis graminum* n. sp. *Nuove Ann. Sci. Natl Bologna* 6: 9-11.

Royer, T. A. 2021. *Management of Insect and Mite Pests in Sorghum. CR-7170*. Stillwater: Oklahoma Cooperative Extension Service.

Royer, T. A., Giles, K. L., Kindler, S. D. and Elliott, N. C. 2001. Developmental response of three geographic isolates of *Lysiphlebus testaceipes* (Cresson) (Hymenoptera: Aphidiidae) to temperature. *Environmental Entomology* 30(4): 637–641.

Royer, T. A., Giles, K. L., Lebusa, M. M. and Payton, M. E. 2008. Preference and suitability of greenbug, *Schizaphis graminum* (Hemiptera: Aphididae) parasitized by *Lysiphlebus testaceipes* (Hymenoptera: Braconidae) as a food source for *Coccinella septempunctata* and Hippodamia convergens (Coleoptera: Coccinellidae). *Biological Control* 47(1): 82–88.

Royer, T. A., Giles, K. L., Kindler, S. D. and Elliott, N. C. 2022. *Cereal Aphid Expert System and Glance-N-Go Sampling for Greenbugs in Winter Wheat: Questions and Answers*. Stillwater, OK: Oklahoma Cooperative Extension Service.

Royer, T. A., Giles, K. L., Nyamanzi, T., Hunger, R. M., Krenzer, E. G., Elliott, N. C., Kindler, S. D. and Payton, M. 2005. Economic evaluation of the effects of planting date and application rate of Imidacloprid for management of cereal aphids and barley yellow dwarf in winter wheat. *Journal of Economic Entomology* 98(1): 95–102.

Royer, T. A., Pendleton, B. B., Elliott, N. C. and Giles, K. L. 2015. Greenbug (Hemiptera: Aphididae) biology, ecology and management in wheat and sorghum. *Journal of Integrated Pest Management* 6(1): 19. DOI: 10.1093/jipm/pmv018.

Ryan, J. D., Johnson, R. C., Eikenbary, R. D. and Dorschner, K. W. 1987. Drought/greenbug interactions: photosynthesis of greenbug resistant and susceptible wheat. *Crop Science* 27(2): 283-288.

Schuster, D. J. and Starks, K. J. 1973. Greenbugs: Components of host-plant resistance in sorghum123. *Journal of Economic Entomology* 66(5): 1131-1134.

Shufran, K. A. 2011. Host race evolution in Schizaphis graminum (Hemiptera: Aphididae): Nuclear DNA sequences. *Environmental Entomology* 40(5): 1317-1322.

Shufran, K. A., Burd, J. D., Anstead, J. A. and Lushai, G. 2000. Mitochondrial DNA sequence divergence among greenbug (Homoptera: Aphididae) biotypes: Evidence for host-adapted races. *Insect Molecular Biology* 9(2): 179-184.

Shufran, K. A., Weathersbee, A. A., III, Jones, D. B. and Elliott, N. C. 2004. Genetic similarities among geographic isolates of *Lysiphlebus testaceipes* (Homenoptera: Aphidiidae) different in cold temperature tolerances. *Environmental Entomology* 33(3): 776-778.

Shufran, R. A., Wilde, G. E., Sloderbeck, P. E. and Morrison, W. P. 1997. Occurrence of insecticide resistant greenbugs (Homoptera: Aphididae) in Kansas, Texas, Oklahoma, and Colorado and suggestions for management. *Journal of Economic Entomology* 90(5): 1106-1116.

Simone, J. C.. and Peccoud, J. 2018. Rapid evolution of aphid pests in agricultural environments. *Current Opinion in Insect Science* 26: 17-24.

Sishodia, R. P., Ray, R. L. and Singh, S. K. 2020. Applications of remote sensing in precision agriculture: a review. *Remote Sensing* 12(19): 3136: https://doi.org/10.3390/rs12193136.

Skendzic, S., Zovko, M., Zivkovic, I. P., Lesic, V. and Lemic, D. 2021. The impact of climate change on agricultural insect pests. *Insects* 12(5): 440. https://doi.org/10.3390/insects12050440.

Starks, K. J., Muniappan, R. and Eikenbary, R. D. 1972. Interaction between plant resistance and parasitism against the greenbug on barley and sorghum. *Annals of the Entomological Society of America* 65(3): 650-655.

Starks, K. J., Wood, E. A., Jr. and Burton, R. L. 1974. Relationship of plant resistance and *Lysiphlebus testaceipes* to population levels of the greenbug on grain sorghum. *Environmental Entomology* 3(6): 950-952

Tan, C. T., Yu, H., Yang, Y., Xu, X., Chen, M., Rudd, J. C., Xue, Q., Ibrahim, A. M. H., Garza, L., Wang, S., Sorrels, M. E. and Liu, S. 2017. Development and validation of KASP markers for the greenbug resistance gene Gb7 and the Hessian fly resistance gene H32 in wheat. *TAG: Theoretical and Applied Genetics. Theoretische und Angewandte Genetik* 130(9): 1867-1884.

Teetes, G. L. and Johnson, J. W. 1973. Damage assessment of the greenbug on grain sorghum. *Journal of Economic Entomology* 66(5): 1181-1186.

Teetes, G. L. and Johnson, J. W. 1974. Assessment of damage by the greenbug in grain sorghum hybrids of different maturities. *Journal of Economic Entomology* 67: 514-516.

Teetes, G. L. and Pendleton, B. 2000. Insect pests of sorghum. In: Smith, C. W. and Fredriksen, R. A. (Eds.), *Sorghum: Origin, History, Technology and Production*. New York, NY: Wiley & Sons, pp. 443-496.

Teetes, G. L., Rosenow, D. T., Fredricksen, R. D. and Johnson, J. W. 1973. *The Predisposing Influence of Greenbugs on Charcoal Rot of Sorghum*. Progress Report. PR-3173. Agricultural Experiment Station, TX.

Teetes, G. L., Schaefer, C. A., Gipson, J. R., McIntyre, R. C. and Latham, E. E. 1975. Greenbug resistance to organophosphorous insecticides on the Texas high plains. *Journal of Economic Entomology* 68: 214-216.

Tyler, J. M., Webster, J. A. and Merkle, O. G. 1987. Designations for genes in wheat germplasm conferring greenbug resistance. *Crop Science* 27(3): 526-527.

Tyler, J. M., Webster, J. A. and Smith, E. L. 1985. Biotype E greenbug resistance in wheat streak mosaic virus-resistant wheat germplasm lines 1. *Crop Science* 25(4): 686-688.

Wadley, F. M. 1931. Ecology of *Toxoptera graminum*, especially as to factors affecting importance in the northern United States. *Annals of the Entomological Society of America* 24(2): 325-395.

Walgenbach, D. D., Elliott, N. C. and Kieckhefer, R. W. 1988. Constant and fluctuating temperature effects on developmental rates and life table statistics of the greenbug (Homoptera: Aphididae). *Journal of Economic Entomology* 81(2): 501-507.

Walker, A. L., Bottrell, D. G. and Cate, J. R., Jr. 1973. Hymenopterous parasites of biotype C greenbug in the High Plains of Texas. *Annals of the Entomological Society of America* 66(1): 173-176.

Washburn, F. L. 1908. Sexual forms of *Toxoptera graminum* Rond. *The Canadian Entomologist* 40(2): 53-54.

Webster, J. A. 1995. *Economic Impact of the Greenbug in the Western United States. 1992-1993*. Volume Publication No. 155. Stillwater, OK: Great Plains Agricultural Council.

Webster, F. M. and Phillips, W. J. 1912. The spring grain aphis or "greenbug." *U. S. Bureau of Entomology Bulletin* 110: 1-153. Available at: https://archive.org/details/springgrainaphis110webs.

Weng, Y., Li, W., Devkota, R. N. and Rudd, J. C. 2005. Microsatellite markers associated with two Aegilops tauschii -derived greenbug resistance loci in wheat. *TAG. Theoretical and Applied Genetics. Theoretische und Angewandte Genetik* 110(3): 462-469.

Wilde, G. E., Shufran, R. A., Kindler, S. D., Brooks, H. L. and Sloderbeck, P. E. 2001. Distribution and abundance of insecticide resistant greenbugs (Homoptera: Aphididae) and validation of a bioassay to assess resistance. *Journal of Economic Entomology* 94(2): 547-551.

Wright, R., Danielson, S. and Mayo, Z. B. 2006. *Management of Greenbugs in Sorghum. G838 NebGuide*. University of Nebraska, Lincoln.

Wu, Y., and Huang, Y. 2008. Molecular mapping of QTLs for resistance to the greenbug Schizaphis graminum (Rondani) in Sorghum bicolor (Moench). Theoretical and Applied Genetics 117: 117-127.

Xu, X., Li, G., Carver, B. F. and Armstrong, J. S. 2020. Gb8: A new gene conferring resistance to economically important greenbug biotypes in wheat. *TAG Theoretical and Applied Genetics. Theoretische und Angewandte Genetik* 133(2): 615-622.

Yang, Z., Rao, M. N., Elliott, N. C., Kindler, S. D. and Popham, T. W. 2005. Using groundbased multispectral radiometry to detect stress in wheat caused by greenbug (Homoptera: Aphididae) infestation. *Computers and Electronics in Agriculture* 47(2): 121-135.

Yunding, Z., Shoudong, B., Qinlei, M., Jiguang, F., Gaochao, C., Gongming, W. and Jialin, L. 1998. Influence of natural enemies on populations of wheat aphid and greenbug. *Chinese Journal of Applied Ecology* 9: 613-616.

Zhu, L. C., Smith, C. M., Fritz, A., Boyko, E. V. and Flinn, M. B. 2004. Genetic analysis and molecular mapping of a wheat gene conferring tolerance to the greenbug

(*Schizaphis graminum* Rondani).*TAG Theoretical and Applied Genetics. Theoretische und Angewandte Genetik* 109(2): 289–293.

Zhu, L. C., Smith, C. M., Fritz, A., Boyko, E., Voothuluru, P. and Gill, B. S. 2005. Inheritance and molecular mapping of new greenbug resistance genes in wheat germplasms derived from Alegilops tauschii. *TAG Theoretical and Applied Genetics. Theoretische und Angewandte Genetik* 111(5): 831-837. DOI: 10.1007/s00122-005-0003-6.

Chapter 8

Greenbug-wheat interactions, pest management and host resistance

L. A. Crespo-Herrera, International Maize and Wheat Improvement Center (CIMMYT), Mexico; J. Huerta-Espino, Instituto Nacional de Investigaciones Forestales Agrícolas y Pecuarias, Mexico; and R. P. Singh, International Maize and Wheat Improvement Center (CIMMYT), Mexico

1 Introduction

Aphids are the most common pests of cereals, distributed globally (CABI-EPPO, 2004, 2022; Blackman and Eastop, 2017). Yield losses due to aphid feeding alone can be up to 30% (Voss et al., 1997; Riedell et al., 1999) and up to 60% when aphid feeding damage is combined with the transmission of viral diseases such as Barley yellow dwarf virus (BYDV) (Riedell et al., 2003). In addition, aphids excrete a sugary compound known as honeydew, which serves as the medium for the development of sooty molds that can also affect plant photosynthesis and consequently yield (Rabbinge et al., 1981).

Schizaphis graminum (Rondani), commonly known as greenbug, is one of the most common cereal aphids with economic importance. *Schizaphis graminum* individuals are light-green with dark-tipped siphunculi and typically have a green longitudinal stripe on the dorsal side of the abdomen. The forewings of winged individuals have the medial vein forked once (Stoetzel,

http://dx.doi.org/10.19103/AS.2022.0114.05

1987) (Aradottir and Crespo-Herrera, 2021). Direct feeding of *Schizaphis graminum* on susceptible plants causes chlorosis and necrotic spots at the feeding site (Fig. 1). Greenbugs are capable of transmitting BYDV, especially the SGV strain (Gray and Gildow, 2003) and the cereal yellow dwarf virus (CYDV).

Possibly having originated from Central Asia (but see Michuad, 2010, who makes a case for a European origin), this species is widely distributed across Asia, Africa, North and South America, Southern Europe and Oceania (Fig. 2). *Schizaphis graminum* feeds on various genera of Poaceae, e.g. *Agropyron*, *Avena*, *Bromus*, *Dactylis*, *Eleusine*, *Festuca*, *Hordeum*, *Lolium*, *Oryza*, *Panicum*, *Poa*, *Sorghum*, *Triticum*, and *Zea* (Blackman and Eastop, 2017), and it has also been found feeding on Triticale, Tritordeo and *Secale* (Clua et al., 2004). Although it has a wide host range, *Schizaphis graminum* populations are composed of specific host-adapted biotypes (Clua et al., 2004; Burd and Porter, 2006). In North America, *Schizaphis graminum* populations had large genetic diversity even before resistant cultivars were released (Porter et al., 1997; Burd and Porter, 2006). Non-cultivated grasses, or green bridges, serve as reservoirs to sustain genetically diverse *Schizaphis graminum* populations and can also act as sources of early crop infestations in wheat (Anstead et al., 2003). Greenbug has been a pest of wheat in parts of Europe, Africa, Asia and North, South and Central America (Van Emden and Harrington, 2007). It has yet to reach Australia and is the subject of vigilance to detect its invasion there (Plant Health Australia, 2021). It has been a pest of wheat in the continental USA for more than a century (Webster and Phillips, 1912). During that time, it also attacked other cultivated grasses and could be collected from more than 30 wild grass species in North America and a few wild species in Europe. In 1968, it was recorded as an outbreaking pest on sorghum in Arizona, Colorado, Kansas, Nebraska, Oklahoma, South Dakota and Texas, USA (Harvey and Hackerott, 1969). It has continued to be problematic in wheat and sorghum, leading to extensive research and research-based guidelines for its management,

Figure 1 *Schizaphis graminum* feeding on susceptible wheat. Source: Photo by L. Crespo-Herrera.

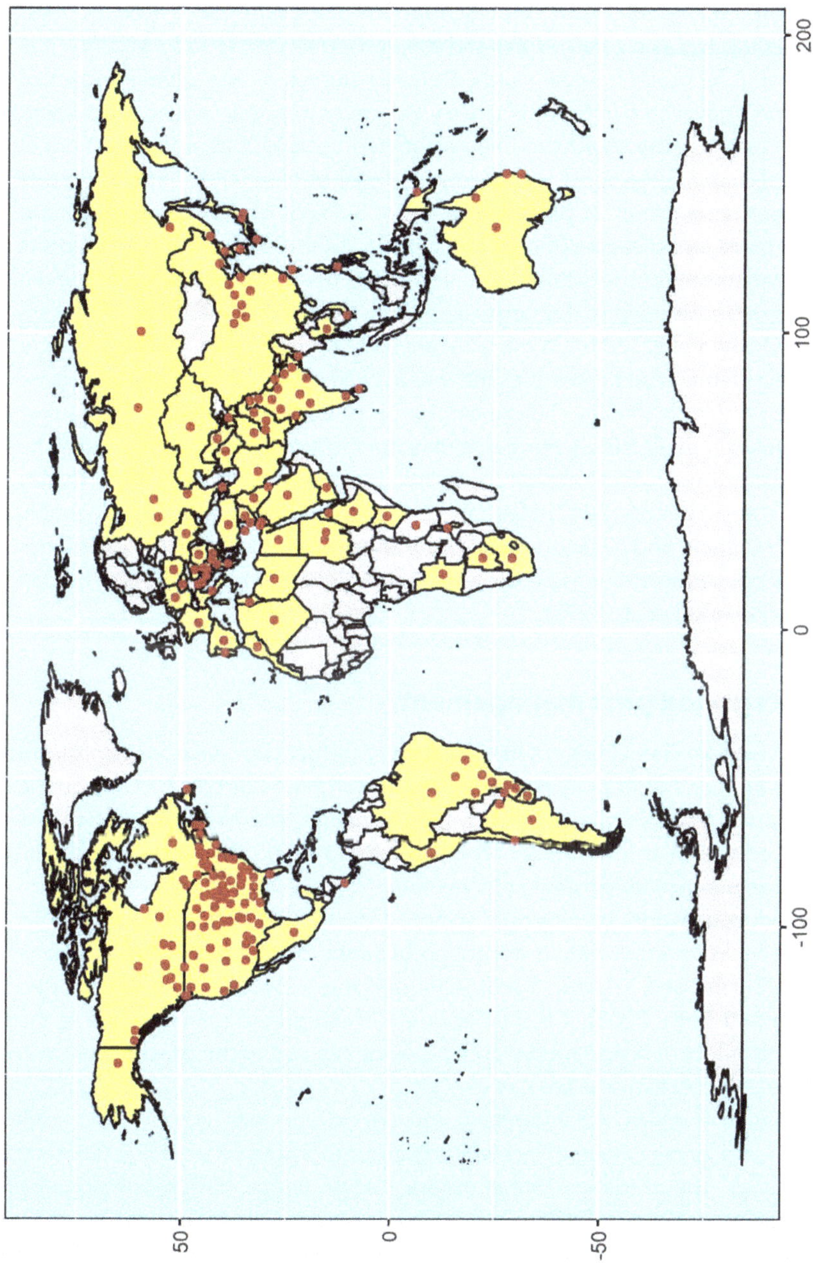

Figure 2 *Schizaphis graminum* world presence. Red dots indicate reports of the species in databases, and regions in yellow indicate report of the species at the country level. Source: Global Biodiversity Facility (2022) and the Commonwealth Agricultural Bureaux International (2022).

especially in the Southern Great Plains where it has been an important pest, and received special focus from researchers in Kansas, Oklahoma and Texas.

Given the yield losses that *Schizaphis graminum* can cause (Kindler et al., 2002; Hussain et al., 2021) and the consequences to food security, it is paramount to protect cereal yields. Detailed studies on the global impact of *Schizaphis graminum* in terms of grain yield losses in cereals are not published. Annual losses in the USA have been estimated to US$100M, and proportional levels of injury likely occur worldwide (Webster et al., 2000). General reports on insects as a group of pests indicate that 8–9% of global wheat losses are due to pests and diseases (Oerke and Dehne, 2004; Oerke, 2005), and reports from some countries indicate that these losses have not decreased and even tend to increase despite the large expenditures on pesticides (Dhaliwal et al., 2010; Oliveira et al., 2014). A more recent assessment of crop yield losses due to insect damage indicates that the wheat grain loss per centigrade degree increase owed to climate change can be up to 25% in relation to the current loss (Deutsch et al., 2018). For these reasons, crop protection is a cornerstone in food production. Present pest control relies primarily on the use of pesticides (Devine and Furlong, 2007; Meehan et al., 2011; Nansen and Ridsdill-Smith, 2013; Meehan and Gratton, 2016). Indiscriminate use of insecticides has serious consequences on the environment (Mitchell et al., 2017) and human health (Pimentel et al., 1992).

2 Integrated pest management

In the Southern Great Plains of the USA during the past few decades, *Schizaphis graminum* management in winter wheat and sorghum has come to rely on the adoption of reduced tillage, conservation of natural enemies, insecticides and improved sampling protocols and thresholds to guide their use, insecticidal seed treatments and resistant crop varieties.

Reduced tillage, increasingly adopted for wheat production worldwide, has been associated with reduced populations of *Schizaphis graminum* in wheat (Burton and Krenzer, 1985). Although this is not consistently reported (Hesler and Berg, 2003), it is a component of integrated management of the pest. The effect of tillage on the aphid is evidently tied to the presence of crop residue, but mechanisms are unclear.

Natural enemies of *Schizaphis graminum* include various coccinellid adult beetles and larvae, larvae of lacewings (Neuroptera) and syrphid flies (Diptera: Syrphidae), parasitoids (Hymenoptera) and fungi (Entomophthorales). *Lysiphlebus testaceipes* (Cresson) (Hymenoptera: Aphidiidae) is the most abundant parasitoid of *Schizaphis graminum* in wheat (Archer et al., 1974) and sorghum (Gilstrap et al., 1984) in the Southern Great Plains. In these regions, parasitism rates of 10–15% can cause *Schizaphis graminum* populations to decline in that region. Fungal entomopathogens (Entomophthorales) can suppress populations of cereal aphids, including *Schizaphis graminum*. Their effectiveness depends on environmental conditions especially relative humidity. Epizootics of these fungi were sporadic and most often associated with the number and timing of rainy days, cumulative precipitation, irrigation regimes and other factors in southwest Idaho (Feng et al., 1992). In nonirrigated cereals in drying climates, infection rates are typically negligible. Management decisions for *Schizaphis graminum* should be taken, considering the potential of natural enemies to suppress its populations below economic injury levels (see below).

Insecticides are important for *Schizaphis graminum* management, and a range of active ingredients are available and registered for use depending on the country or state. The organophosphate insecticides, such as parathion and disulfoton, and carbamates were commonly used for controlling *Schizaphis graminum* in the USA for decades (Teetes et al., 1975) and in Latin America in the 1990s (Gao and Zhu, 2000). These materials continue to be effective, although some resistance has developed to organophosphorus and carbamate insecticides in the USA (Peters et al., 1975; Wilde et al., 2001). These materials are typically applied as foliar sprays or in-furrow, depending upon the material.

Foliar or in-furrow insecticidal treatments should be based on sampling, economic thresholds and decision supports. Sampling plans for *Schizaphis graminum* have been developed in several states in the USA based on within-field distributions (Elliott and Kieckhefer, 1986; Elliott et al., 1990; Feng and Nowierski, 1992 ; Giles et al., 2000). Since the late 1990s, these plans have been binomial sequential plans in which the presence or absence of the aphid on individual plants is scored along a transect and variability during sampling is used to determine the number of samples required for a decision to treat or not treat based on a threshold. This method saves time and resources. Thresholds have been developed for Oklahoma wheat (Kindler et al., 2003) and for sorghum in Texas (Teetes and Johnson, 1973). A typical threshold, based on estimates of grain value and treatment costs, is 3 infested wheat stems out of 15 sampled, with samples of 15 plants taken until it can be concluded whether the infestation exceeds that threshold. Thresholds vary with the time of the sampling and the age of the crop. A useful decision support tool developed in the early 2000s is the online 'Glance 'N Go' (Elliott et al., 2004). In summary, this tool has instructions for implementing a sequential presence-absence sampling plan to determine when aphid populations exceed a specific treatment threshold. Based on the research by Giles (Giles et al., 2000; Giles et al., 2008), the tool also includes quantifying the number of aphid mummies (individuals attacked by *L. testaceipes*) to adjust the threshold upward when they are abundant. The tool is more completely described in Royer et al. (2015, Fig. 7) and is available as a mobile application that is only available and relevant for users in Oklahoma.

Nonetheless, the principle of 'Glance 'N Go' is designed specifically and will be applicable wherever *Schizaphis graminum* is a pest. Those interested in the tool outside of Oklahoma can request access with the authors. Other guides for its management are available in other midwestern states. In Kansas, advice is available under *Schizaphis graminum* here: http://myfields.info/. In addition to using these decision tools, sampling can be implemented based on conditions. For example, *Schizaphis graminum* outbreaks occur more often during a mild winter followed by a cool spring, especially if plants are suffering from lack of moisture (Royer et al., 2020). Although these principles have been delineated for the Great Plains, their adoption for other regions will entail considerable research that is currently lacking.

In addition to its importance as a direct pest, like other cereal aphids, *Schizaphis graminum* is a vector of plant viruses including BYDV (Van Emden and Harrington, 2007). Serotypes of BYDV designated BYDV-SGV are most efficiently transmitted by *Schizaphis graminum*. When the virus is prevalent, direct injury thresholds are not useful. This persistently transmitted virus can be managed by controlling vectors chemically, but treatment can be justified by very low populations since a single feeding event can result in infection and disease. Plants are more susceptible when young, so early-season infestations of the aphid with the presence of the virus can require aggressive vector control to protect the crop.

Neonicotinoid insecticides have been used for seed treatments for cereal aphids, including *Schizaphis graminum*, around the world with good efficacy (Royer et al., 2005; Abd-Ella, 2016; Zhang et al., 2016). These seed treatments can also significantly limit the virus spread by aphids (Perkins et al., 2018). Unfortunately, seed treatments by their nature cannot be applied as-needed but must be applied prophylactically, and the practice is becoming widespread and is contrary to this key principle of integrated pest management (IPM) (Tooker et al., 2017).

3 Host plant resistance

An alternative or complementary control method to insecticides is the use of resistant cultivars. Host resistance (HR) to insects is defined as the set of plant characteristics, genetically governed, that makes the plants to avoid, reduce or endure insect damage. Use of HR is an environmentally friendly method to reduce insect damage to crops. It is easy to use and combine with other control strategies, while it contributes to reducing the use of pesticides. HR is also very advantageous for farmers in situations where they have limited or no access to other control methods. However, sources of HR to insects that render near-immune responses are not always available, which may prevent their use in breeding programs. Nonetheless, HR, if available, along with the determination of economic thresholds that allow growers to attain the desired margin of profit, is considered the base of IPM strategies (Wiseman, 1994). Thus, understanding the underlying mechanisms and interactions between the host and *Schizaphis graminum* is fundamental for more efficient use of HR in breeding and evaluation efforts.

3.1 Schizaphis graminum *biotypes*

Schizaphis graminum populations have been classified based on their ability to feed on cultivated or non-cultivated grasses. In North America, there are 29 designated biotypes differentiated by their host associations and virulence/avirulence patterns in cereals (Burd and Porter, 2006). The biotypes that commonly feed on cultivated grasses are designated with letters A to K (Porter et al., 1997; Berzonsky et al., 2003). The biotype concept is useful to identify the differential damage of pest populations on hosts, but this does not imply that individuals of the same biotype are genetically uniform (Claridge and Den Hollander, 1983). Earlier studies demonstrate that individuals of *Schizaphis graminum* biotypes are diverse and also that such biotypes may have diverged based on host associations (Shufran et al., 1992; Shufran et al., 2000). For instance, based on SSR (simple sequence repeats) markers, biotypes E, I and K are genetically related and placed within one clade of a phylogenetic tree (Weng et al., 2010). However, biotypes I and K are still in a subgroup different from biotype E within that clade. Biotype H is genetically distant from all the other biotypes (Shufran et al., 2000; Weng et al., 2010).

Host associations have a role in this genetic differentiation; biotypes I and K were first found in sorghum, biotype E was identified in wheat, biotype F feeds mainly on bluegrass, biotype G has been collected mainly on *Agropyron* species, and biotype H was found on *Aegilops cylindrica* and *A. intermedium* (Burd and Porter, 2006; Goldman and Springer, 2010; Weng et al., 2010). These host-biotype associations can relate to the display of different proteomic profiles in *Schizaphis graminum* saliva since virulence effectors can be differentially expressed between biotypes. For instance, at the proteomic level, biotypes G and H express higher levels of a particular paralog of glucose dehydrogenase (GDH1) and carbonic anhydrase 7 like, whereas biotypes E and C express higher levels of an abnormal oocyte. These proteins have specific functions to reduce host-produced quinone and phenoxy radicals (GDH1 and other paralogs), binding salicylic acid (SA) and maintaining the pH of the salivary gland cells during salivation (Nicholson and Puterka, 2014). In terms of virulence patterns, biotypes E and C only differ in that the former is virulent to *Gb2*-resistance gene in wheat and sorghum TX 3737, whereas the latter is avirulent. On the other hand, biotypes G and H differ in their virulence from *Gb6* (biotype H) and *Gb3* (biotype G) in wheat (Burd and Porter, 2006; Nicholson and Puterka, 2014).

4 Wheat, environment and *Schizaphis graminum*

Environmental variables, including temperature, water availability and atmospheric CO_2 concentrations, can affect aphids directly or indirectly through their host plant. Soil nutrients can also have indirect effects on aphids via their hosts. There are few publications on the interaction between the environment and *Schizaphis graminum* in comparison with other insect pests.

4.1 Temperature

Most cereal aphids have a thermal window between 4.0°C and 30.0°C, with an optimum of 20.0–25.0°C. The thermal window for *Schizaphis graminum* is 6–29°C (Walgenbach et al., 1988). A temperature rise of 2.0°C would consequently increase the number of aphid generations in regions where temperatures presently are near but below optimal (Hullé et al., 2010; Aljaryian and Kumar, 2016; Deutsch et al., 2018). With global warming, it is expected that *Schizaphis graminum* will increase its geographical presence in the Northern Hemisphere, particularly in Europe and North America, due to the increase of new suitable areas for its establishment (Aljaryian and Kumar, 2016). Whereas in the Southern Hemisphere, the trend is that those suitable areas remain almost the same (Aljaryian and Kumar, 2016).

Global warming generally facilitates pest dispersal beyond their original geographical limits (Bebber et al., 2013). One example of this in wheat is *Sitobion avenae* Fabricius, which was not reported in northern latitudes until the early 1900s, but toward the end of the second half of the twentieth century, it was reported in both the southern latitudes and more northern areas (Bebber et al., 2013).

However, for an insect pest to become invasive in other geographical areas, other factors besides climatic conditions, such as the presence and quality of the host, food trade, and agricultural practices play an important role (Bebber et al., 2014; Skendžić et al., 2021). Nonetheless, climate can set the limits of the dispersal of pests to become invasive, increase earlier infestations of those pests already present and/or increase the presence of new species in new regions (Skendžić et al., 2021). For example, data analysis from the European observation network EXAMINE, which provides information from 58 sites across Europe spanning 2000 km from southern Mediterranean areas to northern Scotland, indicates an increment of 20% in the number of aphid species captured over 3–4 decades, thus increasing the diversity due to migrations and rare species becoming more common (Hullé et al., 2010).

Changing temperatures not only affect the productivity of the crop but also, if resistant, may modify the expression of its defenses. One example is the specific case of wheat and Hessian fly, where the resistance mediated by single major genes is not effective at higher temperatures (Tang et al., 2018). On the other hand, *Schizaphis graminum* resistance in sorghum, a tropical crop, is enhanced at higher temperatures (Schweissing and Wilde, 1978, 1979; Harvey et al., 1994; Thindwa and Teetes, 1994; Pendleton et al., 2009), possibly because higher temperatures contribute to an increase of metabolites associated with plant defenses (Schweissing and Wilde, 1979; Harvey et al., 1994).

However, this cannot be generalized to all crops of the *Poaceae* family as the expression of the resistance is related to the optimal host temperature and specific crop-pest associations (Schweissing and Wilde, 1978; Nechols et al., 2020). In the case of wheat, the situation may be the opposite of that of sorghum because of the observations made on oats, barley and rye, where *Schizaphis graminum* tolerance, measured as biomass loss, remained effective in lower temperatures (10–21°C), and more damage was observed in temperatures exceeding 21°C (Schweissing and Wilde, 1978). This topic of 'HR × temperature' interactions requires further investigation, as it has important consequences under global warming.

4.2 Ambient CO_2 concentration

Elevated CO_2 concentration in the atmosphere, as a result of human activity, will most likely influence aphid performance indirectly through the modification in plant chemistry, physiology and nutritional content (Robinson et al., 2012). The literature indicates that the response to this factor is species-specific (Sun et al., 2009; Sun and Ge, 2011; Oehme et al., 2013; Sun et al., 2016; Wang et al., 2018).

Higher CO_2 increases the carbon:nitrogen (C:N) ratio, causing an increase in total carbohydrates, diluting the N-based metabolites like soluble proteins and free amino acids (Robinson et al., 2012). In wheat plants, the assimilation of NO_3 and its conversion into organic N compounds is inhibited (Bloom et al., 2010). The plant N content correlates positively with the efficiency of herbivores to convert plant tissue into their own biomass (Robinson et al., 2012). Additionally, the photosynthetic capacity of plants is enhanced under higher CO_2 by reducing the rates of photorespiration (Bloom et al., 2010). However, the inhibition of NO_3 is also considered to be responsible for CO_2 acclimation, which occurs when the photosynthetic enhancement declines over time (Bloom et al., 2010).

Research indicates that CO_2 by itself does not negatively affect the reproduction of *Schizaphis graminum* in wheat and could even reduce the interspecific competition with other aphid species under specific fertilization rates, thus potentially increasing the damage caused to the plants by infestations of multiple aphid species (Sun et al., 2009; Long et al., 2019). For example, it has been shown that under higher CO_2, *Schizaphis graminum* populations can significantly increase at 100 kg N/ha and reduce interspecific competition with *Rhopalosiphum padi* L. at 200 kg N/ha (Sun et al., 2009). However, these observations were made under controlled conditions, and the efficiency of N uptake by the plant needs to be considered in real field conditions, where normally the N uptake efficiency can vary depending on various factors such as soil conditions, agricultural practices and plant genotypes (Cormier et al., 2016).

Evidence also suggests that at higher CO_2, there are aphid-induced response changes, particularly in the deposition of callose, which are driven mainly by the calcium-signaling pathways (Sinha and Smith, 2014). Under elevated CO_2, there can be high 'genotype × environment' interactions that can be influenced by aphid biotypes. Thus, the resistance of certain sources may be less effective for certain aphid biotype/resistance source combinations. In the case of *Diuraphis noxia* (Mordvilko), biotype 1 from South Africa can potentially overcome the resistance provided by the resistance gene *Dn5* under higher (50 ppm) than ambient CO_2 (Botha et al., 2017). In the case of *Schizaphis graminum*, the existence of 'HR × biotype' interactions under different CO_2 levels has not been investigated, but if present, it would make the deployment of *Schizaphis Graminum*-resistance genes more complex.

4.3 Host nutritional changes due to abiotic stresses

Additional abiotic factors such as drought, high nighttime temperatures and nitrogen starvation can change the composition of the phloem sap. The effects of these changes on aphids are unpredictable, firstly because sudden changes in the environment can be unpredictable and secondly because the plant may allocate resources in different ways to cope with particular stresses. Thus, aphids must develop strategies to cope with the effects of these factors. Furthermore, it is important to have an integrated understanding of these multiple plant responses to such abiotic stresses and their interactions.

Water stress is one of the most important factors that reduce wheat productivity. A drought event occurs when there is a prolonged, abnormal shortage of water in the soil that makes the plants suffer (Munné-Bosch and Alegre, 2004). Drought stress induces changes in the plants, including osmotic adjustment, photosynthetic activity, gas exchange and elevated reactive oxygen species (Anjum et al., 2011). Amino acids, sugars and organic acids are the main substances to change in concentration under water-deficit conditions in wheat as well as under high nighttime temperatures (Michaletti et al., 2018;

Impa et al., 2019). Amino acids appear to increase in drought-susceptible wheat genotypes exposed to water-deficit conditions (Michaletti et al., 2018). However, the metabolic response of wheat genotypes to drought varies (Guo et al., 2018; Michaletti et al., 2018; Marček et al., 2019). Drought-tolerant genotypes display larger accumulations of sugars (Guo et al., 2018; Nemati et al., 2018; Marček et al., 2019).

In sorghum, barley and wheat, higher *Schizaphis graminum* reproductive rates are observed when plants have not been water stressed (Sumner et al., 1986; Cabrera et al., 1995; Pendleton and Veerabomma, 2008). Despite lower reproductive rates under severe drought conditions (Sumner et al., 1983), the combination of *Schizaphis graminum* feeding and drought results in greater damage to the photosynthetic capacity than either of the stresses alone (Dorschner et al., 1986; Ryan et al., 1987). This is because of the *Schizaphis graminum* density (aphids/plant biomass units) on the smaller stressed plants (Dorschner et al., 1986). An important observation from these earlier studies is that it appears that resistant wheat genotypes continue to express their resistance under drought conditions (Sumner et al., 1986).

Nitrogen is a major nutrient factor for wheat to survive and successfully produce grain, and its deficiency in the soil causes the plants to reduce their growth, development, fertility and photosynthesis, changes the allocation of photoassimilates within the plant and quickly leads to a decline of amino-acid content in the phloem (Caputo and Barneix, 1997; Jeuffroy and Bouchard, 1999; Curci et al., 2017). The nutritional value of the phloem sap will determine the amount of ingested sap and consequently drive aphid performance (Bale et al., 2007; Douglas and Emden, 2007). Sugar levels in the host are critical because they determine the aphid feeding rate, which is inversely related to the concentration of sugars. A phenomenon that is seen as compensatory feeding behavior, i.e. under low levels of sugars, aphids feed more to compensate for the lower nutritious quality of the phloem sap (Douglas, 2003). The relatively more complex sugars, such as sucrose, need to be hydrolyzed to the reducing sugars glucose and fructose to be directly assimilated by the gut wall and then into the hemolymph (Douglas and Emden, 2007). The content of amino acids from the phloem sap is critical for aphid growth since 95% of these are used for protein synthesis, which may account for up to one-fourth of aphid weight (Douglas and Emden, 2007). About 180 nM of total amino-acid content has been found in *Schizaphis graminum* stylet exudate (Sandström et al., 2000). Additionally, *Schizaphis graminum* studies have shown that the concentration of phenylalanine, proline, glutamine and tryptophan is increased upon feeding between two- to eleven-fold (Sandström et al., 2000; Zhang et al., 2019). These amino acids also increase in drought-susceptible wheat genotypes subject to water-deficit conditions (Michaletti et al., 2018).

5 Molecular observations

Plant hormones modulate defense pathways; the most important pathways in this respect are governed by SA and jasmonic acid (JA). During feeding, aphids inject effectors (salivary compounds) that modulate the host plant response. These effectors have not yet been characterized in *Schizaphis graminum*. Recently, Zhang et al. (2022) identified 76 salivary proteins in *Schizaphis graminum*, some of which had similarities with other characterized effectors from *Myzus persicae* (Sultzer) and *Macrosiphum euphorbiae* Thomas. In this study, a salivary protein named Sg2204 was characterized, which suppresses callose deposition, downregulates JA- and SA-mediated defenses and, in parallel, increases *Schizaphis graminum* growth and fecundity (Zhang et al., 2022). Aphid feeding mostly induces SA but also JA upon cell damage. Pathways of SA and JA pathways are regulated at the gene transcription level; these are independent reactions and interact in complex relations, where other plant hormones participate, such as abscisic acid and ethylene which can promote JA-regulated responses and antagonize with SA responses (Smith and Chuang, 2014; Caarls et al., 2015; Züst et al., 2016).

On the other hand, certain salivary compounds (elicitors) may be recognized by resistant plants and hence activate defense reactions that include the induction of secondary metabolites or other resistance reactions such as hypersensitive responses and cell death (Smith and Chuang, 2014; Züst et al., 2016). A potential elicitor identified in *Schizaphis graminum* saliva is the cathepsin B protein (Zhang et al., 2022), which has been shown to elicit tobacco defenses against *Myzus persicae* (Guo et al., 2020). The recognition of elicitors is defined as a response to herbivory by resistant plants (Smith and Boyko, 2007; Smith and Chuang, 2014).

When *Schizaphis graminum* individuals feed on susceptible plants, they reduce the photosynthetic rate via chlorophyll reduction (Nagaraj et al., 2002), which becomes evident when host plants are severely damaged by the aphid (Fig. 2). Indeed, a quick and reliable method to evaluate plant resistance is measuring the chlorophyll content (Deol et al., 1997; Crespo-Herrera et al., 2019).

Schizaphis graminum feeding on susceptible plants also induces degenerative changes in vascular cells within an hour of feeding, leading to severe organelle degeneration in the phloem parenchyma cells (Morgham et al., 1994). Additionally, in the mesophyll cells, there is also structural damage to the chloroplast and cellular membranes, generating an increment in the size and number of plastoglobuli within chloroplasts. Plastoglobuli are plastid lipoprotein particles in various organelles, but in the chloroplast they are associated with thylakoids, the metabolism of prenol lipids, redox and photosynthetic regulation, plastid biogenesis and senescence, including recycling of phytol, remobilization of thylakoid lipids and metabolism of JA (Van Wijk and Kessler, 2017). In contrast, feeding on resistant plants displays less cellular damage on mesophyll cells (Morgham et al., 1994).

Transcriptome analysis of susceptible and resistant wheat plants carrying the *Gb3*-resistance gene identified various differentially expressed transcripts mediating biotic stress responses (Reddy et al., 2013). Resistant plants activate transcripts that encode membrane-bound receptor kinases and calcium signaling (Reddy et al., 2013). In susceptible plants, there is an expression of transcripts mediating the final steps of JA biosynthesis, redox homeostasis, peroxidases, glutathione S-transferases and notably defense-related secondary metabolites, as well as callose and cell wall modifications (Reddy et al., 2013; Donze-Reiner et al., 2017; Rojas et al., 2020). Other transcriptome responses induced by *Schizaphis graminum* feeding are DNA and RNA repairments and ethylene pathways (Zhang et al., 2019; Rojas et al., 2020).

6 Wheat breeding for *Schizaphis graminum* resistance

Breeding for aphid resistance is a challenging task. At least two conditions are required for developing resistant germplasm: (1) genetic variation for resistance is required to be present in either elite crop germplasm or related species from which the genes conferring the resistance can be transferred to elite germplasm and (2) selection and evaluation methods, tools or strategies are available to increase the frequency of positive alleles in the breeding populations.

Various studies have demonstrated the existence of genetic sources for *Schizaphis graminum* resistance in wheat. Resistance has been reported in synthetic hexaploid wheats (SHWs) and SHW-derived germplasm originating from the goat grass, *A. tauschii* Coss. (Smith and Starkey, 2003; Zhu et al., 2005; Crespo-Herrera et al., 2014, 2019). Other reported sources of resistance are *A. speltoides* Tausch and *Secale cereale* L. (Sebesta and Wood, 1978; Tyler et al., 1987; Porter et al., 1994; Lu et al., 2010; Crespo-Herrera et al., 2013). There are 17 resistance genes and quantitative trait loci reported for *Schizaphis graminum* resistance in wheat (Aradottir and Crespo-Herrera, 2021). Although *A. tauschii* is the source of most of these resistance genes, there is one gene from *A. speltoides* and two from *Secale cereale*. All the reported genes act as single dominant in a gene-for-gene fashion, except the *Gb1* gene, which is recessive (Curtis et al., 1960).

Unlike *R. padi* and *Sitobion avenae*, feeding by *Schizaphis graminum* causes visual symptoms (Fig. 3). The severity of these visual symptoms can be used to detect tolerant germplasm. Protocols generally consist of the following: sowing completely randomized row or hill plots of eight to ten seeds in flats; 3 days after emergence, plants are infested by placing infested leaves on the plots

Figure 3 *Schizaphis graminum* biotype E on wheat leaves 15 days after infestation. (a) Feeding damage on a susceptible leaf; (b) intermediate resistance reaction to Schizaphis graminum feeding; (c) resistant leaf of c.v. Largo carrying *Gb3*-resistance gene. Source: Photo by L. Crespo-Herrera.

with an average density of four to five aphids per plant and scores of symptoms in percent of chlorosis are taken 10–14 days after infestation or using a 0–9 damage scale where 0 = no damage and 9 = dead (Berzonsky et al., 2003). Another more quantitative method to evaluate resistance is the measurement of chlorophyll content with Soil Plant Analysis Development (SPAD) meters (Girma et al., 1998; Lage et al., 2003), which has been also used to identify genomic regions associated to *Schizaphis graminum* resistance in wheat by associating the SPAD readings with DNA markers through the implementation of quantitative trait loci mapping methods (Crespo-Herrera et al., 2019).

Traditional methods, such as life tables, have also been used to evaluate *Schizaphis graminum* resistance in wheat germplasm (Smith et al., 2004). However, these methods do not allow the evaluation of large numbers of plant genotypes due to their laborious nature. With high-throughput phenotyping (HTP) methods, it is possible to overcome this challenge by allowing the assessment of larger germplasm sets through spectral/image analysis with high reliability, which permits aphid resistance screening to be more integrated into large breeding programs where breeders need to focus not only on one or few traits but in a set of those that make varieties to cope with the uncertainties of the climate and meet industrial and nutritional quality demands of the market

(Aradottir and Crespo-Herrera, 2021). However, HTP methods need to be also developed for other wheat pests to ensure that the spectral signatures upon aphid feeding can be effectively distinguished from other stresses. Spectral indices have already been suggested to quantify the damage caused by *Schizaphis graminum* in the field (Yang et al., 2009).

Using validated artificial intelligence (AI) models to process large volumes of genotyping and HTP data enable accurate prediction of trait performances such as aphid resistance (Montesinos-López et al., 2021). The information derived from these models can be used to select the individuals with high estimated breeding values and/or discard those with the lowest breeding values and conform the next-generation of parental lines that will give origin to a new breeding cycle.

7 Future trends in research

Schizaphis graminum is expected to expand its presence toward northern latitudes with global warming (Aljaryian and Kumar, 2016). This poses significant risks to wheat-growing areas where this pest was not present before, potentially increasing the current problems due to BYDV and CYDV epidemics since other aphid vectors (*R. padi* and *Sitobion avenae*) of this virus are already present in such geographical areas.

In the context of climate change, there are certain knowledge gaps that need to be addressed. For instance, the response of resistance germplasm to aphid infestation under higher temperatures and limited water conditions needs to be evaluated to allow for effective variety deployment and implementation of pest management strategies.

Efforts should continue to identify new genetic sources of resistance, where germplasm exchange for international phenotyping can be an important component to determine virulence/avirulence patterns in different geographical areas and hence the deployment of resistant cultivars.

Breeding for *Schizaphis graminum* is relatively easier than breeding for other aphid species, such as *R. padi* and *Sitobion avenae*, due to the nature of the resistance (gene-by-gene interactions). Nonetheless, phenotyping and especially field phenotyping remains a challenge to efficiently incorporate aphid resistance as an additional component of breeding programs that already need to deal with critical traits (yield potential and stability, disease resistance, industrial and nutritional quality). Additionally, work on developing gene-based markers for *Schizaphis graminum* resistance needs to be continued, since this can also greatly facilitate breeding with modern breeding technologies and increase the knowledge in functional genomics.

The distinctive chlorosis caused by *Schizaphis graminum* feeding has raised the possibility of using remote sensing to detect its infestations. Certainly, trained scouts can detect its telltale injury in heavy infestations. Substantial efforts to realize this potential have not proven successful, but the future effort appears justified. Additional information is available from multiple temporal samples, more thorough spectral reads, and by harnessing AI to recognize *Schizaphis graminum* related signals still hold promise. *Schizaphis graminum* appears to be one of the most amenable pests to the adoption of this technology.

Where pesticides have been used extensively, *Schizaphis graminum* develops resistance. Integrated pest management that reduces pesticide applications and prudence in the selection of pesticides with different modes of action can reduce this. The effective sampling and decision supports for this aphid depend upon well-established principles but also require investment in research to quantify its potential to cause injury in different regions. To apply these principles, supporting research is needed. The current reliance on neonicotinoid insecticides in wheat production worldwide may be temporary, given the recognized environmental risks of these materials, even when applied as seed treatments that have triggered bans on their use. Research and management planning should be conducted with this eventuality in view.

8 Conclusion

Schizaphis graminum is one of the most common cereal aphids with economic importance, capable of significantly reducing grain yield through direct feeding and the transmission of viral diseases. These aphid populations have been classified based on their ability to colonize cultivated and non-cultivated grasses, for which host associations have a significant role in the differentiation of *Schizaphis graminum* populations. Although aphids as a group of insects are widely studied, there are few reports specific to the case of *Schizaphis graminum* and its interaction with the environment, for which certain knowledge gaps need to be addressed.

With global warming, it is expected that *Schizaphis graminum* will increase its geographical presence in the Northern Hemisphere, particularly in Europe and North America, due to the increase of new suitable areas for its establishment. Host plant resistance is a viable control method against this pest. Various reports have demonstrated the resistance against *Schizaphis graminum* in *A. tauschii*, *A. speltoides* and *Secale cereale*. Genes associated with HR have been identified from these sources. In association with the identification of genetic resources for resistance against this aphid, it is important to develop selection and evaluation methods to breed for this trait. Some of these methods consist of the utilization of SPAD meters and visual assessments and have permitted the identification of genomic regions associated to the resistance as well as the development of molecular markers for wheat breeding.

9 Where to look for further information

Further information can be found in the references cited here:

- Royer, T. A., Pendleton, B. B., Elliott, N. C. and Giles, K. L. (2015). Greenbug (Hemiptera: Aphididae) biology, ecology, and management in wheat and sorghum. *Journal of Integrated Pest Management* 6(1): 19. doi: 10.1093/jipm/pmv1018.
- Brewer, M. J., Peairs, F. B. and Elliott, N. C. (2019). Invasive cereal aphids of North America: Ecology and pest management. *Annual Review of Entomology* 64: 73-93.
- Royer, T. A., Giles, K., Elliott, N. C. and Zarrabi, A. A. (2020). Small grain aphids in Oklahoma and their management - EPP-7099 (ed. O.C.E. Service). Oklahoma Cooperative Extension Service, Stillwater, OK.
- Plant Health Australia. (2021). Greenbug. Plant Health Australia. https://www.planthealthaustralia.com.au/wp-content/uploads/2021/2001/Greenbug-FS.pdf.
- McCornack, B. P. and Whitworth, R. J. (2022). Wheat insect pest management - MF75 (ed. K.S.R.a. Extension). Kansas State Research and Extension, Kansas State University. https://bookstore.ksre.ksu.edu/pubs/MF745.pdf.

10 References

Abd-Ella, A. A. (2016). Evaluation of certain neonicotinoid insecticide seed treatments against cereal aphids on some wheat cultivars, *Journal of Phytopathology and Pest Management* 3(1), 21-33.

Aljaryian, R and Kumar, L. (2016). Changing global risk of invading greenbug *Schizaphis graminum* under climate change, *Crop Protection* 88, 137-148. doi: 10.1016/j.cropro.2016.06.008.

Anjum, S. A., Xie, X., Wang, L. C., Saleem, M. F., Man, C. and Lei, W. (2011). Morphological, physiological and biochemical responses of plants to drought stress, *African Journal of Agricultural Research* 6(9), 2026-2032. doi: 10.5897/ajar10.027.

Anstead, J. A., Burd, J. D. and Shufran, K. A. (2003). Over-summering and biotypic diversity of *Schizaphis graminum* (Homoptera: Aphididae) populations on noncultivated grass hosts, *Environmental Entomology* 32(3), 662-667. doi: 10.1603/0046-225X-32.3.662.

Aradottir, G. I. and Crespo-Herrera, L. (2021). Host plant resistance in wheat to barley yellow dwarf viruses and their aphid vectors: A review, *Current Opinion in Insect Science* 45, 59-68. doi: 10.1016/j.cois.2021.01.002.

Archer, T. L., Cate, R. H., Eikenbary, R. D. and Starks, K. J. (1974). Parasitoids collected from greenbugs and corn leaf aphids in Oklahoma in 1972, *Annals of the Entomological Society of America* 67(1), 11-14.

Bale, J. S., Ponder, K. L. and Pritchard, J. (2007). Coping with stress. In: Van Emden, H. F. and Harrington, R. (Eds.), *Aphids as Crop Pests*. Oxfordshire, UK: CAB International, pp. 287-309. doi: 10.1079/9780851998190.0287.

Bebber, D. P., Holmes, T. and Gurr, S. J. (2014). The global spread of crop pests and pathogens, *Global Ecology and Biogeography* 23(12), 1398–1407. doi: 10.1111/geb.12214.

Bebber, D. P., Ramotowski, M. A. T. and Gurr, S. J. (2013). Crop pests and pathogens move polewards in a warming world, *Nature Climate Change* 3(11), 985–988. doi: 10.1038/nclimate1990.

Berzonsky, W. A., Ding, H., Haley, S. D., Harris, M. O., Lamb, R. J., McKenzie, R. I. H., Ohm, H. W., Patterson, F. L., Peairs, F., Porter, D. R. and Ratcliffe, R. H. (2003). Breeding wheat for resistance to insects, *Plant Breeding Reviews* 22, 221–296.

Blackman, R. L. and Eastop, V. F. (2017). Taxonomic issues. In: *Aphids as Crop Pests*. Wallingford: CABI Publishing, pp. 1–36. doi: 10.1079/9781780647098.0001.

Bloom, A. J., Burger, M., Rubio Asensio, J. S. and Cousins, A. B. (2010). Carbon dioxide enrichment inhibits nitrate assimilation in wheat and Arabidopsis, *Science (New York, N.Y.)* 328(5980), 899–903. doi: 10.1126/science.1186440.

Botha, C. E. J., Sacranie, S., Gallagher, S. and Hill, J. M. (2017). Russian wheat aphids: Breakfast, lunch, and supper. Feasting on small grains in South Africa, *South African Journal of Botany* 109, 154–173. doi: 10.1016/j.sajb.2016.12.006.

Burd, J. D. and Porter, D. R. (2006). Biotypic diversity in Greenbug (Hemiptera : Aphididae): characterizing new virulence and host associations, *Journal of Economic Entomology* 99(3), 959–965.

Burton, R. L. and Krenzer, E. G. (1985). Reduction of greenbug (Homoptera: Aphididae) populations by surface residues in wheat tillage studies, *Journal of Economic Entomology* 78(2), 390–394.

Caarls, L., Pieterse, C. M. J. and Van Wees, S. C. M. (2015). How salicylic acid takes transcriptional control over jasmonic acid signaling, *Frontiers in Plant Science* 6, 170. doi: 10.3389/fpls.2015.00170.

CABI. (2004). *EPPO, Distribution Maps of Plant Pests*. INCOMPLETE.

CABI. (2022). *EPPO, Plantwise Knowledge Bank*. CABI Publishing. Available at: http://www.plantwise.org/ (Accessed: 9 February 2022).

Cabrera, H. M., Argandoña, V. H., Zúñiga, G. E. and Corcuera, L. J. (1995). Effect of infestation by aphids on the water status of barley and insect development, *Phytochemistry* 40(4), 1083–1088. doi: 10.1016/0031-9422(95)00325-2.

Caputo, C. and Barneix, A. J. (1997). Export of amino acids to the phloem in relation to N supply in wheat, *Physiologia Plantarum* 101(4), 853–860. doi: 10.1111/J.1399-3054.1997.TB01073.X.

Claridge, M. F. and Den Hollander, J. (1983). The biotype concept and its application to insect pests of agriculture, *Crop Protection* 2(1), 85–95. doi: 10.1016/0261-2194(83)90028-5.

Clua, A., Castro, A. M., Ramos, S., Gimenez, D. O., Vasicek, A., Chidichimo, H. O. and Dixon, A. F. G. (2004). The biological characteristics and distribution of the greenbug, *Schizaphis graminum*, and Russian wheat aphid, *Diuraphis noxia* (Hemiptera: Aphididae), in Argentina and Chile, *European Journal of Entomology* 101(1), 193–198. doi: 10.14411/eje.2004.024.

Cormier, F., Foulkes, J., Hirel, B., Gouache, D., Moënne-Loccoz, Y. and Le Gouis, J. (2016). Breeding for increased nitrogen-use efficiency: A review for wheat (*T. aestivum* L.), *Plant Breeding* 135(3), 255–278. doi: 10.1111/PBR.12371.

Crespo-Herrera, L., Singh, R. P., Reynolds, M. and Huerta-Espino, J. (2019). Genetics of greenbug resistance in synthetic hexaploid wheat derived germplasm, *Frontiers in Plant Science* 10, 782. doi: 10.3389/fpls.2019.00782.

Crespo-Herrera, L. A., Akhunov, E., Garkava-Gustavsson, L., Jordan, K. W., Smith, C. M., Singh, R. P. and Ahman, I. (2014). Mapping resistance to the bird cherry-oat aphid and the greenbug in wheat using sequence-based genotyping, *TAG. Theoretical and Applied Genetics. Theoretische und Angewandte Genetik* 127(9), 1963–1973. doi: 10.1007/s00122-014-2352-5.

Crespo-Herrera, L. A., Smith, C. M., Singh, R. P. and Åhman, I. (2013). Resistance to multiple cereal aphids in wheat–alien substitution and translocation lines, *Arthropod-Plant Interactions* 7(5), 535–545. doi: 10.1007/s11829-013-9267-y.

Curci, P. L., Aiese Cigliano, R., Zuluaga, D. L., Janni, M., Sanseverino, W. and Sonnante, G. (2017). Transcriptomic response of durum wheat to nitrogen starvation, *Scientific Reports* 7(1), 1176. doi: 10.1038/s41598-017-01377-0.

Curtis, B. C., Schlehuber, A. M. and Wood, E. A., Jr. (1960). Genetics of greenbug (*Toxoptera graminum* (Rond.)) resistance in two strains of common wheat, *Agronomy Journal* 52(10), 599–602.

Deol, G. S., Reese, J. C. and Gill, B. S. (1997). A rapid, nondestructive technique for assessing chlorophyll loss from greenbug (Homoptera: Aphididae) feeding damage on sorghum leaves, *Journal of the Kansas Entomological Society* 70, 305–312.

Deutsch, C. A., Tewksbury, J. J., Tigchelaar, M., Battisti, D. S., Merrill, S. C., Huey, R. B. and Naylor, R. L. (2018). Increase in crop losses to insect pests in a warming climate, *Science* 361(6405), 916–919. doi: 10.1126/SCIENCE.AAT3466.

Devine, G. J. and Furlong, M. J. (2007). Insecticide use: Contexts and ecological consequences, *Agriculture and Human Values* 24(3), 281–306. doi: 10.1007/s10460-007-9067-z.

Dhaliwal, G. S., Jindal, V. and Dhawan, A. K. (2010). Insect pest problems and crop losses: Changing trends, *Indian Journal of Ecology* 37(1), 1–7. Available at: http://www.cabdirect.org/abstracts/20113091848.html (Accessed: 10 April 2016).

Donze-Reiner, T., Palmer, N. A., Scully, E. D., Prochaska, T. J., Koch, K. G., Heng-Moss, T., Bradshaw, J. D., Twigg, P., Amundsen, K., Sattler, S. E. and Sarath, G. (2017). Transcriptional analysis of defense mechanisms in upland tetraploid switchgrass to greenbugs, *BMC Plant Biology* 17(1), 46. doi: 10.1186/s12870-017-0998-2.

Dorschner, K. W., Johnson, R. C., Eikenbary, R. D. and Ryan, J. D. (1986). Insect-plant interactions: Greenbugs (Homoptera: Aphididae) disrupt acclimation of winter wheat to drought stress, *Environmental Entomology* 15(1), 118–121. doi: 10.1093/EE/15.1.118.

Douglas, A. E. (2003). The nutritional physiology of aphids, *Advances in Insect Physiology* 31, 73–140. doi: 10.1016/S0065-2806(03)31002-1.

Douglas, A. E. and van Emden, H. F. (2007). Nutrition and symbiosis. In: Van Emden, H. F. and Harrington, R. (Eds.), *Aphids as Crop Pests*. Oxfordshire, UK: CAB International, pp. 115–134. doi: 10.1079/9780851998190.0115.

Elliott, N. C. and Kieckhefer, R. W. (1986). Cereal aphid populations in winter wheat: Spatial distributions and sampling with fixed levels of precision, *Environmental Entomology* 15(4), 954–958.

Elliott, N. C., Kieckhefer, R. W. and Walgenbach, D. D. (1990). Binomial sequential sampling methods for cereal aphids in small grains, *Journal of Economic Entomology* 83(4), 1381–1387.

Elliott, N. C., Royer, T. A., Giles, K. L., Kindler, S. D., Porter, D. R., Elliott, D. T. and Waits, D. A. (2004). A web-based decision support system for managing greenbugs in wheat, *Crop Management* 3(1), 1–6.

Feng, M. G. and Nowierski, R. M. (1992). Spatial distribution and sampling plans for four species of cereal aphids (Homoptera: Aphididae) infesting spring wheat in southwestern Idaho, *Journal of Economic Entomology* 85(3), 830–837.

Feng, M. G., Nowierski, R. M., Johnson, J. B. and Poprawski, T. I. (1992). Epizootics caused by entomophthoralean fungi (Zygomycetes, Entomophthorales) in populations of cereal aphids (Homoptera: Aphididae) in irrigated small grains of southwestern Idaho, USA, *Journal of Applied Entomology* 113, 376–390.

Gao, J. R. and Zhu, K. Y. (2000). Comparative toxicity of selected organophosphate insecticides against resistant and susceptible clones of the greenbug, *Schizaphis graminum* (Homoptera: Aphididae), *Journal of Agricultural and Food Chemistry* 48(10), 4717–4722.

GBIF Secretariat. (2022). *GBIF Backbone Taxonomy*. doi: 10.15468/39OMEI.

Giles, K. L., Hein, G. and Peairs, F. (2008). Areawide pest management of cereal aphids in dryland wheat systems of the Great Plains, USA. In: Koul, O., Cuperus, G. W. and Elliott, N. (Eds.), *Areawide Pest Management: Theory and Implementation*. Wallingford, Oxfordshire, UK: CAB International, pp. 441–467.

Giles, K. L., Royer, T. A., Elliott, N. C. and Kindler, S. D. (2000). Development and validation of a binomial sequential sampling plan for greenbug (Homoptera: Aphididae) infesting winter wheat in the southern plains, *Journal of Economic Entomology* 93(5), 1522–1530.

Gilstrap, F. E., Kring, T. J. and Brooks, G. W. (1984). Parasitism of aphids (Homoptera: Aphididae) associated with Texas sorghum, *Environmental Entomology* 13(6), 1613–1617.

Girma, M., Kofoid, K. D. and Reese, J. C. (1998). Sorghum germplasm tolerant to greenbug (Homoptera: Aphididae) feeding damage as measured by reduced chlorophyll loss, *Journal of the Kansas Entomological Society* 71(2), 108–115. doi: 10.2307/25085825.

Goldman, J. and Springer, T. (2010). Population dynamics of greenbug biotypes "E" and "F" on Texas bluegrass, *Grassland Science* 56(1), 26-30. doi: 10.1111/j.1744-697X.2009.00169.x.

Gray, S. and Gildow, F. E. (2003). Luteovirus-aphid interactions, *Annual Review of Phytopathology* 41, 539–566.

Guo, H., Zhang, Y., Tong, J., Ge, P., Wang, Q., Zhao, Z., Zhu-Salzman, K., Hogenhout, S. A., Ge, F. and Sun, Y. (2020). An aphid-secreted salivary protease activates plant defense in phloem, *Current Biology* 30(24), 4826–4836.e7. doi: 10.1016/J.CUB.2020.09.020.

Guo, R., Shi, L., Jiao, Y., Li, M., Zhong, X., Gu, F., Liu, Q., Xia, X. and Li, H. (2018). Metabolic responses to drought stress in the tissues of drought-tolerant and drought-sensitive wheat genotype seedlings, *AoB Plants* 10(2), ply016. doi: 10.1093/AOBPLA/PLY016.

Harvey, T. L. and Hackerott, H. L. (1969). Recognition of a greenbug biotype injurious to sorghum1, *Journal of Economic Entomology* 62(4), 776–779.

Harvey, T. L., Wilde, G. E., Kofoid, K. D. and Bramel-Cox, P. J. (1994). Temperature effects on resistance to greenbug (Homoptera: Aphididae) Biotype I in Sorghum, *Journal of Economic Entomology* 87(2), 500–503. doi: 10.1093/jee/87.2.500.

Hesler, L. S. and Berg, R. K. (2003). Tillage impacts cereal aphid (Homoptera: Aphididae) infestations in spring small grains, *Journal of Economic Entomology* 96(6), 1792–1797.

Hullé, M., Cœur d'Acier, A., Bankhead-Dronnet, S. and Harrington, R. (2010). Aphids in the face of global changes, *Comptes Rendus Biologies* 333(6–7), 497–503. doi: 10.1016/j.crvi.2010.03.005.

Hussain, D., Asrar, M., Khalid, B., Hafeez, F., Saleem, M., Akhter, M., Ahmed, M., Ali, I. and Hanif, K. (2021). Insect pests of economic importance attacking wheat crop (*Triticum aestivum* L.) in Punjab, Pakistan, *International Journal of Tropical Insect Science*, 1–12. doi: 10.1007/s42690-021-00574-9.

Impa, S. M., Sunoj, V. S. J., Krassovskaya, I., Bheemanahalli, R., Obata, T. and Jagadish, S. V. K. (2019). Carbon balance and source-sink metabolic changes in winter wheat exposed to high night-time temperature, *Plant, Cell and Environment* 42(4), 1233-1246. doi: 10.1111/PCE.13488.

Jeuffroy, M. H. and Bouchard, C. (1999). Intensity and duration of nitrogen deficiency on wheat grain number, *Crop Science* 39(5), 1385–1393. doi: 10.2135/CROPSCI1999.3951385X.

Kindler, S. D., Elliott, N. C., Giles, K. L. and Royer, T. A. (2003). Economic injury level for the greenbug, Schizaphis graminum in Oklahoma winter wheat, *Southwestern Entomologist* 28, 163–166.

Kindler, S. D., Elliott, N. C., Giles, K. L., Royer, T. A., Fuentes-Granados, R. and Tao, F. (2002). Effect of greenbugs (Homoptera : Aphididae) on yield loss of winter wheat, *Journal of Economic Entomology* 95(1), 89–95. doi: 10.1603/0022-0493-95.1.89.

Lage, J., Skovmand, B. and Andersen, S. B. (2003). Characterization of greenbug (Homoptera: Aphididae) resistance in synthetic hexaploid wheats, *Journal of Economic Entomology* 96(6), 1922–1928.

Long, W., Xiao-Hui, W., Tong, H., Lei, Q., Li-Kun, L. and Fa-Jun, C. (2019). The effect of fertilizer-N on the inter-specific competition among three wheat aphids under elevated CO_2, *Journal of Applied Entomology* 143(9), 1032–1042. doi: 10.1111/jen.12648.

Lu, H., Rudd, J. C., Burd, J. D. and Weng, Y. (2009). Molecular mapping of greenbug resistance genes Gb2 and Gb6 in T1AL.1RS wheat-rye translocations, *Plant Breeding* 129, 472–476. doi: 10.1111/j.1439-0523.2009.01722.x.

Marček, T., Hamow, K. Á, Végh, B., Janda, T. and Darko, E. (2019). Metabolic response to drought in six winter wheat genotypes, *PLoS ONE* 14(2), e0212411. doi: 10.1371/JOURNAL.PONE.0212411.

Meehan, T. D. and Gratton, C. (2016). A landscape view of agricultural insecticide use across the conterminous US from 1997 through 2012, *PLoS ONE* 11(11), e0166724. doi: 10.1371/journal.pone.0166724.

Meehan, T. D., Werling, B. P., Landis, D. A. and Gratton, C. (2011). Agricultural landscape simplification and insecticide use in the Midwestern United States, *Proceedings of the National Academy of Sciences of the United States of America* 108(28), 11500-11505. doi: 10.1073/pnas.1100751108.

Michaletti, A., Naghavi, M. R., Toorchi, M., Zolla, L. and Rinalducci, S. (2018). Metabolomics and proteomics reveal drought-stress responses of leaf tissues from spring-wheat, *Scientific Reports* 8(1), 5710. doi: 10.1038/s41598-018-24012-y.

Michuad, J. P. (2010). Implication of climate change for cereal aphids on the great plains of North America. In: Kindlmann, P., Dixon, A. F. G. and Houdková, K. (Eds.), *Aphid Biodiversity Under Environmental Change*. New York: Springer.

Mitchell, E. A. D., Mulhauser, B., Mulot, M., Mutabazi, A., Glauser, G. and Aebi, A. (2017). A worldwide survey of neonicotinoids in honey, *Science* 358(6359), 109–111. doi: 10.1126/science.aan3684.

Montesinos-López, A., Runcie, D. E., Ibba, M. I., Pérez-Rodríguez, P., Montesinos-López, O. A., Crespo, L. A., Bentley, A. R. and Crossa, J. (2021). 'Multi-trait genomic-enabled prediction enhances accuracy in multi-year wheat breeding trials', *G3 Genes|Genomes|Genetics, G3* 11(10). doi: 10.1093/g3journal/jkab270.

Morgham, A. T., Richardson, P. E., Campbell, R. K., Burd, J. D., Eikenbary, R. D. and Sumner, L. C. (1994). Ultrastructural responses of resistant and susceptible wheat to infestation by greenbug biotype E (Homoptera: Aphididae), *Annals of the Entomological Society of America* 87(6), 908-917. doi: 10.1093/aesa/87.6.908.

Munné-Bosch, S. and Alegre, L. (2004). Die and let live: Leaf senescence contributes to plant survival under drought stress, *Functional Plant Biology* 31(3), 203-216. doi: 10.1071/FP03236.

Nagaraj, N., Reese, J. C., Kirkham, M. B., Kofoid, K., Campbell, L. R. and Loughin, T. M. (2002). Relationship between chlorophyll loss and photosynthetic rate in greenbug (Homoptera: Aphididae) damaged sorghum, *Journal of the Kansas Entomological Society* 75, 101-109.

Nansen, C. and Ridsdill-Smith, J. (2013). The performance of insecticides-a critical review. In: Trdan, S. (Ed.), *Insecticides: Development of Safer and More Effective Technologies*. INTECH Open Access Publisher, pp. 195-231. doi: 10.5772/3356.

Nechols, J. R., Hough, A. R., Margolies, D. C., Ruberson, J. R., McCornack, B. P., Sandercock, B. K. and Murray, L. (2020). Effect of temperature on plant resistance to arthropod pests, *Environmental Entomology* 49(3), 537-545. doi: 10.1093/ee/nvaa033.

Nemati, F., Ghanati, F., Gavlighi, H. and Sharifi, M. (2018). Comparison of sucrose metabolism in wheat seedlings during drought stress and subsequent recovery, *Biologia Plantarum* 62(3), 595-599. doi: 10.1007/S10535-018-0792-5.

Nicholson, S. J. and Puterka, G. J. (2014). Variation in the salivary proteomes of differentially virulent greenbug (*Schizaphis graminum* Rondani) biotypes, *Journal of Proteomics* 105, 186-203. doi: 10.1016/J.JPROT.2013.12.005.

Oehme, V., Högy, P., Zebitz, C. P. W. and Fangmeier, A. (2013). Effects of elevated atmospheric CO_2 concentrations on phloem sap composition of spring crops and aphid performance, *Journal of Plant Interactions* 8(1), 74-84. doi: 10.1080/17429145.2012.736200.

Oerke, E.-C. (2006). Crop losses to pests, *The Journal of Agricultural Science* 144(1), 31-43. doi: 10.1017/S0021859605005708.

Oerke, E.-C. and Dehne, H.-W. (2004). Safeguarding production–Losses in major crops and the role of crop protection, *Crop Protection* 23(4), 275-285. doi: 10.1016/j.cropro.2003.10.001.

Oliveira, C. M., Auad, A. M., Mendes, S. M. and Frizzas, M. R. (2014). Crop losses and the economic impact of insect pests on Brazilian agriculture, *Crop Protection* 56, 50-54. doi: 10.1016/j.cropro.2013.10.022.

Pendleton, B. B., Copeland, A. L. and Michels, G. J. (2009). Effect of biotype and temperature on fitness of greenbug (Hemiptera: Aphididae) on sorghum, *Journal of Economic Entomology* 102(4), 1624-1627. doi: 10.1603/029.102.0429.

Pendleton, B. B. and Veerabomma, S. (2008). Effects of soil water and nitrogen on fitness of greenbug (Hemiptera: Aphididae) on sorghum, *Southwestern Entomologist* 33(4), 281-287. doi: 10.3958/0147-1724-33.4.281.

Perkins, C. M., Steckel, S. J. and Stewart, S. D. (2018). Impact of insecticide seed treatments and foliar insecticides on aphid infestations in wheat, incidence of Barley yellow dwarf, and yield in west Tennessee, *Journal of Economic Entomology* 111(6), 2734-2740.

Peters, D. C., Wood, E. A. J. and Starks, K. J. (1975). Insecticide resistance in selections of the greenbug, *Journal of Economic Entomology* 68(3), 399–340.

Pimentel, D., Acquay, H., Biltonen, M., Rice, P., Silva, M., Nelson, J., Lipner, V., Giordano, S., Horowitz, A. and D'Amore, M. (1992). Environmental and economic costs of pesticide use, *BioScience* 42(10), 750–760. doi: 10.2307/1311994.

Plant Health Australia. (2021). *Greenbug*. Plant Health Australia. Available at: https://www.planthealthaustralia.com.au/wp-content/uploads/2021/2001/Greenbug-FS.pdf.

Porter, D. R., Burd, J. D., Shufran, K. A., Webster, J. A. and Teetes, G. L. (1997). Greenbug (Homoptera: Aphididae) biotypes: Selected by resistant cultivars or preadapted opportunists?, *Journal of Economic Entomology* 90(5), 1055–1065.

Porter, D. R., Webster, J. A. and Friebe, B. (1994). Inheritance of greenbug biotype-G resistance in wheat, *Crop Science* 34(3), 625–628.

Rabbinge, R., Drees, E. M., Graaf, M., Verberne, F. C. M. and Wesselo, A. (1981). Damage effects of cereal aphids in wheat, *Netherlands Journal of Plant Pathology* 87(6), 217–232.

Reddy, S. K., Weng, Y., Rudd, J. C., Akhunova, A. and Liu, S. (2013). Transcriptomics of induced defense responses to greenbug aphid feeding in near isogenic wheat lines, *Plant Science* 212, 26–36. doi: 10.1016/j.plantsci.2013.08.002.

Riedell, W. E., Kieckhefer, R. W., Haley, S. D., Langham, M. A. C. and Evenson, P. D. (1999). Winter wheat responses to bird cherry-oat aphids and barley yellow dwarf virus infection, *Crop Science* 39(1), 158–163.

Riedell, W. E., Kieckhefer, R. W., Langham, M. A. C. and Hesler, L. S. (2003). Root and shoot responses to bird cherry-oat aphids and barley yellow dwarf virus in spring wheat, *Crop Science* 43(4), 1380–1386.

Robinson, E. A., Ryan, G. D. and Newman, J. A. (2012). A meta-analytical review of the effects of elevated CO_2 on plant-arthropod interactions highlights the importance of interacting environmental and biological variables, *New Phytologist* 194(2), 321–336. doi: 10.1111/j.1469-8137.2012.04074.x.

Rojas, L. A., Scully, E., Enders, L., Timm, A., Sinha, D. and Smith, C. M. (2020). Comparative transcriptomics of *Diuraphis noxia* and *Schizaphis graminum* fed wheat plants containing different aphid-resistance genes, *PLoS ONE* 15(5), e0233077. doi: 10.1371/journal.pone.0233077.

Royer, T. A., Giles, K., Elliott, N. C. and Zarrabi, A. A. (2020). *Small Grain Aphids in Oklahoma and Their Management - EPP-7099*. Stillwater, OK: Oklahoma Cooperative Extension Service.

Royer, T. A., Giles, K. L., Nyamanzi, T., Hunger, R. M., Krenzer, E. G., Elliott, N. C., Kindler, S. D. and Payton, M. (2005). Economic evaluation of the effects of planting date and application rate of Imidacloprid for management of cereal aphids and barley yellow dwarf in winter wheat, *Journal of Economic Entomology* 98(1), 95–102.

Ryan, J. D., Johnson, R. C., Eikenbary, R. D. and Dorschner, K. W. (1987). Drought/greenbug interactions: Photosynthesis of greenbug resistant and susceptible wheat, *Crop Science* 27(2), 283–288. doi: 10.2135/cropsci1987.0011183X002700020033x.

Sandström, J., Telang, A. and Moran, N. A. (2000). Nutritional enhancement of host plants by aphids – A comparison of three aphid species on grasses, *Journal of Insect Physiology* 46(1), 33–40. doi: 10.1016/S0022-1910(99)00098-0.

Schweissing, F. C. and Wilde, G. (1978). Temperature influence on greenbug resistance of crops in the seedling stage, *Environmental Entomology* 7(6), 831–834. doi: 10.1093/EE/7.6.831.

Schweissing, F. C. and Wilde, G. (1979). Temperature and plant nutrient effects on resistance of seedling sorghum to the greenbug, *Journal of Economic Entomology* 72(1), 20–23. doi: 10.1093/JEE/72.1.20.

Sebesta, E. E. and Wood, E. A. (1978). Transfer of greenbug resistance from rye to wheat with X-rays, *Agronomy Abstracts* 70, 61–62.

Shufran, K. A., Burd, J. D., Anstead, J. A. and Lushai, G. (2000). Mitochondrial DNA sequence divergence among greenbug (Homoptera: Aphididae) biotypes: Evidence for host-adapted races, *Insect Molecular Biology* 9(2), 179–184. doi: 10.1046/J.1365-2583.2000.00177.X.

Shufran, K. A., Margolies, D. C. and Black, W. C. (1992). Variation between biotype-E clones of Schizaphis-graminum (Homoptera, Aphididae), *Bulletin of Entomological Research* 82(3), 407–416.

Sinha, D. K. and Smith, C. M. (2014). Selection of reference genes for expression analysis in *Diuraphis* noxia (Hemiptera: Aphididae) fed on resistant and susceptible wheat plants, *Scientific Reports* 4, 5059. doi: 10.1038/srep05059.

Skendžić, S., Zovko, M., Živković, I. P., Lešić, V. and Lemić, D. (2021). The impact of climate change on agricultural insect pests, *Insects* 12(5). doi: 10.3390/INSECTS12050440.

Smith, C. M. and Boyko, E. V. (2007). The molecular bases of plant resistance and defense responses to aphid feeding: Current status, *Entomologia Experimentalis et Applicata* 122(1), 1–16. doi: 10.1111/j.1570-7458.2006.00503.x.

Smith, C. M. and Chuang, W. P. (2014). Plant resistance to aphid feeding: Behavioral, physiological, genetic and molecular cues regulate aphid host selection and feeding, *Pest Management Science* 70(4), 528–540. doi: 10.1002/ps.3689.

Smith, C. M., Havlícková, H., Starkey, S., Gill, B. S. and Holubec, V. (2004). Identification of *Aegilops germplasm* with multiple aphid resistance, *Euphytica* 135(3), 265-273.

Smith, C. M. and Starkey, S. (2003). Resistance to greenbug (Heteroptera: Aphididae) biotype I in *Aegilops tauschii* synthetic wheats, *Journal of Economic Entomology* 96(5), 1571–1576. doi: 10.1093/jee/96.5.1571.

Stoetzel, M. B. (1987). Information on and identification of *Diuraphis* noxia (Homoptera: Aphididae) and other aphid species colonizing leaves of wheat and barley in the United States, *Journal of Economic Entomology* 80(3), 696–704.

Sumner, L. C., Dorschner, K. W., Ryan, J. D., Eikenbary, R. D., Johnson, R. C. and Mcnew, R. W. (1986). Reproduction of *Schizaphis graminum* (Homoptera: Aphididae) on resistant and susceptible wheat genotypes during simulated drought stress induced with polyethylene glycol, *Environmental Entomology* 15(3), 756–762. doi: 10.1093/EE/15.3.756.

Sumner, L. C., Need, J. T., Mcnew, R. W., Dorschner, K. W., Eikenbary, R. D. and Johnson, R. C. (1983). Response of *Schizaphis graminum* (Homoptera: Aphididae) to drought-stressed wheat, using polyethylene glycol as a matricum, *Environmental Entomology* 12(3), 919–922. doi: 10.1093/EE/12.3.919.

Sun, Y. and Ge, F. (2011). How do aphids respond to elevated CO_2?, *Journal of Asia-Pacific Entomology* 14(2), 217–220. doi: 10.1016/j.aspen.2010.08.001.

Sun, Y., Guo, H. and Ge, F. (2016). Plant-aphid interactions under elevated CO_2: Some cues from aphid feeding behavior, *Frontiers in Plant Science* 7, 502. doi: 10.3389/fpls.2016.00502.

Sun, Y. C., Chen, F. J. and Ge, F. (2009). Elevated CO_2 changes interspecific competition among three species of wheat aphids: *Sitobion avenae*, *Rhopalosiphum padi*, and *Schizaphis graminum*, *Environmental Entomology* 38(1), 26–34. doi: 10.1603/022.038.0105.

Tang, G., Liu, X., Chen, G. H., Witworth, R. J. and Chen, M. S. (2018). Increasing temperature reduces wheat resistance mediated by major resistance genes to *Mayetiola destructor* (Diptera: Cecidomyiidae), *Journal of Economic Entomology* 111(3), 1433–1438 doi: 10.1093/jee/toy048.

Teetes, G. L. and Johnson, J. W. (1973). Damage assessment of the greenbug on grain sorghum, *Journal of Economic Entomology* 66(5), 1181–1186.

Teetes, G. L., Schaefer, C. A., Gipson, J. R., McIntyre, R. C. and Latham, E. E. (1975). Greenbug resistance to organophosphorous insecticides on the Texas High Plains, *Journal of Economic Entomology* 68(2), 214–216.

Thindwa, H. P. and Teetes, G. L. (1994). Effect of temperature and photoperiod on sorghum resistance to Biotype C and E greenbug (Homoptera: Aphididae), *Journal of Economic Entomology* 87(5), 1366–1372. doi: 10.1093/JEE/87.5.1366.

Tooker, J. F., Douglas, M. R. and Krupke, C. H. (2017). Neonicotinoid seed treatments: Limitations and compatibility with integrated pest management, *Agricultural and Environmental Letters* 2(1), ael2017.2008.0026.

Tyler, J. M., Webster, J. A. and Merkle, O. G. (1987). Designations for genes in wheat germplasm conferring greenbug resistance 1, *Crop Science* 27(3), 526–527. doi: 10.2135/cropsci1987.0011183X002700030020x.

Van Emden, H. and Harrington, R. (2007). *Aphids as Crop Pests*. CAB International, Cromwell Press.

Van Wijk, K. J. and Kessler, F. (2017). Plastoglobuli: Plastid microcompartments with integrated functions in metabolism, plastid developmental transitions, and environmental adaptation, *Annual Review of Plant Biology* 68, 253–289. doi: 10.1146/ANNUREV-ARPLANT-043015-111737.

Voss, T. S., Kieckhefer, R. W., Fuller, B. W., McLeod, M. J. and Beck, D. A. (1997). Yield losses in maturing spring wheat caused by cereal aphids (Homoptera: Aphididae) under laboratory conditions, *Journal of Economic Entomology* 90(5), 1346–1350. doi: 10.1093/jee/90.5.1346.

Walgenbach, D. D., Elliott, N. C. and Kieckhefer, R. W. (1988). Constant and fluctuating temperature effects on developmental rates and life table statistics of the greenbug (Homoptera: Aphididae), *Journal of Economic Entomology* 81(2), 501–507. Available at: http://www.ingentaconnect.com/content/esa/jee/1988/00000081/00000002/art00013.

Wang, L., Li, L., Song, Y., Han, T., Li, Z., Wan, G. and Chen, F. (2018). Elevated CO_2 and temperature alter specific-species population dynamic and interspecific competition of three wheat aphids, *Journal of Applied Entomology* 142(9), 863–872. doi: 10.1111/JEN.12536.

Webster, F. M. and Phillips, W. J. (1912). The spring grain Aphis or "Green- Bug", *United States Bureau of Entomology Bulletin* 110, pp. 1–153.

Webster, J. A., Treat, R., Morgan, L. and Elliott, N. (2000). Economic impacts of the Russian wheat aphid and greenbug in the western United States 1993–1994, 1994–1995, and 1997–1998. *U.S. Department of Agriculture, ARS Service report PSWCRL Rep.* 00–001.

Weng, Y., Perumal, A., Burd, J. D. and Rudd, J. C. (2010). Biotypic diversity in greenbug (Hemiptera: Aphididae): Microsatellite-based regional divergence and host-adapted differentiation, *Journal of Economic Entomology* 103(4), 1454–1463. doi: 10.1603/ec09291.

Wilde, G. E., Shufran, R. A., Kindler, S. D., Brooks, H. L. and Sloderbeck, P. E. (2001). Distribution and abundance of insecticide resistant greenbugs (Homoptera: Aphididae) and validation of a bioassay to assess resistance, *Journal of Economic Entomology* 94(2), 547–551.

Wiseman, B. R. (1994). Plant resistance to insects in integrated pest management, *Plant Disease* 78(9), 927. doi: 10.1094/PD-78-0927.

Yang, Z., Rao, M. N., Elliott, N. C., Kindler, S. D. and Popham, T. W. (2009). Differentiating stress induced by greenbugs and Russian wheat aphids in wheat using remote sensing, *Computers and Electronics in Agriculture* 67(1–2), 64–70. doi: 10.1016/j.compag.2009.03.003.

Zhang, P., Zhang, X., Zhao, Y., Wei, Y., Mu, W. and Liu, F. (2016). Effects of Imidacloprid and clothianidin seed treatments on wheat aphids and their natural enemies on winter wheat, *Pest Management Science* 72(6), 1141–1149.

Zhang, Y., Fan, J., Fu, Y., Francis, F. and Chen, J. (2019). Plant-mediated interactions between two cereal aphid species: Promotion of aphid performance and attraction of more parasitoids by infestation of wheat with phytotoxic aphid *Schizaphis graminum*, *Journal of Agricultural and Food Chemistry* 67(10), 2763–2773. doi: 10.1021/acs.jafc.8b06150.

Zhang, Y., Fu, Y., Fan, J., Li, Q., Francis, F. and Chen, J. (2019). Comparative transcriptome and histological analyses of wheat in response to phytotoxic aphid *Schizaphis graminum* and non-phytotoxic aphid *Sitobion avenae* feeding, *BMC Plant Biology* 19(1), 547. doi: 10.1186/s12870-019-2148-5.

Zhang, Y., Liu, X., Francis, F., Xie, H., Fan, J., Wang, Q., Liu, H., Sun, Y. and Chen, J. (2022). The salivary effector protein Sg2204 in the greenbug *Schizaphis graminum* suppresses wheat defense and is essential for enabling aphid feeding on host plants, *Plant Biotechnology Journal* 20(11), 2187–2201. doi: 10.1111/PBI.13900.

Zhu, L. C., Smith, C. M., Fritz, A., Boyko, E., Voothuluru, P. and Gill, B. S. (2005). Inheritance and molecular mapping of new greenbug resistance genes in wheat germplasms derived from *Alegilops tauschii*, *TAG: Theoretical and Applied Genetics. Theoretische und Angewandte Genetik* 111(5), 831–837. doi: 10.1007/s00122-005-0003-6.

Züst, T. and Agrawal, A. A. (2016). Mechanisms and evolution of plant resistance to aphids, *Nature Plants* 2(1), 15206. doi: 10.1038/nplants.2015.206.

Chapter 9

Fescue aphid (*Metopolophium festucae*)

Sanford D. Eigenbrode and Subodh Adhikari, University of Idaho, USA; and Arash Rashed, Virginia Tech, USA

1 Introduction

Metopolophium festucae Theobald (the fescue aphid or grass aphid) is native to Western Europe and the British Isles (Edwards and Heath, 1964; Blackman and Eastop, 2021). The aphid occurs on species within the family Poaceae throughout its range. *M. festucae sensu stricto* tends to infest primarily wild and cultivated perennial grasses but can also infest cultivated small grains and require management (Edwards and Heath, 1964). A subspecies, *M. festucae cerealium* (herein *M. f. cerealium*), delineated based primarily on morphological traits (Stroyan, 1982), readily colonizes cultivated small grains causing considerable damage (Edwards and Heath, 1964; Vickerman and Wratten, 1978; Dent, 1986). Prior to 1982, this subspecies was not distinguished so the biology in earlier reports presumably is confounded with information about this subspecies and *M. f. sensu stricto*. The species, evidently *M. f. cerealium*, has been reported within the past 30 years in New Zealand, Argentina, Bolivia, Chile, and western North America (Blackman and Eastop, 2021) and is potentially important as a globally invasive pest of wheat and other small grains.

http://dx.doi.org/10.19103/AS.2022.0114.06

1.1 Key issues

The information about *M. f. cerealium* is sparse, and this chapter is included in this volume because of the recent occurrence of *M. festucae* as an invasive pest of small grains outside of its native range. Here, we summarize available knowledge about its biology and management. Ongoing and future research to improve its management will require improving identification and recognition of the species by pest managers, assessing the severity and economic impacts of the feeding injury it causes to wheat and other small grains, determining its competence as a vector of *Barley yellow dwarf virus* (BYDV), documenting its range expansion in North America and elsewhere, determining the genetics and ecology of those invading populations, and projecting its potential global geographic range.

2 Appearance and identification

Apterae of *M. festucae* are shiny, yellowish green to pale green (Fig. 1a,b) and alatae have conspicuous dark dorsal abdominal cross-bands (Fig. 1b). The species frequently co-occurs in wheat with *M. dirhodum* (rose-grass aphid or rose-grain aphid) in its native and invaded ranges (Vickerman and Wratten, 1978; Remaudière et al., 1993; Foottit et al., 2006; Blackman and Eastop, 2021). Alatae of the two species can be distinguished based on the

Figure 1 *M. f. cerealium* - (a) mature aptera and two female alatae, (b) a colony of apterae with alatae on a wheat leaf. Photo credits: (a) Bradley Stokes and (b) Taylor Murphy.

lack of horizontal bands on the abdomen of *M. dirhodum*. Under a hand lens, *M. dirhodum* apterae can be seen to have darkly pigmented antennal joints (Fig. 2a). These are not visible on *M. festucae* (Fig. 2b), which tends to have antennae that darken progressively from the base to apex (Halbert et al., 2013). This difference in antennal pigmentation is also visible with a microscope on alcohol-preserved specimens (Fig. 2c,d). Where *M. f. sensu stricto* and *M. f. cerealium* co-occur, it may be important to distinguish them, given evidence that *M. f. cerealium* can become abundant and injurious in wheat. Apterae can be distinguished from those of *M. f. sensu stricto* (Stroyan, 1982) in that the body is more elongate and legs and antennae are relatively longer in *M. f. cerealium*. *M. f. cerealium* apterae may appear paler yellow than *M. f. sensu stricto*. Alates of the subspecies differ in similar ways. In addition, although both have darkened abdominal bands, those on *M. f. cerealium* are reduced on two segments producing the impression of a patch, never observed on *M. f. sensu stricto* (Stroyan, 1982; Halbert et al., 2013). Host preference is also indicative of *M. f. cerealium* and likely more important for pest management; large populations on cultivated cereal crops in the UK are almost certainly *M.*

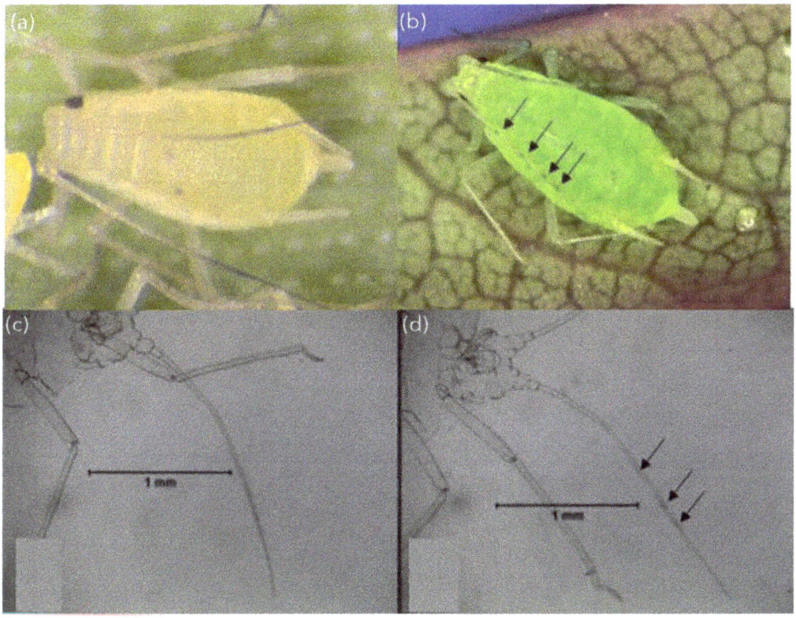

Figure 2 (a) *M. f. cerealium* aptera illustrating a lack of distinct pigmentation at antennal joints. (b) *M. dirhodum* aptera illustrating the prominent pigmentation near the joints of the antennae (black arrows). (c) *M. f. cerealium* antenna of apterous form in EtOH illustrating a lack of distinct pigmentation at antennal joints. (d) *M. dirhodum antenna* of apterous form illustrating the prominent pigmentation at joints of the antennae. Photo credits: (a) Bradley Stokes, (b) Andy Jensen, aphidtrek.org, (c and d) Halbert et al. (2013).

Figure 3 (a) Feeding damage at sites of individual aphids on wheat, (b) a fused necrotic lesion associated with a colony-infesting wheat in Washington state, USA. Photo credits: (a) Bradley Stokes and (b) Bruce Palmer, McGregor Company.

f. cerealium (Stroyan, 1982). *M. f. cerealium* can cause distinctive injury to the host plant visible as red spots localized at aphid feeding sites, which ultimately fuse into larger chlorotic to necrotic patches associated with infestations of the aphid in the field (Fig. 3a,b). This damage can be readily recognized on wild and cultivated hosts and is useful as an aid when scouting. Similar sized infestations of *M. dirhodum* do not produce these symptoms in the field or in laboratory colonies (Eigenbrode, pers. obs.). Whether this type of injury is distinctive to *M. f. cerealium* vs. *M. f. sensu stricto*, however, is uncertain. Caged colonies of what was determined to be *M. f. cerealium* caused 'rapid yellowing and withering' of barley and oats, while colonies of *M. f. sensu stricto* remained small and did not cause yellowing (Stroyan, 1982). Rigorous comparisons of chlorosis caused per capita by *M. dirhodum*, *M. f. cerealium*, and *M. f. sensu stricto*, however, have not been made.

3 Native and invaded ranges

M. festucae (subspecies unspecified) is present throughout Europe and in Kenya (Hille Ris Lambers, 1947; Marzocchi and Nicoli, 1991; Wangai, 1992) and in Iran and Central Asia (Kadyrbekov, 2017, p. 378). *M. festucae*, either *M. f. sensu stricto* or *M. f. cerealium*, or both, has been recorded in the UK since the mid-twentieth century (Edwards and Heath, 1964), and the species has been included in work on aphids and viruses in the UK and Italy (Robert and Rouzejouan, 1976; Vickerman and Wratten, 1978; Chambers et al., 1982; Hand, 1989; Coceano et al., 2009).

Stroyan (1982) described *M. festucae* subsp. *cerealium* based on specimens from sites in southern England, Wales, the Netherlands, Belgium, France, Spain, Switzerland, and Germany. Dent (Dent, 1986; Dent and Wratten, 1986) worked with a laboratory colony of *M. f. cerealium* derived from a collection made in Hampshire, UK. *M. festucae* (no subspecies identified) has been detected in the suction trap network of the Rothamsted Insect Survey (https://www.rothamsted.ac.uk/insect-survey, RIS) since the mid-1960s, initially at the Rothamsted and Newcastle sites and later in other sites; *M. f. cerealium* was detected in four traps in Scotland in the mid-1980s and 1990s; more recently, specimens of *M. festucae* collected in the RIS have not been separated into subspecies (W. Bell and A. Greenslade, RIS, pers. comm.). In the early 1970s, field sites in the UK were sampled within 25 km of suction traps, and *M. festucae* (subspecies not defined) was consistently detected in nearby suction traps weeks before detection in the field (George, 1974). On the other hand, in the West Midlands and South-West Regions, small colonies of *M. festucae* were found in fields in March and April, possibly indicating their having overwintered on grasses (George, 1974). Whether *M. f. cerealium* and *M. f. sensu stricto* occur over the whole Palearctic range of *M. festucae* is unknown. Descriptions either predate the identification of the subspecies or do not attempt to resolve specimens to the subspecies level (Kadyrbekov, 2017, p. 378).

The first report of *M. festucae* (subspecies not specified) in the Western Hemisphere was in California in 1970 (Foottit et al., 2006). *M. festucae* briefly became a pest of fescue lawns in 1994 in Boise Idaho, the specimens most resembling *M. f. sensu stricto,* but no colonization of small grains was found (Halbert et al., 2013). That year, a few specimens resembling *M. f. cerealium* were collected in suction traps in Corvallis, Madras, and Hermiston, Oregon, but no colonization of small grain crops was observed at that time (Halbert and Sandvol, 1995). The species was first detected and definitively identified as *M. f. cerealium* in wheat systems of the inland Pacific Northwest (PNW), USA, in 2011 (Halbert et al., 2013) and since then has continued to be an abundant aphid species on wheat in that region (Adhikari et al., 2022). Ongoing surveys have detected a pattern of apparent spread in the PNW (Table 1). In 2018, it was found for the first time in eastern Idaho and in southern Montana (Eigenbrode, Rashed, unpublished data), and three alatae were detected in Kansas (P. Michaud, pers. comm.). In 2019–2022, *M. f. cerealium* continued to be present in Idaho and Washington State. Monitoring in northwestern wheat-growing states, from Washington to Utah and including Colorado, continues. There have been no further published reports of *M. festucae* in California since 1970. Lack of such reports, however, does not rule out the presence of the aphid elsewhere in North America.

The species has been reported in South America, with populations of *M. f. cerealium* reported in Chile (Remaudière et al., 1993), Bolivia, and Peru

Table 1 Timeline of *M. festucae* invasion in North America

1976–1993	1994–1995	1996–2010	2011–2017	2018	2019–2020	2021	2022
M. festucae Reported in CA in 1970	Specimens at three suction trap locations in Oregon USA; on Fescue lawns, Boise Idaho, USA	No reports	Abundant (*M. f. cerealium*) in wheat, eastern Washington, northern Idaho, and northern Oregon. Present on perennial grasses in PNW	As in 2011–2017; First detected in southeastern Idaho, Montana, Kansas	As in 2018, but no reports from Kansas First detected in southeastern Idaho	As in 2020; No new detections[a]	As in 2021; First detected in western Oregon and Wyoming
Foottit et al. (2006)	Halbert and Sandvol (1995)		Halbert et al. (2013), Jensen, pers. comm.	Unpublished, the authors	Unpublished, the authors	Unpublished, the authors	Unpublished, the authors

[a] Cereal aphids (all species) were rare regionally (53 sites sampled in Washington, Idaho, Montana, Wyoming, and Utah) due to hot dry weather: *M. f. cerealium* detected as previously in Washington and Idaho.

(Remaudière et al., 1991). There are no recent reports of it becoming more abundant in those countries since, but it remained abundant in small grains until 10 years ago when a successful campaign to improve biological control of aphids in small grains greatly reduced abundance of all species in Chile (M.E. Gerding, pers. comm.). Specimens of *M. festucae* (no subspecies reported) were detected in New Zealand (Farrell and Stufkens, 1992), and the species is evidently still present there (David Teulon, pers. comm.).

4 Reproductive biology and host range

Both subspecies of *M. festucae* are monoecious and facultatively (but rarely) holocyclic on Poaceae year around (Stroyan, 1982; Blackman and Eastop, 2021). In North America (Eigenbrode et al., unpublished data) and Chile (Remaudière et al., 1993), the species is evidently anholocyclic. In the native range, hosts of *M. festucae* (subspecies not always specified) include *Avena sativa*, *Festuca pratense*, *F. arundinacea*, *F. rubra*, *Holcus lanatus*, *Lolium perenne* and *L. multiflorum*, *Nardus stricta*, *Phleum pratense*, *Poa pratensis*, other *Poa spp.*, and *Triticum aestivum* (Stroyan, 1982; Dent and Wratten, 1986). In North America, hosts of *M. f. cerealium* include *Alopecurus pratensis* (Sadeghi et al., 2016), *Elymus glaucus*, *Festuca scabrella*, *Hordeum vulgare*, *Pseudoroegneria spicata*, *T. aestivum* (Davis et al., 2014), and likely other species because no comprehensive host range testing has been conducted. It has not been recorded anywhere in the world on *Zea mays* or *Sorghum bicolor*, and in the laboratory, *M. f. cerealium* will not settle on or reproduce on *Z. mays* (Davis et al., 2014). The species host range may therefore be taxonomically constrained within Poaceae, perhaps to subfamily Pooideae, or to C3 grasses.

5 Pest status and damage

As a direct pest: In the UK, *M. f. cerealium* can be a significant pest of cereal crops (Dent and Wratten, 1986). Its feeding on wheat and oats produces distinct chlorosis that progresses to necrosis of leaf tissue (Fig. 3) reminiscent of injury caused by *Schizaphis graminum* (Rondani) (Plumb, 1974). This injury presumably leads to substantial per capita injury to the crop, with potential losses to yield. In a greenhouse experiment, with a colony established from populations in the PNW, USA, biomass of wheat seedlings was reduced more by *M. f. cerealium* than by bird cherry-oat aphid, *Rhopalosiphum padi* (L.), while biomass was reduced to a similar degree by Russian wheat aphid, *Diuraphis noxia* (Mordvilko ex Kurdjumov) (Fig. 4). This suggests that *M. f. cerealium* can reduce yield as strongly as does *D. noxia*, which is one of the most destructive aphid pests of wheat worldwide (Ward et al., 2020). Controlled studies are needed in the field to determine the yield loss relationship for *M. f. cerealium*

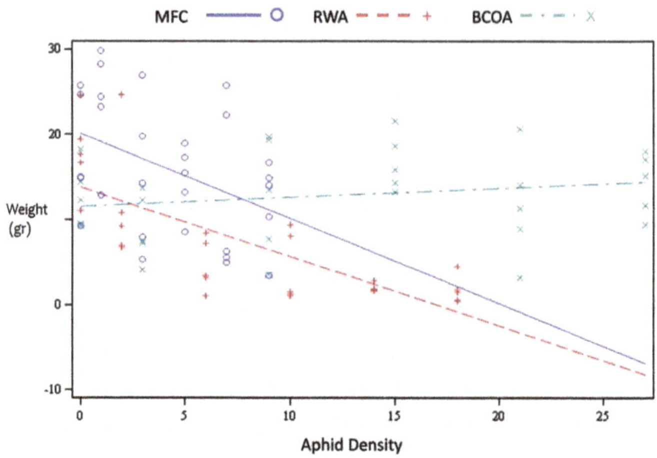

Figure 4 Fresh weight (g) of potted wheat seedlings on which aphid colonies had developed for 2 weeks following inoculation at the indicated aphid densities (*x*-axis). Slopes of the linear fitted regressions do not differ between *M. f. cerealium* (MFC) and *Diuraphis n*oxia, Russian wheat aphid (RWA), while both differ from *Rhopalosiphum padi*, bird cherry-oat aphid (BCOA).

as a basis for estimating economic injury levels for the pest and the direct injury it causes. The distinctive discoloration of leaves fed upon by *M. f. cerealium* is consistent with phytotoxicity, which may result from salivary proteins, peptides, or small molecules, as occurs in saliva of other aphid species (Nicholson et al., 2012; Nicholson and Puterka, 2014). Damage by *M. f. cerealium* superficially resembles phytotoxic reactions caused by *S. graminum*, which has varying salivary proteome related to its virulence (Nicholson and Puterka, 2014), suggesting further studies of the phenomenon could involve comparing the salivary proteome of *M. f. cerealium* with closely related taxa that produce less elicit evident phytotoxic symptoms.

As a virus vector: *M. festucae* is a reported vector of one or more of the barley yellow dwarf complex of cereal viruses (Plumb, 1974, 1983). Blackman et al. (1990) and Stroyan (1982) suggested that these records should refer to *M. f. cerealium*. Sadeghi et al. (2016) reported that *M. f. cerealium* from a colony established from specimens collected in the PNW, USA, is not a vector for BYDV-PAV (the most abundant serotype of BYDV worldwide). That report, however, was based on a greenhouse experiment using one clonal colony of the aphid and a single isolate of the virus, leaving uncertain whether other populations of *M. f. cerealium* can transmit BYDV-PAV and/or other strains of BYDV. Given the severity of yield losses to BYDV globally (Plumb, 1983; McKirdy et al., 2002; Royer et al., 2005; Bockus et al., 2010; Pscheidt and Ocamb, 2013) and the recent occurrence of severe outbreaks of the disease in southern Idaho (Marshall and Rashed, 2014), the role of *M. f. cerealium* as a vector needs to be more rigorously investigated. That should be done taking care to use representative regional genotypes of the aphid and virus isolates where BYDV is prevalent, including South America (Araya, 1990; Parizoto et al., 2013). In addition to serving as a vector of BYDV, *M. f. cerealium* may indirectly promote virus spread by facilitating growth in populations of *R. padi*, a known vector of BYDV-PAV (Plumb, 1974). That species reproduction is increased on plants co-infested with *M. f. cerealium* in greenhouse studies (Foote et al., 2017; Kaisley and Eigenbrode, unpublished data). In field surveys of Washington, Idaho, and Oregon across 160 site-years (2011–2014), positive correlations between *M. f. cerealium* and *R. padi* are consistent with such facilitation in the field but positive correlations are also present with other aphid species, either aspects of sites affect all aphids, because facilitation occurs for other species, or both.

6 Potential for host plant resistance

Although *M. f. cerealium* has a relatively wide host range within Poaceae, there are differences in the suitability among its hosts. Relative resistance in two cultivated small grains and four noncultivated grass host species of *M. f. cerealium* were assessed in terms of the percentage of alatae flying from the host and their reproductive rates (Dent, 1986). *Festuca rubra* and *F. arundinacea* exhibited highest levels of resistance, inducing a greater proportion of the alatae when placed on these plants to abandon the plants after producing only a few nymphs compared with *L. multiflorum*, oats, *L. perenne*, and wheat. A laboratory colony of *M. f. cerealium* from Idaho differed in its performance on, and preference for, native grasses from that region and cultivated grass species (Davis et al., 2014). Although there is no known resistance to the pest in wheat, genetic variation in susceptibility to its phytotoxic effects may exist. In greenhouse experiments in which individual *M. f. cerealium* alates were caged on wheat seedlings, differences were detected in reductions in chlorophyll content as estimated with a Soil Plant Analysis Development (SPAD) chlorophyll meter (Minolta Soil-Plant Analyses Development) among five widely grown wheat varieties (Fig. 5), suggesting there may be sources of tolerance in well-adapted germplasm suitable for introgression to other wheat varieties.

7 Management

There are no formal recommendations for management of *M. f. cerealium*, reflecting the lack of information about its direct injury to yield in any crop and its capacity as a vector of BYDV. Nonetheless, given that its injury is potentially similar to that caused by *D. noxia*, it would be prudent at least to use a nominal threshold similar to the one often used for that species (University of California

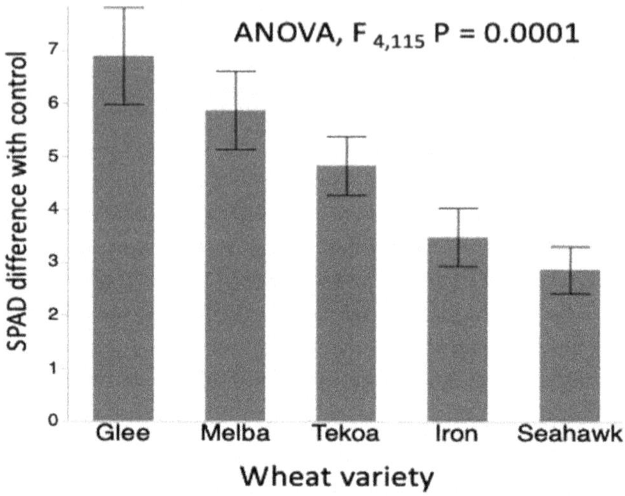

Figure 5 Reduction in chlorophyll after feeding by individual *M. f. cerealium* apterae, as measured by SPAD among spring wheat varieties. Positive values = greater injury, defined as a reduction in chlorophyll signal.

Pest Management Guidelines): fall-seedlings (1 tiller) - 10% of plants infested; fall-larger plants - treat if plants are stressed or there is danger of winter kill; spring-winter grain green-up to the appearance of first node - 5% of plants with reproducing populations and fresh damage; spring-winter grain appearance of first node to head emergence - 10% of tillers infested; spring-spring grain emergence to head emergence - 10% of tillers infested; spring-head emergence to soft dough - treat only if heavy populations (i.e. more than 20 aphids per plant) develop on 10% to 20% of flag leaves or stems; after the soft dough stage, no treatment recommended. This might be modified depending on the degree of injury observed, which could differ depending upon the wheat variety (Fig. 5). If *M. f. cerealium* is confirmed as a BYDV vector and risk of virus presence is high, controls should be implemented when the aphid is detected.

8 Research needs and future trends

As is evident from this comprehensive review, relatively little is known about the biology and ecology of *M. f. cerealium*, its capacity to injure wheat, and strategies for its management. Since the taxonomic resolution of *M. festucae* into two subspecies (Stroyan, 1982), work in the UK has been limited to delineation of the host range for *M. f. cerealium* (Dent, 1986; Dent and Wratten, 1986) and records of the two subspecies captured in suction traps (Rothamsted Insect Survey, https://www.rothamsted.ac.uk/insect-survey). With its occurrence as an adventive species on three continents (Farrell and Stufkens, 1992; Remaudière et al., 1993; Halbert et al., 2013; Blackman and Eastop, 2021), it is important to improve on understanding of the biology of the species and *M. f. cerealium* to aid in its management. Given the lack of information, the scope of needed research is broad:

1 Research is needed to monitor the extent and spread of *M. f. cerealium* in its invaded range in North America; its occurrence, abundance, and distribution in South America and New Zealand; and whether it has spread to other wheat-producing areas.
2 In its invaded range, host range in cultivated and noncultivated grassy hosts should be identified. Currently, its overwintering biology and migration patterns are unknown outside of the UK. The aphid can be found in natural habitats in the PNW that could serve as reservoirs for populations that subsequently colonize crops.
3 The per capita injury the aphid can inflict to crops should be quantified and used to develop and validate treatment thresholds.
4 The basis of the phytotoxic effects caused by *M. f. cerealium* and implications for injury to wheat should be examined. This should be coupled with an assessment of responses of host species and varieties, with an eye to potentially identifying sources of host plant resistance.
5 Feeding alters grain chemistry and quality, as reported with other phytotoxic aphids (Gray et al., 1990; Royer et al., 2015) (other chapters in this volume), should be determined.
6 Whether various populations of the aphid are effective vectors of multiple strains of BYDV must be assessed and this information used to develop additional management recommendations.
7 The subspecies prevalence genetic structure of populations of *M. festucae* in its invaded ranges should be assessed to determine whether they are uniform genetically and phenotypically or variable, which could require different management approaches.
8 In the event of genotypic variability associated with differences in biology and management, molecular tools will be needed to facilitate identification of strains or biotypes.
9 Several parasitoids have been reared from field-collected individuals of *M. f. cerealium* in North America (Eigenbrode, unpublished data), and others have been reported from the species in its native range (Arthur, 1944; Monadjemi, 1972; Vorley and Wratten, 1987; Levie et al., 2005), but the full suite of its natural enemies and potential to suppress its populations is unknown.
10 As with most of the pests covered in this volume, ongoing climate change in its invaded and native ranges has the potential to affect its geographic extent and severity as a pest. Future research should

address each of these aspects to anticipate its continuing prevalence in wheat systems.

9 Conclusion

Given the potential of this species to remain a member of cereal aphid complex in much of the globe, research to improve understanding of its biology and approaches to management can contribute to sustainable production of cultivated small grains, at least wheat, barley, and oats, for smallholders and large-scale commercial production.

10 Where to look for further information

The cited references in this chapter are most of the sources of information about *M. festucae* and *M. f. cerealium* as a pest of wheat. If further discoveries of the invaded range or biology of this species and its subspecies are made, they may appear in Blackman, R. L. and Eastop, V. F. (2021) (cited) an online resource that is periodically updated.

For additional information about aphid biology and diversity, readers can consult "Influential Points", another excellent on-line resource (https://influentialpoints.com/Gallery/Aphid_genera.htm).

Information about the diversity and biology of aphids in the Pacific Northwest of the USA can be found at Aphidtrek (https://aphidtrek.org), an online resource maintained by Andrew Jensen.

11 References

Adhikari, S., Seamon, E., Wu, Y., Sadeghi, S. E. and Eigenbrode, S. D. (2022). Do invasive and naturalized aphid pest populations respond differently to climatic and landscape factors? *J. Econ. Entomol.*, toac044. doi: 10.1093/jee/toac044.

Araya, I. R. (1990). A review of *Barley yellow dwarf virus* in the southern cone countries of South America, In: Burnett, P. (Ed.), *World Perspectives on Barley Yellow Dwarf*, pp. 29–33. México-Veracruz, El Batán: CIMMYT.

Arthur, D. R. (1944). *Aphidus granarius*, marsh, in relation to its control of *Myzus kaltenbachi Schout*, *Bull. Entomol. Res.* 35, 257–270.

Blackman, R. L. and Eastop, V. F. (2021). *Aphids on the World's Plants: An Online Identification and Information Guide*. Available at: http://www.aphidsonworldsplants.info.

Blackman, R. L., Eastop, V. F. and Brown, P. A. (1990). The biology and taxonomy of the aphids transmitting barley yellow dwarf virus. In: Burnett, P. A. (Ed.), *World Perspectives on Barley Yellow Dwarf*, pp. 197–214. México-Veracruz, El Batán: CIMMYT.

Bockus, W. W., Bowden, R. L., Hunger, R. M., Morrill, W. L., Murray, T. D. and Smiley, R. W. (2010). *Compendium of Wheat Diseases and Pests* (3rd edn.). St. Paul: APS Press.

Chambers, R. J., Sunderland, K. D., Stacey, D. L. and Wyatt, I. J. (1982). A survey of cereal aphids and their natural enemies in winter-wheat in 1980, *Ann. Appl. Biol.* 101(1), 175–178.

Coceano, P. G., Peressini, S., Bianchi, G. L. and Caciagli, P. (2009). Long-term changes of aphid vectors of Barley yellow dwarf viruses in north-eastern Italy (Friuli-Venezia Giulia), *Ann. Appl. Biol.* 155(1), 37–50.

Davis, T. S., Wu, Y. and Eigenbrode, S. D. (2014). *Metopolophium festucae* subsp *cerealium*, *Environ. Entomol.* 43(3), 682–688.

Dent, D. R. (1986). Resistance to the aphid *Metopolophium festucae cerealium*: Effects of the host plant on flight and reproduction, *Ann. Appl. Biol.* 108(3), 577–583.

Dent, D. R. and Wratten, S. D. (1986). The host-plant relationships of apterous virginoparae of the grass aphid *Metopolophium festucae cerealium*, *Ann. Appl. Biol.* 108, 567–576.

Edwards, C. A. and Heath, G. W. (1964). The grass aphid. In: Edwards, C. A. and Heath, G. W. (Eds.), *Principles of Agricultural Entomology*, p. 182. London: Chapman and Hall Ltd.

Farrell, J. A. and Stufkens, M. W. (1992). Cereal aphid flights and barley yellow dwarf virus-infection of small grains in Canterbury, New Zealand, *N. Z. J. Crop Hortic. Sci.* 20(4), 407–412.

Foote, N. E., Davis, T. S., Crowder, D. W., Bosque-Perez, N. A. and Eigenbrode, S. D. (2017). Plant water stress affects interactions between an invasive and a naturalized aphid species on cereal crops, *Environ. Entomol.* 46(3), 609–616.

Foottit, R. G., Halbert, S. E., Miller, G. L., Maw, E. and Russell, L. M. (2006). Adventive aphids (Hemiptera: Aphididae) of America north of Mexico, *Proc. Entomol. Soc. Wash.* 108, 583–610.

George, K. S. (1974). Damage assessment aspects of cereal aphid attack in autumn-sown and spring-sown small grains, *Ann. Appl. Biol.* 77(1), 67–74.

Gray, M. E., Hein, G. L., Walgenbach, D. D. and Elliott, N. C. (1990). Effects of Russian wheat aphid (Homoptera: Aphididae) on winter and spring wheat infested during different plant growth stages under greenhouse conditions, *J. Econ. Entomol.* 83(6), 2434–2442.

Halbert, S. and Sandvol, L. (1995). New or unusual pests in Idaho in 1994, *Oreg. State Univ. Res. Rep.* 71.

Halbert, S. E., Wu, Y. and Eigenbrode, S. D. (2013). *Metopolophium festucae cerealium* (Hemiptera: Aphididae), a new addition to the aphid fauna of North America, *Insecta Mundi* 301, 1–6.

Hand, S. C. (1989). The overwintering of cereal aphids on Gramineae in southern England, 1977–1980, *Ann. Appl. Biol.* 115(1), 17–29.

Hille Ris Lambers, D. (1947). Contributions to a monograph of the Aphididae of Europe, *Temminckia*, 179–319.

Kadyrbekov, R. K. (2017). Aphids (Hemiptera: Aphidoidea, Phylloxeroidea) of Kazakhstan, a checklist. Almaty, Kazakhstan, 583 pp. (cited in: Blackman, R. L. and Eastop, V. F. (2021), *Aphids on the World's Plants: An Online Identification and Information Guide*. Available at: http://www.aphidsonworldsplants.info).

Levie, A., Legrand, M. A., Dogot, P., Pels, C., Baret, P. V. and Hance, T. (2005). Mass releases of Aphidius rhopalosiphi (Hymenoptera: Aphidiinae), and strip management to control of wheat aphids, *Agric. Ecosyst. Environ.* 105(1–2), 17–21.

Marshall, J. M. and Rashed, A. (2014). *Barley Yellow Dwarf Virus in Idaho Cereal Crops, CIS 810*. Mosow: University of Idaho Extension ID.

Marzocchi, L. and Nicoli, G. (1991). The principal pests of wheat. (I principali fitofagi del frumento.), *Inf. Fitopatologico* 41(2), 29–33.

McKirdy, S. J., Jones, R. A. C. and Nutter, F. W. (2002). Quantification of yield losses caused by Barley *yellow dwarf virus* in wheat and oats, *Plant Dis.* 86(7), 769–773.

Monadjemi, N. (1972). Variations of fecundity of *Aphelinus asychis* (Hym. Aphelinidae) according to aphid species available at different periods of his imaginal life, *Ann. Soc. Entomol. Fr.* 8(2), 451.

Nicholson, S. J., Hartson, S. D. and Puterka, G. J. (2012). Proteomic analysis of secreted saliva from Russian wheat aphid (Diuraphis noxia Kurd.) biotypes that differ in virulence to wheat, *J. Proteomics* 75, 2252–2268.

Nicholson, S. J. and Puterka, G. J. (2014). Variation in the salivary proteomes of differentially virulent greenbug (Schizaphis graminum Rondani) biotypes, *J. Proteomics* 105, 186–203.

Parizoto, G., Rebonatto, A., Schons, J. and Lau, D. (2013). Barley yellow dwarf virus-PAV in Brazil: Seasonal fluctuation and biological characteristics, *Trop. Plant Pathol.* 38(1), 11–19.

Plumb, R. T. (1974). Properties and isolates of barley yellow dwarf virus, *Ann. Appl. Biol.* 77(1), 87–91.

Plumb, R. T. (1983). Barley yellow dwarf virus-a global problem. In: Plumb, R. T. and Thresh, J. M. (Eds.), *Plant Virus Epidemiology*, pp. 185–198. Oxford: Blackwell Publishing.

Pscheidt, J. W. and Ocamb, C. M. (2013). *Pacific Northwest Plant Disease Management Handbook*. Oregon State University. Available at: http://pnwhandbooks.org/plantdisease.

Remaudière, G., Stary, P. and Gerding, M. (1993). *Sitobion fragarie* (Walker) y *Metopolophium festucae cerealium* Stroyan (Homoptera: Aphididae), dos nuevos afidos de los cereales, en Chile, *Agric. Tec.* 53(1), 91–92.

Remaudière, G., Weemaels, N. and Nicolas, J. (1991). Contribution to the knowledge of the aphid fauna of Bolivia (Homoptera: Aphididae), *Parasitica* 47(1), 19–46.

Robert, Y. and Rouzejouan, J. (1976). Trapping small grains aphids, *Acyrthosiphon* (*Metopolophium*) *dirhodum* Wlk, *A* (*M*) *festucae* Wlk, *Macrosiphum* (*Sitobion*) *avenae*-f *M* (*S*) *fragariae* Wlk and *Rhopalosiphum padi* over 9 years in Brittany, *Rev. Zool. Agric. Pathol. Veg.* 75(2), 67–80.

Royer, T. A., Giles, K. L., Nyamanzi, T., Hunger, R. M., Krenzer, E. G., Elliott, N. C., Kindler, S. D. and Payton, M. (2005). Economic evaluation of the effects of planting date and application rate of Imidacloprid for management of cereal aphids and barley yellow dwarf in winter wheat, *J. Econ. Entomol.* 98(1), 95–102.

Royer, T. A., Pendleton, B. B., Elliott, N. C. and Giles, K. L. (2015). Greenbug (Hemiptera: Aphididae) biology, ecology, and management in wheat and sorghum, *J. Integr. Pest Manag.* 6(1), 19. DOI: 10.1093/jipm/pmv018.

Sadeghi, S. E., Bjur, J., Ingwell, L., Unger, L., Bosque-Pérez, N. A. and Eigenbrode, S. D. (2016). Interactions between *Metopolophium festucae cerealium* (Hemiptera: Aphididae) and *Barley yellow dwarf virus* (BYDV-PAV), *J. Insect Sci.* 16(1), 1–6.

Stroyan, H. L. G. (1982). Revisionary notes on the genus *Metopolophium* Mordvilko, 1914, with keys to European species and descriptions of 2 new taxa (Homoptera, Aphidoidea), *Zool. J. Linn. Soc.* 75(2), 91–140.

University of California pest management guidelines. Available at: http://ipm.ucanr.edu/PMG/r730300211.html.

Vickerman, G. P. and Wratten, S. D. (1978). The biology and pest status of cereal aphids in Europe: A review, *Bull. Entomol. Res.* 69, 1–32.

Vorley, V. T. and Wratten, S. D. (1987). Migration of parasitoids (Hymenoptera, Braconidae) of cereal aphids (Hemiptera, Aphididae) between grassland, early-sown small grains and late-sown small grains in southern England, *Bull. Entomol. Res.* 77(4), 555–568.

Wangai, A. (1992). Barley yellow dwarf virus research in Kenya. Barley yellow dwarf in West Asia and North Africa. In: Comeau, A. and Makkouk, K. M. (Eds.), *Proceedings of a Workshop Held at Rabat*, Morocco, 19–21 November 1989, pp. 109–111. International Centre for Agricultural Research in the Dry Areas.

Ward, S., van Helden, M., Heddle, T., Ridland, P. M., Pirtle, E. and Umina, P. A. (2020). Biology, ecology and management of *Diuraphis noxia* (Hemiptera: Aphididae) in Australia, *Austral. Entomol.* 59(2), 238–252.

Chapter 10

The English grain aphid *Sitobion avenae*

Deguang Liu, College of Plant Protection, Northwest A&F University, China

1 Introduction

The English grain aphid, *Sitobion avenae* (Fabricius), is an economically important insect pest of cereal crops worldwide. This aphid has extremely complex life-history traits (e.g. possessing different life-cycle variants, symbiont-induced phenotypic plasticity, and rapid adaptation under changing environments), which may be conducive to their survival under rapidly changing environments (Huang et al., 2013, 2015; Dai et al., 2014, 2015; Ahmed et al., 2017). There has been increasing evidence for *S. avenae*'s evolving and differentiating populations (i.e. geographic and biotype differentiation), which are closely related to this aphid's population growth, evolutionary ecology, dispersal, and management. Damage by this aphid can be destructive for cereal crops due to its ability to transmit yellow dwarf viruses (YDV), and its control often relies on routine application of synthetic pesticides. However, the emergence of insecticide resistance for *S. avenae* has been reported by many studies (Lu and Gao, 2009; Foster et al., 2014; Gong et al., 2021). In addition, some neonicotinoid insecticides (e.g. imidacloprid, acetamiprid, clothianidin, and thiamethoxam), which are effective against this pest, have been banned in Europe (Mc Namara et al., 2020) and are threatened with the regulation in other parts of the world. Climate change can also pose a significant challenge for the management of *S. avenae* (Li

http://dx.doi.org/10.19103/AS.2022.0114.07

et al., 2022). These issues potentially make the control of aphids in cereal fields increasingly difficult in the future. In this chapter, we provide a review of recent research progress on the above-mentioned issues (e.g. geographic divergence, biotype development, insecticide resistance, climate change, etc.) and highlight the opportunities for improving integrated management of *S. avenae*.

2 Distribution and economic importance

Sitobion avenae is widely distributed throughout Europe, Asia, North America, South America, the Mediterranean region, and east Africa (Fig. 1; https://www.cabi.org/isc/datasheet/51737). This aphid can colonize and become a pest on various cereal crops [e.g. wheat (*Triticum aestivum*), barley (*Hordeum vulgare*), oat (*Avena sativa*), two-rowed barley (*Hordeum distichon*), rice (*Oryza sativa*), rye (*Secale cereale*), and maize (*Zea mays*)], and pasture grasses of temperate climates (e.g. cocksfoot *Dactylis glomerata*); its hosts also include many other monocots (i.e. wild grasses), as well as certain dicots (e.g. *Capsella bursa-pastoris* (L.) and *Spergula arvensis* L.) (Blackman and Eastop, 2006). It can cause damage to cereal production directly by extraction of phloem sap and indirectly by transmitting the yellow dwarf virus (YDV) disease (resulting in yield reductions of up to 80%) (Mc Namara et al., 2020). Annual worldwide losses of cereal crops due to infestations by this aphid are estimated to be in the hundreds of millions of dollars (Mc Namara et al., 2020; Gao et al., 2014).

In China, cereal aphids like *S. avenae* are among the most dominant and destructive grain pests and have been included in the 'list of the most

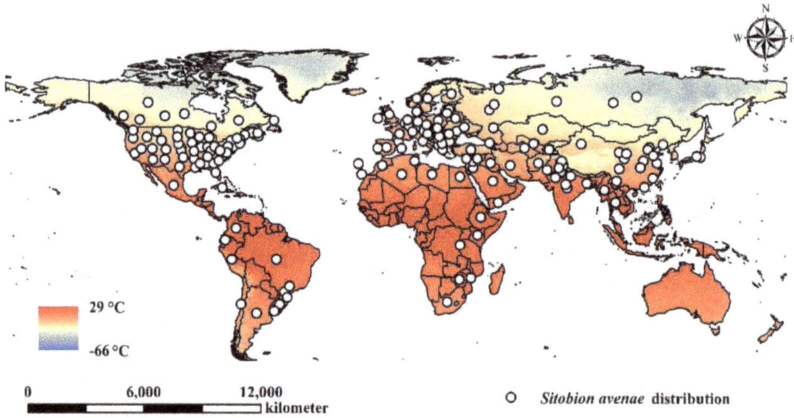

Figure 1 Worldwide distribution of *Sitobion avenae* in relation to mean temperature of coldest quarter of the year (open circles indicate known occurrence points; data of the expertly compiled and validated occurrence points from https://www.cabi.org; temperature data from the PaleoClim database http://www.paleoclim.org/) illustrating the predominance of records in north temperate regions.

important crop pests and diseases' by the Ministry of Agriculture and Rural Affairs of China (http://www.zzys.moa.gov.cn/). Total agricultural land of wheat cultivation in China is about 23.4 million hectares, in which 10–15 million hectares are infested by cereal aphids, resulting in about 10% annual crop yield losses (Chinese National Agro-Tech Extension and Service Center [NATESC] database, http://www.natesc.moa.gov.cn/). This aphid is sometimes referred to as *Sitobion miscanthi* in China (Wang et al., 2009a). Based on morphology and mtDNA sequences of *S. avenae*, *S. miscanthi*, and *Sitobion akebiae*, Choe et al. (2006) concluded that the latter two should be considered as the same species as *S. avenae*. However, it is still controversial to treat *S. miscanthi* as a synonym of *S. avenae* in China (Hales et al., 2010), and more research is still needed to resolve this issue by using samples from more geographic areas and molecular markers of more genomic regions.

2.1 Life cycle and reproduction strategies

Sitobion avenae is a monoecious species overwintering on grasses and cereals as diapausing eggs or active aphids (i.e. parthenogenetic individuals). Each year, this aphid usually appears in the field when plants are in the late tillering or stem-elongation stage and reaches maximum numbers at the heading/flowering stage (Dean, 1974; Wiktelius and Ekbom, 1985). It has a complex life cycle, consisting of four kinds of reproductive variants (i.e. holocyclic, anholocyclic, androcyclic, and intermediate) (Dedryver et al., 1998). Holocyclic clones are cyclical parthenogens, which reproduce parthenogenetically in the summer but sexually in the autumn; in response to short day-lengths and low temperatures, they produce mating females (oviparae) and males in the autumn, which mate and produce overwintering eggs. Anholocyclic clones are obligate parthenogens, which reproduce parthenogenetically all year round. Androcyclic clones can give birth to only males in the sexual phase. Intermediate clones can produce both males and mating females in the autumn, in addition to parthenogenetic females.

In different geographic areas, populations of *S. avenae* can be a mixture of the above-mentioned life cycle variants with different investments in sexual reproduction (Helden and Dixon, 2002). For example, intermediate clones accounted for a relatively large proportion in French oceanic areas, and androcyclic clones presented a relatively higher frequency in the United Kingdom (Dedryver et al., 1998; Helden and Dixon, 2002). However, in North Europe (e.g. Scandinavian countries like Finland and Sweden), *S. avenae* has only holocyclic lifecycles, and overwinters as eggs (Hansen, 2006). This makes sense, because in areas with cold winters, it is advantageous for *S. avenae* to overwinter as eggs, and the occurrence of holocyclic clones can be favored, while anholocyclic clones are supposed to be favored in areas with mild winters (Dedryver et al., 1998). It is believed that winter severity, instead of summer fitness, could be the most important factor affecting the occurrence frequency of *S. avenae* clones of different reproductive modes in a particular area (Helden and Dixon, 2002). Therefore, the geographical variation in life cycles and reproductive strategies can be an adaptation of *S. avenae* to different climatic scenarios. Although a high amount of genetic differentiation among different clones of life cycle variants of *S. avenae* have been identified using microsatellite markers (Dedryver et al., 1998), the underlying genetic and molecular mechanisms still remain elusive.

3 Factors affecting population growth and spread

It is of great importance to clarify the divergence and dispersal of insect pest populations, as well as their evolutionary implications. Divergence of insect populations can occur due to combined effects of selection, mutation, gene flow, and genetic drift (Huang et al., 2015; Wang et al., 2019). This often involves mechanisms of isolation by distance (i.e. geographic isolation) and/or isolation by adaptation to host plants (Vidal et al., 2019). In addition, climate change potentially has significant impacts on the adaptability and dispersal of aphids.

3.1 Climatic factors

Developmental and biological processes of insects are temperature-dependent, so temperature often can determine the geographical distribution limits of an insect and the timing of outbreaks. The developmental threshold of *S. avenae* nymphs was found to be about 4°C, and the thermal units required for development from birth to adult was about 150 degree days (Kieckhefer et al., 1989; Salman, 2006). Based on comparisons of fitness parameters (e.g. fecundity, growth rate, and longevity) in life-table tests at different constant temperatures (i.e. 12–28°C) in the laboratory, the optimal temperature range for population growth and development of *S. avenae* is 18–25°C (Salman, 2006; Xu et al., 2014). This is consistent with its global distribution which is concentrated in the north temperate regions (Fig. 1).

The mean temperature of coldest quarter (i.e. winter) was found to have the highest contribution to the ecological niche modeling in the software MaxEnt with the current occurrence records of *S. avenae*, suggesting its critical role in determining the distribution of *S. avenae* (Li et al., 2022). Indeed, an analysis of 24-year time-series field data of aphid abundance on winter wheat showed that the abundance of *S. avenae* each year was closely associated with temperature and could be predicted from temperature data preceding the aphid population growth (Brabec et al., 2014). Temperature from mid of February to March had strongest effects (meaning highest contribution) in the model, suggesting that temperature data of this period can be used to predict *S. avenae* outbreaks and plan aphid management strategies (Brabec et al., 2014).

A combination of laboratory experiments, field experiments, and analysis of multiyear field observations revealed that extreme high temperature (EHT) had stronger effects on demographic rates and fitness of *S. avenae* than mean temperature, and large frequency and amplitude of EHTs could significantly reduce the fitness of *S. avenae* (Ma et al., 2015). This may explain the high prevalence of *S. avenae* during summer in temperate regions, but not in tropical regions.

3.2 Water

Water availability is also among the most important abiotic variables influencing the abundance and distribution of insects. Indeed, for *S. avenae*, precipitation of wettest quarter (i.e. summer in China) was shown to have the second largest contribution (after mean winter temperature only) to the models (constructed by ecological niche modeling with MaxEnt), which can be used to predict the occurrence and suitability of this aphid in different areas of China under various climatic conditions (Fig. 2; Li et al., 2022). In addition, high relative humidity favors population growth of *S. avenae*, and this aphid develops best at a relative humidity of 65–80% (Ahmad et al., 2016; Gao and Liu, 2013; Dai et al., 2015). On the other hand, abundance of *S. avenae* is negatively correlated with relative humidity and rainfall in the field (Ahmad et al., 2016), meaning that excess water is also unfavorable for the development and survival of this aphid.

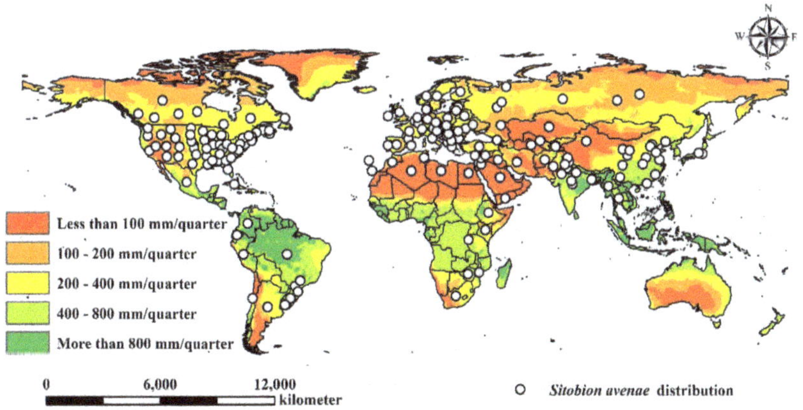

Figure 2 Worldwide distribution of *Sitobion avenae* in relation to precipitation in the wettest quarter of the year (open circles indicate known occurrence points; data of the expertly compiled and validated occurrence points from https://www.cabi.org; rainfall data from the PaleoClim database http://www.paleoclim.org/).

Water deficit can have particularly strong effects on sap-sucking insects like aphids because they feed dominantly or exclusively on plant phloem sap. The survival, growth, and reproduction of *S. avenae* can be significantly and negatively affected by water-deficit stress, despite that environmental stresses can increase nutritional quality (e.g. higher concentration of amino acids) of host plants (White, 1969, 2009). For example, the developmental time of *S. avenae* nymphs showed a tendency to increase under treatments with increasing water-deficit stress, and with increasing lengths of water-deficit exposure time (one to five generations), and the effects of severe water-stress on *S. avenae*'s fitness was negative in terms of fecundity and weight (Dai et al., 2015; Liu et al., 2018).

Nonetheless, detailed comparisons of *S. avenae* samples from northwestern China showed that some clones had higher fitness on water-deficit stressed plants, suggesting that they could be somewhat adapted to water-deficit conditions (Liu et al., 2018). Based on comparisons of desiccation-related traits (e.g. survival time under desiccation stress), Yang et al. (2020) found the occurrence of desiccation-resistant clones of *S. avenae* in northwestern China and that winged aphids exhibited lower desiccation resistance than wingless aphids. The desiccation resistant clones were found to have suppressed water loss rates, further indicating their adaptation to dry environments (Yang et al., 2020). In addition, after their exposure to moderate water-deficit stress for only five generations, adult weight and fecundity tended to increase for some *S. avenae* clones, suggesting rapid shifts in their vital life-history characters and adaptation potential of this aphid under continuous water-deficit stress for long term (Liu et al., 2018). Clones of *S. avenae* appeared to be more likely to become adapted to moderate water-deficit stress than severe water-deficit stress (Liu et al., 2018). Five mechanisms of adaptation to water deficit in insects have been proposed, and they include enhanced water conservation, behavioral avoidance, desiccation tolerance, high water turnover, and water sorption dependence (Addo-Bediako et al., 2001). Based on the study of Yang et al. (2020), desiccation-resistant clones of *S. avenae* had enhanced water conservation. Further studies are needed to determine how *S. avenae* clones reduce water loss under desiccation, and if other mechanisms are also at play. Such studies could have significant implications for prediction of evolutionary dynamics and outbreak risks of aphids in semiarid and arid areas with widely planted cereals.

3.3 Geographic divergence and local adaptation

Several studies have used microsatellite DNA markers to assess genetic divergence of *S. avenae* populations (Simon et al., 1999; Fievet et al., 2007; Jensen et al., 2008; Xu et al., 2011a). For example, in Europe, *S. avenae* populations of Romania (with severe winters) were found to have higher genotypic diversity and lower linkage disequilibrium than those of France (Papura et al., 2003). In China, populations of *S. avenae* from western provinces (i.e. Gansu, Qinghai, Shaanxi, and Xinjiang) had higher genetic differentiation than those from eastern provinces (i.e. Anhui, Hubei, Henan, Jiangsu, and Zhejiang), shown by comparisons of pairwise F_{ST} values and gene diversity between both areas (Wang et al., 2020b). In both studies, genetic differentiation among *S. avenae* populations could be explained by different proportions of sexual reproduction (recombination of alleles). In China, western provinces had lower winter temperatures, which could lead to more occurrences of sexual reproduction for *S. avenae* there than in eastern provinces (Yang et al., 1994); similarly, Romania with severe winters possessed a higher proportion of sexual forms than France with milder winters (Dedryver et al., 2001; Delmotte et al., 2002). Long-distance dispersal of *S. avenae* between eastern and western provinces of China was found to be very limited in both directions, contributing to the significant geographic divergence between both areas (Wang et al., 2020b). So, another possible explanation for divergence of *S. avenae* population in the above-mentioned studies is isolation by distance (IBD). It has long been considered as an important mechanism underlying evolutionary changes and divergence in many insect populations, since gene flow is often found to have negative correlations with geographic distance (Vidal et al., 2019; He et al., 2018). Mantel tests detected significant positive relationships between genetic and geographic distances, indicating a significant IBD effect among different geographic populations of *S. avenae* (Wang et al., 2020b). Thus, geographic distance plays a significant role in the genetic structuring and divergence of *S. avenae* populations.

Another important geographic factor in the differentiation of *S. avenae* populations is mountain ranges. For example, comparisons between *S. avenae* populations from areas north and south of the Qinling Mountains showed that northern populations had reduced developmental time for first instar nymphs, but prolonged for third and fourth instar nymphs, whereas southern populations showed longer post-reproductive time, longer adult lifespan, and higher fecundities (lifetime or daily), clearly showing the divergence of *S. avenae* populations from both sides of the Qinling Mountains. Similar structuring effects of geographical barriers were found for Russian wheat aphid (*Diuraphis noxia*) (Zhang et al., 2012). Although *S. avenae* alatae are able to disperse across the Qinling Mountains, this movement is apparently limited based on evidence of low gene flow between populations on both sides of the mountains identified with microsatellite markers (Huang et al., 2015). The low gene flow in combination with differing selective pressures north and south of the mountains has evidently contributed to genetic differentiation and local adaptation. Based on significant differences between genetic profiles of migrants and those of local populations,

Vialatte et al. (2007) concluded that local populations, instead of migrants from distant sources, were important in determining outbreak risks of *S. avenae*. So, local adaptation appears to be a common phenomenon for this pest.

In summary, evidence of geographic isolation coupled with varying selective pressures across the broad geographic range of *S. avenae* has evidently led to considerable local adaptation by this pest, with implications for its pest status and outbreak potential. Patterns of genetic divergence can depend on geographic scale (Vidal et al., 2019) necessitating additional study of the population structure of *S. avenae* at multiple scales to better understand the drivers of population divergence in this species and implications for pest management.

3.4 Climate change

Climate warming is well known to have significant impacts on a species' phenology and abundance, and its interactions with other species as well. Due to increasing rates of warming in the twenty-first century, global mean surface temperature is predicted to increase by 1.5–4.5°C at the end of this century (IPCC, 2021). Cereal aphids like *S. avenae* are characterized by their small body sizes and short life cycles, thereby making them relatively more sensitive to shifts in ambient temperatures. Maximum temperatures in wheat fields are about 38–40°C on hot days during the growing season in the major wheat-growing regions of China, and *S. avenae* mainly infests wheat spikes with higher temperatures than surrounding air (Ma and Ma, 2012). So, climate change will increase exposure opportunities of *S. avenae* to high temperatures exceeding its upper physiological limits.

In addition to changes in average temperature, extreme temperature events are also predicted to increase in both frequency and severity under future climate change scenarios (Ma et al., 2015). *S. avenae* can avoid potential heat injuries through escaping behaviors (e.g. moving to cool microhabitats like soil crevices) (Ma and Ma, 2012). In addition, to cope with thermal stress, aphids may need to produce and accumulate heat shock proteins and increase the abundance of secondary symbionts (Montllor et al., 2002; Ma and Ma, 2012). Thus, in the future, it may take lots of time and resources for aphids to avoid heat stress and prevent heat injuries, influencing their reproduction and population growth. Despite the negative effects of high temperatures on *S. avenae*, the mean fecundity was significantly higher in infrared heated plots compared to control ones, and the intrinsic growth rates, reproductive rates, and finite increase rates of *S. avenae* were also significantly higher in the warmed plots (Han et al., 2019). The phenomenon could be at least partly attributed to top-down effects, since warming significantly decreased the rate of parasitism by parasitoids (Han et al., 2019).

Increasing frequency and intensities of drought is also expected in many grain-growing areas under climate change. Liu et al. (2018) found that *S. avenae* populations from moist, semiarid, and arid areas had both adaptive and non-adaptive changes in development time, fecundity, and adult weight, providing evidence of shifts in life history traits and potential divergence under variable drought conditions in the future. The adaptability of arid-area clones of *S. avenae* could be increased after their continuous exposure to water-deficit stress for only five generations, suggesting that some *S. avenae* clones could become adapted to particular drought conditions and spread quickly in the future with the increasing pace of the climate warming trend (Liu et al., 2018). Furthermore, Ahmed et al. (2017) found that water-deficit-stressed *S. avenae* could have indirect and negative impacts on the development of a third trophic level parasitoid (i.e. *Aphidius gifuensis*), suggesting that the vulnerability of parasitoids could be potentially increased in future warming scenarios.

Climate niche modeling of *S. avenae* recently showed that under high greenhouse gas emissions (CMIP-6, SSP585), the potential suitable area for *S. avenae* will expand by nearly 37% of China's total land area with its range centroid moving northward by 3.57 degrees in China by 2100. The model also projected increased damage due to this aphid in several of China's main grain-growing areas (e.g. Huang-Hui-Hai, Northeast, Along the Great Wall, and Loess Plateau). These projections are consistent with historical records of increasing frequency of aphid outbreaks on cereal crops coincident with the climate warming trend in China (Cao et al, 2006; Deng et al., 2010).

Taken together, projected climate change will potentially influence distribution, abundance, and evolutionary ecology of *S. avenae*, exacerbating its severity as a pest of cereals. Future work should aim to better delineate these projections to anticipate challenges and improve pest management.

4 Host plants

Sitobion avenae can feed and survive on various plants of the family Poaceae. Some of them have resistant properties, which can negatively affect the growth, reproduction, or behavior of this aphid. The use of resistant cultivars of cereal crops is probably the most economically and environmentally sound way of controlling this aphid. Thus, many researchers have focused on identifying sources of resistance to *S. avenae*. For example, using glasshouse tests with immature *S. avenae*, winter wheat cv. Kador, and barley cv. CI16145 were found to be resistant to this aphid (Lowe, 1980). Moderate levels of resistance to *S. avenae* were found to exist quite widely in spring (26 out of 91 tested lines) and winter wheat breeding lines in Britain (Lowe, 1982, 1984). *Triticum monococcum* (L.) lines Tm44 and Tm46 of France were identified to be resistant to *S. avenae* (Caillaud et al., 1995). Based on both field tests and chemical analyses, the winter triticale cultivar 'Lasko' with relatively high contents of non-protein amino acids was found to be moderately resistant to this aphid, and the cultivar 'Grado' was found to be susceptible (Ciepiela et al., 1999).

In China, wheat varieties of Fengchan No.3, Zhong 4 wumang, Xiaobaidongmai, Hongmanghong, Xiaoyan 22, 186-TM12-34, Liying No.5, Liying No.6, Nongda 198, Shan 167, Yanda 1817, PI262260, and Shannong 82(2)-12 have been found to present different levels of resistance to *S. avenae* (Xu et al., 2011b; Hu et al., 2011; Liu et al., 2014, 2015). Wang et al. (2020a) screened 33 wheat varieties and 25 barley varieties and found that high levels of resistance occurred in wheat varieties of Zhong 4 wumang, Zhengmai 366, and 186-TM12-34, as well as barley varieties of Yanyin2129-56, Dulihuang, Zhudamai No.4, Zaoshu No.3, Zangqing 25, Ganpi No.6, and Xiyin No.2. Antixenosis, antibiosis, or tolerance (Painter, 1951; Horber, 1980) can all be the possible mechanism underlying the resistance of these varieties against *S. avenae*.

Thus, it is highly likely for widely planted cereal plants (e.g. wheat and barley) to exert some levels of selective pressures on *S. avenae* populations. Indeed, compared with barley clones, wheat clones of this aphid have shorter developmental durations, higher fecundities, and higher population growth rates on the plant of origin (Gao and Liu, 2013). Similarly, significant differences in developmental durations and fecundities were found between *S. avenae* clones from oat and barley, suggesting these clones could have diverged to a certain extent on both plants (Gao et al., 2014). In another study, some clones of *S. avenae* were found to be non-virulent on barley, whereas some other clones were characterized by their high virulence on barley (e.g. Zaoshu No.3), clearly showing the potential effects of host plants on the divergence of *S. aveane* clones (Wang et al., 2020a). These recent studies with laboratory bioassays have confirmed the occurrence of divergence among *S. avenae* clones on widely planted cereal crops. Thus, the pattern of natural selection can vary greatly on different host plants, contributing to adaptation of *S. avenae* clones to a particular plant and their genetic differentiation.

Since the 1990s, research has identified *S. avenae* populations adapted to different host plant species including but not limited to cereal crops. Host plant-associated genetic differentiation was first found among *S. avenae* clones in Europe by using RAPD (random amplified polymorphic DNA)-PCR (De Barro et al., 1995). Using additional molecular markers (i.e. microsatellites and mitochondrial DNA markers), host plant associated differentiation among *S. avenae* clones were found to be substantial in Europe, and some clones were found to associate with only a single plant species (i.e. cocksfoot or wheat) (Sunnucks et al., 1997). Using microsatellite markers, *S. avenae* populations were found to be diverged to a certain extent not only on cereal crops like wheat and barley but also on non-cultivated grass hosts (Wang et al., 2020a,b; Gao, 2014). Non-cultivated grass hosts have been found to be important in the differentiation of the greenbug, *Schizaphis graminum* (Rondani), and Russian wheat aphid (*D. noxia* Kurdjumov) populations (Burd and Porter, 2006; Weiland et al., 2008). It remains to be determined if non-cultivated hosts in different environments can contribute to the divergence of *S. avenae* populations.

Overall, *S. avenae* appears to be prone to develop variable host plant associated populations (Wang et al., 2019, 2020b). This has significant implications not only for its evolutionary ecology but also for plant breeding programs.

5 Biotype development

5.1 Discovery of S. avenae *biotypes*

Different biotypes of *S. avenae* on wheat were first found in China from a very limited number of clones (i.e. seven) from wheat by using a small number of wheat varieties (e.g. Zhong 4 wumang, Amigo and Fengchan No.3) (Xu et al., 2011b). Gao and Liu (2013) found that clones of *S. avenae* from wheat were more specialized to their host plant than those from barley based on comparisons of fitness parameters (e.g. fecundity) in host transfer experiments. Comparisons of *S. avenae* clones from barley and oat showed that barley clones presented significantly higher extents of specialization than oat clones (Gao et al., 2014). These studies suggest that clonal specialization of *S. avenae* into biotypes may be more common than previously recognized. So, a follow-up study by Wang et al. (2020a) screened 33 wheat varieties and 25 barley varieties, identified a panel of *S. avenae*-resistant varieties of wheat and barley, and detected six biotypes based on performance from among 200 genotypes (distinguishable based on microsatellite markers) of the aphid collected in nine provinces in China.

Among the biotypes detected by Wang et al. (2020a), biotype 1 was predominant accounting for over 82% of the total samples across the provinces sampled, while others were more restricted geographically. Furthermore, three biotypes were collected only on barley, while biotype 4 was collected only on wheat, indicating the importance of species-specific plant traits for biotype development. Wang et al. (2020a) collected *S. avenae* only from cultivated cereals and not from non-cultivated wild grasses. Since wild grasses are thought to be a key force in evolution and maintenance of biotypes in other cereal aphids (greenbug, Weiland et al., 2008; Russian wheat aphid, Burd and Porter, 2006), additional samples from wild grasses that are alternative hosts for *S. avenae* in production landscapes in China would likely detect more biotypes.

5.2 Mechanisms underlying biotype development

The mechanisms underlying the divergence of variable biotypes in aphids has long been an intriguing question for researchers. First, significant genetic differentiation among *S. avenae* biotypes was found in the study of Wang et al. (2020b). Furthermore, based on characterization of transcriptomes of two distinct biotypes (i.e. biotypes 1 and 3) of *S. avenae* fed on wheat and barley, some important defense-related DEGs (differentially expressed genes) were found (e.g. three peroxidases, two UDP-glycosyltransferases, and one glutathione *S*-transferase), suggesting that such DEGs could play key roles in the divergence of *S. avenae* biotypes (Wang et al., 2020c,d). Thus, genetic differentiation of *S. avenae* biotypes can involve many genes, instead of rare genes often assumed in the gene-for-gene model (Lapitan et al., 2007; Smith and Clement, 2012). Aphid salivary effectors (e.g. C002 and Armet in *Acyrthosiphon pisum*, and Mp1, Mp2, and Mp55 in *Myzus persicae*) have also been shown to be closely related to aphid virulence and plant resistance, which may contribute to evolution of biotypes in aphids (Rodriguez and Bos, 2013; Rodriguez et al., 2017; Cui et al., 2019). Future studies are still needed to determine key genes involved in both the aphid and the cereal varieties, and their exact functions.

Although some genetic differentiation is detectable among *S. avenae* biotypes, it is not sufficient to explain phenotypic divergence among them in some cases (Wang et al., 2020a,c). This raises the possibility that certain *S. avenae* biotypes also differ in their adaptive transcriptional plasticity (Wang et al., 2020c). Such a mechanism has been detected in other aphid species (Gorur et al., 2007). Indeed, plasticity in life history traits in response to different plant hosts has been documented in *S. avenae* (Dai et al., 2014), and greater plasticity is associated with specialization in the species (Dai et al., 2016).

A third possibility is that different *S. avenae* biotypes can be associated with particular secondary symbionts, because some aphid biotypes have been found to be associated with unique secondary symbionts (Tsuchida et al., 2004; Ferrari et al., 2007; Simon et al., 2003; Wang et al., 2016). It appears that there can be multiple mechanisms at work during the evolution of *S. avenae* biotypes, and a challenging task for their clarification is still ahead.

6 Options for integrated management

Although host plant resistance is important as part of management for cereal aphids (Glen, 2000) including *S. avenae* (Gao and Liu, 2013), its use has been rare due to a lack of effective resistance in agronomically adapted cultivars. Furthermore, the genetic variability of *S. avenae* biotypes (Wang et al., 2020a) could undermine durability of host plant resistance to this species. Nonetheless, even low levels of resistance can support integrated control of *S. avenae* because of its compatibility with cultural and biological control measures (Glen, 2000). When the population density of *S. avenae* is high, insecticide treatments are often justified and applied. Here, we describe and discuss potentially additive and synergistic effects of different measures and opportunities for integrating different kinds of options: cultural, biological, and chemical.

6.1 Host plant resistance

Resistance to *S. avenae* has been found to widely occur in many cultivars or lines of wheat and barley, such as diploid wheat (*T. monococcum* L.), tetraploid wheat (*Triticum durum* Desf.), barley 'Dulihuang', and barley 'Xiyin No.2' (Sotherton and Lee, 1988; Bohidar et al., 1986; Wang et al., 2020a). Host plant secondary metabolites, such as hydroxamic acids (Hx) (e.g. DIMBOA, 2,4-dihydroxy-7-methoxy-l,4-benzoxazin-3-one), phenolic compounds and alkaloids, were all shown to be involved in the resistance of cereals against *S. avenae* (Thackray et al., 1990; Corcuera, 1993; Tuvesson and Johansson, 2000; Yu et al., 2001). Isolation and location of resistant genes in host plants have also been attempted in quite a few studies (Li and Peng, 2014). It appears that many plant genes, instead of single genes, are associated with resistance to *S. avenae* identified so far (Li and Peng, 2014; Wang et al., 2020c,d). Some studies have started to examine the sufficiency of plant resistance improvement by knocking out or down plant susceptibility (*S*) genes with CRISPR/Cas9 or RNAi techniques, providing a new resistance breeding technique (Åhman et al., 2019). Nonetheless, there's still a huge gap in knowledge of resistance (or susceptibility) genes and molecular mechanisms involved, and a challenging task for aphid resistance to be deliberately bred into a cereal crop variety.

6.2 Cultural control

Adjusting the sowing date is a common cultural method for *S. avenae* management and YDV control. Compared to late plantings, early plantings can lead to earlier colonization and larger numbers of cereal aphids like *S. avenae* and a relatively higher risk of YDV on winter wheat (Royer et al., 2005). In contrast, late sowing of autumn cereals was found to allow a lower number of, or even zero, insecticide applications for cereal aphid management (Kennedy and Connery, 2001).

Tillage intensity also affects colonization of cereal crops by cereal aphids like *S. avenae*. Cereal aphid populations are less abundant when surface crop residues (serving as reflective mulch against aphid colonization) are preserved (Burton and Krenzer, 1985). Furthermore, parasitism rate of aphids can be increased by reduced tillage, likely due to limited disturbance by farming practices in such systems (Tamburini et al., 2016) or an increase in weed abundance and diversity (providing more floral nectar resources for parasitoids) (Soane et al., 2012; Araj et al., 2011). However, the benefits of reduced tillage may change depending on local soil types and climates, and the benefits can even be offset by several downsides such as increased weed abundance (Holland, 2004; Soane et al., 2012).

Habitat diversification has also been attempted in management of *S. avenae*. Several studies found that *S. avenae* abundance was reduced, and predators and aphid parasitism were increased in intercrops of wheat with oilseed rape or alfalfa compared to monocultures (Ma et al., 2006; Wang et al., 2009b, 2011). Beyond the field margin, there is potential to enhance biological control of aphids through diversification of non-crop (e.g. hedgerows, tree lines, and semi-natural habitats) and crop habitats within the farming landscape (Andow, 1991; Brewer and Elliott, 2004; Tamburini et al., 2016). Thus, both local management and landscape composition should be considered in order to maximize natural biological control services in agro-ecosystems.

6.3 Biological control

Sitobion avenae has numerous natural enemies that can potentially suppress its colonization and abundance on cereal crops. These natural enemies fall into mainly three categories (Elliott et al., 2006; Van Veen et al., 2008; Barczak et al., 2014): (1) koinobiont parasitoids: Aphidiinae (Ichneumonoidea: Braconidae) and Aphelinidae (Chalcidoidea); (2) predators: lady beetles (Coccinellidae), syrphid fly larvae (Syrphidae), lacewing larvae (Chrysopidae), rove beetles (Staphylinidae), ground beetles (Carabidae), spiders (Araneae), predatory midge larvae (Cecidomyiidae), and predatory bugs (Nabidae and Anthocoridae); and (3) fungi of Entomophthorales (subclass Zygomycetes).

Specialist parasitoids are often believed to be highly effective in aphid control. Among all natural enemies, aphid parasitoids were the most common (followed by coccinellids) in wheat fields of northern China, and the most common parasitoid was *A. gifuensis* Ashmead (Hymenoptera: Braconidae) with the highest mean parasitism up to 98 mummies per 100 tillers (Yang et al., 2009; Ali et al., 2018). In Europe, the most numerous and most frequently occurring parasitoids are *Aphidius ervi*, *Aphidius uzbekistanicus*, or *Praon volucre* depending on location (Chambers et al., 1986; Barczak et al., 2014). These studies show the potential of using *Aphidius* parasitoids in conservation and augmentative biological control for management of *S. avenae*.

Generalist predators can also contribute significantly to natural control of cereal aphids like *S. avenae* (Rice and Wilde, 1991). Predators can account for between 5% and 30% aphid mortality in the field in France (Hewson and Sagenmuller, 2000). In wheat fields of northern China, most common generalist predators include coccinlellids (e.g. *Coccinella septempunctata* L., *Harmonia axyridis* Pallas, and *Propylaea japonica* Thunberg), syrphids, and lacewings (e.g. *Chrysoperla sinica* Tjeder), and they play a particularly large role in reducing *S. avenae* infestations (Ali et al., 2018; Zou et al., 1998). In a wheat-cotton relay intercropping system of northern China, the most abundant generalist predators were found to be coccinellids like *C. septempunctata* and *P. japonica* (Ma et al., 2006). Li et al. (2022) found that the aphid suppression capacity of the lady beetle *H. axyridis* would increase with climate change, while it would weaken for the syrphid fly *Episyrphus balteatus* and the parasitoid *A. gifuensis*, showing that climate change can favor generalist over specialist enemies. Therefore, generalist predators, particularly lady beetles, can be used through conservation biological control in integrated management of *S. avenae*, especially under climate change in the future.

The abundance of generalist predators can be significantly influenced by tillage. Indeed, conservation tillage in cereal fields can increase the abundance of generalist predators (e.g. web spiders, ladybirds, ground beetles, and syrphid fly larvae) and the aphid predation (16% higher than in the fields under conventional tillage) (Tamburini et al., 2016; Holland, 2004; Soane et al., 2012). An increase of semi-natural habitats in the landscape can also increase the abundance of generalist predators and enhance aphid predation in the crop fields (Tamburini et al., 2016), showing the potential of *S. avenae* management through adjustments of landscape composition. Furthermore, cis-jasmone was found to be directly attractive to the predacious seven-spot ladybird, *C. septempunctata* L., and to the aphid parasitoid, *A. ervi* Haliday (Birkett et al., 2000). Thus, this semiochemical may be applied in the field to enhance *S. avenae* control by both generalist and specialist enemies.

Aphids, including cereal aphids like *S. avenae*, are vulnerable to entomopathogenic fungi (e.g. *Pandora neoaphidis* and *Beauveria bassiana*), which under certain conditions can reduce their populations. Several species have been reported to infect *S. avenae* in the field and were virulent to the aphid in the laboratory (Feng et al., 1990a,b). Cereal aphids including *S. avenae* can be killed in large numbers in the field, as was reported for a field survey in the United Kingdom (Chambers et al., 1986). Microclimatic factors including humidity, precipitation, and others influence whether infections by these fungi act to suppress aphid populations (Chambers et al., 1986). Thus, although their epizootics are rare in natural conditions, some entomopathogens (e.g. *B. bassiana*) may have the potential to be included in integrated pest management (IPM) programs when humidity, precipitation, and other related factors are appropriate for their application in the field.

6.4 Chemical control

Insecticides are still important tools for integrated management of *S. avenae* on different cereal crops. During the past decades, four main classes of insecticides have been used, including carbamates, organophosphates, pyrethroids, and neonicotinoids (e.g. imidacloprid).

Seed treatments with neonicotinoids (e.g. clothianidin and imidacloprid) can control aphids on cereal crops as soon as the seedlings emerge from the soil and for approximately 6 weeks after sowing (Dewar, 2007). Recently, imidacloprid has been one of the main neonicotinoid insecticides included in seed treatments to control wheat aphids in China (Miao et al., 2014). In another study, seed dressing treatments with imidacloprid had good management effects for *S. avenae* and posed little threat to coccinellids, shown by low aphid populations and high coccinellid populations particularly during the booting and flowering stages of wheat (Nyaanga et al., 2014). This is probably because imidacloprid is a systemic insecticide, and lady beetles do not directly feed on wheat leaf contents where seed dress toxins are concentrated. However, decreased sensitivity to imidacloprid has been identified for *S. avenae* populations from different provinces of China (Wu et al., 2020; Gong et al., 2021). In another study, *S. avenae* populations were shown to have the potential to develop high levels of resistance to imidacloprid under continuous selection in the greenhouse, suggesting a risk of imidacloprid cross resistance to similar neonicotinoid insecticides (Zhang et al., 2020).

Populations of *S. avenae* also appear to have developed variable levels of resistance to other insecticides like carbamates, organophosphates, and pyrethroids. Pirimicarb was shown to have little adverse effects on polyphagous predators (Powell et al., 1985). Pirimicarb was introduced into China for wheat aphid control in early 1990s, but differential susceptibilities of *S. avenae* populations to pirimicarb have been observed there (Lu and Gao, 2009). In China, pyrethroids and organophosphates have also been widely employed to control wheat aphids in the field over the last 20 years, and beta-cypermethrin and thiamethoxam are particularly popular and widely used in recent years (Deng et al., 2019). In Shanxi province of China, *S. avenae* populations were found to have low to moderate resistance against pirimicarb and chlorpyrifos, but they showed no resistance to acetamiprid or beta-cypermethrin (Wu et al., 2020). A large scale study covering 15 provinces of China showed that most *S. avenae* populations were moderately resistant to pyrethroids, one population highly resistant to beta-cypermethrin, and one highly resistant to bifenthrin; but over half of the populations exhibited only low resistance to organophosphates and macrolides (e.g. avermectin) (Gong et al., 2021). In Europe, pyrethroid insecticides are the main option for controlling aphids on cereals, and their widespread use has led to the selection of resistance in *S. avenae* (Foster et al., 2014). Thus, pyrethroid resistance in *S. avenae* has been linked to control failures and yellow virus disease outbreaks in the United Kingdom (Dewar and Foster, 2017). Pesticide detoxification and target site mutations are the main mechanisms underpinning resistance in aphids to pyrethroids, carbamates, and organophosphates (Chen et al., 2007). In China, these studies all lack assessments in the field, and studies in this respect need to be conducted in different areas in order to make appropriate management decisions and accurately evaluate the significance of the resistance impact on crop yields.

In Europe, since 2020, the authorization for use of chlorpyriphos has been withdrawn, and the use of neonicotinoids has also been banned on cereal crops for protection of pollinators (Mc Namara et al., 2020). This can exacerbate selective pressures exerted by a limited number of insecticides on cereal aphid populations (Mc Namara et al., 2020). While it appears that numbers of highly insecticide-resistant *S. avenae* populations are still low, growers should be encouraged to use insecticide resistance management strategies such as economic threshold-based chemical sprays, the use of more selective insecticides, and rotation of insecticides with different modes of action (Sparks and Nauen, 2015).

Threshold-based chemical control should be used in IPM programs to help growers make management decisions that prevent excessive pesticide applications. The use of economic and damage thresholds in barley and wheat are now widely accepted in making crop protection decisions regarding cereal aphid control in Europe and the United States (Elliott et al., 1990; Winder et al., 1999; Mc Namara et al., 2020). Economic thresholds have been suggested in several countries (e.g. England, Germany, Denmark, Sweden, and China), varying from three to five aphids per tiller or ear around flowering (Larsson, 2005). High populations of natural enemies could occur during booting and flowering stages of cereal crops, and hence foliar insecticide treatments should be avoided during this period (Nyaanga et al., 2014). Spraying early on winter cereals can eliminate beneficial insects (Mc Namara et al., 2020). Thus, it might be better to also incorporate natural enemy abundance into insecticide treatment decisions, and insecticides should be carefully selected based on their specific properties regarding efficiency, potential effects on beneficial insects, potential environmental hazard, and residual activity (Gong et al., 2021; Nyaanga et al., 2014).

7 Future trends in research

In the last few decades, significant advances have been made in understanding driving forces (e.g. geographic factors and host plants) for *S. avenae*'s population expansion, dispersal, and adaptive evolution, providing useful information for its integrated management. Recent research finds that *S. avenae* has developed at least six biotypes in China, and biotype 1 is predominant (accounting for over 82% of the total samples) there (Wang et al., 2020a), meaning that the majority of resistant cereal cultivars identified so far can still be incorporated in integrated management programs for this aphid. Although still rare, it is highly desirable to plant cereal crops with moderate levels of resistance to *S. avenae* in areas where natural biological control can play an important role, in order to enhance integrated management of this aphid. Breeding for better resistance in cereals through modern biotechnology (e.g. transgenic crops and RNAi) can provide an economically and environmentally sound way of controlling this aphid in the future. However, the acceptance of transgenic cereals can be a major issue that has to be addressed in the process. Due to climate change, *S. avenae* will become increasingly highly suitable in China's main grain-growing areas (e.g. Huang-Hui-Hai, Northeast, Along the Great Wall, and Loess Plateau), and its sustainable management will present an even bigger challenge in the future. Compared to specialist enemies, climate change will favor generalist predators (particularly lady beetles), which can be used to maximize conservation biological control in cereal-producing systems through local management and landscape composition adjustments.

It is necessary to have a diversity of viable control options for a sustainable pest control management program. For *S. avenae* control, options are limited, and insecticide sprays are still the mainstay. This can make insecticide resistance an increasingly serious problem worldwide. Future management of *S. avenae* needs a strategy that includes careful integration of aphid monitoring, cultural practices, biological controls, and chemical control methods, and improved cereal cultivars as well. For this to happen, there are still significant knowledge gaps. First, biotype evolution in *S. avenae* is not well understood. Genetic, genomic, and evolutionary factors involved in the development and spread of its different biotypes need more clarification. Increasing efforts should be made for assessment of effects of non-cultivated grass hosts, identification of resistant germplasms, characterization of resistance at the molecular level, and identification of resistance genes. Second, future management of *S. avenae* may require the inclusion of aphid transgenic control by using improved cereal varieties with biotechnology. New plant breeding techniques (e.g. CRISPR/Cas9 or RNAi) targeting host plant susceptibility (*S*) genes may be used to expedite this effort (Åhman et al., 2019). Third, knowledge of the migratory behavior of *S. avenae* is still insufficient, and the factors determining timing and size of aphid migration into cereal crops are poorly understood. This information is important to improve aphid monitoring and decision-support tools in order for accurate identification of risk periods and risk areas, and for decisions of threshold-based chemical applications. Fourth, with climate change, management of *S. avenae* can face increasing challenges. Ecological niche modeling studies show that warming climate will increase *S. avenae*'s damage through enhancing its population growth, aiding its range expansion, and decreasing the top-down effects of parasitoids (Li et al., 2022). Further studies need to consider additional bioclimatic variables, and complex species interactions as well. In addition, the projected ecological changes in related studies need to be further explored and verified with field warming experiments in the future. Such studies will be of significance for developing more sophisticated models for prediction of *S. avenae*'s changing dynamics on cereals in future warming scenarios and development of corresponding managing strategies.

8 Where to look for further information

The following literature provides a good overview of the subject:

- Bell, J. R., Alderson, L., Izera, D., Kruger, T., Parker, S., Pickup, J., Shortall, C. R., Taylor, M. S., Verrier, P. and Harrington, R. 2015. 'Long-term phenological trends, species accumulation rates, aphid traits and climate: five decades of change in migrating aphids', J Anim Ecol 84: 21–34.
- Bosem Baillod, A., Tscharntke, T., Clough, Y. and Batáry, P. 2017. 'Landscape-scale interactions of spatial and temporal cropland heterogeneity drive biological control of cereal aphids', J Appl Ecol 54: 1804–1813.
- Dewar, A. M., Ferguson, A., Pell, J. K., Nicholls, C. and Watts, J. 2016. Research review no. 86, 'A review of pest management in cereals and oilseed rape in the UK', AHDB Cereals & Oilseeds, Kenilworth. https://projectblue.blob.core.windows.net/media/Default/Research%20Papers/Cereals%20and%20Oilseed/rr86.pdf.
- Jactel, H., Verheggen, F., Thiéry, D., Escobar-Gutiérrez, A. J., Gachet, E., Desneux, N. and Neonicotinoids Working Group 2019. 'Alternatives to neonicotinoids', Environ Int 129: 423–429.
- Poehling, H. M., Thieme, T. and Heimbach, U. 2017. 'IPM case studies: grain', In: aphids as crop pests, 2nd edn, ed. by van Emden, H. F. and Harrington, R. CAB International, Wallingford.

Key research in this area can be found at the following organizations:

- Web of Science (www.webofscience.com).
- United States Department of Agriculture (USDA) (pubag.nal.usda.gov).
- CAB International (CABI) (www.cabi.org).
- International Maize and Wheat Improvement Center (www.cimmyt.org).
- Agronet (www.agronet.com.cn) (in Chinese).

9 Acknowledgements

I would like to acknowledge Dexian Li of Northwest A&F University for making the worldwide distribution maps of *S. avenae*.

10 References

Addo-Bediako, A., Chown, S. L. and Gaston, K. J. 2001. Revisiting water loss in insects: A large scale view. *J. Insect Physiol.* 47(12): 1377–1388.

Ahmad, H., Mir, I. A., Sharma, D., Srivastava, K., Sharma, S. A. and Sharma, S. 2016. Seasonal incidence and management of wheat aphid, *Sitobion avenae* (f.). *Indian J. Entomol.* 78(2): 148–152.

Åhman, I., Kim, S. Y. and Zhu, L. H. 2019. Plant genes benefitting aphids' potential for exploitation in resistance breeding. *Front. Plant Sci.* 10: 1452.

Ahmed, S. S., Liu, D.-G. and Simon, J. C. 2017. Impact of water-deficit stress on tritrophic interactions in a wheat-aphid-parasitoid system. *PLoS ONE* 12(10): e0186599.

Ali, A., Desneux, N., Lu, Y. H. and Wu, K. M. 2018. Key aphid natural enemies showing positive effects on wheat yield through biocontrol services in northern China. *Agric. Ecosyst. Environ.* 266: 1–9.

Andow, D. A. 1991. Vegetational diversity and arthropod population response. *Annu. Rev. Entomol.* 36(1): 561–586.

Araj, S. E., Wratten, S., Lister, A., Buckley, H. and Ghabeish, I. 2011. Searching behavior of an aphid parasitoid and its hyperparasitoid with and without floral nectar. *Biol. Control* 57(2): 79–84.

Barczak, T., Debek-Jankowska, A. and Bennewicz, J. 2014. Primary parasitoid and hyperparasitoid guilds (hymenoptera) of grain aphid (*Sitobion avenae* F.) in northern Poland. *Arch. Biol. Sci.* 66(3): 1141–1148.

Birkett, M. A., Campbell, C. A. M. , Chamberlain, K., Guerrieri, E., Hick, A. J., Martin, J. L., Matthes, M., Napier, J. A., Pettersson, J., Pickett, J. A., Poppy, G. M., Pow, E. M., Pye, B. J., Smart, L. E., Wadhams, G. H., Wadhams, L. J. and Woodcock, C. M. 2000. New roles for cis-jasmone as an insect semiochemical and in plant defense. *Proc. Natl Acad. Sci. U. S. A.* 97(16): 9329–9334.

Blackman, R. and Eastop, V. F. 2006. Aphids on the World's herbaceous plants and shrubs. Available at: http://www.aphidsonworldsplants.info/index.htm.

Bohidar, K., Wratten, S. D. and Niemeyer, H. M. 1986. Effects of hydroxamic acids on the resistance of wheat to the aphid *Sitobion avenae*. *Ann. Appl. Biol.* 109(1): 193–198.

Brabec, M., Honek, A., Pekar, S. and Martinkova, Z. 2014. Population dynamics of aphids on cereals: Digging in the time-series data to reveal population regulation caused by temperature. *PLoS ONE* 9(9): e106228.

Brewer, M. J. and Elliott, N. C. 2004. Biological control of cereal aphids in North America and mediating effects of host plant and habitat manipulations. *Annu. Rev. Entomol.* 49: 219–242.

Burd, J. D. and Porter, D. R. 2006. Biotypic diversity in greenbug (Hemiptera: Aphididae): Characterizing new virulence and host associations. *J. Econ. Entomol.* 99(3): 959–965.

Burton, R. L. and Krenzer Jr., E. G. 1985. Reduction of greenbug (Homoptera: Aphididae) populations by surface residues in wheat tillage studies. *J. Econ. Entomol.* 78(2): 390–394.

Caillaud, C. M., Dedryver, C. A., Di Pietro, J. P., Simon, J. C., Fima, F. and Chaubet, B. 1995. Clonal variability in the response of *Sitobion avenae* (Homoptera: Aphididae) to resistant and susceptible wheat. *Bull. Entomol. Res.* 85(2): 189–195.

Cao, Y.-Z., Li, K.-B., Yin, J. and Zhang, K.-C. 2006. Occurring dynamics and sustainable management strategies and practices of wheat major insect pests. *China Plant Prot.* 26: 11–14.

Chambers, R. J., Sunderland, K. D., Stacey, D. L. and Wyatt, I. J. 1986. Control of cereal aphids in winter wheat by natural enemies: Aphid-specific predators, parasitoids and pathogenic fungi. *Ann. Appl. Biol.* 108(2): 219–231.

Chen, M., Han, Z., Qiao, X. and Qu, M. 2007. Resistance mechanisms and associated mutations in acetylcholinesterase genes in *Sitobion avenae* (Fabricius). *Pestic. Biochem. Physiol.* 87(3): 189–195.

Choe, H. J., Lee, S. H. and Lee, S. 2006. Morphological and genetic indiscrimination of the grain aphids, *Sitobion avenae* complex (Hemiptera: Aphididae). *Appl. Entomol. Zool.* 41(1): 63–71.

Ciepiela, A. P., Sempruch, C. and Chrzanowski, G. 1999. Evaluation of natural resistance of winter triticale cultivars to grain aphid using food coefficients. *J. Appl. Entomol.* 123(8): 491–494.

Corcuera, L. J. 1993. Biochemical basis for the resistance of barley to aphids. *Phytochemistry* 33(4): 741–747.

Cui, N., Lu, H., Wang, T., Zhang, W., Kang, L. and Cui, F. 2019. Armet, an aphid effector protein, induces pathogen resistance in plants by promoting the accumulation of salicylic acid. *Phil. Trans. R. Soc. B Biol. Sci.* 374(1767): 20180314.

Dai, P., Liu, D.-G. and Shi, X.-Q. 2015. Impacts of water deficiency on life history of *Sitobion avenae* clones from semi-arid and moist areas. *J. Econ. Entomol.* 108(5): 2250–2258.

Dai, P., Shi, X., Liu, D., Ge, Z., Wang, D., Dai, X., Yi, Z. and Meng, X. 2016. Life-history trait plasticity and its relationships with plant adaptation and insect fitness: A case study on the aphid *Sitobion avenae*. *Sci. Rep.* 6: 29974.

Dai, X.-J., Gao, S.-X. and Liu, D.-G. 2014. Genetic basis and selection for life-history trait plasticity on alternative host plants for the cereal aphid *Sitobion avenae*. *PLoS ONE* 9(9): e106179.

De Barro, P. J., Sherratt, T. N., David, O. and Maclean, N. 1995. An investigation of the differential performance of clones of the aphid *Sitobion avenae* on two host species. *Oecologia* 104(3): 379–385.

Dean, G. J. 1974. The four dimensions of cereal aphids. *Ann. Appl. Biol.* 77(1): 74–78.

Dedryver, C.-A., Hullé, M., Le Gallic, J. F., Caillaud, M. C. and Simon, J.-C. 2001. Coexistence in space and time of sexual and asexual populations of the cereal aphid *Sitobion avenae*. *Oecologia* 128(3): 379–388.

Dedryver, C.-A., Le Gallic, J. F., Gauthier, J. P. and Simon, J.-C. 1998. Life-cycle of the cereal aphid *Sitobion avenae* F. polymorphism and comparison of life history traits associated with sexuality. *Ecol. Entomol.* 23(2): 123–132.

Delmotte, F., Leterme, N., Gauthier, J. P., Rispe, C. and Simon, J. C. 2002. Genetic architecture of sexual and asexual populations of the aphid *Rhopalosiphum padi* based on allozyme and microsatellite markers. *Mol. Ecol.* 11(4): 711–723.

Deng, D., Duan, W., Wang, H., Zhang, K., Guo, J., Yuan, L., Wang, L. and Wu, S. 2019. Assessment of the effects of lethal and sublethal exposure to dinotefuran on the wheat aphid *Rhopalosiphum padi* (Linnaeus). *Ecotoxicology* 28(7): 825–833.

Deng, Z.-Y., Wang, Q., Zhang, Q., Qing, J.-Z., Yang, Q.-G., Yuan, Z.-P., Liu, W.-J. and Xu, J.-F. 2010. Impact of climate warming and drying on food crops in northern China and the countermeasures. *Acta Ecol. Sin.* 30: 6278–6288. (in Chinese).

Dewar, A. M. 2007. Chemical control. In: van Emden, H. and Harrington, R. (Eds.), *Aphids as Crop Pests*. Wallingford: CABI Publishing, pp. 391–422.

Dewar, A. M. and Foster, S. P. 2017. Overuse of pyrethroids may be implicated in recent BYDV epidemics in cereals. *Outlooks Pest Manag.* 28(1): 7–12.

Elliott, N. C., Kieckhefer, R. W. and Walgenbachi, D. D. 1990. Binomial sequential sampling methods for cereal aphids in small grains. *J. Econ. Entomol.* 83(4): 1381–1387.

Elliott, N. C., Tao, F. L., Fuentes-Granados, R., Giles, K. L., Elliott, D. T., Greenstone, M. H., Shufran, K. A. and Royer, T. A. 2006. D-vac sampling for predatory arthropods in winter wheat. *Biol. Control* 38(3): 325–330.

Feng, M. G., Johnson, J. B. and Kish, L. P. 1990a. Survey of entomopathogenic fungi naturally infecting cereal aphids (Homoptera: Aphididae) of irrigated grain crops in southwestern Idaho. *Environ. Entomol.* 19(5): 1534–1542.

Feng, M. G., Johnson, J. B. and Kish, L. P. 1990b. Virulence of *Verticil/ium lecanii* and an aphid-derived isolateof *Beauveria bassiana* (Fungi: Hyphomycetes) for six species of cereal aphids (Homoptera: Aphididae). *Environ. Entomol.* 19(3): 815–820.

Ferrari, J., Scarborough, C. L. and Godfray, H. C. J. 2007. Genetic variation in the effect of a facultative symbiont on host-plant use by pea aphids. *Oecologia* 153(2): 323–329.

Fievet, V., Dedryver, C.-A., Plantegenest, M., Simon, J.-C. and Outreman, Y. 2007. Aphid colony turn-over influences the spatial distribution of the grain aphid *Sitobion avenae* over the wheat growing season. *Agric. For. Entomol.* 9(2): 125–134.

Foster, S. P., Paul, V. L., Slater, R., Warren, A., Denholm, I., Field, L. M. and Williamson, M. S. 2014. A mutation (l1014f) in the voltage-gated sodium channel of the grain aphid, *Sitobion avenae*, is associated with resistance to pyrethroid insecticides. *Pest Manag. Sci.* 70(8): 1249–1253.

Gao, S.-X. 2014. *Population differentiation and its genetic basis for Sitobion avenae on different host plants*. Dissertation, Northwest A&F University, Yangling, Shaanxi Province, China.

Gao, S.-X. and Liu, D.-G. 2013. Differential performance of *Sitobion avenae* clones from wheat and barley with implications for its management through alternative cultural practices. *J. Econ. Entomol.* 106(3): 1294–1301.

Gao, S.-X., Liu, D.-G., Chen, H. and Meng, X. X. 2014. Fitness traits and underlying genetic variation related to host plant specialization in the aphid *Sitobion avenae*. *Insect Sci.* 21(3): 352-362.

Glen, D. M. 2000. The effects of cultural measures on cereal pests and their role in integrated pest management. *Integr. Pest Manag. Rev.* 5(1): 25–40.

Gong, P., Li, X., Gao, H., Wang, C., Li, M., Zhang, Y., Li, X., Liu, E. and Zhu, X. 2021. Field evolved resistance to pyrethroids, neonicotinoids, organophosphates and macrolides in *Rhopalosiphum padi* (Linnaeus) and *Sitobion avenae* (Fabricius) from China. *Chemosphere* 269: 128747.

Gorur, G., Lomonaco, C. and Mackenzie, A. 2007. Phenotypic plasticity in host choice behavior in black bean aphid, *Aphis fabae* (Homoptera: Aphididae). *Arthropod Plant Interact.* 1(3): 187–194.

Hales, D., Foottit, R. G. and Maw, E. 2010. Morphometric studies of the genus *Sitobion* Mordvilko 1914 in Australia (Hemiptera: Aphididae). *Aust. J. Entomol.* 49(4): 341–353.

Han, Z., Tan, X., Wang, Y., Xu, Q., Zhang, Y., Harwood, J. D. and Chen, J.-L. 2019. Effects of simulated climate warming on the population dynamics of *Sitobion avenae* (Fabricius) and its parasitoids in wheat fields. *Pest Manag. Sci.* 75(12): 3252–3259.

Hansen, L. M. 2006. Models for spring migration of two aphid species *Sitobion avenae* (f.) and *Rhopalosiphum padi* (L.) infesting cereals in areas where they are entirely holocyclic. *Agric. Forest Entomol.* 8(2): 83–88.

He, Y., Liu, D.-G., Dai, P., Wang, D. and Shi, X.-Q. 2018. Genetic differentiation and structure of *Sitobion avenae* (Hemiptera: Aphididae) populations from moist, semiarid and arid areas in northwestern China. *J. Econ. Entomol.* 111(2): 603–611.

Helden, A. J. and Dixon, A. F. G. 2002. Life-cycle variation in the aphid *Sitobion avenae*: Costs and benefits of male production. *Ecol. Entomol.* 27(6): 692–701.

Hewson, R. T. and Sagenmuller, A. 2000. Some contributions to integrated crop management in Europe. *Pest Manag. Sci.* 56(11): 954–956.

Holland, J. M. 2004. The environmental consequences of adopting conservation tillage in Europe: Reviewing the evidence. *Agric. Ecosyst. Environ.* 103(1): 1–25.

Horber, E. 1980. Types and classification of insect resistance. In: Maxwell, F. G. and Jennings, P. R. (Eds.), *Breeding Plants Resistant to Insects*. New York: Wiley, p. 15D21.

Hu, X.-S., Liu, X.-F. and Hu, Z.-Q. 2011. Resistance of 10 wheat varieties to *Sitobion avenae* at wheat seedling stage. *Plant Prot.* 37: 81–85. (in Chinese).

Huang, X.-L., Liu, D.-G., Gao, S.-X. and Chen, H. 2013. Differential performance of *Sitobion avenae* populations from both sides of the Qinling Mountains under common garden conditions. *Environ. Entomol.* 42(6): 1174–1183.

Huang, X.-L., Liu, D.-G., Wang, D., Shi, X.-Q. and Simon, J. C. 2015. Molecular and quantitative genetic differentiation in *Sitobion avenae* populations from both sides of the Qinling Mountains. *PLoS ONE* 10(3): e0122343.

IPCC. 2021. 2021: Synthesis report; contribution of working groups I, II and III to the sixth assessment report of the Intergovernmental Panel on Climate Change (IPCC), Geneva, Switzerland. *Clim. Change*.

Jensen, A. B., Hansen, L. M. and Eilenberg, J. 2008. Grain aphid population structure: No effect of fungal infections in a 2-year field study in Denmark. *Agric. For. Entomol.* 10(3): 279–290.

Kennedy, T. and Connery, J. 2001. Barley yellow dwarf virus in winter barley in Ireland: Yield loss and timing of autumn aphicides in controlling the MAV-strain. *Ir. J. Agric. Food Res.* 40: 55–70.

Kieckhefer, R. W., Elliot, N. C. and Walagenbach, D. D. 1989. Effects of constant and fluctuating temperatures on development rates and demographic statistics of the English grain aphid (Homoptera: Aphididae). *Ann. Entomol. Soc. Am.* 82(6): 701–706.

Lapitan, N. L., Li, Y. C., Peng, J. and Botha, A. M. 2007. Fractionated extracts of Russian wheat aphid eliciting defense responses in wheat. *J. Econ. Entomol.* 100(3): 990–999.

Larsson, H. 2005. A crop loss model and economic thresholds for the grain aphid, *Sitobion avenae* (f.), in winter wheat in southern Sweden. *Crop Prot.* 24(5): 397–405.

Li, D.-X., Li, Z.-X., Liu, Z.-W., Yang, Y.-J., Khoso, A. G., Wang, L. and Liu, D.-G. 2022. Climate change simulations revealed potentially drastic shifts in insect community structure and crop yields in China's farmland. *J. Pest Sci.* https://doi.org/10.1007/s10340-022-01479-3.

Li, F. Q. and Peng, J. H. 2014. Identification of associations between SSR markers and *Sitobion avenae* tolerance trait in bread wheat. *Cereal Res. Communic.* 42(1): 111–118.

Liu, D.-G., Dai, P., Li, S.-R., Ahmed, S. S., Shang, Z.-M. and Shi, X.-Q. 2018. Life history responses of insects to water-deficit stress: A case study with the aphid *Sitobion avenae*. *BMC Ecol.* 18(1): 17.

Liu, X., Wang, C., Wang, Y., Zhang, H. and Ji, W. 2014. Screening and evaluation of different wheat varieties for resistance to English grain aphid *Sitobion avenae* at seedling and adult-plant stages. *Acta Phytophylacica Sin.* 41: 216–224. (in Chinese).

Liu, X., Wang, C., Wang, Y., Zhang, H. and Ji, W. 2015. Screening and evaluation of different wheat cultivars for resistance to *Sitobion avenae* at seedling and adult-plant stages. *Agri. Sci. Technol.* 16: 1686. (in Chinese).

Lowe, H. J. B. 1980. Resistance to aphids in immature wheat and barley. *Ann. Appl. Biol.* 95(1): 129-135.

Lowe, H. J. B. 1982. Some observations on susceptibility and resistance of winter wheat to the aphid *Sitobion avenae* in Britain. *Crop Prot.* 1(4): 431-440.

Lowe, H. J. B. 1984. Development and practice of a glasshouse screening technique for resistance of wheat to the aphid *Sitobion avenae*. *Ann. Appl. Biol.* 104(2): 297-305.

Lu, Y. and Gao, X. 2009. Multiple mechanisms responsible for differential susceptibilities of *Sitobion avenae* (Fabricius) and *Rhopalosiphum padi* (Linnaeus) to pirimicarb. *Bull. Entomol. Res.* 99(6): 611-617.

Ma, G. and Ma, C. S. 2012. Effect of acclimation on heat-escape temperatures of two aphid species: Implications for estimating behavioral response of insects to climate warming. *J. Insect Physiol.* 58(3): 303-309.

Ma, G., Rudolf, V. H. and Ma, C. S. 2015. Extreme temperature events alter demographic rates, relative fitness, and community structure. *Glob. Chang. Biol.* 21(5): 1794-1808.

Ma, X. M., Liu, X. X., Zhang, Q. W., Zhao, J. Z., Cai, Q. N., Ma, Y. A. and Chen, D. M. 2006. Assessment of cotton aphids, Aphis gossypii, and their natural enemies on aphid resistant and aphid susceptible wheat varieties in a wheat cotton relay intercropping system. *Entomol. Exp. Appl.* 121(3): 235-241.

Mc Namara, L., Gauthier, K., Walsh, L., Thébaud, G., Gaffney, M. and Jacquot, E. 2020. Management of yellow dwarf disease in Europe in a post-neonicotinoid agriculture. *Pest Manag. Sci.* 76(7): 2276-2285.

Miao, J., Du, Z. B., Wu, Y. Q., Gong, Z. J., Jiang, Y. L., Duan, Y., Li, T. and Lei, C. L. 2014. Sub-lethal effects of four neonicotinoid seed treatments on the demography and feeding behavior of the wheat aphid *Sitobion avenae*. *Pest Manag. Sci.* 70(1): 55-59.

Montllor, C. B., Maxmen, A. and Purcell, A. H. 2002. Facultative bacterial endosymbionts benefit pea aphids *Acyrthosiphon pisum* under heat stress. *Ecol. Entomol.* 27(2): 189-195.

Nyaanga, J. G., Kamau, A. W., Pathak, R. S. and Tuey, R. K. 2014. Effect of insecticide application regimes on cereal aphids and their coccinellid predators on wheat. *Int. J. Sci. Natl.* 5: 658-662.

Painter, R. H. 1951. *Insect Resistance in Crop Plants*. New York: The Macmillan Company.

Papura, D., Simon, J.-C., Halkett, F., Delmotte, F., Le Gallic, J.-F. and Dedryver, C.-A. 2003. Predominance of sexual reproduction in Romanian populations of the aphid *Sitobion avenae* inferred from phenotypic and genetic structure. *Heredity (Edinb)* 90(5): 397-404.

Powell, W., Dean, D. J. and Bardner, R. 1985. Effects of pirimicarb, dimethoate and benomyl on natural enemies of cereal aphids in winter wheat. *Ann. Appl. Biol.* 106: 235-242.

Rice, M. E. and Wilde, G. E. 1991. Aphid predators associated with conventional-and conservation-tillage winter wheat. *J. Kans. Entomol. Soc.* 64: 245-250.

Rodriguez, P. A. and Bos, J. I. B. 2013. Toward understanding the role of aphid effectors in plant infestation. *Mol. Plant. Microbe Interact.* 26(1): 25-30.

Rodriguez, P. A., Escudero-Martinez, C. and Bos, J. I. B. 2017. An aphid effector targets trafficking protein VPS52 in a host-specific manner to promote virulence. *Plant Physiol.* 173(3): 1892-1903.

Royer, T. A., Giles, K. L., Nyamanzi, T., Hunger, R. M., Krenzer, E. G., Elliott, N. C., Kindler, S. D. and Payton, M. 2005. Economic evaluation of the effects of planting date and application rate of Imidacloprid for management of cereal aphids and barley yellow dwarf in winter wheat. *J. Econ. Entomol.* 98(1): 95–102.

Salman, A. M. A. 2006. Influence of temperature on the developmental rate and reproduction of the English grain aphid, *Sitobion avenae* (Fabricius) (Homoptera: Aphidadae). *Arab Univ. J. Agric. Sci.* 14: 789–801.

Simon, J. C., Baumann, S., Sunnucks, P., Hebert, P. D. N., Pierre, J. S., Le Gallic, J. F. and Dedryver, C. A. 1999. Reproductive mode and population genetic structure of the cereal aphid *Sitobion avenae* studied using phenotypic and microsatellite markers. *Mol. Ecol.* 8(4): 531–545.

Simon, J.-C. , Carré, S., Boutin, M., Prunier-Leterme, N., Sabater-Muoz, B., Latorre, A. and Bournoville, R. 2003. Host-based divergence in populations of the pea aphid: Insights from nuclear markers and the prevalence of facultative symbionts. *Proc. R. Soc. Lond. B* 270(1525): 1703–1712.

Smith, C. M. and Clement, S. L. 2012. Molecular bases of plant resistance to arthropods. *Annu. Rev. Entomol.* 57: 309–328.

Soane, B. D., Ball, B. C., Arvidsson, J., Basch, G., Moreno, F. and Roger-Estrade, J. 2012. No-till in northern, western and south-Western Europe: A review of problems and opportunities for crop production and the environment. *Soil Till. Res.* 118: 66–87.

Sotherton, N. W. and Lee, G. 1988. Field assessments of resistance to the aphids, *Sitobion avenae* and *Metopolophium dirhodum* in old and modern spring-sown wheats. *Ann. Appl. Biol.* 112(2): 239–248.

Sparks, T. C. and Nauen, R. 2015. IRAC: Mode of action classification and insecticide resistance management. *Pestic. Biochem. Physiol.* 121: 122–128.

Sunnucks, P., De Barro, P. J., Lushai, G., Maclean, N. and Hales, D. 1997. Genetic structure of an aphid studied using microsatellites: Cyclic parthenogenesis, differentiated lineages and host specialization. *Mol. Ecol.* 6(11): 1059–1073.

Tamburini, G., Simone, S. D., Sigura, M., Boscutti, F. and Marini, L. 2016. Conservation tillage mitigates the negative effect of landscape simplification on biological control. *J. Appl. Ecol.* 53(1): 233–241.

Thackray, D. J., Wrattent, S. D., Edwards, P. J. and Niemeyer, H. M. 1990. Resistance to the aphids *Sitobion avenae* and *Rhopalosiphum padi* in Gramineae in relation to hydroxamic acid levels. *Ann. Appl. Biol.* 116(3): 573–582.

Tsuchida, T., Koga, R. and Fukatsu, T. 2004. Host plant specialization governed by facultative symbiont. *Science* 303(5666): 1989.

Tuvesson, S. and Johansson, M. 2000. Does indole alkaloid gramine confer resistance in barley to aphid Rhopalosiphum padi? *J. Chem. Ecol.* 26(1): 233–255.

Van Veen, F. J. F., Müller, C. B., Godfray, J. K. and Godfray, H. C. 2008. Food web structure of three guilds of natural enemies: Predators, parasitoids, and pathogens of aphids. *J. Anim. Ecol.* 77(1): 191–200.

Vialatte, A., Plantegenest, M., Simon, J.-C. and Dedryver, C.-A. 2007. Farm-scale assessment of movement patterns and colonization dynamics of the grain aphid in arable crops and hedgerows. *Agric. For. Entomol.* 9(4): 337–346.

Vidal, M. C., Quinn, T. W., Stireman, J. O., Tinghitella, R. M. and Murphy, S. M. 2019. Geography is more important than host plant use for the population genetic structure of a generalist insect herbivore. *Mol. Ecol.* 28(18): 4317–4334.

Wang, D., Liu, D.-G., Zhai, Y.-T., Zhang, R.-F. and Shi, X.-Q. 2019. Clonal diversity and genetic differentiation of *Sitobion avenae* (Hemiptera: Aphididae) from wheat and barley in China. *J. Econ. Entomol.* 112(3): 1217–1226.
Wang, D., Shi, X.-Q., Dai, P., Liu, D.-G., Dai, X.-J., Shang, Z.-M., Ge, Z.-H. and Meng, X.-X. 2016. Comparison of fitness traits and their plasticity on multiple plants for *Sitobion avenae* infected and cured of a secondary endosymbiont. *Sci. Rep.* 6: 23177.
Wang, D., Zhai, Y.-T., Liu, D.-G., Zhang, N., Li, C.-B. and Shi, X.-Q. 2020a. Identification and genetic differentiation of *Sitobion avenae* (Hemiptera: Aphididae) biotypes in China. *J. Econ. Entomol.* 113(1): 407–417.
Wang, D., Shi, X.-Q., Liu, D.-G., Yang, Y.-J. and Shang, Z.-M. 2020b. Genetic divergence of two *Sitobion avenae* biotypes on barley and wheat in China. *Insects* 11(2): 117.
Wang, D., Liu, D.-G., Shi, X.-Q., Yang, Y.-J., Zhang, N. and Shang, Z.-M. 2020c. Transcriptome profiling revealed potentially important roles of defensive gene expression in the divergence of insect biotypes: A case study with the cereal aphid *Sitobion avenae*. *BMC Genomics* 21(1): 546.
Wang, D., Shi, X.-Q., Liu, D., Yang, Y., Shang, Z., Yang, Y.-J. and Shang, Z.-M. 2020d. Transcriptome profiling revealed potentially critical roles for digestion and defense-related genes in insects' use of resistant host plants: A case study with *Sitobion avenae*. *Insects* 11(2): 90.
Wang, G., Cui, L.-L., Dong, J., Francis, F., Liu, Y. and Tooker, J. 2011. Combining intercropping with semiochemical releases: Optimization of alternative control of *Sitobion avenae* in wheat crops in China. *Entomol. Exp. Appl.* 140(3): 189–195.
Wang, Z., Shen, Z. R., Song, Y., Liu, H. Y. and Li, Z. X. 2009a. Distribution and diversity of *Wolbachia* in different populations of the wheat aphid *Sitobion miscanthi* (Hemiptera: Aphididae) in China. *Eur. J. Entomol.* 106(1): 49–55.
Wang, W., Liu, Y., Chen, J., Ji, X., Zhou, H. and Wang, G. 2009b. Impact of intercropping aphid-resistant wheat cultivars with oilseed rape on wheat aphid (*Sitobion avenae*) and its natural enemies. *Acta Ecol. Sin.* 29(3): 186–191.
Weiland, A. A., Peairs, F. B., Randolph, T. L., Rudolph, J. B., Haley, S. D. and Puterka, G. J. 2008. Biotypic diversity in Colorado Russian wheat aphid (Hemiptera: Aphididae) populations. *J. Econ. Entomol.* 101(2): 569–574.
White, T. 1969. An index to measure weather-induced stress of trees associated with outbreaks of psyllids in Australia. *Ecology* 50: 905–909.
White, T. C. R. 2009. Plant vigour versus plant stress: A false dichotomy. *Oikos* 118(6): 807–808.
Wiktelius, S. and Ekbom, B. S. 1985. Aphids in spring sown cereals in central Sweden: Abundance and distribution 1980–1983. *Z. Angew. Entomol.* 100(1–5): 8–16.
Winder, L., Perry, J. N. and Holland, J. M. 1999. The spatial and temporal distribution of the grain aphid *Sitobion avenae* in winter wheat. *Entomol. Exp. Appl.* 93(3): 275–288.
Wu, Y., Cao, Y., Fan, S., Zhao, F. and Xing, K. 2020. Path analysis of insecticide resistance and related detoxification enzymes of *Sitobion avenae* in different geographical population. *Acta Agric. Shanghai* 36: 74–78. (in Chinese).
Xu, L.-Y., Mi, Y., Lu, H., Sun, L., Chen, Z.-Z., Yu, J.-F. and Xu, J. 2014. The age-stage life tables of *Sitobion avenae* (Fabricius) at different temperatures. *J. Plant Prot.* 41: 673–679. (in Chinese).
Xu, Z.-H., Chen, J.-L., Cheng, D.-F., Liu, Y. and Francis, F. 2011a. Genetic variation among the geographic population of the grain aphid, *Sitobion avenae* (Hemiptera: Aphididae)

in China inferred from mitochondrial COI gene sequence. *Agric. Sci. China* 10(7): 1041–1048. (in Chinese).

Xu, Z.-H., Chen, J.-L., Cheng, D.-F., Sun, J. R., Liu, Y. and Francis, F. 2011b. Discovery of English grain aphid (Hemiptera: Aphididae) biotypes in China. *J. Econ. Entomol.* 104(3): 1080–1086.

Yang, S., Yang, S. Y., Zhang, C. P., Wei, J.-N. and Kuang, R. P. 2009. Population dynamics of *Myzus persicae* on tobacco in Yunnan Province, China, before and after augmentative releases of *Aphidius gifuensis*. *Biocontrol Sci. Technol.* 19(2): 219–228.

Yang, Y., Lu, Z., Han, J. and Chen, X. 1994. Overwintering eggs of *Sitobion avenae* in Yangzhou area. *Entomol. Knowl.* 31: 207–209. (in Chinese).

Yang, Y.-J., Liu, D.-G., Liu, X.-M., Wang, B.-Y. and Shi, X.-Q. 2020. Divergence of desiccation-related traits in *Sitobion avenae* from northwestern China. *Insects* 11(9): 626.

Yu, J., Vasanthan, T. and Temelli, F. 2001. Analysis of phenolic acids in barley by high-performance liquid chromatography. *J. Agric. Food Chem.* 49(9): 4352–4358.

Zhang, B., Edwards, O. R., Kang, L. and Fuller, S. J. 2012. Russian wheat aphid (*Diuraphis noxia*) in China: Native range expansion or recent introduction? *Mol. Ecol.* 21(9): 2130–2144.

Zhang, B.-Z., Su, X., Xie, L.-F., Zhen, C.-A., Hu, G.-L., Jiang, K., Huang, Z. Y., Liu, R., Gao, Y., Chen, X. and Gao, X. 2020. Multiple detoxification genes confer Imidacloprid resistance to *Sitobion avenae* Fabricius. *Crop Prot.* 128: 105014. (in Chinese).

Zou, Y.-D., Bi, S.-D., Meng, Q.-L., Geng, J.-G., Chen, G.-C., Wang, G.-M. and Li, J.-L. 1998. Influence of natural enemies on populations of wheat aphid and greenbug. *Chin. J. Appl. Ecol.* 9: 613–616. (in Chinese).

Chapter 11

Wheat curl mite ecology and epidemiology of its associated wheat viruses

Gary L. Hein, Anthony J. McMechan and Lindsay Overmyer, University of Nebraska-Lincoln, USA

1 Introduction

The wheat curl mite (WCM), *Aceria tosichella* Keifer, is a serious pest of wheat (*Triticum aestivum* L.) and other cereals in many parts of the world. Although WCM can cause damage to wheat plants through their feeding alone, its primary impact is through the viruses it transmits. Worldwide WCM transmits four virus species to wheat: *Wheat streak mosaic virus* (WSMV) (Slykhuis, 1955), *High Plains wheat mosaic virus* (HPWMoV) (Seifers et al., 1997), *Triticum mosaic virus* (TriMV) (Seifers et al., 2009) and *Brome streak mosaic virus* (BrSMV) (Goetz and Maiss, 1995). Interactions between these viruses are not well understood, but infections by more than one of these viruses can increase the severity of the resulting disease (Byamukama et al., 2014; Tatineni et al., 2010).

Across the Great Plains of North America, WSMV, HPWMoV and TriMV are present in mixed populations, creating a virus complex that results in mixed

http://dx.doi.org/10.19103/AS.2022.0114.11

symptomology for infected plants (Burrows et al., 2009; Byamukama et al., 2013). In that region, the wheat-mite-virus complex regularly results in serious yield losses (Velandia et al., 2010; Wegulo et al., 2008; Hein et al., 2023). The best available data on the annual presence and impact of this disease complex is from Kansas where the estimated average yield loss over the last 10 years is 1.34% with a range of 0.05–5.6% (Hollandbeck et al., 2020). This annual average loss in Kansas alone would equal approximately $20 million at a wheat price of $5/bushel. Losses across the Great Plains are variable, but loss estimates for Kansas would be representative of potential loss in surrounding states. Perhaps the most impactful aspect of this disease complex is that losses are often spatially concentrated so that complete yield loss in particular fields (Fig. 1) within these serious epidemic areas is common (Hein et al., 2023). In addition to these losses in the Great Plains, losses occur on a more sporadic basis in other winter wheat-producing regions of North America (Townsend et al., 1996; Rashad et al., 2019).

The impact of these mite-transmitted viruses around the world is sporadic. WSMV has been reported in several European and Asian countries, but no reports of severe yield impacts have been recorded (Navia et al., 2013). WSMV was detected in Australia in 2002 (Ellis et al., 2003), and since then, it has had serious but sporadic effects on wheat yield in some areas of the country (Coutts et al., 2014; Murray and Brennan, 2009). At about this same time, WSMV was reported in Argentina where it also has occasionally caused serious injury (Truol et al., 2004; Alemandri et al., 2017).

Figure 1 Images of 2017 winter wheat field in western Nebraska showing complete yield loss due to wheat-mite-virus complex.

HPWMoV has been reported across North America (Burrows et al., 2009; Byamukama et al., 2013). Its first detection outside North America was from Argentina in 2006, and since then mixed infections with WSMV have been observed (Alemandri et al., 2017). HPWMoV has recently been found in Australia (Coutts et al., 2014) and Ukraine (Snihur et al., 2020). TriMV has been reported across the Great Plains in the USA, but in surveys it does not reach the incidence level of WSMV (Burrows et al., 2009; Byamukama et al., 2013). BrSMV has only been reported from some European countries with no associated yield losses reported (Navia et al., 2013).

Recent reviews have addressed the WCM-virus relationship from different perspectives (Hadi et al., 2011; Navia et al., 2013; Stenger et al., 2016; Singh et al., 2018; Skoracka et al., 2018b; Tatineni and Hein, 2018; Tatineni and Hein, 2021a; Tatineni and Hein, 2021b; Nachappa et al., 2021). The focus of this chapter will be to provide a review and update from recent literature on mite ecology and virus epidemiology that is most relevant to the management of the WCM and its associated viruses.

2 Mite identification

The WCM is a white, cigar-shaped mite with only two pairs of legs near the head end (Fig. 2a and b). They are very small, measuring up to ~0.25 mm in length as adults (del Rosario and Sill, 1965). The mouthparts are composed of a two-lobed rostrum with a pair of slightly curved, needle-like stylets (Fig. 2c; Orlob, 1966). Only about a third of a stylet (i.e. 5–7μ) is able to penetrate the plant when feeding (Orlob, 1966); therefore, their feeding is restricted to epidermal cells. On their grass hosts, mites will most often be found feeding deep in the leaf whorl of the youngest leaves and readily move to the whorl of new leaves as these appear (Gillespie et al., 1997).

The WCM is a member of the family Eriophyidae, one of three families in the superfamily Eriophyoidea (Oldfield and Proeseler, 1996). The taxonomic history of the WCM has suffered from considerable confusion. Mites found on wheat in North America were first identified in 1938 as *Aceria tulipae* Keifer because of morphological similarities to the dry bulb mite, a mite that infests tulips (Keifer, 1939). This misidentification continued into the 1950s when the connection to virus transmission in wheat was first made (Slykhuis, 1955). In 1969, Keifer described WCM from wheat in Yugoslavia as *A. tosichella* and separated *A. tulipae* and *A. tosichella* into two distinct species (Keifer, 1969; Amrine and Stasny, 1994). However, in 1971 Newkirk and Keifer (1971) reassigned mites from the genus *Aceria* to the genus *Eriophyes*. This change was not universally adopted, but more importantly, the proper naming of WCM as *A. tosichella* was lost in this naming confusion. In 1989, the WCM was restored to the genus

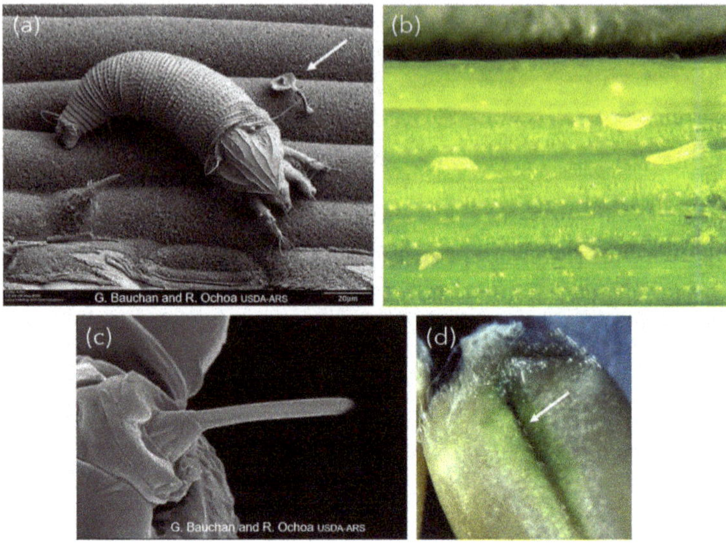

Figure 2 Wheat curl mite feeding on wheat leaf with arrow identifying spermatophore attached to the leaf (a); wheat curl mite adults, immature and eggs on wheat leaf (b); wheat curl mite needle-shaped stylet and mouthparts (c) and wheat curl mite infesting developing wheat kernel (d).

Aceria (Amrine and Stasny, 1994), and later, Amrine and Stasny (1994) clearly re-established the proper naming of the WCM as *A. tosichella* Keifer. As a result of this extended confusion, the naming of the WCM in the literature has been inconsistent through the years: *A. tulipae* (1950s–1990s), *Eriophyes tulipae* (1970s–1990s) and *A. tosichella* (mid-1990s-present). References to WCM on wheat using any of these names prior to 2000 are likely *A. tosichella*.

Although the proper scientific name of WCM is now widely used, there is evident genetic diversity within the taxon. Harvey et al. (1999) collected five regionally distinct populations of WCM from across the Great Plains that differentially survived on wheat varieties with different mite-resistant genes, indicating the presence of mite biotypes. Genetic sequencing techniques used by Carew et al. (2009) targeting the mitochondrial 16S rRNA and the nuclear internal transcribed spacer 1 (ITS1), identified two distinct genotypes of WCM in Australia and designated them Type 1 and Type 2. Hein et al. (2012), using ITS1 and cytochrome oxidase I and II (COI, COII), characterized the five populations collected by Harvey et al. (1999) and placed them into either the Type 1 (Kansas, South Dakota, Texas, and Montana) or Type 2 (Nebraska) lineages found in Australia. These five populations have been used in several major studies through the last two decades. Using these now-established molecular methods, researchers should identify the mite types they are working with to facilitate comparative studies.

Additional work in Poland has identified greater genetic diversity for the mite than in other areas of the world. Skoracka et al. (2012) used molecular sequencing of the mtCOI and rDNA regions D2 and ITS to identify several distinct lineages and concluded that WCM is a complex of cryptic lineages. Using molecular techniques along with plant host data, Skoracka et al. (2013) found these lineages range from polyphagous to host specific. Further WCM genetic diversity was identified through additional sampling in Turkey by Szydło et al. (2015). WCM diversity was characterized by Skoracka et al. (2014) by identifying three genetic lineages that have spread throughout the world.

1 MT-1 (same as Type 2 from North America and Australia; also present in South America and Asia);
2 MT-7 (also present in Asia and Australia); and
3 MT-8 (same as Type 1 from North America and Australia; also present in Asia).

These three European lineages are the most polyphagous lineages in their studies, and it was hypothesized that the broader host range of these three groups contributed to their invasive success (Skoracka et al., 2014).

The spread of WCM throughout the world was comprehensively summarized by Navia et al. (2013) and Skoracka et al. (2014). Based on host usage patterns and genetic structure, Skoracka et al. (2018a) concluded that the lineage diversity found within WCM pre-dates agricultural practices and began with the diversification of C4 grasses.

The mite was originally identified and associated with WSMV impacting wheat in North America in the 1950s (Slykhuis, 1955; Keifer, 1953), although the presence of WSMV in North America has been traced back to the observation of 'yellow mosaic' in 1922 (Haskell and Wood, 1923). The mite had spread to Australia by the mid-1990s (Frost, 1995; Ellis et al., 2003) and to South America by the 2000s (Navia et al., 2006). It has not been reported in Africa, but reports of the presence of WSMV have appeared from Algeria (Benmokhtar and Yahia, 2009) and Zambia (Kapooria and Ndunguru, 2004), suggesting the mite's presence.

3 Wheat curl mite life cycle

WCM completes its life cycle in 8–10 days at 24–25°C with four stages: egg, first instar (nymph), second instar (larva) and adult (Staples and Allington, 1956). A recent study by Kuczyński et al. (2016) established the effect of temperature on the intrinsic population growth rates for Type 1 and Type 2 WCM. They found the temperature ranges for positive population growth rates were 10.4–35.7°C for Type 1 and 12.2–40.0°C for Type 2. Population growth rates peaked at 32°C for Type 1 and 35°C for Type 2, dropping dramatically at higher temperatures with no population growth by 40°C. They discussed the potential impact these differences could have on geographical distribution between the two types.

After a 1-2-day pre-oviposition period, WCMs reproduce rapidly with a female producing 12–25 eggs in her lifetime (Staples and Allington, 1956; del Rosario and Sill, 1965). It is estimated that a single female mite could have over three million descendants in 60 days under ideal conditions (Somsen and Sill, 1970). The mites reproduce by arrhenotokous parthenogenesis and with unfertilized eggs developing into males (Helle and Wysoki, 1983). Fertilization is indirect by males depositing spermatophores on the leaf surface to be picked up later by females (Fig. 2a). Even after fertilization, females produce both diploid females and haploid males (Helle and Wysoki, 1983). This reproductive system enables unfertilized females to disperse and still be able to start a new colony.

Because of the WCM's high reproductive capacity, mites can increase to very large populations when conditions are favorable. Their primary economic host is winter wheat. Winter wheat provides the benefits of being an excellent host, but even more importantly, it enables them to survive through the winter on a living host. The mites have a continuous life cycle while on winter wheat over-wintering in all life stages.

Two major population bottlenecks exist in the WCM life cycle. The first of these occurs as wheat nears harvest and begins to dry down. WCM can survive off green plants for only a few hours to a few days depending on temperature and humidity (Wosula et al., 2015). Under hot and dry conditions in the summer, they will likely desiccate and die within 12 h. As wheat is approaching maturity, mite populations build up rapidly to very high levels (Gillespie et al., 1997; McMechan et al., 2017) that disperse on wind currents to find a 'green bridge' host on which to over-summer. There are several potential alternate hosts, but by far the optimum host is volunteer wheat that is present at harvest.

The second bottleneck in the mite's life cycle occurs in the fall as a new crop of winter wheat is emerging. Mites have built up on their green bridge hosts (e.g. volunteer wheat, corn, other hosts) and must find a new host to survive the winter. The movement of mites from the summer hosts to winter wheat enables mites to continue their life cycle, but this can also cause a serious impact on the winter wheat crop. If enough mites move into the wheat crop and transmit viruses, serious infection and yield impacts will occur. The buildup of mite populations during the green bridge period is determined by the available bridge hosts, the environmental conditions during this time and the length of the bridge period.

4 Plant damage and response to mite feeding

Wheat curl mites are most often found on developing leaves within the whorl, and this feeding causes the edges of these leaves to remain tightly curled

Figure 3 Curling of the wheat leaf from wheat curl mite feeding trapping subsequent developing leaf (a) and showing curled leaf edges after leaf emerged from the whorl (b).

inward. This curl creates a beneficial micro-environment for the mites and enhances their survival. As this leaf curl is exposed with the extension of the leaf, the mites move back into the whorl to colonize the next developing leaf. As the plant grows, subsequent leaves or awns on the head can be trapped in the previous leaf curl causing distorted leaves (Fig. 3) or curled heads.

As the plants progress through the heading stages, mite populations can build dramatically (Fig. 2d). If mite populations build to high levels, these mites can have an impact on wheat yield. Their feeding during the heading stages can negatively impact yields up to 15% (Harvey et al., 2000; Harvey et al., 2002). The greatest impact from the WCM remains its ability to transmit its associated viruses; however, the development of more effective virus resistance in wheat (see Section 9.3) may allow greater mite populations to develop on the healthier wheat and increase the importance of direct injury from the mite (Wosula et al., 2018).

5 Impact of virus complex

In the early stages of virus symptom development, infected plants exhibit a subtle yellow-green mosaic pattern of parallel discontinuous streaks on the uppermost leaf (Fig. 4). If infection occurs prior to or early in tillering, severe stunting and spraddling (i.e. tillers spreading out close to the ground) will occur (Fig. 5a) (Hein et al., 2023; Wosula et al., 2018). Wheat infected after tillering will primarily show only leaf symptoms (Fig. 5b). As symptoms progress, the leaves become much yellower, and it can become hard to distinguish WSMV symptoms from other causes, such as barley yellow dwarf virus or nitrogen deficiency.

Mixed infections of two or even three viruses often occur (Burrows et al., 2009; Byamukama et al., 2013). However, it is not possible to distinguish between single or multiple infections based on symptoms. The extent of impact from multiple infections is not clear, but infections of more than one virus do increase the impact on the plant (Tatineni et al., 2014; Byamukama et al., 2012).

Figure 4 Progression of leaf symptoms from wheat streak mosaic virus infection from mottled mosaic pattern (a) to heavier discoloration (b) as virus symptoms develop.

Figure 5 Impact of mite-transmitted viruses on susceptible (S) and resistant (R) winter wheat varieties. Pre-tillering stage infection with severe stunting and yellowing with prostrate tiller growth (a). Post-tillering infection showing severe discoloration but upright growth (b).

Viral disease severity is greatly influenced by the time of infection. Early fall infection of a susceptible winter wheat variety by WSMV resulted in the greatest impact and yield reduction compared to late fall or early spring infections (Hunger et al., 1992; Wosula et al., 2018). This pattern of symptom development was also documented for single and double infection by TriMV and WSMV (Byamukama et al., 2014). In addition to the timing of infection, temperature also affects virus development within the plant, specifically virus replication, systemic movement, and symptom development (Wosula et al., 2017). For a WSMV-susceptible variety, WSMV was limited within the plant to the point of inoculation at 10°C, but as temperatures increased from 15°C to 25°C systemic movement of the virus and symptom development increased.

In the central Great Plains, a serious infection of mite-transmitted viruses in winter wheat typically begins in the fall, but symptoms do not appear until spring after optimal temperatures occur. However, with extended warm fall weather or in areas further south with warmer fall conditions, symptom development can begin in the fall. Projections of increased fall and winter temperatures due to climate change are likely to increase the risk of virus spread and development (Vose et al., 2017).

5.1 Wheat streak mosaic virus

It is the type species of the genus *Tritimovirus* in the family *Potyviridae* (Stenger et al., 2016; Tatineni and Hein, 2021a). WSMV is a single-stranded positive-sense RNA virus (Choi et al., 2002). It has a wide host range that includes wheat, other cereals and many grasses. Navia et al. (2013) and McMechan et al. (2016) provide summaries of the host range of WSMV.

WSMV must be acquired by immature mites, and these mites are able to carry the virus through each molt and into the adult stage (Slykhuis, 1955; Orlob, 1966; Siriwetwiwat, 2006). This virus is not transmitted through eggs. Orlob (1966) observed low transmission rates in as little as 15 min of feeding on WSMV-infected plants, although transmission rates increased with WCM feeding time up to 16 h. Transmission by adults persisted for at least seven days, but transmission efficiency for adults declined with time. Both Type 1 and Type 2 WCM are able to efficiently transmit WSMV, but the transmission efficiency of Type 2 mites is greater than that of Type 1 mites (45% vs 25%; Wosula et al., 2016). These differences varied with WSMV strain. Even though the genetics of the Type 1 mites in North America and Australia are comparable, Schiffer et al. (2009) found that Type 1 mites were unable to transmit the WSMV strain they used. WSMV can also be transmitted via wheat seed at low levels (~0.5–1.5%; Jones et al., 2005), and this pathway has been considered important in the introduction of WSMV to Australia from the United States (Dwyer et al., 2007).

5.2 High Plains wheat mosaic virus

It was first reported in the Great Plains during the 1993–1994 growing season from corn (*Zea mays* L.) and wheat (Jensen et al., 1996). In the literature, HPWMoV has been referred to as High Plains virus, wheat mosaic virus, maize red stripe virus and wheat mosaic High Plains virus. This virus is classified in the family *Fimoviridae*, genus *Emaravirus* (Tatineni and Hein, 2021b). HPWMoV is a multipartite, negative-sense RNA virus composed of eight genomic segments

(Tatineni et al., 2014). Seifers (1997) identified WCM as a vector of HPWMoV causing infection in barley and wheat. A partial host range of HPWMoV includes downy brome (*Bromus tectorum* L.), corn, barley (*Hordeum vulgare* L.), oats (*Avena sativa* L.), rye (*Secale cereale* L.), green foxtail (*Setaria viridis* (L.) P. Beauv.), yellow foxtail (*Setaria pumila* L.) and wheat (Seifers et al., 1998). Unlike WSMV and TriMV, HPWMoV cannot be mechanically transmitted, which has limited studies describing its transmission characteristics. Seifers et al. (2002) demonstrated differential transmission of HPWMoV with Type 2 mites being most efficient, but transmission rates between the two mite types varied with virus isolate and co-infection with WSMV. The economic losses associated with HPWMoV are unknown, but co-infections of these two viruses have been observed to have greater symptomatic expression and impact on the plant.

5.3 *Triticum mosaic virus*

It was first discovered in Kansas in 2006 expressing symptoms in wheat similar to WSMV (Seifers et al., 2008). TriMV is a positive-sense, single-stranded RNA virus in the genus *Poacevirus*, family *Potyviridae* (Tatineni et al., 2009). Seifers et al. (2009) determined the WCM to be the vector of TriMV but with poor efficiency. Later studies found that Type 2 mites transmitted TriMV very well (efficiency of 40.3%), whereas Type 1 mites were only able to transmit TriMV under very high mite populations (McMechan et al., 2014). Adult WCMs are unable to acquire TriMV; therefore, mites must acquire it as nymphs and retain the virus through molting (Knoell, 2018). TriMV can be acquired within an hour of feeding on infected plants with the acquisition peaking after 24 h of feeding (Knoell, 2018). Adult mites removed from infected plants are able to transmit TriMV at a high efficiency for the first day, but efficiency declines to low levels by day 6 (Knoell, 2018).

The host range of TriMV includes wheat, barley, oats, rye and triticale, plus several grass species (Seifers et al., 2010; Tatineni et al., 2010). Knoell (2018) found that unlike WSMV and HPWMoV, mite populations had very limited ability to acquire TriMV from corn. TriMV has been shown to reduce wheat yields (Seifers et al., 2011). Greenhouse (Byamukama et al., 2012) and field studies (Byamukama, 2014) both demonstrated that WSMV-TriMV co-infections caused a synergistic reduction in wheat growth and yield. Tatineni et al. (2010) showed that TriMV is synergistic in co-infections with WSMV, with TriMV exceeding the titer of WSMV late in the infection process. Virus interactions with co-infections are largely unknown, but in an extensive survey across the Great Plains, 91% of all TriMV-positive samples were co-infected with WSMV (Byamukama et al., 2013).

6 Alternative hosts and green bridge species for the mite-virus complex

Winter wheat serves to support WCM through a good part of the year, but especially for survival of the mite through the winter. After wheat harvest, WCM must find an alternate host for survival through this 'green bridge' period until the emergence of winter wheat in the fall. Christian and Willis (1993) established five characteristics that determine a host's risk to serve as a green bridge to carry mites and viruses through the summer to fall-planted winter wheat.

6.1 Host acceptability

The reproductive success of WCM on its host is critical to determining the potential risk of that host serving as an important green bridge host. Reviews by Navia et al. (2013) and McMechan (2016) provide a list of nearly 90 plant species identified as hosts for the WCM through field observations or lab experiments. Much of this information was developed in the 1950s and 1960s. Most studies on the reproductive potential of WCM were conducted as short-term (typically 7 days) exposures under controlled conditions using early, vegetative stages of plant development and collecting non-quantitative (i.e. classification) data. Determining a list of potential WCM hosts is difficult due to these limitations, but approximately 71 grass species show at least some level of survival of WCM over a short-term period. To establish the true risk of a plant species serving as a green bridge host, studies must quantify the reproductive potential of WCM and its performance across all growth stages of the host that occur during the green bridge period. This reproductive potential has been quantified for very few species.

Field observations have detected WCM on 66 of 90 plant species sampled (McMechan, 2016; Navia et al., 2013). The abundance of WCM on these species is highly variable because it will be affected not only by the suitability of the host but synchrony of winter wheat and alternative host phenology. For example, even though volunteer wheat is an excellent host, infestation rates can range from 100% to 0% (Staples and Allington, 1956; Brey et al., 1998). This is likely due to the variable emergence date of volunteer wheat (Staples and Allington, 1956) with high populations resulting when volunteer wheat emerges before harvest and much reduced or absent populations when volunteer wheat emerges after wheat harvest.

Another uncertainty with these historical accounts is the accuracy of species identification since other eriophyid mites can be found on grassy plants (Nault and Styer, 1969), and most studies provide limited information to verify mite identification. Additionally, only a few recent studies account for the genotype differences within the WCM species complex noted by Hein et al. (2012) and Skoracka et al. (2013) because these differences had not been identified when the studies occurred. For example, Harvey et al. (2001) evaluated 29 grass species and found differential survival of WCM on rye depending on the mite source, with a population collected from Kansas increasing on rye over a 7-day period, whereas a population collected from Nebraska declined in the same time period. These two populations were found to differ genetically (Kansas = Type 1; Nebraska = Type 2; Hein et al., 2012), but Type 1 and Type 2 mites also have different virus transmission characteristics (McMechan et al., 2014; Wosula et al., 2016) and reproductive rates on virus-infected plants (Siriwetwiwat, 2006; McMechan, 2014).

6.2 Host seasonality

The most important green bridge hosts will be summer annuals that are growing as wheat harvest approaches and continue growth until after the fall wheat crop has emerged. Several recorded hosts for the mite match this seasonality (e.g. corn, foxtail millet (*Setaria italica* (L.) P. Beauv.), barnyardgrass (*Echinochloa crus-galli* (L.) P. Beauv.)). Hosts that emerge after wheat harvest could be important. But because mite dispersal activity is very low after harvest, these hosts will need to grow long enough to become infested and enable significant mite buildup.

Several winter annual species listed as WCM hosts, such as jointed goatgrass (*Aegilops cylindrica* Host), possess a life cycle comparable to winter wheat. These hosts may serve as an additional source of mites in the agroecosystem and may enhance the survival of maintenance populations of mites and perhaps viruses in the agroecosystem. However, because these hosts' life cycles are in synchrony with wheat, they are not a threat as green bridge hosts and will not be a factor in establishing epidemics in winter wheat. Where spring wheat is grown, however, these winter annual hosts could enable mite populations to survive the winter to infest spring wheat.

6.3 Host availability

Extensive mite dispersal and subsequent virus spread that would result in field epidemics require significant mite density at the mite source, which in turn depends on the density of the host. McMechan (2016) found that many of the potential grass weed hosts for the WCM-virus complex are seldom present in significant densities in the west-central Great Plains, or they are present only in sporadic patches that limit mite population buildup. Under favorable conditions, volunteer wheat stands are dense enough to allow considerable mite population buildup (Fig. 6).

Figure 6 Pre-harvest volunteer wheat resulting from severe hail. Plants growing from hail-lodged head (a); plants growing in and between wheat stubble rows following hail (b); plants growing within understory of hail-damaged wheat stubble (c); mite-infested and virus-infected volunteer wheat growing in sunflower field after sunflower harvest (d).

6.4 Virus susceptibility

A literature review of over-summering hosts indicates that approximately 197 plant species have been tested for WSMV susceptibility by using mechanical inoculation with 91 species testing positive for WSMV, but only 30 of those species have been tested by more than one author (McMechan, 2016). Mechanical inoculation with WSMV provides a good estimation of a potential background source for WSMV in the landscape. In comparison, field detection of WSMV has been conducted on 44 plant species with 18 testing positive for WSMV.

The involvement of two additional viruses complicates the determination of risk for green bridge hosts. Volunteer wheat will host all three viruses, but several other hosts differ in whether they can be infected by each virus and the susceptibility of others is unknown. For example, mites moving off corn will remain viruliferous for WSMV and HPWMoV but not TriMV (Knoell, 2018).

Seifers et al. (1998) provide a listing of known HPWMoV hosts, and Seifers et al. (2009) provide background on TriMV hosts.

6.5 Mite adaptation to wheat and other hosts

There are almost no studies that have addressed the adaptation of the mites as they move from host to host. Skoracka et al. (2013) first documented reduced WCM reproduction in wheat when transferring mites from other hosts back to wheat. WCM transferred from wheat to wheat had a population growth rate (PGR) of 50 progeny per female, whereas WCM transferred from quackgrass (*Elymus repens* (L.) Gould) to wheat had a PGR of 0.2–4 depending on the mite source. McMechan (2016) also demonstrated that the movement of mites between hosts requires adaptation to the new hosts in some situations. Mites are efficient enough at transmitting viruses that even if mites do not survive well on wheat, it is likely that virus transmission will still result even from limited mite feeding and survival. An important consideration is that this effect will mostly impact the secondary spread of mites and viruses in wheat.

In 1980 and 1981, Shahwan and Hill (1984) tracked 11 fields that were severely impacted by WSMV and attempted to correlate disease severity with the adjacent fields' cropping and environmental history. Only one field was planted adjacent to corn, and another field had been planted adjacent to foxtail millet implicating these hosts as sources of the virus. However, 9 of the 11 fields had late-season hail, resulting in the presence of pre-harvest volunteer wheat.

In the central Great Plains, there is documented evidence (Staples and Allington, 1956; Wegulo et al., 2008; Nguy-Robertson et al., 2016; Hein et al., 2023) that pre-harvest volunteer wheat arising from hail serves as an optimal green bridge host. Hail events just prior to harvest shatter seed to the ground and typically provide moisture for germination of this seed (Fig. 6). Pre-harvest volunteer wheat also includes volunteer growing at this time in summer crops (e.g. corn, sunflower and proso millet) (Fig. 6d). The prevalence of mites during this time insures rapid infestation of this volunteer wheat.

Volunteer wheat is a problem after wheat harvest as rainfall events bring on a flush of new volunteer wheat. This volunteer wheat can also serve as a bridge host, but low mite dispersal activity at this time causes it to become infested much more slowly and sporadically. The risk of post-harvest volunteer wheat serving as a bridge host increases to the south in the Great Plains where wheat harvest is earlier, planting is often later and warm weather extends much later in the fall.

The risk of other grasses serving as green bridge hosts is highly variable and difficult to establish because of sporadic data in the historical literature. Webb (2018) established a green bridge host risk assessment by determining the proportion of studies in the literature where a host is listed as a positive or

Table 1 Risk of grass species serving as green bridge host for mite-virus complex. Ten high-risk species from the literature (Webb 2018) are compared against five criteria for green bridge hosts proposed by Christian and Willis (1993). Entries in italics designate factors that limit or mitigate risk posed by the host

Host	Host acceptability	Seasonal host availability	Host density	Virus susceptibility	Adaptation effects on return to wheat	Potential green bridge risk
Wheat *Triticum aestivum* L.	0.96[a] T1:+++; T2:+++[b]	Yes (volunteer)	Yes (crop)	0.96[a]	None	Very high
Corn *Zea mays* L.	0.86[a]	Crop - Yes (warm season)	Yes (crop)	0.91[a]	Unknown	Moderate
Rye *Secale cereale* L.	0.83[a]	***Crop - No (winter annual)***	Yes (crop)	0.92[a]	Unknown	***None (crop)***
Barley *Hordeum vulgare* L.	1.0[a]	***Crop - No (cool season)***	Yes (crop)	0.88[a]	Unknown	***None (crop)***
Jointed goatgrass *Aegilops cylindrica* host	1.0[a] T1:+++; T2:+++[b]	***Weed - No (winter annual)***	Poor–fair (weed)	0.88[a]	Unknown	***None***
Barnyardgrass *Echinochloa crus-galli* (L.) P. Beauv.	0.8[a] T1:++; T2:+++[b]	Weed - Yes (warm season)	***Poor (weed)***	0.93[a]	T1: yes/T2: no[b]	***Low***
Green foxtail *Setaria viridis* (L.) P. Beauv.	0.85[a] T1:+; T2:+[b]	Weed - Yes (warm season)	***Fair (weed)***	1.0[a]	Unknown	***Low***
Stinkgrass *Eragrostis cilianensis* (All.) Vignolo ex Janchen	0.64[a]	Weed - Yes (warm season)	***Poor–fair (weed)***	0.78[a]	Unknown	***Low***

(*Continued*)

Table 1 (*Continued*)

Host	Host acceptability	Seasonal host availability	Host density	Virus susceptibility	Adaptation effects on return to wheat	Potential green bridge risk
Field sandbur *Cenchrus pauciflorus* Benth.	0.75[a]	Weed – Yes (warm season)	***Poor–fair (weed)***	1.0[a]	Unknown	***Low***
Canada wildrye *Elymus canadensis* L.	0.7[a]	Yes (perennial grass)	***Poor (wild)***	0.89[a]	Unknown	***Low***

[a] Data from Webb (2018). Ratio of the number of studies with positive host designation for WCM or WSMV divided by the total studies for that host.
[b] Data from McMechan (2016). Extensive evaluation of Type 1 (T1) and Type 2 (T2) mite reproduction on host across a 42-day period. +, ++ and +++ represent the degree of reproductive success on the host for that mite type.

negative host for WCM and WSMV. Ten hosts with a high proportion of studies listing them as hosts were identified as 'high-risk' (Table 1). Table 1 includes the green bridge criteria for each host as identified by Christian and Willis (1993) listed above. Using these criteria, jointed goatgrass, wheat, rye and barley can be eliminated as 'high-risk' because as winter or cool season annuals their growing cycle largely mimics wheat. However, if these hosts are growing out of cycle as a volunteer or perhaps as cover crops, they could still be a concern. Corn is a host of concern because it is growing through all or most of the green bridge period, and mites and viruses survive through this host (Knoell, 2018). Canada wild rye is a native grass that is not found in dense plantings across most of the Great Plains unless in scattered pasture or prairie areas. Thus, its importance is limited. The four remaining grass weeds (barnyardgrass, green foxtail, field sandbur and stinkgrass) are common in the central Great Plains (McMechan, 2016) but rarely in large dense patches.

The quantitative data on mite reproduction for field sandbur and stinkgrass are limited; thus, assigning risk for these grasses is difficult. McMechan (2016) provided extensive quantitative mite reproductive data that showed barnyardgrass to be a very good host for WCM and green foxtail to be a rather poor host. McMechan (2016) also showed green foxtail in the west-central Great Plains to be widespread but not in great density, while barnyardgrass occurred mainly in isolated patches. Considering these factors, neither barnyardgrass nor green foxtail is a high-risk green bridge host.

7 Mite dispersal

Different potential modes of WCM dispersal have been proposed including walking (active), phoresy and wind (passive) dispersal (Sabelis and Bruin, 1996; Michalska et al., 2010). Most short distance plant-to-plant movement of the WCM is by walking from leaf to leaf, and WCMs have been shown to walk at a speed of 4–5 cm per h (del Rosario and Sill, 1965; Kuczyński et al., 2020). Majer et al. (2021a) found that WCMs have much higher dispersal success and subsequent colonization when walking plant to plant than when dispersing aerially. Dispersal by walking though would be a factor in plant-to-plant secondary virus spread (Overmyer, 2020).

Gibson and Painter (1957) documented phoresy with WCMs hitching a ride on aphids in a greenhouse setting. Recent laboratory studies have tried to simulate the potential phoretic transport of mites with very limited success, and the authors have concluded that this dispersal method is possible but quite rare (Kuczyński et al., 2020; Majer et al., 2021b). Populations of WCM and the transport vector would need to be very high for effective phoretic mite dispersal and virus spread.

Mites do not produce silk for ballooning as do some arthropods, and they have no specialized structures for aerial transport; therefore, they must rely on passive aerial transport for longer range dispersal. Slykhuis (1955) and Pady (1955) both identified mite dispersal to occur via air currents. Staples and Allington (1956) found a correlation between wind speed and the number of mites collected on sticky glass slides placed in and around wheat and volunteer wheat fields. Later, Nault and Styer (1969), Coutts et al. (2008), Umina et al. (2016) and Stilwell et al. (2019) using various field techniques (e.g. sticky slides, direct plant counts of mites, and estimates of virus spread) obtained directional mite movement and/or virus spread data and concluded that wind was the primary factor in WCM dispersal.

Although WCMs have no morphological adaptations for wind dispersal, there is evidence for behavioral adaptations for this mode of dispersal. WCMs have been found to alter their behavior in the presence of wind by moving upward on the plant to exposed areas and standing vertically on the plant (Nault and Styer, 1969; Kiedrowicz et al., 2017) and crawling over each other to form multiple-mite chains that may increase take-off potential (Laska et al., 2019). Dispersal behavior was mostly limited to adults as nearly all the mites collected on sticky glass slides were adults (Nault and Styer, 1969). Although Kiedrowicz et al. (2017) found that a low proportion of mites exhibited these dispersal behaviors, given the high reproductive potential of WCM, dispersal of a few individuals could be important for the spread of the population.

Nault and Styer (1969) demonstrated field-to-field mite movement in a wheat-corn agroecosystem as wheat was approaching maturity in mid-summer and again in the fall when corn was maturing. They concluded that the declining quality of the hosts drove mite dispersal. Other studies have demonstrated that mites are at their greatest density as the crop matures in both winter wheat (McMechan et al., 2017) and corn (Knoell, 2018). The relative importance of host quality and mite density to mite dispersal is important during the critical period of dispersal from volunteer winter wheat in the fall.

Between declining host quality and mite density, mite density has been documented to be the most important factor determining mite dispersal. Jiang et al. (2005) found that mite abundance increased with increased total green leaf area on source plants treated with glyphosate. This suggests that improved plant condition would lead to an increased reproductive rate for WCM and an increase in subsequent dispersal potential. Wind tunnel studies showed that over time mite movement off plants was greater from plants less impacted by glyphosate because of the greater mite densities remaining on those plants (Thomas and Hein, 2003). In the field, Thomas and Hein (2003) used various stress levels to modify the plant condition of simulated volunteer winter wheat and showed that mite density on the volunteer wheat was the most important factor influencing the number of WCM dispersing. In a wind tunnel study, Overmyer (2020) used differential virus infection to establish varied plant conditions and mite densities on wheat and found that mite density was more important than plant condition in determining mite dispersal.

Majer et al. (2021a) in determining the factors important to mite colonization resulting from dispersal via wind found that propagule pressure (i.e. the number of mites dispersing) was the most important factor for colonization success. Stilwell et al. (2019), using virus spread to evaluate mite dispersal in the field, found that the highest mite densities in volunteer wheat corresponded to the greatest virus spread into surrounding areas. Thus, mite density has been found to be the best predictor of mite dispersal and virus spread. This distinction is important because mite dispersal from volunteer winter wheat in the fall occurs over an extended period of time. The longer the quality of the volunteer is maintained through this period, the greater the buildup in mite density and the greater the potential mite dispersal and virus spread. There is still much to learn about mite dispersal, but the risk of virus spread and the spatial extent of that spread into surrounding wheat fields appear to depend primarily on mite density in the source field.

Time of day, air temperature, wind direction and wind speed have been examined for their effects on WCM dispersal, but the work has been limited. Staples and Allington (1956) found that mites primarily dispersed during the daytime. Laska et al. (2019) in wind tunnel studies found an increase in dispersal behavior for mites up to a wind speed of about 4 m/s. However, Stilwell et al. (2019) in monitoring the spatial movement of WCM through the fall found that the greatest distances of mite dispersal into surrounding winter wheat correlated best with wind directions when wind speeds were greater than 9 m/s.

In addition to dispersal behavior that leads to increased virus risk locally, long-range mite dispersal could also affect the pest status of WCM. McNeil et al. (1996) found that WSMV genetic diversity among counties across Nebraska (~600 km distance) was comparable to the genetic diversity within sampled fields, concluding that dispersal of WSMV isolates by mites over these distances would be needed to account for this pattern. The risk that this long-range movement of WCM will result in epidemic populations is very low because the number of mites moving these distances will be low. However, this movement could be important in the gene flow of mite and virus populations that could result in the spread of resistance-breaking mite biotypes or virus isolates. A better understanding of the potential for this long-range dispersal will be important in establishing the genetic stability of both the mite and its associated viruses.

8 Case study: spatial characterization of mite dispersal and virus spread

Most winter wheat farmers across the western Great Plains have experienced first-hand or seen the devastating impact of the wheat-mite-virus complex. A common question posed by these farmers is: How far can WCM move and spread the virus into my field? Research into mite dispersal has enabled us to begin to answer this question.

Mites can move at any time during the year, but WCM dispersal has the greatest impact when mites move from maturing wheat to their over-summering host and when mites move from their over-summering host to the new winter wheat crop and initiate virus infection. These movements depend primarily on passive aerial dispersal.

During the vegetative growth stages of winter wheat, WCM feeds within the whorls, moving into the whorl of the newest leaves as each leaf emerges. When wheat heads emerge, the mites move into them, utilizing countless protected niches to survive and reproduce (Fig. 2d). As a result, WCM populations increase steadily during the heading stages of winter wheat (Gillespie et al., 1997; McMechan et al., 2017), reaching high densities by the hard dough stage across the western Great Plains (Mahmood et al., 1998; McMechan et al., 2017; Byamukama et al., 2016). All fields in these three studies were healthy wheat fields (i.e. not impacted by virus), but the WCM density in these studies averaged over 500 mites per head or conservatively 2.5 billion mites per hectare. Thus, across the Great Plains where winter wheat is grown extensively, there is potential for significant background WCM populations in all winter wheat fields as they approach maturity, and the mites seek to find alternate hosts. These mite populations typically carry the viruses from this mite-virus complex (Byamukama et al., 2016), so virus inoculum is present every year during this dispersal cycle.

Wheat curl mite survival is limited when off a host plant (Wosula et al., 2015), helping to explain the high WCM dispersal events that occur as the wheat crop matures and dries down (Nault and Styer, 1969; McMechan et al., 2017). During this 'mite rain' period, mites can infest available hosts across the landscape (Nault and Styer, 1969; Gillespie et al., 1997; McMechan et al., 2017). In western Nebraska, dispersing mites have been detected more than 30 km from the nearest wheat fields during this time (GLH unpublished). Mite activity drops substantially after harvest, lowering the risk of mite colonization of newly emerging bridge hosts.

As mentioned previously, the most important of these bridge hosts is volunteer wheat that has emerged before harvest. McMechan (2016) studied the development of volunteer wheat from maturing wheat. Pre-maturity germination rates were found to differ by wheat variety and were related to the seed dormancy requirement for that variety. Potential for germination increased as the seed matured through the seed fill stages, but a window of risk for significant germination and the formation of pre-harvest volunteer following hail events was determined to be about a 3-week period before harvest, corresponding to the soft to hard dough stages of wheat.

From a farmer's perspective, mite dispersal from green bridge hosts to newly emerged winter wheat is the most important dispersal event. Understanding mite dispersal during this fall period is critical because the most serious virus infections will result from fall mite infestations and virus infections beginning during the early growth stages (Hunger et al., 1992; Wosula et al., 2018). Because of declining temperatures during the fall and wheat dormancy through the winter, the expression of serious virus symptoms occurs at least several weeks and in most cases several months (i.e. fall to spring) after the mite dispersal period and initiation of virus infection. This delay makes it difficult to connect the two in the farmers' minds. Studies discussed earlier on mite dispersal shed light on the factors that affect WCM dispersal activity and address directional movement resulting from wind and the extent of mite dispersal.

To characterize the spatial spread of mites and viruses, Stilwell et al. (2019) used small (10 m × 10 m), simulated field plots of mite-infested and virus-infected volunteer wheat as a mite source in the middle of large wheat fields. They monitored mite populations as they moved into the surrounding wheat crop through the fall. The following spring, aerial remote sensing data and virus infection rates from plant sampling were used to quantify the spatial spread of the virus surrounding these simulated volunteer source plots (Fig. 7). The red edge vegetative index was found to be the best predictor of virus symptoms created in the surrounding wheat (Stilwell et al. 2013). Ground truth data included Soil Plant Analysis Development (SPAD) meter readings that measured relative chlorophyll (i.e. yellowing symptoms) (Yadava, 1986; Byamukama et al., 2014) taken at numerous spatially defined points in the surrounding wheat. Cokriging of these data and virus infection values provided a map of spatial virus spread into high, medium and low infection risk zones surrounding the mite source.

Mite dispersal occurs throughout the fall, and due to variable wind directions through this period, mite movement and virus spread extended in all directions from the source (Stilwell et al., 2019). Although wind direction is variable during this period, winds greater than 9 m/s tended to occur more frequently from the northwest accounting for mite distribution and virus spread in an oval pattern around the mite source but extending farther to the southeast from source fields (see Fig. 7). This pattern is the basis for predicting potential WCM dispersal from green bridge source fields and virus spread

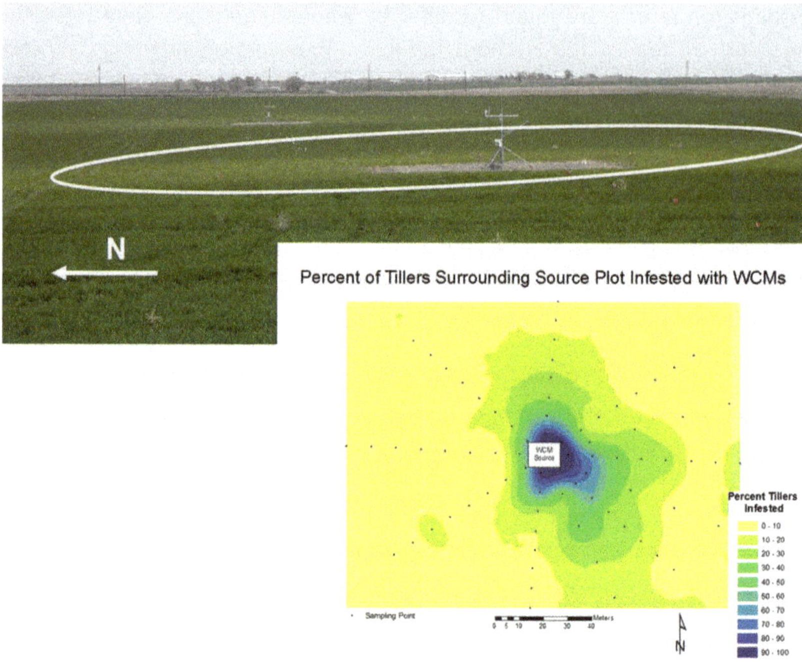

Figure 7 Oval-shaped pattern of symptomatic plants showing virus spread from central mite/virus source (light areas). Inset: Cokriged analysis of mite infestation isolines surrounding the central mite source. Source: From Stilwell (2009).

into surrounding production fields in the Great Plains (Fig. 8). Stilwell et al. (2019) used the spatial dimensions of virus spread from these small plots into the surrounding wheat to estimate the potential sphere of influence for virus spread around a typical size field. Based on these dimensions, they estimated mite dispersal significant enough to produce serious virus infection could extend up to 3.3 km. These estimates need to be further validated at this larger scale, but observations made from serious virus epidemics indicate that these estimates are reasonable.

Mite dispersal and risk of virus spread depend upon the wind direction patterns that generate oval-shaped spheres of influence of virus risk around source fields (Fig. 8). The extent of significant mite dispersal and virus spread will depend on the source field area and the mite abundance in those source fields. For wheat fields immediately adjacent to a source field, minimal mite presence will result in a limited gradient of virus symptoms appearing just at the margin of the field. Increasing mite presence will increase the gradient across the adjacent field, and extensive mite presence will result in serious virus symptoms extending across the entire field and beyond (Fig. 9).

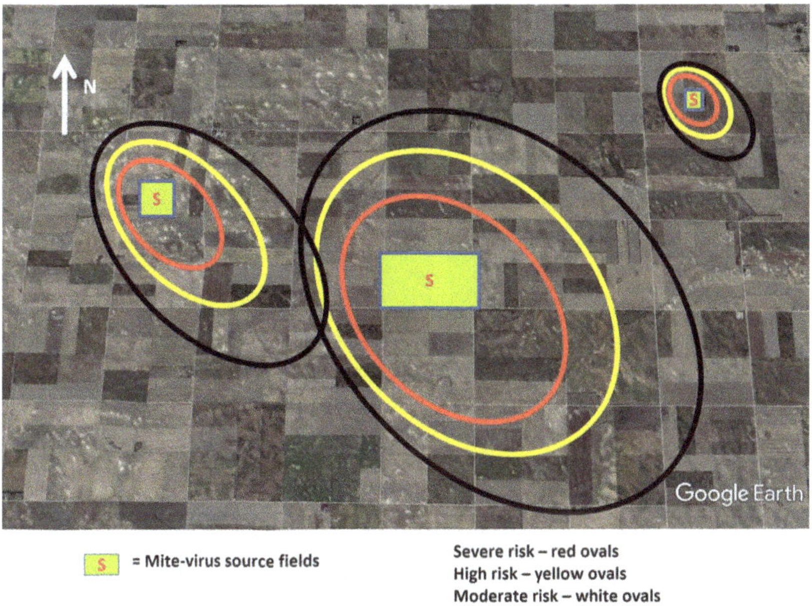

Figure 8 The sphere of influence for wheat curl mite dispersal and virus spread extending in all directions in an oval pattern around mite/virus source fields. The long axis of the spread pattern extends with the direction of prevailing strong winds, and the different color ovals indicate decreasing virus spread risk levels (red to yellow to black). Source: From Hein et al. (2023).

In wheat production landscapes, mites and viruses are prevalent every year as wheat approaches maturity. This results in significant risk if green bridge hosts are present at this time. Reducing virus risk depends on eliminating these green bridge sources. To protect the surrounding fields from virus, management of source fields needs to be promoted as a 'good neighbor' practice. Managing virus risk is complicated further when large pre-harvest hail events result in multiple foci of mite source volunteer wheat fields that are insufficiently controlled throughout the neighborhood (Fig. 8). If the density of these mite source fields is great enough, the entire neighborhood can act as a single mite source to infest all fields within the neighborhood and well beyond. This problem has been demonstrated numerous times after a large hail event one year results in an extensive virus epidemic the following year over large acreages. These large hail streaks can cover dozens of square miles, and if they occur during the 3-week window of risk prior to harvest, it is even more important that effective management involves an areawide emphasis to eliminate significant mite/virus source fields.

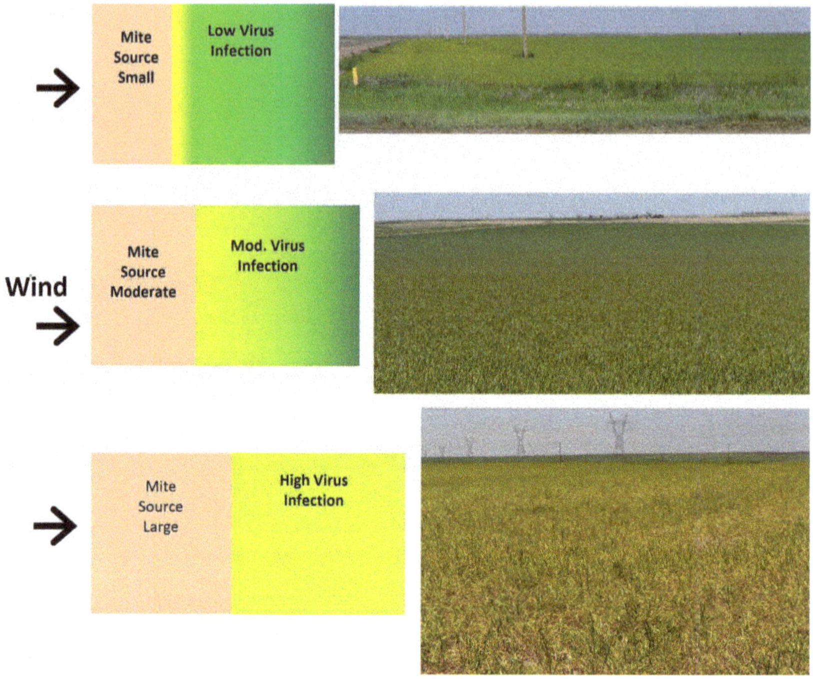

Figure 9 Relationship between mite source and extent of mite dispersal and virus spread into adjoining wheat. 'Size' of the mite source is determined by mite and green bridge host density and area of the source field. The virus spread increases with increasing 'size' of the mite source. Source: Revised from Hein et al. (2023).

9 Management of mite-virus complex

WCM is a pest almost exclusively because of its role as a virus vector. The elements of WCM management include assessing the risk of virus infection, eliminating the high-risk green bridge hosts as sources of infestation and infection, avoiding early planting and planting mite- and virus-resistant varieties. Available insecticides are ineffective because the mites are protected within the whorl of the plant or not susceptible to available systemic materials. Some biological control effects have been identified (Stilwell, 2009), but their overall impact on mite populations is deemed minimal.

Although management efforts cannot alter the environmental factors that influence risk, they must be considered (Fig. 10; Hein et al., 2023). Increased prevalence of winter wheat acreage in the landscape will increase the mite-virus risk for a growing area. Cooler summers with timely precipitation can increase the growth of volunteer wheat and increase the potential for mite buildup. Extended warm conditions through the fall into the winter will increase the period of mite dispersal and reproduction on the plants and virus replication in

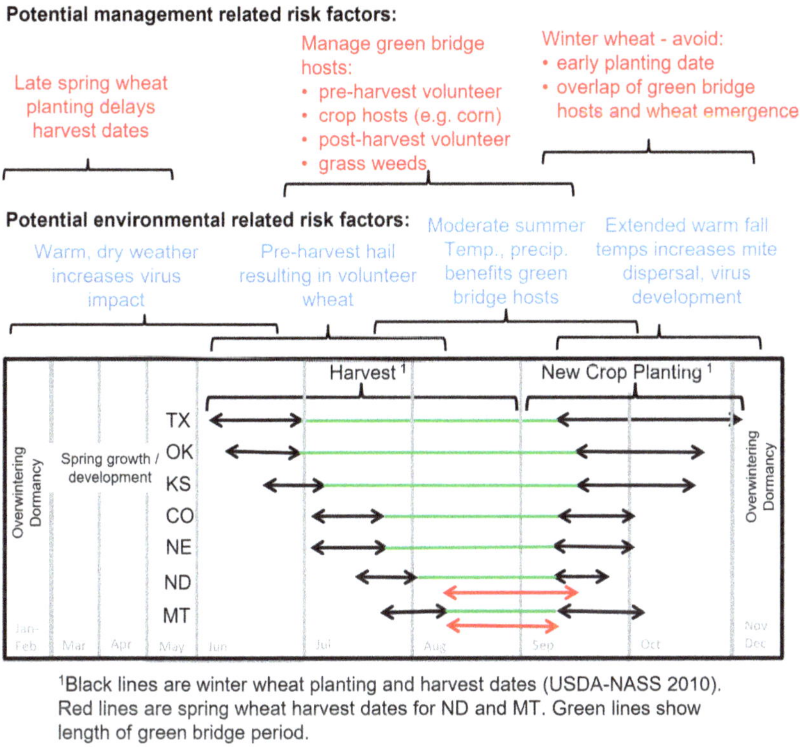

Figure 10 Wheat-mite-virus cycle in winter wheat across the Great Plains, showing varied lengths of green bridge period due to varied harvest and planting dates and associated risk factors during the cycle.

the plants. Warm conditions occurring earlier in the spring when the plants are at earlier growth stages will increase the impact of virus infection on the plants.

9.1 Green bridge management

The presence of WCM during the summer bridge period is a primary risk factor. If the green bridge for the mites is broken and mites are not able to survive the over-summering period, virus risk is eliminated. Therefore, management must focus on breaking this green bridge, preferably for at least 2 weeks (Hein et al., 2023). High-risk bridge hosts, especially pre-harvest volunteer wheat, should be controlled completely before the emergence of the fall wheat crop. Controlling green bridge hosts can be effectively accomplished with tillage or herbicides, but environmental and application variables can impact the effectiveness of control measures (Thomas et al., 2004; Jiang et al., 2005).

Another source of WCM in the crop landscape is corn. Mites move into corn with the flush of WCM moving off winter wheat just prior to wheat harvest (Knoell, 2018). Mite populations increase in corn, especially in the ears, and they move off corn as the ears are drying down (Knoell, 2018). Dryland corn often dries down before wheat planting in the fall and carries a lower risk for virus problems. Irrigated corn will stay green longer, and WCM can remain on that crop, posing an additional risk of virus infection if it overlaps with winter wheat emergence.

Other potential green bridge hosts in the surrounding agroecosystem include post-harvest volunteer wheat and several other grass hosts for the mites. The risk from post-harvest volunteer wheat increases with the length of time that it is growing, allowing time for mites to increase to significant levels. The ability of mites to reproduce on other grass hosts is much less than it is on wheat; therefore, the risk from the presence of these hosts is much lower than that for volunteer wheat. However, it is important to avoid the overlap of newly emerging winter wheat with those alternate grass hosts that pose a significant risk (e.g. foxtail millet and spring cereals).

9.2 Adjust planting date

A delay in the planting date of winter wheat has been shown to reduce the risk of serious virus infection in winter wheat (Hunger et al, 1992; McMechan et al., 2016). However, McMechan et al. (2016) also demonstrated that even with delayed planting dates, this benefit can be overridden by high mite populations and virus prevalence. Thus, an important practice that reduces, but does not eliminate, the risk of serious disease development is to avoid planting winter wheat before the recommended window for the production area.

9.3 Mite and virus resistance

Early efforts to identify resistance to the WCM in wheat had limited success, necessitating the identification and development of resistant genes from close wheat relatives that could be introduced into wheat. Thus far, four WCM colonization genes (*Cmc1*-*Cmc4*) have been transferred to and described in wheat (Table 2). However, recent research has demonstrated that the *Cmc1* and *Cmc4* genes are the same gene (Silva, 2021). Additional mite resistance genes have been identified, but they have not yet been characterized and named. The value of mite resistance lies not only in the potential for reduced mite populations and virus transmission and spread within the crop but also to reduce mite buildup in volunteer wheat that serves as a summer bridge host (Thomas et al., 2004; Harvey et al., 2003).

Table 2 Summary of resistance genes identified for mite-transmitted viruses in wheat and resistance to the wheat curl mite

Resistance genes	Wheat gene	Source	Released varieties	Primary literature citations
WCM resistance (curl mite colonization) genes				
Cmc1[a]	6DS	*Aegilops tauschii*	Radient	Thomas and Conner, 1986; Whelan and Thomas ,1989; Cox et al., 1999; Thomas et al., 2012
Cmc2	6DL	*Triticum ponticum*	-	Whelan and Hart, 1988
Cmc3	1AL	*Secale cereale*	TAM 107	Martin et al., 1983; Porter et al., 1987; Harvey et al., 1995; Harvey et al., 1997
Cmc4[a]	6DS	*A. tauschii*	OK 05312, Byrd, TAM 112	Cox et al., 1999; Malik et al., 2003; Carver et al., 2016; Zhao et al., 2021
WSMV resistance genes				
Wsm-1	4DS	*Thinopyrum intermedium*	Mace	Gill et al., 1995; Seifers et al., 1995; Graybosch et al., 2009; Kumssa et al., 2017
Wsm-2	3BS	*Triticum aestivum*	RonL, Snowmass, Clara CL, Oakley CL	Haley et al., 2002; Lu et al., 2011
Wsm-3	7BS	*Triticum intermedium*		Friebe et al., 2009; Liu et al., 2011; Danilova et al., 2017; Kumssa et al., 2017

[a] Recent publication by Silva (2021) demonstrates that *Cmc1* and *Cmc4* are the same gene.

Concern about the stability of resistance genes was demonstrated by the adaptation of mite populations to *Cmc3* after extensive use (Harvey et al., 1995; Harvey et al., 1997). Murugan et al. (2011), Carrera et al. (2012) and Khalaf et al. (2019) have begun to characterize the mechanisms of resistance and their effect on mite genotypes. Kiani et al. (2021) have described the initial response of the wheat transcriptome to mite feeding for one variety containing the *Cmc4* gene. It will be important to improve our understanding of the mechanisms of these and future resistance genes to predict and manage the stability of these genes and deploy them for maximum effectiveness.

Resistance to WSMV in wheat has been pursued since the identification of the virus connection to the disease in the 1950s with limited success in identifying highly effective resistance. However, effective sources of resistance have been identified in close relatives of wheat and introgressed into wheat. A summary of the genes identified is listed in Table 2. The *Wsm1* gene in the variety 'Mace' has shown strong levels of resistance to both WSMV and TriMV (Graybosch et al., 2009), but its incorporation has resulted in yield drag that has limited this variety and the broader use of this gene. A second gene, *Wsm2*, has been more widely incorporated into commercially released lines; however, this gene only provides resistance to WSMV. There are several challenges with developing virus-resistant varieties. The stability of these genes is unknown, but a strain of WSMV has recently been found that overcomes the resistance in *Wsm2* (Kumssa et al., 2019). Genetic resistance in these varieties may provide resistance to only a single virus, as with *Wsm2*, and little is known about the resistance levels of these genes to HPWMoV. Virus resistance in both *Wsm1* and *Wsm2* is temperature sensitive and begins to break down as temperatures rise above about 20°C (Seifers et al., 1995; Seifers et al., 2006). Thus, these varieties may limit fall and early spring virus establishment, and they are likely to be more effective where fall temperatures are cooler. Additional genes are being evaluated and may provide more consistent resistance and temperature stability (Liu et al., 2011; Seifers et al., 2013).

An additional need for mite resistance in wheat has become apparent from the recent development of more effective virus resistance in wheat. The potential for greater usage of virus-resistant wheat cultivars may alter the impact of the WCM. Because strong virus resistance will dramatically lessen the impact of the virus, the ability of the mites to build up to very high levels will become a more important factor. Harvey et al. (2002) demonstrated the impact of non-viruliferous mites can approach 15%, and significant mite buildup and impact on yield in the absence of the virus have been observed for Mace (Wosula et al., 2018; GLH unpublished). This demonstrates the potential importance of pyramiding mite resistance with virus resistance as it is more broadly deployed in wheat.

10 Conclusion

The WCM can be found worldwide and exerts its primary impact through the transmission of four different viruses. The relatively recent spread of components of this wheat-mite-virus complex to new areas around the world increases the importance of understanding the aspects of managing this complex. Variation in cropping systems can dramatically alter the risk for this mite-virus complex, but understanding mite ecology and virus epidemiology is critical to developing the best management options for each area. Because plant resistance for both the mite and the virus has been difficult to establish in commercial wheat varieties, management of this wheat-mite-virus complex has relied primarily on cultural practices. Some reduction of risk and impact can be obtained by avoiding early planting, but the primary management practices rely on the control of over-summering hosts for the mite and virus. Management through this green bridge period requires a good understanding of host relationships and dispersal characteristics of the mite. Two periods of mite dispersal are important. The first occurs as wheat is approaching maturity when populations are very high, and mites need to find green bridge hosts to survive the summer. The optimal host at this time is volunteer wheat, and if present, it will become readily infested with mites. The second important dispersal period is when mites move from their green bridge host to the emerging winter wheat in the fall. The density of mites in source fields at this time and the wind conditions through the fall are critical factors in determining the degree of dispersal and the spatial patterns of virus spread surrounding these mite source fields. Additional management tools (e.g. improved mite and virus resistance) would be advantageous to enhance the durability of management programs for this mite-virus complex.

11 Future trends in research

Recent reviews cited have addressed research needs in various areas. Additionally, many areas in mite ecology and management need to be addressed to optimize management for this mite-virus complex. It will be important that more diverse and effective management options are available as we move into the future.

- Quantification of the mite's relationship to potential green bridge hosts is needed. Most available mite-host data provide little quantification of the relationship and make risk predictions difficult. Further research is needed for:
    - a greater understanding of the reproductive potential of WCM on alternate hosts across various host growth stages;
    - quantification of adaptation requirements when WCM changes hosts;
    - impact of mite haplotypes (e.g. Type 1, Type 2 and potentially others) on these host relationships; and
    - a more complete understanding of the impact of virus infection, both single and in combination, on these host relationships and on the epidemiology of the virus surrounding the green bridge hosts.
- Wheat curl mite dispersal largely drives the risk for this virus complex. Because of the importance of mite density to dispersal, risk assessment efforts will require:
    - modeling mite buildup on and dispersal from green bridge hosts in the field; and
    - quantification of the factors impacting dispersal and colonization success.

- Most of the emphasis on WCM resistance in wheat has focused primarily on antibiotic resistance. However, this resistance strategy is very prone to adaptation due to the strong selection pressure involved. Research needs include:
  - a more complete understanding of the physiological interactions between WCM feeding and wheat for both WCM-resistant and susceptible wheat lines;
  - an understanding of the mechanisms of WCM resistance genes in wheat and its close relatives, especially resistance mechanisms that would minimize selection pressure (e.g. tolerance sources);
  - an understanding of the variability in WCM response to mite resistance genes across the known mite haplotypes and the ability of these populations to adapt to these genes;
  - a more complete understanding of mite genetic diversity and gene flow will better enable predictions of the stability of mite resistance and the potential for effective pyramiding of resistance genes; and
  - identifying the impact of virus presence on wheat-mite dynamics and the interactions between WCM and the presence of virus resistance genes.

12 Where to look for further information

More information can be found in the following reviews that provide an exposure to the literature and past and current research activities on this wheat-mite-virus complex:

- Navia et al. (2013) - comprehensive review of the mite and its transmitted viruses from a world perspective as it relates to occurrence, impact and management;
- Hein et al. (2023) - general extension review focusing on managing the mite-virus complex in the Great Plains of the USA; and
- other reviews address more specific mite-virus interactions (Stenger et al. 2016; Tatineni and Hein, 2021a), WSMV transmission by WCM (Tatineni and Hein, 2018), HPWMoV (Tatineni and Hein, 2021b) and relationships between WCM and wheat, its primary host (Skoracka et al., 2018b; Nachappa et al., 2021).

Research activities around the world have been limited to a few concentrated research groups with some additional research groups addressing recent introductions of either the mites or one of the viruses. Primary research groups that have focused on WCM and/or its virus complex include:

- University of Nebraska - Lincoln, Department of Entomology, Vector Ecology Lab (https://entomology.unl.edu/);

- USDA-ARS, Wheat, Sorghum and Forage Research Unit, Lincoln, NE; Virology and Wheat Genetics expertise (https://www.ars.usda.gov/plains-area/lincoln-ne/wheat-sorghum-and-forage-research/);
- several private and public wheat breeding programs across the Great Plains of the USA have efforts to improve mite and/or virus resistance in wheat; and
- Adam Mickiewicz University, Poznań, Poland, Population Ecology Lab (http://popecol.home.amu.edu.pl/).

13 References

Alemandri, V., Mattio, M. F., Rodriguez, S. M. and Truol, G. (2017). Geographical distribution and first molecular detection of an Emaravirus, high plains wheat mosaic virus in Argentina, *Eur. J. Plant Pathol.* 149(3), 743–750.

Amrine, J. W. and Stasny, T. A. (1994). *Catalog of the Eriophyoidea (Acarina: Prostigmata) of the World*. Indira Publishing House, West Bloomfield, MI, p. 798.

Benmokhtar, K. and Yahia, A. A. (2009). Contribution to the study of cereal viruses by the biological characterization of wheat and barley mosaic viruses: WSMV, WSSMV and BMSV in the central zone of Algeria. In: *9ème conférence international sur les maladies des plantes*. Association FrançAise de Protection Des Plantes, Tours, France, pp. 48–57.

Brey, C. W., Johnson, G. D. and Blodgett, S. L. (1998). Survey of Montana grasses for wheat curl mite (Acari: Eriophyidae), the vector of Wheat streak mosaic virus, *J. Agric. Entomol.* 15(3), 173–181.

Burrows, M., Franc, G., Rush, C., Blunt, T., Ito, D., Kinzer, K., Olson, J., O'Mara, J., Price, J., Tande, C., Ziems, A. and Stack, J. (2009). Occurrence of viruses in wheat in the Great Plains region, 2008, *Plant Health Prog.* 10(1), 14. https://doi.org/10.1094/PHP-2009-0706-01-RS.

Byamukama, E., Seifers, D. L., Hein, G. L., De Wolf, E., Tisserat, N. A., Langham, M. A. C., Osborne, L. E., Timmerman, A. and Wegulo, S. N. (2013). Occurrence and distribution of Triticum mosaic virus in the central great plains, *Plant Dis.* 97(1), 21–29.

Byamukama, E., Tatineni, S., Hein, G. L., Graybosch, R. A., Baenziger, P. S., French, R. and Wegulo, S. N. (2012). Effects of single and double infections of winter wheat by Triticum mosaic virus and Wheat streak mosaic virus on yield determinants, *Plant Dis.* 96(6), 859–864.

Byamukama, E., Tatineni, S., Hein, G. L., McMechan, J. and Wegulo, S. N. (2016). Incidence of Wheat streak mosaic virus, Triticum mosaic virus, and Wheat mosaic virus in wheat curl mites recovered from maturing winter wheat spikes, *Plant Dis.* 100(2), 318–323. http://doi.org/10.1094/PDIS-06-15-0692-RE.

Byamukama, E., Wegulo, S. N., Tatineni, S., Hein, G. L., Graybosch, R. A., Baenziger, P. S. and French, R. (2014). Quantification of yield loss caused by Triticum mosaic virus and Wheat streak mosaic virus in winter wheat under field conditions, *Plant Dis.* 98(1), 127–133.

Carew, M., Schiffer, M., Umina, P., Weeks, A. and Hoffmann, A. (2009). Molecular markers indicate that the wheat curl mite, *Aceria tosichella* Keifer, may represent a species complex in Australia, *Bull. Entomol. Res.* 99(5), 479–486.

Carrera, S. G., Davis, H., Aguirre-Rojas, L., Murugan, M. and Smith, C. M. (2012). Multiple categories of resistance to wheat curl mite (Acari: Eriophyidae) expressed in accessions of *Aegilops tauschii*, *J. Econ. Entomol.* 105(6), 2180–2186.

Carver, B. F., Smith, C. M., Chang, W., Hunger, R. M., Edwards, J. T., Yan, L., Brown-Guedira, G., Gill, B. S., Bai, G. and Bowden, R. L. (2016). Registration of OK05312, a high-yielding hard winter wheat donor of Cmc4 for wheat curl mite resistance, *J. Plant Reg.* 10(1), 75–79.

Choi, I. R., Horken, K. M., Stenger, D. C. and French, R. (2002). Mapping of the P1 proteinase cleavage site in the polyprotein of Wheat streak mosaic virus (genus Tritimovirus), *J. Gen. Virol.* 83(2), 443–450.

Christian, M. L. and Willis, W. G. (1993). Survival of wheat streak mosaic virus in grass hosts in Kansas for wheat harvest to fall wheat emergence, *Plant Dis.* 77(3), 239–242.

Coutts, B. A., Banovic, M., Kehoe, M. A., Severtson, D. L. and Jones, R. A. C. (2014). Epidemiology of Wheat streak mosaic virus in wheat in a Mediterranean-type environment, *Eur. J. Plant Pathol.* 140(4), 797–813.

Coutts, B. A., Strickland, G. R., Kehoe, M. A., Severtson, D. L. and Jones, R. A. C. (2008). The epidemiology of wheat streak mosaic virus in Australia: Case histories, gradients, mite vectors, and alternative hosts, *Aust. J. Agric. Res.* 59(9), 844–853.

Cox, T. S., Bockus, W. W., Gill, B. S., Sears, R. G., Harvey, T. L., Leath, S. and Brown-Guedira, G. L. (1999). Registration of KS96WGRC40 hard red winter wheat germplasm resistant to wheat curl mite, Stagonospora leaf blotch, and Septoria leaf blotch, *Crop Sci.* 39(2), 597.

Danilova, T. V., Zhang, G., Liu, W., Friebe, B. and Gill, B. S. (2017). Homoeologous recombination-based transfer and molecular cytogenetic mapping of a wheat streak mosaic virus and Triticum mosaic virus resistance gene Wsm3 from *Thinopyrum intermedium* to wheat, *Theor. Appl. Genet.* 130(3), 549–556.

del Rosario, M. S. and Sill, W. H. (1965). Physiological strains of *Aceria tulipae* and their relationships to transmission of wheat streak mosaic virus, *Phytopathology* 55, 1168–1175.

Dwyer, G. I., Gibbs, M. J., Gibbs, A. J. and Jones, R. A. C. (2007). Wheat streak mosaic virus in Australia: Relationship to isolates from the Pacific Northwest of the USA and its dispersion via seed transmission, *Plant Dis.* 91(2), 164–170.

Ellis, M. H., Rebetzke, G. J., Mago, R. and Chu, P. (2003). First report of Wheat streak mosaic virus in Australia, *Australas. Plant Pathol.* 32(4), 551–553.

Friebe, B., Qi, L. L., Wilson, D. L., Chang, Z. J., Seifers, D. L., Martin, T. J., Fritz, A. K. and Gill, B. S. (2009). Wheat-*Thinopyrum intermedium* recombinants resistant to Wheat streak mosaic virus and Triticum mosaic virus, *Crop Sci.* 49(4), 1221–1226.

Frost, W. E. (1995). *The ecology of cereal rust mite Abacarus hystrix (Nalepa) in irrigated perennial pastures in South Australia*. Ph.D. Dissertation, University of Adelaide, Adelaide.

Gibson, W. W. and Painter, R. H. (1957). Transportation by aphids of the wheat curl mite, *Aceria tulipae* (K.), a vector of the Wheat streak mosaic virus, *J. Kans. Entomol. Soc.* 30(4), 147–153.

Gill, B. S., Friebe, B., Wilson, D. L., Martin, T. J. and Cox, T. S. (1995). Registration of KS93WGRC27 wheat streak mosaic virus resistant T4DL-4Ai# 2S wheat germplasm, *Crop Sci.* 35(4), 1236–1237.

Gillespie, R. L., Roberts, D. E. and Bentley, E. M. (1997). Population dynamics and dispersal of wheat curl mite (Acari: Eriophyidae) in north central Washington, *J. Kans. Entomol. Soc.* 70(4), 361–364.

Goetz, R. and Maiss, E. (1995). The complete nucleotide-sequence and genomic organization of the mite-transmitted brome streak mosaic Rymovirus in comparison with those of potyviruses, *J. Gen. Virol.* 76(8), 2035–2042.

Graybosch, R. A., Peterson, C. J., Baenziger, P. S., Baltensperger, D. D., Nelson, L. A., Jin, Y., Kolmer, J., Seabourn, B., French, R. C., Hein, G. L., Martin, T. J., Beecher, B., Schwarzacher, T. and Heslop-Harrison, P. (2009). Registration of "mace" hard red winter wheat, *J. Plant Reg.* 3(1), 51–56.

Hadi, B. A. R., Langham, M. A. C., Osborne, L. and Tilmon, K. J. (2011). Wheat streak mosaic virus on wheat: Biology and management, *J. Integr. Pest Manag.* 2(1), J1–J5. https://doi.org/10.1603/IPM10017.

Haley, S. D., Martin, T. J., Quick, J. S., Seifers, D. L., Stromberger, J. A., Clayshulte, S. R., Clifford, B. L., Peairs, F. B., Rudolph, J. B., Johnson, J. J., Gill, B. S. and Friebe, B. (2002). Registration of C0960293-2 wheat germplasm resistant to wheat streak mosaic virus and Russian wheat aphid, *Crop Sci.* 42(4), 1381–1382.

Harvey, T. L., Martin, T. J. and Seifers, D. L. (1995). Survival of five wheat curl mite, *Aceria tosichilla* Keifer (Acari: Eriophyidae), strains on mite resistant wheat, *Exp. Appl. Acarol.* 19(8), 459–463.

Harvey, T. L., Martin, T. J. and Seifers, D. L. (2000). Effect of nonviruliferous wheat curl mites on yield of winter wheat, *J. Agric. Urban Entomol.* 17(1), 9–13.

Harvey, T. L., Martin, T. J. and Seifers, D. L. (2003). Resistance to the wheat curl mite (Acari: Eriophyidae) prevents loss in wheat yield, *J. Agr. Urban Entomol.* 20(1), 7–10.

Harvey, T. L., Martin, T. J., Seifers, D. L. and Sloderbeck, P. E. (1997). Change in virulence of wheat curl mite detected on TAM 107 wheat, *Crop Sci.* 37(2), 624–625.

Harvey, T. L., Seifers, D. L. and Martin, T. J. (2001). Host range differences between two strains of wheat curl mites (Acari: Eriophyidae), *J. Agric. Urban Entomol.* 18(1), 35–41.

Harvey, T. L., Seifers, D. L. and Martin, T. J. (2002). Wheat yield reduction due to wheat curl mite (Acari: Eriophyidae) infestations, *J. Agr. Urban Entomol.* 19(1), 9–13.

Harvey, T. L., Seifers, D. L., Martin, T. J., Brown-Guedira, G. and Gill, B. S. (1999). Survival of wheat curl mites on different sources of resistance in wheat, *Crop Sci.* 39(6), 1887–1889.

Haskell, R. J. and Wood, J. I. (1923). Diseases of cereal and forage crops in the United States in 1922, *USDA Bur. Plant Ind. Plant Dis. Bul. Sup.* 27, 164–226.

Hein, G. L., McKelvy, U., Wegulo, S., Hunger, R., Burrows, M., Rush, C., Workneh, F., Brorsen, W., Fiskop, A. and Byamukama, E. (2023). *Mite-Transmitted Virus Disease Complex of Wheat in the Great Plains of the United States*. Ext. Circular 3063, University Nebraska Extension, Lincoln.

Hein, G. L., French, R., Siriwetwiwat, B. and Amrine, J. W. (2012). Genetic characterization of North American populations of the wheat curl mite and dry bulb mite, *J. Econ. Entomol.* 105(5), 1801–1808.

Helle, W. and Wysoki, M. (1983). The chromosomes and sex-determination of some actinotrichid taxa (Acari), with special reference to Eriophyidae, *Int. J. Acarol.* 9(2), 67–71.

Hollandbeck, G. F., Onofre, K. A., DeWolf, E. and Todd, T. (2020). Preliminary 2020 Kansas wheat disease loss estimates, *Kans. Coop. Plant Dis. Rep., Kansas Dept. Agriculturists*, October 9, p. 6.

Hunger, R. M., Sherwood, J. L., Evans, C. K. and Montana, J. R. (1992). Effects of planting date and inoculation date on severity of wheat streak mosaic in hard red winter wheat cultivars, *Plant Dis.* 76(10), 1056–1060.

Jensen, S. G., Lane, L. C. and Seifers, D. L. (1996). A new disease of maize and wheat in the high plains, *Plant Dis.* 80(12), 1387–1390.

Jiang, W., Garrett, K. A., Peterson, D. E., Harvey, T. L., Bowden, R. L. and Fang, L. (2005). The window of risk for emigration of Wheat streak mosaic virus varies with host eradication method, *Plant Dis.* 89(8), 853–858.

Jones, R. A. C., Coutts, B. A., Mackie, A. E. and Dwyer, G. I. (2005). Seed transmission of Wheat streak mosaic virus shown unequivocally in wheat, *Plant Dis.* 89(10), 1048–1050.

Kapooria, R. G. and Ndunguru, J. (2004). Occurrence of viruses in irrigated wheat in Zambia, *EPPO Bull.* 34(3), 413–419.

Keifer, H. H. (1939). Eriophyid studies, *Bull. Calif. Dep. Agric.* 28, 329.

Keifer, H. H. (1953). Eriophyid studies XXI, *Bull. Dep. Agric. State Calif.* 42, 65-79.

Keifer, H. H. (1969). *Eriophyid Studies C-3 (Special Publication).* Agricultural Research Service, United States Department of Agriculture.

Khalaf, L., Chuang, W., Aguirre-Rojas, L. M., Klein, P. and Smith, C. M. (2019). Differences in *Aceria tosichella* population responses to wheat resistance genes and wheat virus transmission, *Arthropod Plant Interact.* 13(6), 807-818.

Kiani, M., Bryan, B., Rush, C. and Szczepaniac, A. (2021). Transcriptional responses of resistant and susceptible wheat exposed to wheat curl mite, *Int. J. Mol. Sci.* 22(5), 2703. https://doi.org/10.3390/ijms22052703.

Kiedrowicz, A., Kuczyński, L., Lewandowski, M., Proctor, H. and Skoracka, A. (2017). Behavioural responses to potential dispersal cues in two economically important species of cereal-feeding eriophyid mites, *Sci. Rep.* 7(1), 3890. https://doi.org/10.1038/s41598-017-04372-7.

Knoell, E. A. (2018). *Transmission characteristics of Triticum mosaic virus by the wheat curl mite, Aceria tosichella Keifer, and ecology of the wheat-mite-virus complex on field corn.* MS Thesis, University Nebraska, Lincoln, p. 103.

Kuczyński, L., Radwańska, A., Karpicka-Ignatowska, K., Laska, A., Lewandowski, M., Rector, B. G., Majer, A., Raubic, J. and Skoracka, A. (2020). A comprehensive and cost-effective approach for investigating passive dispersal in minute invertebrates with case studies of phytophagous eriophyid mites, *Exp. Appl. Acarol.* 82(1), 17–31.

Kuczyński, L., Rector, B. G., Kiedrowicz, A., Lewandowski, M., Szydło, W. and Skoracka, A. (2016). Thermal niches of two invasive genotypes of the wheat curl mite *Aceria tosichella*: Congruence between physiological and geographical distribution data, *PLoS ONE* 11(4), e0154600.

Kumssa, T. T., Rupp, J. S., Fellers, M. C., Fellers, J. P. and Zhang, G. (2019). An isolate of Wheat streak mosaic virus from foxtail overcomes Wsm2 resistance in wheat, *Plant Pathol.* 68(4), 783–789.

Kumssa, T. T., Zhao, D., Bai, G. and Zhang, G. (2017). Resistance to Wheat streak mosaic virus and Triticum mosaic virus in wheat lines carrying Wsm1 and Wsm3, *Eur. J. Plant Pathol.* 147(3), 709–712.

Laska, A., Rector, B. G., Skoracka, A. and Kuczyński, L. (2019). Can your behaviour blow you away? Contextual and phenotypic precursors to passive aerial dispersal in phytophagous mites, *Anim. Behav.* 155, 141–151.

Liu, W., Seifers, D. L., Qi, L. L., Friebe, B. and Gill, B. S. (2011). A compensating wheat-*Thinopyrum intermedium* Robertsonian translocation conferring resistance to *Wheat steak mosaic virus* and *Triticum* mosaic virus, *Crop Sci.* 51(6), 2382-2390.

Lu, H., Price, J., Devkota, R., Rush, C. and Rudd, J. (2011). A dominant gene for resistance to in winter wheat line CO960293-2, *Crop Sci.* 51(1), 5-12.

Mahmood, T., Hein, G. L. and Jensen, S. G. (1998). Mixed infection of hard red winter wheat with high plains virus and wheat streak mosaic virus from wheat curl mites in Nebraska, *Plant Dis.* 82(3), 311-315.

Majer, A., Laska, A., Hein, G. L., Kuczynski, L. and Skoracka, A. (2021a). Propagule pressure rather than population growth determines colonization ability: A case study using two phytophagous mite species differing in their invasive potential, *Ecol. Entomol.* 46(5), 1136-1147.

Majer, A., Laska, A., Hein, G. L., Kuczynski, L. and Skoracka, A. (2021b). Hitchhiking or hang gliding? Dispersal strategies of two cereal-feeding eriophyoid mite species, *Exp. Appl. Acarol.* 85(2-4), 131-146.

Malik, R., Brown-Guedira, G. L., Smith, C. M., Harvey, T. L. and Gill, B. S. (2003). Genetic mapping of wheat curl mite resistance genes and in common wheat, *Crop Sci.* 43(2), 644-650.

Martin, T. J., Harvey, T. L., Bender, C. G., Seifers, D. L. and Hatchett, J. H. (1983). Wheat curl mite resistant wheat germplasm, *Crop Sci.* 23(4), 809.

McMechan, A. J. (2016). *Over-summering ecology of the wheat curl mite (Aceria tosichella Keifer)*. PhD Dissertation, University Nebraska, Lincoln, p. 238.

McMechan, A. J. and Hein, G. L. (2016). Planting date and variety selection for management of viruses transmitted by the wheat curl mite (Acari: Eriophyidae), *J. Econ. Entomol.* 109(1), 70-77.

McMechan, A. J. and Hein, G. L. (2017). Population dynamics of the wheat curl mite (Acari: Eriophyidae) during the heading stages of winter wheat, *J. Econ. Entomol.* 110(2), 355-361.

McMechan, A. J., Tatineni, S., French, R. and Hein, G. L. (2014). Differential transmission of Triticum mosaic virus by wheat curl mite populations collected in the Great Plains, *Plant Dis.* 98(6), 806-810.

McNeil, J. E., French, R., Hein, G. L., Baenziger, P. S. and Eskridge, K. M. (1996). Characterization of genetic variability among natural populations of wheat streak mosaic virus, *Phytopathology* 86(11), 1222-1227.

Michalska, K., Skoracka, A., Navia, D. and Amrine, J. W. (2010). Behavioural studies on eriophyoid mites: An overview, *Exp. Appl. Acarol.* 51(1-3), 31-59.

Murray, G. M. and Brennan, J. P. (2009). *The Current and Potential Costs From Diseases of Wheat Tin Australia*. Grains Research and Development Corp, Kingston, ACT, p. 69.

Murugan, M., Cardona, P. S., Duraimurugan, P., Whitfield, A. E., Schneweis, D., Starkey, S. and Smith, C. M. (2011). Wheat curl mite resistance: Interactions of mite feeding with wheat streak mosaic virus infection, *J. Econ. Entomol.* 104(4), 1406-1414.

Nachappa, P., Haley, S. and Pearce, S. (2021). Resistance to the wheat curl mite and mite-transmitted viruses: Challenges and future directions, *Curr. Opin. Insect Sci.* 45, 21-27.

Nault, L. R. and Styer, W. E. (1969). The dispersal of *Aceria tulipae* and three other grass-infesting eriophyid mites in Ohio, *Ann. Entomol. Soc. Am.* 62(6), 1446-1455.

Navia, D., de Mendonça, R. S., Skoracka, A., Szydło, W., Knihinicki, D., Hein, G. L., da Silva Pereira, P. R. V., Truol, G. and Lau, D. (2013). Wheat curl mite, *Aceria tosichella*, and transmitted viruses: An expanding pest complex affecting cereal crops, *Exp. Appl. Acarol.* 59(1-2), 95-143.

Navia, D., Truol, G., Mendoça, R. S. and Sagadin, M. (2006). *Aceria tosichella* Keifer (Acari: Eriophyidae) from wheat streak mosaic virus-infected wheat plants in Argentina, *Int. J. Acarol.* 32(2), 189-193.

Newkirk, R. A. and Keifer, H. H. (1971). Revision of types of *Eriophyes* and *Phytoptus*, *USDA-ARS Spec. Publ. Eriophyid Stud.* C5, 1-10.

Nguy-Robertson, A. L., Zygielbaum, A. I., McMechan, A. J., Hein, G. L., Wegulo, S. N., Stilwell, A. R. and Smith, T. M. (2016). Developing the framework for a risk map for mite vectored viruses in wheat resulting from pre-harvest hail damage, *Crop Prot.* 89, 21-31.

Oldfield, G. N. and Proeseler, G. (1996). Eriophyoid mites as vectors of plant pathogens. In: Lindquist, E. E., Sabelis, M. W. and Bruin, J. (Eds.), *Word Crop Pests Eriophyoid Mites: Their Biology, Natural Enemies and Control* (Vol. 6). Elsevier, Amsterdam, pp. 259-273.

Orlob, G. B. (1966). Feeding and transmission characteristics of *Aceria tulipae* keifer as vector of wheat streak mosaic virus, *J. Phytopathol.* 55(3), 218-238.

Overmyer, L. M. (2020). *Factors influencing wheat curl mite, Aceria tosichella Keifer, dispersal*. MS Thesis, University Nebraska, Lincoln, p. 100.

Pady, S. M. (1955). The occurrence of the vector of wheat streak mosaic, *Aceria tulipae*, on slides exposed to the air, *Plant. Dis. Report.* 39(4), 296-297.

Porter, K. B., Worrall, W. D., Gardenhire, J. H., Gilmore, E. C., McDaniel, M. E. and Tuleen, N. A. (1987). Registration of TAM 107 wheat, *Crop Sci.* 27, 818-819.

Rashad, A., Liang, X. and Marshall, J. M. (2019). *Wheat Streak Mosaic Virus Management in Idaho Cereals, BUL 941*. University Idaho Ext., Moscow.

Sabelis, M. W. and Bruin, J. (1996). Evolutionary ecology: Life history patterns, food plant choice and dispersal. In: Lindquist, E. E., Sabelis, M. W. and Bruin, J. (Eds.), *Word Crop Pests Eriophyoid Mites: Their Biology, Natural Enemies and Control* (Vol. 6). Elsevier, Amsterdam, pp. 329-366.

Schiffer, M., Umina, P., Carew, M., Hoffmann, A., Rodoni, B. and Miller, A. (2009). The distribution of wheat curl mite (*Aceria tosichella*) lineages in Australia and their potential to transmit Wheat streak mosaic virus, *Ann. Appl. Biol.* 155(3), 371-379.

Seifers, D. L., Harvey, T. L., Louie, R., Gordon, D. T. and Martin, T. J. (2002). Differential transmission of isolates of the High Plains virus by different sources of wheat curl mites, *Plant Dis.* 86(2), 138-142.

Seifers, D. L., Harvey, T. L., Martin, T. J. and Jensen, S. G. (1997). Identification of the wheat curl mite as the vector of the High Plains virus of corn and wheat, *Plant Dis.* 81(10), 1161-1166.

Seifers, D. L., Harvey, T. L., Martin, T. J. and Jensen, S. G. (1998). A partial host range of the High Plains virus of corn and wheat, *Plant Dis.* 82(8), 875-879.

Seifers, D. L., Martin, T. J. and Fellers, J. P. (2011). Occurrence and yield effects of wheat infected with Triticum mosaic virus in Kansas, *Plant Dis.* 95(2), 183-188.

Seifers, D. L., Martin, T. J. and Fellers, J. P. (2010). An experimental host range for Triticum mosaic virus, *Plant Dis.* 94(9), 1125-1131.

Seifers, D. L., Martin, T. J. and Haber, S. (2013). Temperature-sensitive resistance to Wheat streak mosaic virus in CO960333 and KS06HW79 wheat, *Plant Dis.* 97(7), 983-987.

Seifers, D. L., Martin, T. J., Harvey, T. L. and Gill, B. S. (1995). Temperature sensitivity and efficacy of *Wheat streak mosaic virus* resistance derived from *Agropyron intermedium*, *Plant Dis.* 79(11), 1104-1106.

Seifers, D. L., Martin, T. J., Harvey, T. L., Fellers, J. P. and Michaud, J. P. (2009). Identification of the wheat curl mite as the vector of Triticum mosaic virus, *Plant Dis.* 93(1), 25-29.

Seifers, D. L., Martin, T. J., Harvey, T. L., Fellers, J. P., Stack, J. P., Ryba-White, M., Haber, S., Krokhin, O., Spicer, V., Lovat, N., Yamchuk, A. and Standing, K. G. (2008). Triticum mosaic virus: A new virus isolated from wheat in Kansas, *Plant Dis.* 92(5), 808-817.

Seifers, D. L., Martin, T. J., Harvey, T. L., Haber, S. and Haley, S. D. (2006). Temperature sensitivity and efficacy of Wheat streak mosaic virus resistance derived from CO960293 wheat, *Plant Dis.* 90(5), 623-628.

Shahwan, I. M. and Hill, J. P. (1984). Identification and occurrence of wheat streak mosaic virus in winter wheat in Colorado and its effects on several wheat cultivars, *Plant Dis.* 68(7), 579-581.

Silva, P. (2021). *Breeding strategies for improving pest and disease resistance in wheat*. PhD Dissertation, Kansas State University, Manhattan, p. 188.

Singh, K., Wegulo, S. N., Skoracka, A. and Kundu, J. K. (2018). Wheat streak mosaic virus: A century old virus with rising importance worldwide, *Mol. Plant Pathol.* 19(9), 2193-2206.

Siriwetwiwat, B. (2006). *Interactions between the wheat curl mite, Aceria tosichella Keifer (Eriophyidae), and wheat streak mosaic virus and distribution of wheat curl mite biotypes in the field*. PhD Dissertation, University Nebraska, Lincoln, p. 172.

Skoracka, A., Kuczyński, L., de Mendonca, R., Dabert, M., Szydło, W., Knihinicki, D., Truol, G. and Navia, D. (2012). Cryptic species within the wheat curl mite *Aceria tosichella* (Keifer) (Acari, Eriophyoidea) revealed by mitochondrial, nuclear and morphometric data, *Invert. Syst.* 26(4), 417-433.

Skoracka, A., Kuczyński, L., Szydło, W. and Rector, B. (2013). The wheat curl mite *Aceria tosichella* (Acari: Eriophyoidea) is a complex of cryptic lineages with divergent host ranges: Evidence from molecular and plant bioassay data, *Biol. J. Linn. Soc. Lond.* 109(1), 165-180.

Skoracka, A., Lopes, L. F., Alves, M. J., Miller, A., Lewandowski, M., Szydło, W., Majer, A., Różańska, E. and Kuczyński, L. (2018a). Genetics of lineage diversification and the evolution of host usage in the economically important wheat curl mite, *Aceria tosichella* Keifer, 1969, *BMC Evol. Biol.* 18(1), 122. https://doi.org/10.1186/s12862-018-1234-x.

Skoracka, A., Rector, B. G. and Hein, G. L. (2018b). The interface between wheat and the wheat curl mite, *Aceria tosichella*, the primary vector of globally important diseases, *Front. Plant Sci.* 9, 1098. https://doi.org/10.3389/fpls.2018.01098.

Skoracka, A., Rector, B., Kuszynski, L., Szydło, W., Hein, G. L. and French, R. (2014). Global spread of wheat curl mite by its most polyphagous and pestiferous lineages, *Ann. Appl. Biol.* 165(2), 222-235.

Slykhuis, J. T. (1955). *Aceria tulipae* Keifer (Acarina: Eriophyidae) in relation to the spread of wheat streak mosaic, *Phytopathology* 45(3), 116-128.

Snihur, H., Pozhylov, I., Budzanivska, I. and Shevchenko, O. (2020). First report of High Plains wheat mosaic virus on different hosts in Ukraine, *J. Plant Pathol.* 102(2), 545-546.

Somsen, H. W. and Sill, W. H. (1970). *The Wheat Curl Mite, Aceria Tulipae Keifer, in Relation to Epidemiology and Control of Wheat Streak Mosaic*. Kansas State University, Agricultural Experiment Station, Manhattan, p. 28.

Staples, R. and Allington, W. B. (1956). Streak mosaic of wheat in Nebraska and its control, *Res. Bull.* 178, 40.

Stenger, D. C., Hein, G. L., Tatineni, S. and French, R. (2016). Eriophyid mite vectors of plant viruses. In: Brown, J. K. (Ed.), *Vector-Mediated Transmission of Plant Pathogens*. The American Phytopathological Society, St. Paul, pp. 263–274.

Stilwell, A. R. (2009). *Remote sensing to detect the movement of wheat curl mites through the spatial spread of virus symptoms, and identification of thrips as predators of wheat curl mites*. PhD Dissertation, University Nebraska, Lincoln, p. 179.

Stilwell, A. R., Hein, G. L., Zygielbaum, A. I. and Rundquist, D. C. (2013). Proximal sensing to detect symptoms associated with wheat curl mite-vectored viruses, *Int. J. Remote Sens.* 34(14), 4951–4966.

Stilwell, A. R., Rundquist, D. C., Marx, D. B. and Hein, G. L. (2019). Differential spatial gradients of Wheat streak mosaic virus into winter wheat from a central mite-virus source, *Plant Dis.* 103(2), 338–344.

Szydło, W., Hein, G. L., Denizhan, E. and Skoracka, A. (2015). Exceptionally high levels of genetic diversity in wheat curl mite (Acari: Eriophyidae) populations from Turkey, *J. Econ. Entomol.* 108(4), 2030–2039.

Tatineni, S., Graybosch, R. A., Hein, G. L., Wegulo, S. N. and French, R. (2010). Wheat cultivar-specific disease synergism and alteration of virus accumulation during co-infection with Wheat streak mosaic virus and Triticum mosaic virus, *Phytopathology* 100(3), 230–238.

Tatineni, S. and Hein, G. L. (2018). Genetics and mechanisms underlying transmission of Wheat streak mosaic virus by the wheat curl mite, *Curr. Opin. Virol.* 33, 47–54.

Tatineni, S. and Hein, G. L. (2021a). Tritimoviruses and rymoviruses. In: Bamford, D. H. and Zuckerman, M. (Eds.), *Encyclopedia of Virology* (4th edn., Vol. 3). Academic Press, Oxford, pp. 797–804.

Tatineni, S. and Hein, G. L. (2021b). High plains wheat mosaic virus: An enigmatic disease of wheat and corn causing the high plains disease, *Mol. Plant Pathol.* 22(10), 1167–1179.

Tatineni, S., McMechan, A. J., Wosula, E. N., Wegulo, S. N., Graybosch, R. A., French, R. and Hein, G. L. (2014). An eriophyid mite-transmitted plant virus contains eight genomic RNA segments with unusual heterogeneity in the nucleocapsid protein, *J. Virol.* 88(20), 11834–11845.

Tatineni, S., Ziems, A. D., Wegulo, S. N. and French, R. (2009). Triticum mosaic virus: A distinct member of the family Potyviridae with an unusually long leader sequence, *Phytopathology* 99(8), 943–950.

Thomas, J. A. and Hein, G. L. (2003). Influence of volunteer wheat plant condition on movement of the wheat curl mite, *Aceria tosichella*, in winter wheat, *Exp. Appl. Acarol.* 31(3–4), 253–268.

Thomas, J. A., Hein, G. L. and Lyon, D. J. (2004). Spread of wheat curl mite and Wheat streak mosaic virus is influenced by volunteer wheat control methods, *Plant Health Prog.* 5(1). https://doi.org/10.1094/PHP-2004-1206-01-RS.

Thomas, J. B. and Conner, R. L. (1986). Resistance to colonization by the wheat curl mite in *Aegilops squarrosa* and its inheritance after transfer to common wheat 1, *Crop Sci.* 26(3), 527–530.

Thomas, J. B., Conner, R. L. and Graf, R. J. (2012). Radiant hard red winter wheat, *Can. J. Plant Sci.* 92(1), 169–175.

Townsend, L., Johnson, D. and Hershman, D. (1996). *Wheat Streak Mosaic Virus and the Wheat Curl Mite, ENTFACT-117*. University Kentucky Ext., Lexington.

Truol, G., French, R., Sagadin, M. and Arneodo, J. (2004). First report of Wheat streak mosaic virus infecting wheat in Argentina, *Australas. Plant Pathol.* 33(1), 137–138.

Umina, P. A., Schiffer, M., Parker, P. and Hoffmann, A. A. (2016). Distribution and influence of grazing on wheat curl mites (*Aceria tosichella* Keifer) within a wheat field, *J. Appl. Entomol.* 140(6), 426–433.

United States Department of Agriculture – Nationals Agricultural Statistics Service. (2010). Field crops: Usual planting and harvesting dates, USDA-NASS Agric. Handbook No. 628, October, p. 51.

Velandia, M., Rejesus, R. M., Jones, D. C., Price, J. A., Workneh, F. and Rush, C. M. (2010). Economic impact of Wheat streak mosaic virus in the Texas High Plains, *Crop Prot.* 29(7), 699–703.

Vose, R., Easterling, D., Kunkel, K., LeGrande, A. and Wehner, M. (2017). Temperature changes in the United States. In: *Clim. Sci. Spec. Rep. Fourth Natl. Clim. Assess,* U.S., Global Change Research Program, Washington, DC, pp. 185–206.

Webb, C. A. (2018). *Epidemiological factors impacting the development of Wheat streak mosaic virus outbreaks*. MS Thesis, Kansas State University, Manhattan, p. 88.

Wegulo, S. N., Hein, G. L., Klein, R. N. and French, R. C. (2008). Managing wheat streak mosaic, *EC1871*. University Nebraska Ext., Lincoln.

Whelan, E. D. P. and Hart, G. E. (1988). A spontaneous translocation that transfers wheat curl mite resistance from decaploid *Agropyron elongatum* to common wheat, *Genome* 30(3), 289–292.

Whelan, E. D. P. and Thomas, J. B. (1989). Chromosomal location in common wheat of a gene (Cmc1) from *Aegilops squarrosa* that conditions resistance to colonization by the wheat curl mite, *Genome* 32(6), 1033–1036.

Wosula, E. N., McMechan, A. J. and Hein, G. L. (2015). The effect of temperature, relative humidity, and virus infection status on off-host survival of the wheat curl mite (Acari, Eriophyidae), *J. Econ. Entomol.* 108(4), 1545–1552.

Wosula, E. N., McMechan, A. J., Knoell, E., Tatineni, S., Wegulo, S. N. and Hein, G. L. (2018). Impact of timing and method of virus inoculation on the severity of wheat streak mosaic disease, *Plant Dis.* 102(3), 645–650.

Wosula, E. N., McMechan, A. J., Oliveira-Hofman, C., Wegulo, S. N. and Hein, G. L. (2016). Differential transmission of two isolates of Wheat streak mosaic virus by five wheat curl mite populations, *Plant Dis.* 100(1), 154–158.

Wosula, E. N., Tatineni, S., Wegulo, S. N. and Hein, G. L. (2017). Effect of temperature on wheat streak mosaic disease development in winter wheat, *Plant Dis.* 101(2), 324–330.

Yadava, U. L. (1986). A rapid and nondestructive method to determine chlorophyll in intact leaves, *Hort. Sci.* 21(6), 1449–1450.

Zhao, L., Liu, S., Abdelsalam, N. R., Carver, B. F. and Bai, G. (2021). Characterization of wheat curl mite resistance gene *Cmc4* in OK05312, *Theor. Appl. Genet.* 134(4), 993–1005.

Chapter 12

Advances in managing wireworms in cereal crops: challenges and future directions

Arash Rashed, Virginia Tech, USA; and Erik J. Wenninger, University of Idaho, USA

1 Introduction

There are approximately 100 species of click beetles (Coleoptera: Elateridae) that are known to be pests of agricultural crops during their larval stage, in which these insects are commonly referred to as 'wireworms' (Vernon and van Herk, 2022). The adult beetles can be easily distinguished by their slender and elongated bodies with almost parallel lateral sides (Fig. 1). The side corners of the posterior end of their pronotum are pointed toward the abdomen (Fig. 1). Click beetles can vary in size and color depending on the species; most species are dark in color and do not exceed 2.5 cm in length. Some species may have spots or lines on their pronotum and elytra. Click beetles get their name from their unique defensive mechanism triggered by the forceful release of a ventral peg which locks into the mesosternal lip. In response to stimuli associated with any threat, the beetle flips up into the air making a "click" sound, which may enable them to evade the potential danger. 'Death feigning', or thanatosis, is also used by click beetles as a behavioral defensive mechanism against predation.

Adult click beetles emerge during the spring to mate, after which females deposit their eggs in the soil, near the soil surface (Andrews et al., 2008). In the Pacific Northwest region of the United States, the peak emergence time for adults appears to be between April and June (Andrews et al., 2008). After egg hatching, the emerged larvae can live in the soil for several years before

http://dx.doi.org/10.19103/AS.2022.0114.16

Figure 1 Different developmental stages of the click beetle *Limonius californicus* (Coleoptera: Elateridae); click beetle larva ('wireworm') feeding on a wheat plant (left), pupa (center), and the adult click beetle (right).

they pupate (Becker et al., 1991); the larval stage of the North American wireworm species can range from more than one generation per year to 5 years (Vernon and van Herk, 2022). Depending on the species and quality of the environment, larval development can take more than a decade (Vernon et al., 2009).

Wireworms are protected by their thick integument and high tolerance to adverse environmental conditions. Wireworms can evade cold winter temperatures by moving deeper into the soil profile; in the spring they return to more shallow depths to feed on emerging plants (Traugott et al., 2015). Wireworms may be found in a variety of cultivated and non-cultivated terrestrial habitats where they feed on belowground tissues of a wide range of host plants, which include crops such as potato, legumes, onion, lettuce, carrot, and cereals. Wilting, delayed growth, and seedling death are examples of the general foliar symptoms which may indicate wireworm feeding (Traugott et al., 2015) (Fig. 2). Damage in small grains is commonly reflected in failed emergence and seedling death. The damage in larger seedlings can result in wilting and death of the central leaf, a symptom also known as 'deadheart' (Rashed et al., 2017). Stand thinning reduces yield and allows for the establishment of weeds (Traugott et al., 2015).

Both irrigated and dryland production systems can be affected by wireworms. Because some of the most damaging wireworm species (e.g. *Limonius* spp., *Agriotes* spp.) prefer soil with relatively higher moisture levels, cultural practices such as no-till farming are expected to generate favorable soil conditions for their prolonged activity and damage (Milosavljević et al., 2016b; Vernon and van Herk, 2022). Their multi-year larval stage and

Figure 2 Wireworm (*Limonius californicus*) feeding at the base of a wheat seedling (left), and plant thinning and delayed maturation in a field infested with *Limonius californicus* (right).

wide host ranges mean that wireworms can damage entire rotational systems once they become established in small grain fields.

Decades-long reliance on broad-spectrum insecticides in conventional farming not only derailed research on developing environmentally sensitive control measures but also likely contributed to the current resurgence of this devastating pest, once these chemistries were discontinued (Vernon et al., 2008). Now, with limited effective chemical control replacements (see Section 3.4), recent research has focused on developing IPM protocols to manage wireworms (Andrews et al., 2020; Knodel and Shrestha, 2018; Vernon and Van Herk, 2017).

Another critical issue that complicates wireworm management is that multiple species are often erroneously considered as a single pest. Wireworms are a complex of different species with distinct differences in their ecology and potential to cause damage (Horton, 2006). Differences in species composition and prevalence can be related to cropping systems and other environmental factors (Milosavljević et al., 2016b), as discussed later in this chapter. Among-species differences in wireworm ecology and phenology may explain the observed variations in the efficacy of the applied pest management practices (Esser et al., 2015). Therefore, identifying wireworm species complexes within fields and across regions can help with establishing location-specific management approaches because the makeup of species complexes in a given area tends to remain consistent over time.

The success of any IPM program relies on effective monitoring, pest identification, management guidelines and thresholds, integration of various control practices, and evaluation of our success in protecting crops. Following these principles, this book chapter will provide an overview of the most recent advances in our knowledge of wireworm biology and ecology, which are expected to improve effective monitoring and species identification and facilitate the decision-making process to implement the most appropriate IPM strategies that are species-specific and tailored to geographical locations. The existing challenges and areas needing further research are presented and discussed.

2 Decision-making in wireworm management

'The importance of understanding insect activity to make decisions in insect pest management cannot be overstated. Indeed, this knowledge is indispensable to the paramount principle of pest management: to take no action against a pest unless that pest is known to be present and is an actual or potential threat.' - Pedigo et al. (1996); *Entomology and Pest Management*, Chapter 6.

Accurate pest identification, effective monitoring, and economic thresholds are management decision tools, which, combined with the understanding of environmental variables that influence wireworm infestation and crop damage, can allow for a precise assessment of the risk to the crop. This, however, is not an easy task to achieve, in part, due to wireworms' subterranean habitat, difficulty of surveillance, and our limited species-specific knowledge of their biology and ecology.

2.1 Wireworm species and ecology

2.1.1 Identification

Identifying the species composition of wireworms in any given agroecosystem is a prerequisite for the implementation of an effective management strategy. With more than 10 000 described species worldwide, the click beetle family is one of the most diverse insect taxa. There are several identification keys based on larval morphological traits (e.g. Becker et al., 1991; Glen, 1950; Wilkinson, 1963), which allow identification to genus and species levels. Accurate identification of larvae is important because the larva is the only damaging stage of click beetles. Unfortunately, species identification based on morphological traits of the immature wireworms is challenging and, in many cases, ambiguities remain. Limitation to reliable species identification include: the limited development of some features in early larval instars (Emden, 1945), lack of distinct traits to distinguish some species, and loss or wearing away of some morphological traits as larvae move through soil (Traugott et al., 2015). Recent advances in molecular identification have helped with some of the ambiguities in species identification and have made available powerful tools to explore new areas of investigation such as within-species population structures and dispersal patterns (e.g. Andrews et al., 2020).

Species within the *Agriotes* complex, which are serious pests of agricultural crops in Europe and Canada (Blackshaw and Vernon, 2008), have been successfully identified from the cytochrome oxidase I (COI) and 16S regions of the mitochondrial DNA (Ellis et al., 2009; Staudacher et al., 2011), using terminal restriction fragment length polymorphism (T-RFLP) (Ellis et al., 2009) and multiplex polymerase chain reaction (Staudacher et al., 2011) approaches. Ellis et al. (2009) identified the species *A. obscurus*, *A. sputator* and *A. lineatus*, for which the presence of several haplotypes within species was also detected. Although these three species are known to have behavioral differences, they are not easily morphologically distinguishable, especially in their early larval instars. COI sequences are available for the species in the genera *Aeolus*, *Agriotes*, *Athous*, *Conoderus*, *Dalopius*, *Hadromorphus*, *Hemicrepidius*, *Hypnoidus*, *Melanotus*, *Pseudanostirus* (Etzler et al., 2014; Lindroth and Clark, 2009), and *Limonius* (Gerritsen et al., 2016). The mitochondrial 16S rRNA has been used to distinguish species in the genera *Aeolus*, *Agriotes*, *Hypnoidus*, *Limonius*, *Melanotus*, and *Selatosomus* (Benefer et al., 2013).

Molecular identification methods have been critical in identifying 'cryptic' species that are not distinguishable based on morphological traits. DNA barcoding allowed detection of cryptic species within *Hadromorphus glaucus* Germar and the reemerging *Hypnoidus bicolor* Eschscholtz species complexes (Benefer et al., 2013; Etzler et al., 2014). Most recently, using COI and 16S rRNA sequences along with nuclear genome-wide single nucleotide polymorphisms (SNPs) generated from restriction site associated DNA, cryptic species complexes were detected within *Limonius californicus* and *L. infuscatus* species groups (Andrews et al., 2020). Andrews et al (2020) also identified genetically distinct populations within *L. californicus* and *L. infuscatus* and quantified potential genomic responses to broad-spectrum insecticide treatments, albeit using a limited number of samples in a relatively small geographical range. Future long-term studies with more site-specific samples across populations are warranted to determine the effect of habitat variability on population genetic diversity and identify ecological variables that predict the evolution of adaptations and patterns of dispersal (Balkenhol et al., 2015). Despite their effectiveness, the use of genomic tools in wireworm management has yet to gain traction among researchers in the field. Barriers to widespread adoption of these tools include insufficient genomic resources by researchers and the need for annotated whole genome assemblies of the pestiferous species of wireworms.

2.1.2 Species distribution

There is more to be done to delineate the global distribution of wireworms, especially pest species. Species complexes tend to vary across regions. For example, *Agriotes* spp. are frequently encountered across Europe and eastern and western coastal Canada, whereas the genus *Melanotus* has been reported from Japan and eastern and central United States. *Limonius* spp., *H. bicolor*, and *Agriotes* spp. are examples of species complexes that commonly occur in the western United States and Canada (Andrews et al., 2020; Benefer et al., 2013). Co-infestations by multiple wireworm species are not uncommon (Parker and Howard, 2001; Poggi et al., 2021). Although one species is often the predominant species within each field (Rashed, unpublished data), shifts in species dominance over time may occur due to management practices or other ecological changes.

The distribution of wireworms within fields is often patchy (Benefer et al., 2010; Blackshaw and Vernon, 2008; Cherry and Stansly, 2008; Seal et al., 1997), reflected in the observed patchiness of the damage at the field scale. Due to this nonrandom distribution some sampling approaches are likely not to accurately reflect the existing population density, especially where relatively low, yet still damaging, numbers are present in a field (Yates and Finney, 1942). Even sampling plans designed to overcome the problem of aggregation through trapping or soil core sampling do not necessarily predict the extent of crop damage (Parker, 1996) as the wireworm activity is driven by various environmental factors. The patchiness of wireworm distribution can be influenced by factors such as female dispersal and oviposition site preference, environmental constraints (Blackshaw and Vernon, 2008), as well as the limited mobility of wireworms in the soil after egg hatching (Schallhart et al., 2011).

Adult females are known to be attracted to grasses, a preference that explains wireworm presence in grasslands, including natural vegetation surrounding fields, and/or cereal fields, especially those lacking rotations with non-cereal crops (Blackshaw and Vernon, 2008; Cherry and Stansly, 2008; Milosavljević et al., 2016b; Traugott et al., 2015). Grass pastures and grassy patches within fields (e.g. Conservation Reserve Program [CRP]) and around field margins serve as reservoirs for wireworms, focal points from which wireworms, especially more mobile later instars, can disperse into fields (Vernon and van Herk, 2022; Vernon and van Herk, 2017; Vernon et al., 2014). There, soil factors can influence the extent of wireworm dispersal within its subterranean habitat (Blackshaw and Vernon, 2008).

2.1.3 Phenology

In addition to the pest density, a clear understanding of pest development and seasonal activity in relation to the growth stage of the crop enables the development of phenological models for improving IPM (Johnson et al., 2007; Logan et al., 2006). The phenology of wireworms is species-dependent, and risk of damage also varies according to the developmental stages of crops. In a review of literature on several wireworm species, Traugott et al. (2015) stated that the semivoltine wireworm species of the Holarctic generally have a spring and a fall activity peak. In many regions, these peak activity times may correspond with spring planting (e.g. small grains, corn, wide variety of vegetables) or fall planting (e.g. small grains) of crops. In the case of fall-harvested root crops such as sugar beet or especially potato (Wenninger et al., 2020), a wireworm activity peak during early fall can result in significant quality loss through direct feeding or secondary pathogenic infections.

There are a few variables that should be considered when wireworm activity or presence is monitored and/or predicted in agroecosystems. Wireworm activity, i.e. detectability through various sampling methods (see below), is influenced by molting, food availability, and soil conditions such as moisture levels and temperature (Parker and Howard, 2001; Traugott et al., 2015). Moreover, because wireworm populations belonging to more than one species may be present within a field, multiple peak activity times may occur; one or two samplings, timed for capturing the status of a single species may not provide a complete picture of the extent of the problem. Some wireworm species may maintain their activity throughout the season (Milosavljević et al., 2016b). Several surveys in the Pacific Northwest (PNW) region of the United States, have identified *Limonius* spp. as the most common wireworm genera, among which *L. californicus* is known as the most damaging species, causing severe economic damage especially to small grains (Esser et al., 2015; Milosavljević et al., 2017; Rashed et al., 2017, 2015). In the state of Washington, optimal temperatures for *L. canus* and *L. californicus* larval activity are 21-23°C. Wireworm species in the inland PNW also differ in their moisture needs; *L. californicus* inhabits mostly irrigated soils, whereas *S. pruininus* is an obligate dryland species (Andrews et al., 2008; Milosavljević et al., 2017). Although neonicotinoid seed treatment can result in yield increases of up to 30% in fields infested with *L. californicus*, they do not have the same effect for fields infested with *L. infuscatus* (Esser et al., 2015), and fail to provide crop protection when wireworm populations are very high (van Herk et al., 2018). Differences in phenology between the two species could explain the inconsistency in the efficacy of the neonicotinoid seed treatments; *L. infuscatus* is active and predominantly found in traps in April and May, whereas *L. californicus* is active throughout the season in the inland PNW (Milosavljević et al., 2017). Despite its activity throughout the season, the number of *L. californicus* found in bait traps still appears to be relatively higher between April and May than later in September (Rashed, unpublished data). Inconsistencies in the number of peak activity times have also been reported from Saskatchewan, Canada in *Ctenicera destructor* - an observation that was explained by agricultural practices (i.e. fallow rotation) (Doane, 1981).

2.2 Monitoring

Effective monitoring of wireworm populations within an agroecosystem is arguably one of the most important components of any IPM approach as it provides critical information needed for risk assessment and decision-making. Examples of the types of information that an effective monitoring program is expected to generate include: accurate data on wireworm population and distribution within fields and across landscapes, species composition, seasonal activity patterns (of each species), and larval age structure. One challenge in developing and implementing an effective monitoring protocol is the extreme variability in wireworm species' response to biotic and abiotic variables in their subterranean habitat, as discussed in Section 2.1.3. Monitoring should be ongoing in fields at high risk of wireworm infestations, including those lacking proper rotation (see Section 3.1), with a history of wireworm infestation, or in transition from perennial use (e.g. CRP) to annual cropping (see Section 2.3.1).

Soil sampling can be done with a soil core sampler or a shovel. A review of the available literature by Barsics et al. (2013) concluded that this method of sampling is not favored by growers and other field persons due to the difficulties of removing adequate number of samples, sorting through the collected soil, and the patchy wireworm distribution within fields (Barsics et al., 2013). Even though soil sampling may be a laborious approach, it can be effective because with adequate human power and tools (e.g. a four-wheeler equipped with a core sampler) numerous spots within a field can be sampled in a relatively short period of time. Collected samples can be run through Berlese or Tullgren funnel traps to extract and count wireworms. In our experience, however, at times funnel traps could fail to extract all individuals from samples, at least in the case of *Limonius* spp.

Larval bait traps come in different versions and if installed at the right time (see Section 2.1.3), they may provide a relatively accurate estimate of wireworm presence (Catton et al., 2021; Kirfman et al., 1986; Parker, 1996, 1994). Wireworms use CO_2 and other volatile compounds released from germinating seeds and plant roots to orient toward the food source (Section 3.1.2). Their directional response to these plant signals and preference for adequate soil temperature and moisture have been exploited to make bait traps that are effective in attracting wireworms and reducing dispersal from the traps over the days between trap deployment and retrieval. Solar bait traps use germinating cereal seed (Doane, 1981) or other attractive substances (e.g. oatmeal, carrot, potato) as sources of CO_2 and volatiles to attract wireworms (Horton and Landolt, 2002). The bait (e.g. one cup of presoaked cereal grain) is buried 15 cm deep in the soil and covered with dark plastic on the soil surface which absorbs heat from the sun, retains soil moisture, and traps and concentrates the released volatiles (Rashed et al., 2015). Due to their design and function, the bait traps are most effective on bare ground prior to crop planting. Setting up the solar bait trap, however, can be somewhat time consuming, which could be challenging in the large-scale farming of North America, specifically the PNW, where approximately one bait per acre is recommended for effective field scouting (Andrews et al., 2008). Although sorting through the collected traps is time consuming, this method is more accurate than soil core sampling because the traps remain deployed for up to 2 weeks, attracting wireworms that are present in the vicinity of the trap.

Adult click beetle trapping using pheromone-baited lures are effective tools in detecting the presence and estimating the abundance and the distribution of the beetles within landscapes (Blackshaw and Vernon, 2008). The female sex pheromones have been identified and synthesized for a few species of click beetles, mostly those native to Europe (Furlan and Tóth, 2007). Recently, progress has been made in developing effective sex pheromone lures for some of the most damaging north American wireworm species, including *Cardiophorus tenebrosus* (Serrano et al., 2018), *Limonius* spp. (Gries et al., 2021), *Selatosomus aeripennis* destructor (Gries et al., 2022), and *Melanotus communis* (Williams et al., 2019). Limoniic acid is shown to be highly effective in attracting adult *L. californicus*, *L. infuscatus*, and *L. canus* males (Van Herk et al., 2021a), three of the most damaging species in the PNW. Limoniic acid lures can facilitate monitoring for the species complex rather than a single species. Pheromone traps can also be used to lure the males to entomopathogens, facilitating the pathogen spread within the population. This 'attract-and-kill' method can not only increase the amount of acquired inoculum per beetle (i.e. conidia dose/beetle), but also can significantly reduce the application rate of the entomopathogenic fungus (i.e. *Metarhizium brunneum*) (Kabaluk et al., 2015). Pheromone traps potentially can be deployed to trap out adult populations, but on a large scale this could become prohibitively expensive (Van Herk et al., 2022; Vernon et al., 2014). Pheromone traps can potentially be used to develop wireworm distribution maps, in relation to landscape, meteorological, agricultural, and soil characteristics (Barsics et al., 2013). To achieve this, more data are needed to determine the relationship between numbers of adults taken in pheromone traps and wireworms taken in bait traps at the appropriate spatial scale.

2.3 Wireworm risk and management thresholds

2.3.1 Damage risk

The variables that determine the risk of crop damage are not limited to wireworm species, abundance, seasonal activity, phenology, crop growth stage, adult dispersal, and female oviposition site preference. Other contributors include location-specific factors, such as continuous grass coverage (Hermann et al., 2013; Miles, 1942; Parker and Seeney, 1997), availability of alternative food sources, climatic conditions (e.g. temperature and moisture), and soil properties (e.g. organic matter, texture) (Hermann et al., 2013; Rashed et al., 2017). Models have been developed to map the risk of wireworm activity and damage in relation to soil moisture, climate data, and soil type (Jung et al., 2014; Lefko et al., 1998). Lafrance (1968) reported that a soil temperature of 13°C at a depth of 10 cm was optimal for wireworm (mostly *Dalopius pallidus*) activity. Decades later, in a simulation model for *Agriotes* spp. larvae, Jung et al. (2014) determined the risk of wireworm damage to field crops on a binary scale (i.e. risk vs no risk), in relation to soil moisture and temperature in a loamy soil type. The 'damage zone' for *Agriotes* spp. was associated with soil temperatures ranging between 8°C and 14°C and soil moisture between 30% and 32% in 15 cm soil depth. Jung et al. (2014) also demonstrated the preferred soil moisture was significantly influenced by soil type. Soil type can also affect the ability of wireworms to cause damage. It was demonstrated through a greenhouse study that feeding damage by *L. californicus* to wheat and barley occurred more frequently in a sand-dominated medium compared to a peatmoss-dominated medium or 1:1 mix of sand and peatmoss (Rashed et al., 2017). This observation was consistent with Hermann et al. (2013), who found a positive association between sandy soil and potato tuber damage by *Agriotes* sp. in the field through landscape analyses. Rapid changes in moisture levels in sandy soil have been proposed as a factor triggering wireworms to burrow into succulent tuber tissue in search of moisture (Hermann et al., 2013). Soils with higher sand contents usually have relatively lower bulk density, which is associated with wireworm infestations in the field (Hermann et al., 2013; Rashed, unpublished data; Parker and Seeney, 1997). Soil pH and organic matter also can influence the risk of wireworm presence and damage. Salt and Hollick (1946) detected positive associations between wireworm numbers and pH in topsoil. The same authors reported negative associations between wireworms and soil organic matter and soil nitrogen (Salt and Hollick, 1949). However, wireworm response to soil pH may be species-dependent; *Agriotes* and *Melanotus* species usually prefer acidic soils (Milosavljević et al., 2016a), whereas *Limonius* spp. can survive a considerable range of soil pH and can thus cause significant damage to the crops in both alkaline and acidic soils if other factors are favorable (Milosavljević et al., 2016a). Soil organic matter was thought to be utilized by wireworms as a food source, potentially reducing damage to crops (Traugott et al., 2015, and references within). Based on a study using an isotope nitrogen signature ($\delta^{15}N$), it is now clear that dead vegetation and organic matter are unlikely to serve as a food source for wireworms (Traugott et al., 2008, 2007), and a later long-term survey by Furlan et al. (2017) determined soil organic matter as a risk factor for damage by *Agriotes* spp. (Furlan et al., 2017). The role of soil organic content in relation to soil type, wireworm species, age, and crop warrants further studies.

2.3.2 Action thresholds and economic injury levels

When an effective monitoring protocol is in place, damage thresholds are needed as guidelines as to whether or when to act and what control tactics to use, if necessary. Damage thresholds can be influenced by the wireworm species, developmental instar, crop susceptibility (both plant and agronomic susceptibility) (Furlan, 2005, 1996; Griffiths, 1974), monitoring method (Furlan, 2014), and geographical location. Investigating all combinations would be a tremendous task and, to date, there have only been a handful of species-specific studies that have determined thresholds in two different crops. Cherry et al. (2013), estimated 9 wireworms per 25 soil samples (each ca. 30 × 30 × 30 cm in volume), as the economic injury level in sugarcane fields, primarily infested with *Melanotus* and *Conoderus* wireworms (Cherry et al., 2013). Furlan (2014) detected substantial differences in damage severity among three *Agriotes* spp.; although *A. ustulatus* did not appear to cause considerable economic damage to corn, *A. sordidus* and *A. brevis* proved to be damaging to the seedlings. Economic injury levels of 1, 2, and 5 wireworm per trap were estimated for *A. brevis*, *A. sordidus*, and *A. ustulatus*, respectively. In potatoes, at-plant treatments are recommended when an average of more than two *L. californicus* or *L. canus* are found per bait trap (Bechinski et al., 1994) when soil temperature in the 15 cm depth is higher than 7°C (Wright et al., 2002).

Wireworms are increasingly becoming a problem in small grain production in the western United States. Although small grains are not considered high-value crops, nearly all growers apply prophylactic insecticides as a preventive measure. Such applications are costly, and developing accurate economic thresholds are expected to reduce production input especially in dryland farming.

Furlan (2005) stated that a 'rational IPM strategy' depends on identifying high-risk areas, confirming wireworm presence and numbers, and making appropriate management decisions if control is warranted. Predicting the risk is only possible with explicit knowledge of species-specific (and location-specific) ecology as influenced by an array of biotic and abiotic environmental factors. We summarized some of these studies in this chapter; a detailed review of environmental parameters that can influence the risk of wireworm damage has been covered in Poggi et al. (2021). Most recent advances in arial imaging (e.g. improved remote sensing through satellite and drone technologies) combined with predictive models and larval bait traps can facilitate identifying damaged areas and predicting yield loss at harvest. Since most damaging wireworm species live for several years and tend to be found year after year in a given field, the history of wireworm infestation and their damage within specific fields should weigh heavily in any risk assessment model. Pheromone traps can also be practical tools in assessing wireworm risk. We discussed the most recent advances in this area, especially those in the *Limonius* complex, and echoed the need for understanding the relationship between adult male catches and larval population and distribution and quantifying the effective range of the sex pheromones and the environmental factors that influence the effectiveness of the approach. Finally, as previously mentioned, simple economic injury thresholds have been estimated for only a few species of wireworms in two crop types. Further research is needed to develop economic injury levels for more pestiferous wireworm species in small grains.

3 Control tactics

In response to the resurgence of wireworms across agroecosystems in recent decades and the limited number of effective synthetic insecticides to reduce wireworm numbers, there has been growing interest in evaluating and developing new ecological methods to control wireworms. Recent advances in our knowledge of specific-specific ecology are enabling development of new approaches that can be integrated with other control methods to minimize the risk of wireworm damage. In this section, we review the existing cultural, biological, and chemical control methods, emphasizing the latest developments in each area.

3.1 Host plant resistance

Using resistant and tolerant plant genotypes may offer an effective method of management for wireworms because their presence, activity pattern, and damage risk might not be known to allow for timely application of 'rescue' treatments. Provided their subterranean habitat and seasonality, at least in theory, host resistance and tolerance may be an effective approach in reducing the risk of damage by wireworms. Recognizing the importance of identifying resistant and tolerant crops, several studies to date examined the susceptibility of several crops to different species of wireworms. Potato, corn, and sweet potato are better-studied crops with respect to variations in susceptibility to wireworm damage. Johnson et al. (2008) studied the susceptibility of 12 potato varieties from several classes to *A. lineatus* and *A. obscurus*, which were considered as a single species due to their morphological similarity. While glycoalkaloid concentrations were discussed as a potential predictor of the observed differences in tuber susceptibility to *Agriotes* spp. feeding, the relationship was not strong enough to suggest glycoalkaloid content as the only underlying mechanism of resistance (Johnson et al., 2008). More recently, Langdon and Abney (2017) confirmed variation in susceptibility in current commercial varieties of potato planted in south and southeastern United States to *M. communis* and *Glyphonyx bimarginatus*. In that study, *M.*

communis appeared to be more damaging than *G. bimarginatus*, although the experimental design prevented comparing the varieties statistically (Fig. 1, in Langdon and Abney, 2017).

Differences in susceptibility to *A. sordidus*, measured as the percentage of infested plants in field trials, were also reported among corn varieties (La Forgia et al., 2020). This variation was attributed, in part, to differences in the diversity of the released compounds and the concentration of root volatiles such as hexanal and heptanal and 2, 3-octenanedione, with higher concentrations and diversity repellent to the wireworms (La Forgia et al., 2020). Hexanal and several other aldehydes have also been detected in barley roots (Barsics et al., 2017). However, since a synthetic blend of these volatiles was not as attractive as the live barley plants, it is conceivable that wireworms use root volatile blends in combination with CO_2 to detect the host plants' presence and suitability of the host plants (Barsics et al., 2017).

Several studies have focused on the search for host plant resistance and tolerance in small grains. In a greenhouse study with the wireworm *L. californicus*, using barley (cv. Idagold) and wheat (cv. Klasic), barley was more tolerant of damage by wireworms (Rashed et al., 2017). Overall, the ability of the wheat to tolerate wireworm feeding (measured in the form of foliar biomass) was influenced by soil type, also demonstrating the importance of environmental variables in wireworm host plant interactions (Rashed et al., 2017). A field study also confirmed barley as a crop tolerant of *L. californicus* damage, where despite wireworm presence, the application of insecticides did not result in significant economic return (Milosavljević et al., 2019). While barley may be more tolerant of wireworm damage than wheat, the reports of fewer wireworms in oat, compared to wheat and barley, might suggest host plant resistance in this crop (Milosavljević et al., 2019). Indeed, oats are known for their resistance to several pests due to their vigorous root development and production of defensive steroidal and triterpenoid saponins (Osbourn, 2003).

Variation in tolerance has also been observed across wheat genotypes. In the only published screening study conducted in fields infested with *L. californicus* and *L. infuscatus* (Higginbotham et al., 2014), 4 out of 163 evaluated wheat genotypes were consistently tolerant of wireworm feeding. One of these wheat cultivars, cv. Hollis, is also known for its resistance against the Hessian fly *Mayetiola destructor* larva (Sadeghi et al., 2021), which has a similar feeding mechanism as wireworms, where the larvae ingest plant sap after piercing through the tissue with their mandibles. Higginbotham et al. (2014) suggested that the observed reduced damage may reflect wireworms' preferential feeding resulting from structural differences in crown depth, root mass, and/or the rate of seedling growth (see also Miles and Petherbridge, 1927). Although these variables could potentially influence the extent of wireworm damage, they are relatively less likely to directly affect wireworm preference or directional response as these are driven by olfactory signals, i.e. root volatiles.

Although genetically modified small grains have yet to be accepted by consumers, transgenic plants expressing double-stranded RNA (dsRNA) that can target vital wireworm genes (RNA interference technology) potentially offer novel tools to reduce wireworm populations and damage. Efforts to identify and construct RNAi trigger sequences in wireworms are ongoing in multiple laboratories, recognizing that the approach should be species-specific to minimize the impact on non-target organisms.

3.2 Cultural control

Cultural control involves modifying the way a crop is grown to make it less suitable for the pest; such tactics may reduce wireworm damage by providing a competitive edge to the crop to evade serious damage by the pest.

3.2.1 Trap crops and intercropping

Wireworms feed on a wide range of crops, but also show preferences for some over others. As discussed earlier their host preference can be characterized by the initial directional response to CO_2 and root volatiles (e.g. Barsics et al., 2012) and the subsequent retention period, determined by feeding and sap ingestion. Wireworm host preference and differential response to subterranean plant cues have been explored as a basis for intercropping to redirect wireworms away from the main crop toward roots of a trap crop (Schallhart et al., 2011). Trap crops have been used successfully to reduce wireworm damage in field crops (Landl and Glauninger, 2013; Miles and Petherbridge, 1927; Sharma et al., 2019; Vernon et al., 2000), including spring wheat (Adhikari and Reddy, 2017; Sharma et al., 2019). Intercropping with lentil and pea resulted in a significant reduction in wireworm numbers in the collected soil samples and damage in spring wheat in both silt clay and sandy loam soils infested with *H. bicolor* and *L. californicus*, respectively (Adhikari and Reddy, 2017). Work has yet to be done to determine cost-to-benefit ratio of trap cropping especially at large scale, required to encourage adoption by growers. Further, Adhikari and Reddy (2017) demonstrated that *L. californicus* was attracted to the legumes they tested, a response shown in the laboratory to be triggered by CO_2 and root volatile cues (e.g. the aldehyde, hexanal) (Nikoukar, Sadeghi, Rashed, unpublished data). Identifying compounds that are attractive to wireworms can facilitate the integration of management tactics such as bait traps as part of an attract-and-kill strategy (Poggi et al., 2021), in which the wireworm is attracted into contact with a killing agent (e.g. Vernon,

2005). This approach was successfully employed to reduce the number of damaged tubers in potatoes, where wireworms attracted to CO_2, produced from encapsulated yeast and starch, were exposed to the entomopathogenic fungus (EPF) *Metarhizium brunneum* (Brandl et al., 2017). Encapsulated potato extract as the attractant in combination with the entomopathogenic nematodes (EPN), *Steinernema carpocapsae* and *Heterorhabditis bacteriophora*, increased their efficacies against *A. sordidus* by 2- and 2.5-fold, respectively, compared to application of the EPN without the attractant (La Forgia et al., 2020).

3.2.2 Crop rotation

Selecting an appropriate rotation is a critical decision in wireworm management. Diversification may include both crops and cover crops in rotations in which the most susceptible crops follow crops that are least supportive of wireworm populations (Poggi et al., 2021). Griffiths (1974) lists the following hosts as less susceptible than wheat to damage by *Agriotes* spp.: yellow mustard (*Sinapis alba*), cabbage (*Brassica oleracea*), flax (*Linum usitatissimum*), clover (*Trifolium pratense*), and French marigold (*Tagetes patula nana*). The small size of seeds and shallow planting depth, and/or the glucosinolate contents in the *Brassica* spp. were suggested as potential factors resulting in reduced wireworm damage. Pea (*Pisum sativum*) and bean (*Vicia faba*) are tolerant to *Agriotes* spp. (Griffiths, 1974) and *L. californicus* (Rashed, unpublished data), but neither of these crops appear to result in a reduction in wireworm populations (Griffiths, 1974).

Defatted brown mustard (*Brassica juncea*) seed meal can suppress both wireworm numbers and feeding in laboratory assays (Brown et al., 1991; Williams et al., 1993). Glucosinolates differ among mustard species, which may at least partly explain differences among mustards in toxicity to wireworms. The major glucosinolates in yellow mustard (*Sinapsis alba*) and brown mustard (*Brassica juncea*) are sinalbin and sinigrin, respectively. When exposed to moisture and through enzymatic degradation, sinalbin produces SCN^-, which is phytotoxic (Borek and Morra, 2005). While sinalbin is not efficacious against wireworms, sinigrin, which hydrolyzes to 2-propenyl isothiocyanate, is known for its toxicity to *L. californicus* (Williams et al., 1993). The effectiveness of the brown mustard *B. juncea* against wireworms was also shown when chopped plant tissue was incorporated into the soil (Furlan et al., 2009). However, it is important to note that evidence suggests that these biofumigants could potentially interfere with the function of beneficial soil organisms (Björkman et al., 2011; Henderson et al., 2009), which could otherwise contribute to effective management of *L. californicus* (see Section 3.2). Moreover, lipid profiles that are reflective of abundance, structure, and physiological status of soil microbial communities have been shown to be sensitive to introduction of *Brassica* crops into cereal grain rotations (Hansen et al., 2019). Although *Brassica* crops have been grown for many years, little is known about effects of these crops on soil health and non-target organisms and additional studies will help to improve our understanding of their role and importance in IPM and production sustainability.

It has been long known that alfalfa may reduce wireworms (Shirck, 1945) possibly due to the dense root structure of alfalfa that can facilitate soil drying and compaction over the years (Shirck, 1946). Observations (Rashed, personal observations) suggest that the effectiveness of alfalfa in rotation to reduce wireworm pressure also depends on the species of wireworms. Although alfalfa may help in reducing wireworm populations, this requires maintaining an alfalfa crop with low weed incidence for up to 4 years; this may be impractical because alfalfa harvest declines as the crop ages, and weed management becomes more difficult as the stand declines. Moreover, in much of the inland PNW, alfalfa may not be economically competitive with small grains even if used for a shorter duration.

Crop rotation options may be very limited in dryland production areas, especially places with limited precipitation. Winter wheat followed by fallow in the inland PNW dryland has been shown to reduce wireworm populations by 50% compared to continuous spring wheat systems, suggesting that incorporating fallow into rotations will provide benefits for wireworm control (Esser et al., 2015). However, the loss of revenue from a field left fallow would have to be balanced against the value of this approach in reducing wireworm populations. In wheat production systems, oats and barley are generally not considered to be ideal rotation crops from an agronomic standpoint; however, the higher tolerance in oats (Milosavljević et al., 2019) and relative tolerance in barley (Milosavljević et al., 2019; Rashed et al., 2017) may provide growers with options to replace the susceptible wheat crops in wireworm-infested fields. Location- and species-specific studies are needed to evaluate and identify appropriate rotations for implementation as a component of IPM against wireworms. The use of proper rotations may also enable producers to apply insecticides that are registered for application in the rotation crop, but not wheat. This would help to knock down wireworm numbers for wheat planting in the subsequent season. The choice of appropriate rotation crops compatible with other control tactics is perhaps the most important consideration in IPM against semivoltine wireworms.

3.2.3 Tillage

Adult wireworms oviposit and larvae pupate within the upper levels of the soil profile; larvae actively forage in that stratum, especially early in the season.

Therefore, it is likely that the disturbance caused by tillage would cause direct mortality and increase exposure to various biotic (e.g. natural enemies) and abiotic (e.g. heat, low moisture) factors that reduce wireworm survival (Andrews et al., 2008; Seal et al., 1992).

Seal et al. (1992) showed that repeated plowing could significantly reduce the number of wireworms based on monitoring with bait traps. However, repeated plowing (up to three events per season) may not be feasible or economically viable in large-scale farming. Tillage is most effective against early-instar larvae because the later instars are capable of resisting desiccation to a greater degree and can move back to suitable soil depth following disturbance (Shirck, 1946). Species-specific knowledge of the adult and larval seasonal activity and movement can help to time tillage such that the negative impacts on eggs, larvae, and pupae are maximized. Furlan et al. (2020) showed that *Agriotes* spp. damage in plots meadow-ploughed before corn planting in the spring was consistently below the assumed 15% damage threshold. There, it was suggested that incorporating plant material into the soil before corn seeding would increase CO_2 emission and provide alternate food sources, which *Agriotes* spp. can utilize, subsequently reducing damage to the field crop (Furlan et al., 2020).

No-tillage farming, or direct seeding, is a water- and soil-conserving practice used by almost all dryland (and rainfed) producers in the inland PNW, United States, and elsewhere. This practice, however, can increase wireworm populations compared to conventional tillage operations given that the higher soil moisture favors survival of eggs and the higher organic matter favors survival of larvae by providing alternative food sources.

3.2.4 Other cultural control measures

Because click beetle eggs and wireworms are generally susceptible to variations in the soil moisture (Poggi et al., 2021), water manipulation in the forms of *soil drying*, *flooding*, or both (Genung, 1970) have been explored as tactics to reduce their populations. As mentioned earlier, alfalfa can promote soil drying and is potentially most effective in sandy soil textures (Shirck, 1946). Since alfalfa is an irrigated crop, a well-planned irrigation schedule is needed to maximize the effectiveness of the soil drying method. Like many other cultural practices, the effectiveness of the moisture manipulation approach also depends on the species of the targeted wireworm. For example, *S. pruininus*, a species associated with dryland farms in the western United States, may be less susceptible than *Limonius* spp. to dry conditions (Andrews et al., 2008). The effectiveness of flooding as a control method also appears to be influenced by temperature. Fully saturated soil can cause 100% *L. californicus* mortality at

30°C but only 26% mortality at 10°C (Lane and Jones, 1936). This finding was supported by Van Herk and Vernon (2006) who also demonstrated that *Argiotes* spp. in flooded soil died more quickly at higher temperatures (i.e. 20°C). The efficacy of the soil flooding approach is also influenced by soil salinity, with more saline soils expected to lead to increased absorption of salt by wireworms and to increased mortality (Van Herk and Vernon, 2006).

Proper seedbed preparation and soil packing can support plant growth by maximizing soil-to-seed contact and hindering the ability of wireworms to disperse through the soil and locate plants (Coghill and Bristol, 1978). Some other cultural practices, such as early seeding and seeding at higher densities to compensate for damage, have been recommended but have little scientific support. Although it has been suggested that shallow seeding may reduce wireworm damage (Griffiths, 1974), this approach did not reduce *L. californicus* damage to wheat in greenhouse assays (Rashed et al., 2017).

Although the various cultural control tactics reviewed here have been studied for decades, results are often not consistent; the wireworm species' complex, climate, crop(s) of interest, field history and topography, soil characteristics, and interactions among these factors may influence the success of many of these tactics.

3.3 Natural enemies and biological control

The opening sentence of Parker and Howard (2001) on the biological control of wireworms, '*Published data on natural enemies of wireworms is sparse*', still holds true more than two decades later.

3.3.1 Predators and parasitoids

Predators that feed on wireworms include rove beetles (Coleoptera: Staphylinidae), predatory ground beetles (Coleoptera: Carabidae) (Fox and MacLellan, 1956), larvae of some darkling beetles (Coleoptera: Tenebrionidae), and various species of birds. Van Herk et al. (2015) reported feeding by larvae of the generalist predator, *Thereva nobilitata* (Diptera: Therevidae), on the wireworm *A. obscurus* in the field (Van Herk et al., 2015). There are also several species of hymenopteran parasitoids (reviewed in Poggi et al., 2021) that attack wireworms. *Paracodrus apterogynus* (Hymenoptera: Proctotrupidae) and *Pristocera depressa* (Hymenoptera: Bethylidae) are two examples, in which the wingless females actively search for wireworm larvae within the soil (Poggi et al., 2021).

Nonetheless, the ability of predators and parasitoids to reduce wireworm populations effectively is limited due to the patchy distribution of wireworms, their vertical movement through the soil profile, and cryptic subterranean habitat. Within the soil, however, wireworms are exposed to soil-dwelling entomopathogenic organisms, including bacteria, fungi, and nematodes (Barsics et al., 2013; Furlan et al., 2009; Traugott et al., 2015).

3.3.2 Entomopathogenic parasites

3.3.2.1 Bacteria

A few examples of free-living entomopathogenic soil-dwelling bacteria with the potential to reduce wireworms include *Rickettsiella* sp. (Kleespies et al., 2013; Leclerque et al., 2011), *Pseudomonas* spp., *Bacillus* spp., (Krieg, 1962; Lacey et al., 2007), and *Rahnella aqualitis* (Lacey et al., 2007). *Serratia* spp. is another bacterial group, reported from *L. canus*, that is known for expressing toxins (Lacey et al., 2007). *Serratia* sp., *Bacillus* spp., and *Pseudomonas* spp. have also been detected in *L. californicus* (Nikoukar, Andrews, Rashed, unpublished data) in the inland PNW; however, studies on the functional roles of these detected entomopathogenic bacteria and endosymbionts are needed to evaluate their suitability for wireworm management.

3.3.2.2 Fungi

Among entomopathogenic fungi (EPF), *Metarhizium* spp. (Kabaluk et al., 2007; Kabaluk and Ericsson, 2014, 2007; Reddy et al., 2014) (Fig. 3) and *Beauveria* spp. (Ester and Huiting, 2007; Ladurner et al., 2009; Sufyan et al., 2017) are the two most studied genera.

Figure 3 The sugar beet wireworm, *Limonius californicus*, infected with the entomopathogenic fungus *Metarhizium brunneum*.

Ansari et al. (2009) showed that *Agriotes lineatus* mortality occurs when inoculated with *Metarhizium anisopliae*; however, the rate of mortality was greatly impacted by the EPF isolate, ranging between 10% and 100% (Ansari et al., 2009). A later study by Reddy et al. (2014) showed reduced number of wireworms (predominantly, *L. californicus* and *H. bicolor*) collected in plots treated with *Metarhizium brunneum* F52, *M. robertsii* DWR346, or *B. bassiana* GHA relative to untreated control, with no differences among EPF treatments. However, in-furrow and soil-drench applications of EPF were more effective than the seed-coating method in reducing wireworm populations (Reddy et al., 2014). EPF isolates from field-collected wireworms have shown promise in causing wireworm mortality in field plots. For example, an *M. anisopliae* that was isolated from a field-collected *A. obscurus* sample proved effective against that species in field trials, albeit when applied at the very high rate of 4 million conidia/cm^3 (Kabaluk et al., 2007). *Beauveria bassina* has been isolated from genera including *Agriotes*, *Conoderus*, and *Hypnoidus* (Traugott et al., 2015) and successfully applied to reduce *Agriotes* spp. damage to potato (Ladurner et al., 2009). In-furrow application of *B. bassiana* was also successful in reducing the number of *Limonius* spp. in bait traps in organic fields (Nikoukar and Rashed, unpublished data). However, in a study by Kölliker et al (2011) comparing efficacies of *M. anisoplaie* isolate ART-2825 (a Swiss strain) and the commercial *B. bassiana* against *Agriotes* spp., the two EPF isolates differed significantly in their effectiveness against the wireworm, with *B. bassiana* being ineffective. This finding supported that of Ansari et al. (2009), in which *B. bassiana* was deemed non-pathogenic against *A. lineatus*. Although the efficacy of *M. anisopliae* reached as high as 97% against *A. obscurus*, it was considerably less virulent against *A. lineatus* and *A. sputator* (Kölliker et al., 2011). Inconsistencies in the outcome of these studies suggest wireworm species differ substantially in the degree of their susceptibility to different EPF species and isolates.

3.3.2.3 Nematodes

EPNs, especially those belonging to the families Steinernematidae and Heterorhabditidae (Grewal et al., 2005), are known to infect wireworms. Under laboratory conditions, *Steinernema feltiae* (Filipjev), *S. carpocapsae* (Weiser), and *S. riobrave* Cabanillas, Poinar and Raulston, have been effective in reducing *L. californicus* populations (Nikoukar et al., 2021; Fig. 4; Sandhi et al., 2020b; Toba et al., 1983). Despite its effectiveness against *L. californicus*, *S. feltiae* appeared ineffective against *Agriotes* spp., whereas *Heterorhabditis bacteriophora* Poinar proved effective against this genus (Ansari et al., 2009; Ester and Huiting, 2007; Morton and Garcia-del-Pino, 2017). However, it has been shown consistently that *H. bacteriophora* has very low efficacy against *L. californicus* (Sandhi et al., 2020a, 2020b; Nikoukar et al., 2021). Differences

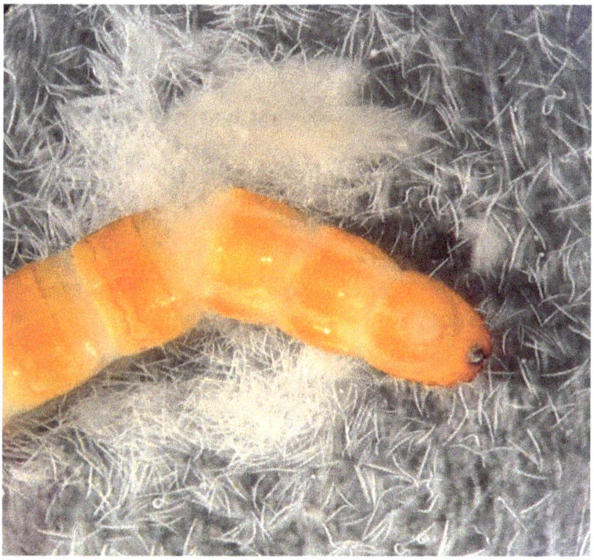

Figure 4 Infective juveniles of *Steinernema feltiae* emerging from the sugar beet wireworm, *Limonius californicus*. Photo by Atoosa Nikoukar.

in EPN efficacies might be related to variation in their foraging strategies (Lewis, 2002; Lewis et al., 2006), habitat preferences (e.g. soil depth; Kaya and Gaugler, 1993), ecologies (Milosavljević et al., 2017), as well as physical and/or behavioral defenses of the wireworm host (Eidt and Thurston, 1995).

The abiotic environment also influences the efficacy of EPN (Campos-Herrera and Gutiérrez, 2009; Kaspi et al., 2010; Kaya and Gaugler, 1993; Kung et al., 1991; Lewis et al., 2006). For example, soil type is shown to influence EPN efficacy since it can impact their survival and movement (Campos-Herrera and Gutiérrez, 2009; Kaya and Gaugler, 1993; Kung et al., 1990). Ensafi et al. (2018) demonstrated that the *S. carpocapsae* was more effective in protecting wheat seedlings from *L. californicus* in sand-dominated compared to peat moss-rich soil (Ensafi et al., 2018).

The above examples show variability in EPN efficacies in relation to the wireworm species, the species and strain of the EPN, and/or environmental variables such as soil type, temperature, and moisture. The commercial EPN are lab-reared for generations and may not be as versatile as the naturally occurring EPN in the field, because the naturally occurring EPN may be adapted to the adverse soil environments and the different endemic hosts within, due to constant exposure through generations (Campos-Herrera and Gutiérrez, 2014; Nikoukar et al., 2021; Rojht et al., 2009; Sandhi et al., 2020a). In the laboratory, two field isolates of EPN from Montana resulted in up to 50% mortality in *L. californicus* (Sandhi et al., 2020a). This mortality rate, however, dropped to 20–25% when the experiment was carried out in different soil types in an outdoor 'shade house'. Most recently, in a laboratory assay, Nikoukar et al. (2021) showed that the efficacy against *L. californicus* of *S. feltiae* isolated from two different wheat fields differed significantly. The *S. feltiae* from a heavily infested field with *L. californicus* and *H. bicolor* caused 56.7% wireworm mortality, 2.1-fold higher than the *S. feltiae* isolate from a field with very low mixed-species infestations. Although the Nikoukar et al. (2021) study was not conclusive, it provided support for the hypothesis that weak EPN efficacy can sometimes be attributed to poor adaptation to the target insect (Kaya and Gaugler, 1993; Koppenhöfer and Fuzy, 2006). Future studies to extract naturally occurring EPN populations from wireworm-infested fields and screen for their efficacies are needed to confirm their superior performance to the commercially available alternatives and determine their suitability as a component of a species-specific wireworm IPM approach.

To improve the efficacy of biological control, studies have also explored the combined use of biological control agents against wireworms and other insects. The outcome of mixed applications of different entomopathogenic organisms on a target pest can be influenced by the choice of the mixed entomopathogens, the rate and timing of applications, the larval instar of the pest (Acevedo et al., 2007; Ansari et al., 2006; Barberchek and Kaya, 1990; Shapiro-Ilan et al., 2004; Wakil et al., 2017), and abiotic environmental variables (Barbercheck and Kaya, 1991). Some additive effects were observed for the combination of the EPN *Heterorhabditis indica* and the EPF *M. anisoplaie* against *Curculio caryae* (Coleoptera: Curculionidae); however, it was concluded that the evaluated pathogen combinations in the study would not consistently improve efficacies beyond that of the most virulent entomopathogen in the mix (Shapiro-Ilan et al., 2004). This conclusion was supported by Ensafi et al. (2018), who also did not detect increased *L. californicus* mortality when they were treated with a combination of *M. anisopliae* and *S. carpocapsae* at planting, compared to the applications of only one of the entomopathogens.

3.4 Insecticides

The recent reemergence of wireworms as pests in cereal crops is thought to be due in part to the elimination of registration of organochlorine insecticides with long residual activity that had been widely used for wireworm control (Van Herk et al., 2021b; Vernon and van Herk, 2022; Vernon et al., 2009). Several organochlorines that were banned during the 1970s may have continued to provide some wireworm control. More recently, the organochlorine seed treatment lindane was banned in Canada (2004) and the United States (2006). The loss of these highly efficacious compounds due to concerns regarding their effects on the environment has driven strong interest in the development of alternative chemistries.

In the years following the deregulation of lindane, neonicotinoid seed treatments have become the industry standard for wireworm control. Whereas organochlorine insecticides provided both crop protection and population reduction, often providing control for multiple years after application as both neonate (newly hatched) and older wireworms were killed, neonicotinoid insecticides generally fail to reduce wireworm populations and typically require application each year. Neonicotinoids have been shown to provide cereal stand and yield protection from low to moderate populations of several different wireworm species (Van Herk et al., 2018; Vernon et al., 2013, 2009; Zhang et al., 2017); however, this protection generally is not a result of wireworm mortality (Milosavljević et al., 2019; Morales-Rodriguez and Wanner, 2015; Van Herk et al., 2018; Vernon et al., 2013, 2009). Rather, wireworms exposed to neonicotinoids exhibit temporary intoxication and morbidity, with inhibition of mobility and feeding behaviors (Van Herk et al., 2015, 2008; Vernon et al., 2008). Affected wireworms may exhibit full recovery later during the growing season (Van Herk et al., 2015, 2008; Vernon et al., 2008) and may readily damage subsequent crops.

Broflanilide, an insecticide in a group with a new mode of action, the meta-diamides (Nakao and Banba, 2016), was recently reported to provide crop protection and wireworm reduction at levels that were comparable to the former industry standard, lindane (Van Herk et al., 2021b). Broflanilide was registered in wheat in Canada in 2020 and in the United States in 2021 and is already being widely used in these growing areas. This new chemistry offers a promising new tool in wireworm management in cereals given its ability to protect yields and reduce field populations – at least for *A. obscurus*. More research is needed to clarify the efficacy of broflanilide against other species.

In recent years, several other chemistries have been evaluated for the management of wireworms, with varying degrees of efficacy. Of these, the most studied groups include the pyrethroids, diamides, and phenyl pyrazoles.

Pyrethroid insecticides have been shown to guard against wheat damage from *A. obscurus* (Vernon et al., 2009) and *L. californicus* (Van Herk et al., 2018). As with neonicotinoids, protection of stand and yield with pyrethroids is achieved without reducing wireworm populations. However, the mechanism appears to be repellency that results in reduced feeding (Van Herk et al., 2015, 2008; Van Herk and Vernon, 2013; Van Herk and Vernon, 2007; Van Herk and Vernon, 2007).

The diamides cyantraniliprole and chlorantraniliprole, on the other hand, are not repulsive but likely to induce temporary morbidity after feeding (Van Herk et al., 2015), similar to the effects of neonicotinoid exposure. Low mortality and short recovery times after exposure to these compounds in the laboratory suggested that they may not be effective for crop protection (Van Herk et al., 2015). However, at least cyantraniliprole has been demonstrated to reduce cereal damage in the field without reducing populations (Van Herk et al., 2018).

The phenyl pyrazole fipronil has been shown to provide excellent stand and yield protection and – unlike the neonicotinoids, pyrethroids, and diamides – high wireworm mortality in both laboratory (Van Herk et al., 2015, 2008; Vernon et al., 2008) and field trials (Van Herk et al., 2018; Vernon et al., 2013, 2009). The mode of action of fipronil is similar to that of lindane, and high mortality has been demonstrated with fipronil against several wireworm species at far lower doses than for lindane (Vernon et al., 2013). Fipronil, however, is not registered for use on cereals in North America. Another phenyl pyrazole, ethiprole, was found to induce morbidity of short duration against *Agriotes obscurus*, but no mortality nor evidence of stand or yield protection was observed (Van Herk et al., 2015). Should fipronil be registered in wheat, it could provide another useful tool for protecting the crop while reducing wireworm densities.

In the development of management approaches, the importance of knowledge of the wireworm species present in a field is underscored by different responses among species to a given chemistry. The neonicotinoid thiamethoxam provides a useful, well-studied example. Though neonicotinoids have generally provided good crop protection, thiamethoxam was reported to be ineffective against very high densities of *L. californicus* (Van Herk et al., 2018), despite providing acceptable control against high densities of *A. obscurus* (Vernon et al., 2013, 2009). Moreover, repeated exposure to thiamethoxam-treated wheat was reported to reduce *L. infuscatus* populations, but not populations of *L. californicus* (Esser et al., 2015; Milosavljević et al., 2019, 2016b). The possibility of such species-specific differences in susceptibility needs to be considered in making insecticide decisions for wireworm management.

Combining insecticides of different modes of action has been explored in several studies e.g. (Morales-Rodriguez and Wanner, 2015; Van Herk et al., 2021b, 2018; Vernon et al., 2013). In addition to the potential for delaying the development of insecticide resistance, combining modes of action may allow for acceptable control while reducing the total active ingredient applied. Insecticide blends may also be necessary to protect wheat from other insect pests in addition to wireworms. Of key importance when combining insecticides against wireworms is the degree to which a particular chemistry may repel wireworms and/or inhibit feeding, thereby reducing their exposure to chemistries that could otherwise contribute to crop protection or population reduction. For example, in laboratory assays, reduced contact time or mortality rates have been exhibited for several wireworm species and several seed treatment blends, including a pyrethroid + thiamethoxam (Van Herk et al., 2008), thiamethoxam + fipronil (Morales-Rodriguez and Wanner, 2015; Van Herk et al., 2015) and thiamethoxam + chlorpyrifos (Van Herk et al., 2015). However, it should be cautioned that apparent antagonistic effects in the laboratory do not necessarily translate into reduced field efficacy (Catton et al., 2021; Morales-Rodriguez and Wanner, 2015; Van Herk et al., 2015).

4 Conclusion and future trends

Historically, decades-long reliance on environmentally persistent insecticides for wireworm control hampered efforts to develop effective IPM. This has also contributed to the resurgence of the pest, as the lingering effects of those chemistries have been fading away following their removal from the market (Vernon et al., 2009). Nearly two decades ago, Furlan (2005) listed the following requirements for developing and implementing an effective IPM strategy against wireworms: (1) improved species identification; (2) established species distribution maps; (3) knowledge of species-specific ecology and biology; (4) improved tools to monitor and estimate population densities; (5) defined economic injury thresholds; and (6) new and improved control tactics compatible to one another. The first five proposed priority areas are primarily associated with the decision-making process of IPM, which are prerequisite to the development of new and improved integrated control tactics that are environmentally friendly and 'climate-smart', compatible with one another, and designed to sustainably reduce the input costs of management to maximize productivity. In this chapter, we provided a review of the progress made over the past two decades, which contributed to a better understanding of species-specific wireworm ecology and management. Gaps in knowledge still exist and we describe those we consider most important below.

- The recent advances in molecular approaches have helped greatly with clarifying some of the ambiguities in species identification and provide powerful tools to further explore within-species population structure, adaptations, and dispersal (Andrews et al., 2020). Although region-specific distribution maps have been developed for parts of Europe (Furlan and Tóth, 2007) and North America (Van Herk et al., 2021c), efforts are needed to pool the existing information and further solicit or collect data from other regions to enable developing a global distribution map for pestiferous species of wireworms and click beetles in general, but see the overview of economic species in the Holarctic in Vernon and van Herk (2022).
- There is still a substantial lack of knowledge of basic biology and ecology for many species of wireworms. Most of the studies within the past two decades have been species-specific, but comparisons among these studies underscore considerable among-species ecological differences. Location-specific field studies must continue and should include controlled laboratory assays, to enable identifying specific factors that can influence the outcomes of IPM.
- Wireworms are difficult to monitor, and considerable damage can be done to crops before their presence is recognized (Esser et al., 2015;

Milosavljević et al., 2017, 2016a). Furlan (2005) raised the need to improve trapping and monitoring of wireworms (i.e. efficiency of bait traps) and adult click beetles (particularly, effective sex pheromones). Trapping and monitoring of wireworms remain challenging; however, significant progress has been made in developing and employing effective sex pheromones for the click beetle species that impact agroecosystems in Europe and North America (see Section 2.2). While such efforts are continuing for several other click beetle species in the United States, future work is much needed to allow for accurate interpretation of the adult catch data for predicting distribution and abundance of wireworms within and across fields as well as risk of injury to crops.

- To use wireworm count data in the decision-making process there is a need for economic thresholds and economic injury levels, which are lacking for the primary wireworm pests affecting small grain production. To our knowledge no objective evaluation of yield loss in relation to wireworm numbers has been performed in wheat. Coordinated efforts are needed to monitor yield (i.e. use of GPS-driven yield monitor technology) and wireworm ground data to develop thresholds based on the production market value and the input cost associated with control. In the absence of objective thresholds, the current management strategy used in cereal crops by most growers is to prophylactically treat seeds with insecticides. Prophylactic seed treatment is costly, especially in dryland production where there is a relatively lower margin for economic return.
- Research into identifying effective control methods has been ongoing and will continue to develop. Broflanilide is a new class of insecticide which was made available to the US and Canadian wheat and barley producers, in the spring of 2021, to reduce the populations of wireworms. This new insecticide has no systemic activity and is therefore not expected to be effective against some of the other common pests of wheat and barley, such as aphids. This means that the growers may continue to prophylactically use neonicotinoids and add broflanilide in a 'stack-on' approach. Such an approach, however, would not be species-specific, sustainable, or protective of non-target organisms. Moreover, the compatibility of new chemistries - alone and in combination with existing insecticides - with alternative control methods (e.g. biological control) and with beneficial soil macro- and micro-organisms is unknown. As wireworm management history has taught us, the sole reliance on insecticides is not sustainable, and development and evaluation of both species- and location-specific cultural and biological control methods must continue.
- Some of the existing recommendations, such as early seeding and seeding at higher densities to compensate for damage, have little scientific support and research is needed to validate the effectiveness of these tactics.

- Cost-to-benefit analyses for recommended IPM practices are needed to provide grain producers with strategies that are scientifically grounded and economically sound.
- In addition to evaluating various alternative control methods, studies are also needed to evaluate the potential interaction between tactics and/or their impact(s) on the soil microbial community and the rhizosphere (Nikoukar and Rashed, 2022). For example, whether incorporating mustard products into the soil or adding mustard cover crop could interfere with the functions of soil micro-organisms (Hansen et al., 2018) is an area which needs further investigation. Such information is critical to assure whether such IPM practices will contribute to the sustainability of small grain production systems (Nikoukar and Rashed, 2022).
- Once location- and species-specific IPM protocols are developed, effective communication with stakeholders is needed to successfully disseminate and implement the IPM strategies. Throughout the process, the success of the implemented programs needs to be evaluated on short-term (e.g. knowledge gain), mid-term (e.g. reduced risk of infestation/damage, improved productivity), and long-term (e.g. socioeconomic benefits at the regional community level) bases.

5 Where to look for further information

Barsics, F., Haubruge, E. and Verheggen, F. J. (2013). 'Wireworms' Management: An Overview of the Existing Methods, with Particular Regards to *Agriotes* spp. (Coleoptera: Elateridae)', *Insects*, 4(1), 117–152.

Poggi, S., Le Cointe, R., Lehmhus, J., Plantegenest, M. and Furlan, L. (2021). 'Alternative Strategies for Controlling Wireworms in Field Crops: A Review', *Agriculture*, 11(5), 436.

Traugott, M., Benefer, C. M., Blackshaw, R. P., Herk, W. G. van and Vernon, R. S. (2015). 'Biology, Ecology, and Control of Elaterid Beetles in Agricultural Land', *Annu. Rev. Entomol.*, 60, 313–334.

Vernon, R. S. and van Herk, W. G. (2022). 'Wireworms as pests of potato'. In A. Alyokhin, S. I. Silva, and Y. Gao (eds.), Insect pests of potato: global perspectives on biology and management (Pp. 501), 2nd edition. Amsterdam, The Netherlands: Academic Press, Elsevier.

6 References

Acevedo, J. P. M., Samuels, R. I., Machado, I. R. and Dolinski, C. (2007). Interactions between isolates of the entomopathogenic fungus Metarhizium anisopliae and the entomopathogenic nematode Heterorhabditis bacteriophora JPM4 during infection of the sugar cane borer Diatraea saccharalis (Lepidoptera: Pyralidae). *J. Invertebr. Pathol.* 96(2), 187–192.

Adhikari, A. and Reddy, G. V. P. (2017). Evaluation of trap crops for the management of wireworms in spring wheat in Montana. *Arthropod Plant Interact.* 11(6), 755-766.

Andrews, K. R., Gerritsen, A., Rashed, A., Crowder, D. W., Rondon, S. I., van Herk, W. G., Vernon, R., Wanner, K. W., Wilson, C. M. and New, D. D. (2020). Wireworm (Coleoptera: Elateridae) genomic analysis reveals putative cryptic species, population structure, and adaptation to pest control. *Commun. Biol.* 3(1), 489, 1-13.

Andrews, N., Ambrosino, M. D., Fisher, G. C. and Rondon, S. I. (2008). *Wireworm: Biology and Nonchemical Management in Potatoes in the Pacific Northwest.* Grants Pass, OR: Oregon State University Extension Service.

Ansari, M. A., Evans, M. and Butt, T. M. (2009). Identification of pathogenic strains of entomopathogenic nematodes and fungi for wireworm control. *Crop Prot.* 28(3), 269-272.

Ansari, M. A., Shah, F. A., Tirry, L. and Moens, M. (2006). Field trials against *Hoplia philanthus* (Coleoptera: Scarabaeidae) with a combination of an entomopathogenic nematode and the fungus Metarhizium anisopliae CLO 53. *Biol. Control* 39(3), 453-459.

Balkenhol, N., Cushman, S., Storfer, A. and Waits, L. (2015). *Landscape Genetics: Concepts, Methods, Applications.* Hoboken, NJ: John Wiley & Sons.

Barberchek, M. E. and Kaya, H. K. (1990). Interactions between *Beauveria bassiana* and the entomogenous nematodes, *Steinernema feltiae* and *Heterorhabditis heliothidis.* *J. Invertebr. Pathol.* 55(2), 225-234.

Barbercheck, M. E. and Kaya, H. K. (1991). Effect of host condition and soil texture on host finding by the entomogenous nematodes *Heterorhabditis bacteriophora* (Rhabditida: Heterorhabditidae) and *Steinernema carpocapsae* (Rhabditida: Steinernematidae). *Environ. Entomol.* 20(2), 582-589.

Barsics, F., Delory, B. M., Delaplace, P., Francis, F., Fauconnier, M.-L., Haubruge, É. and Verheggen, F. J. (2017). Foraging wireworms are attracted to root-produced volatile aldehydes. *J. Pest Sci.* 90(1), 69-76.

Barsics, F., Haubruge, E. and Verheggen, F. J. (2013). 'Wireworms' management: An overview of the existing methods, with particular regards to *Agriotes* spp. (Coleoptera: Elateridae). *Insects* 4(1), 117-152.

Barsics, F., Latine, R., Gfeller, A., Laloux, M., Lognay, G., Wathelet, P., Haubruge, E. and Verheggen, F. J. (2012). Do root-emitted volatile organic compounds attract wireworms? *Commun. Agric. Appl. Biol. Sci.* 77(4), 561-565.

Bechinski, E., Sandvol, L., Carpenter, G. and Homan, H. (1994). *Integrated Pest Management Guide to Wireworms in Potatoes.* University of Idaho, College of Agriculture, Cooperative Extension System.

Becker, E. C., Dogger, J. R. and Stehr, F. W. (1991). Elateridae (Elateroidea) (including Dicronychidae, Lissomidae). *Immature Insects* 2, 410-418.

Benefer, C., Andrew, P., Blackshaw, R., Ellis, J. and Knight, M. (2010). The spatial distribution of phytophagous insect larvae in grassland soils. *Appl. Soil Ecol.* 45(3), 269-274.

Benefer, C. M., Van Herk, W. G., Ellis, J. S., Blackshaw, R. P., Vernon, R. S. and Knight, M. E. (2013). The molecular identification and genetic diversity of economically important wireworm species (Coleoptera: Elateridae) in Canada. *J. Pest Sci.* 86(1), 19-27.

Björkman, M., Klingen, I., Birch, A. N., Bones, A. M., Bruce, T. J., Johansen, T. J., Meadow, R., Mølmann, J., Seljåsen, R., Smart, L. E. and Stewart, D. (2011). Phytochemicals of Brassicaceae in plant protection and human health - Influences of climate, environment and agronomic practice. *Phytochemistry* 72(7), 538-556.

Blackshaw, R. P. and Vernon, R. S. (2008). Spatial relationships between two Agriotes click-beetle species and wireworms in agricultural fields. *Agric. For. Entomol.* 10(1), 1-11.
Borek, V. and Morra, M. J. (2005). Ionic thiocyanate (SCN-) production from 4-hydroxybenzyl glucosinolate contained in *Sinapis alba* seed meal. *J. Agric. Food Chem.* 53(22), 8650-8654.
Brandl, M. A., Schumann, M., Przyklenk, M., Patel, A. and Vidal, S. (2017). Wireworm damage reduction in potatoes with an attract-and-kill strategy using *Metarhizium brunneum*. *J. Pest Sci.* 90(2), 479-493.
Brown, P. D., Morra, M. J., McCaffrey, J. P., Auld, D. L. and Williams, L. (1991). Allelochemicals produced during glucosinolate degradation in soil. *J. Chem. Ecol.* 17(10), 2021-2034.
Campos-Herrera, R. and Gutiérrez, C. (2009). Screening Spanish isolates of Steinernematid nematodes for use as biological control agents through laboratory and greenhouse microcosm studies. *J. Invertebr. Pathol.* 100(2), 100-105.
Campos-Herrera, R. and Gutiérrez, C. (2014). *Steinernema feltiae* intraspecific variability: Infection dynamics and sex-ratio. *J. Nematol.* 46(1), 35-43.
Catton, H., van Herk, W., Saguez, J. and Svendsen, E. (2021). *Guide to Pest Wireworms in Canadian Prairie Field Crop Production*. Lethbridge, Alberta, Canada: Agriculture and AgriFood Canada.
Cherry, R., Grose, P. and Barbieri, E. (2013). Validation of a sequential sampling plan for wireworms (Coleoptera: Elateridae) at sugarcane planting. *J. Pest Sci.* 86(1), 29-32.
Cherry, R. and Stansly, P. (2008). Abundance and spatial distribution of wireworms (Coleoptera: Elateridae) in Florida sugarcane fields on muck versus sandy soils. *Fla. Entomol.* 91(3), 383-387.
Coghill, K. J. and Bristol, A. D. A. S. (1978). Wireworms. *Minist. Agric. Fish. Food ADP* 51.
Doane, J. F. (1981). Evaluation of a larval trap and baits for monitoring the seasonal activity of wireworms in Saskatchewan. *Environ. Entomol.* 10(3), 335-342.
Eidt, D. C. and Thurston, G. S. (1995). Physical deterrents to infection by entomopathogenic nematodes in wireworms (Coleoptera: Elateridae) and other soil insects. *Can. Entomol.* 127(3), 423-429.
Ellis, J. S., Blackshaw, R., Parker, W., Hicks, H. and Knight, M. E. (2009). Genetic identification of morphologically cryptic agricultural pests. *Agric. For. Entomol.* 11(1), 115-121.
Ensafi, P., Crowder, D. W., Esser, A. D., Zhao, Z., Marshall, J. M. and Rashed, A. (2018). Soil type mediates the effectiveness of biological control against *Limonius californicus* (Coleoptera: Elateridae). *J. Econ. Entomol.* 111(5), 2053-2058.
Esser, A. D., Milosavljević, I. and Crowder, D. W. (2015). Effects of neonicotinoids and crop rotation for managing wireworms in wheat crops. *J. Econ. Entomol.* 108(4), 1786-1794.
Ester, A. and Huiting, H. (2007). Controlling wireworms (*Agriotes* spp.) in a potato crop with BioLogicals. *IOBC WPRS Bull.* 30(1), 189-196.
Etzler, F. E., Wanner, K. W., Morales-Rodriguez, A. and Ivie, M. A. (2014). DNA barcoding to improve the species-level management of wireworms (Coleoptera: Elateridae). *J. Econ. Entomol.* 107(4), 1476-1485.
Furlan, L. (1996). The biology of *Agriotes ustulatus* Schäller (Col., Elateridae). I. Adults and oviposition. *J. Appl. Entomol.* 120(1-5), 269-274.
Furlan, L. (2005). An IPM approach targeted against wireworms: What has been done and what has to be done. *IOBC WPRS Bull.* 28(2), 91-100.
Furlan, L. (2014). IPM thresholds for Agriotes wireworm species in maize in Southern Europe. *J. Pest Sci.* 87(4), 609-617.

Furlan, L., Benvegnù, I., Chiarini, F., Loddo, D. and Morari, F. (2020). Meadow-ploughing timing as an integrated pest management tactic to prevent soil-pest damage to maize. *Eur. J. Agron.* 112, 125950.

Furlan, L., Bonetto, C., Costa, B., Finotto, A. and Lazzeri, L. (2009). Observations on natural mortality factors in wireworm populations and evaluation of management options. *IOBC WPRS Bull.* 45, 436–439.

Furlan, L., Contiero, B., Chiarini, F., Colauzzi, M., Sartori, E., Benvegnù, I., Fracasso, F. and Giandon, P. (2017). Risk assessment of maize damage by wireworms (Coleoptera: Elateridae) as the first step in implementing IPM and in reducing the environmental impact of soil insecticides. *Environ. Sci. Pollut. Res. Int.* 24(1), 236–251.

Furlan, L. and Tóth, M. (2007). Occurrence of click beetle pest spp. (Coleoptera, Elateridae) in Europe as detected by pheromone traps: Survey results of 1998–2006. *IOBC WPRS Bull.* 30(7), 19–25.

Genung, W. G. (1970). Flooding experiments for control of wireworms attacking vegetable crops in the Everglades. *Fla. Entomol.* 53(2), 55–63.

Gerritsen, A. T., New, D. D., Robison, B. D., Rashed, A., Hohenlohe, P., Forney, L., Rashidi, M., Wilson, C. M. and Settles, M. L. (2016). Full mitochondrial genome sequence of the sugar beet wireworm *Limonius californicus* (Coleoptera: Elateridae), a common agricultural pest. *Genome Announc.* 4(1), e01628-15.

Glen, R. (1950). *Larvae of the Elaterid Beetles of the Tribe Lepturoidini (Coleoptera: Elateridae)* (Vol. 111). Smithson, Washington, DC. *Misc.* Collect, pp. 1-246.

Grewal, P. S., Ehlers, R.-U. and Shapiro-Ilan, D. I. (2005). *Nematodes as Biocontrol Agents.* Wallingford, UK: CABI Publishing.

Gries, R., Alamsetti, S. K., van Herk, W. G., Catton, H. A., Meers, S., Lemke, E. and Gries, G. (2021). Limoniic acid-major component of the sex pheromones of the click beetles *Limonius canus* and *L. californicus*. *J. Chem. Ecol.* 47(2), 123–133.

Gries, R., van Herk, W., Alamsetti, S. K., Catton, H., Meers, S., Otani, J. and Gries, G. (2022). (Z, E)-α-Farnesene – Sex pheromone component of female click beetle Selatosomus aeripennis destructor with intra- and inter-sexual communication function. *Entomol. Exp. Appl.* 170(4), 344–351.

Griffiths, D. C. (1974). Susceptibility of plants to attack by wireworms (*Agriotes* spp.). *Ann. Appl. Biol.* 78(1), 7–13.

Hansen, J. C., Schillinger, W. F., Sullivan, T. S. and Paulitz, T. C. (2018). Rhizosphere microbial communities of canola and wheat at six paired field sites. *Appl. Soil Ecol.* 130, 185–193.

Hansen, J. C., Schillinger, W. F., Sullivan, T. S. and Paulitz, T. C. (2019). Soil microbial biomass and fungi reduced with canola introduced into long-term monoculture wheat rotations. *Front. Microbiol.* 10, 1488.

Henderson, D. R., Riga, E., Ramirez, R. A., Wilson, J. and Snyder, W. E. (2009). Mustard biofumigation disrupts biological control by *Steinernema* spp. nematodes in the soil. *Biol. Control* 48(3), 316–322.

Hermann, A., Brunner, N., Hann, P., Wrbka, T. and Kromp, B. (2013). Correlations between wireworm damages in potato fields and landscape structure at different scales. *J. Pest Sci.* 86(1), 41–51.

Higginbotham, R. W., Froese, P. S. and Carter, A. H. (2014). Tolerance of wheat (Poales: Poaceae) seedlings to wireworm (Coleoptera: Elateridae). *J. Econ. Entomol.* 107(2), 833–837.

Horton, D. R. (2006). Quantitative relationship between potato tuber damage and counts of Pacific coast wireworm (Coleoptera: Elateridae) in baits: Seasonal effects. *J. Entomol. Soc. B.C.* 103, 37-48.

Horton, D. R. and Landolt, P. J. (2002). Orientation response of Pacific coast wireworm (Coleoptera: Elateridae) to food baits in laboratory and effectiveness of baits in field. *Can. Entomol.* 134(3), 357-367.

Johnson, S. N., Anderson, E. A., Dawson, G. and Griffiths, D. W. (2008). Varietal susceptibility of potatoes to wireworm herbivory. *Agric. For. Entomol.* 10(2), 167-174.

Johnson, S. N., Zhang, X., Crawford, J. W., Gregory, P. J. and Young, I. M. (2007). Egg hatching and survival time of soil-dwelling insect larvae: A partial differential equation model and experimental validation. *Ecol. Modell.* 202(3-4), 493-502.

Jung, J., Racca, P., Schmitt, J. and Kleinhenz, B. (2014). SIMAGRIO-W: Development of a prediction model for wireworms in relation to soil moisture, temperature and type. *J. Appl. Entomol.* 138(3), 183-194.

Kabaluk, J. T. and Ericsson, J. D. (2007a). *Metarhizium anisopliae* seed treatment increases yield of field corn when applied for wireworm control. *Agron. J.* 99(5), 1377-1381.

Kabaluk, J. T. and Ericsson, J. D. (2007b). Environmental and behavioral constraints on the infection of wireworms by *Metarhizium anisopliae*. *Environ. Entomol.* 36(6), 1415-1420.

Kabaluk, J. T., Lafontaine, J. P. and Borden, J. H. (2015). An attract and kill tactic for click beetles based on *Metarhizium brunneum* and a new formulation of sex pheromone. *J. Pest Sci.* 88(4), 707-716.

Kabaluk, T., Vernon, R. and Goettel, M. S. (2007). Mortality and infection of wireworm, *Agriotes obscurus* [Coleoptera: Elateridae], with inundative field applications of *Metarhizium anisopliae*. *Phytoprotection* 88(2), 51-56.

Kaspi, R., Ross, A., Hodson, A. K., Stevens, G. N., Kaya, H. K. and Lewis, E. E. (2010). Foraging efficacy of the entomopathogenic nematode *Steinernema riobrave* in different soil types from California citrus groves. *Appl. Soil Ecol.* 45(3), 243-253.

Kaya, H. K. and Gaugler, R. (1993). Entomopathogenic nematodes. *Annu. Rev. Entomol.* 38(1), 181-206.

Kirfman, G. W., Keaster, A. J. and Story, R. N. (1986). An improved wireworm (Coleoptera: Elateridae) sampling technique for midwest cornfields. *J. Kans. Entomol. Soc.*, 37-41.

Kleespies, R. G., Ritter, C., Zimmermann, G., Burghause, F., Feiertag, S. and Leclerque, A. (2013). A survey of microbial antagonists of Agriotes wireworms from Germany and Italy. *J. Pest Sci.* 86(1), 99-106.

Knodel, J. J. and Shrestha, G. (2018). Pulse crops: Pest management of wireworms and cutworms in the Northern Great Plains of United States and Canada. *Ann. Entomol. Soc. Am.* 111(4), 195-204.

Kölliker, U., Biasio, L. and Jossi, W. (2011). Potential control of Swiss wireworms with entomopathogenic fungi. *IOBC WPRS Bull.* 66, 517-520.

Koppenhöfer, A. M. and Fuzy, E. M. (2006). Effect of soil type on infectivity and persistence of the entomopathogenic nematodes *Steinernema scarabaei*, *Steinernema glaseri*, *Heterorhabditis zealandica*, and *Heterorhabditis bacteriophora*. *J. Invertebr. Pathol.* 92(1), 11-22.

Krieg, A. (1962). *Bacillus thuringiensis* Berliner. Über seine Biologie, Pathogenie und Anwendung in der biologischen Schädlingsbekämpfung. *Anz. Schädlingskunde* 35(4), 61-61.

Kung, S., Gaugler, R. and Kaya, H. K. (1991). Effects of soil temperature, moisture, and relative humidity on entomopathogenic nematode persistence. *J. Invertebr. Pathol.* 57(2), 242-249.

Kung, S.-P., Gaugler, R. and Kaya, H. K. (1990). Soil type and entomopathogenic nematode persistence. *J. Invertebr. Pathol.* 55(3), 401-406.

La Forgia, D., Thibord, J.-B., Larroudé, P., Francis, F., Lognay, G. and Verheggen, F. (2020). Linking variety-dependent root volatile organic compounds in maize with differential infestation by wireworms. *J. Pest Sci.* 93(2), 605-614.

Lacey, L. A., Unruh, T. R., Simkins, H. and Thomsen-Archer, K. (2007). Gut bacteria associated with the Pacific Coast wireworm, *Limonius canus*, inferred from 16s rDNA sequences and their implications for control. *Phytoparasitica* 35(5), 479-489.

Ladurner, E., Quentin, U., Franceschini, S., Benuzzi, M. and Ehlers, R. (2009). Efficacy evaluation of the entomopathogenic fungus *Beauveria bassiana* strain ATCC 74040 against wireworms (*Agriotes* spp.) on potato. *IOBC WPRS Bull.* 45, 445-448.

Lafrance, J. (1968). Seasonal movements of wireworms (Coleoptera-Elateridae) in relation to soil moisture and temperature in organic soils of southwestern Quebec. *Can. Entomol.* 100(8), 801-807.

Landl, M. and Glauninger, J. (2013). Preliminary investigations into the use of trap crops to control *Agriotes* spp. (Coleoptera: Elateridae) in potato crops. *J. Pest Sci.* 86(1), 85-90.

Lane, M. C. and Jones, E. W. (1936). Flooding as a means of reducing wireworm infestations. *J. Econ. Entomol.* 29(5), 842-850.

Langdon, K. W. and Abney, M. R. (2017). Relative susceptibility of selected potato cultivars to feeding by two wireworm species at two soil moisture levels. *Crop Prot.* 101, 24-28.

Leclerque, A., Kleespies, R. G., Ritter, C., Schuster, C. and Feiertag, S. (2011). 'Genetic and electron-microscopic characterization of *'Rickettsiella agriotidis'*, a new Rickettsiella pathotype associated with wireworm, *Agriotes* sp.(Coleoptera: Elateridae). *Curr. Microbiol.* 63(2), 158-163.

Lefko, S. A., Pedigo, L. P., Batchelor, W. D. and Rice, M. E. (1998). Spatial modeling of preferred wireworm (Coleoptera: Elateridae) habitat. *Environ. Entomol.* 27(2), 184-190.

Lewis, E. E. (2002). Behavioural ecology. In: *Entomopathogenic Nematology*. New York, NY: CABI Publishing, pp. 205-220.

Lewis, E. E., Campbell, J., Griffin, C., Kaya, H. and Peters, A. (2006). Behavioral ecology of entomopathogenic nematodes. *Biol. Control* 38(1), 66-79.

Lindroth, E. and Clark, T. L. (2009). Phylogenetic analysis of an economically important species complex of wireworms (Coleoptera: Elateridae) in the Midwest. *J. Econ. Entomol.* 102(2), 743-749.

Logan, J. D., Wolesensky, W. and Joern, A. (2006). Temperature-dependent phenology and predation in arthropod systems. *Ecol. Modell.* 196(3-4), 471-482.

Miles, H. W. (1942). Wireworms and agriculture, with special reference to *Agriotes obscurus* L. *Ann. Appl. Biol.* 29(2), 176-180.

Miles, H. W. and Petherbridge, F. R. (1927). Investigations on the control of wireworms. *Ann. Appl. Biol.* 14(3), 359-387.

Milosavljević, I., Esser, A. D., Bosque-Pérez, N. A. and Crowder, D. W. (2016a). The identity of belowground herbivores, not herbivore diversity, mediates impacts on plant productivity. *Sci. Rep.* 6, 1.

Milosavljević, I., Esser, A. D. and Crowder, D. W. (2016b). Effects of environmental and agronomic factors on soil-dwelling pest communities in cereal crops. *Agric. Ecosyst. Environ.* 225, 192-198.

Milosavljević, I., Esser, A. D. and Crowder, D. W. (2017). Seasonal population dynamics of wireworms in wheat crops in the Pacific Northwestern United States. *J. Pest Sci.* 90(1), 77-86.

Milosavljević, I., Esser, A. D., Murphy, K. M. and Crowder, D. W. (2019). Effects of Imidacloprid seed treatments on crop yields and economic returns of cereal crops. *Crop Prot.* 119, 166-171.

Morales-Rodriguez, A. and Wanner, K. W. (2015). Efficacy of thiamethoxam and fipronil, applied alone and in combination, to control *Limonius californicus* and *Hypnoidus bicolor* (Coleoptera: Elateridae). *Pest Manag. Sci.* 71(4), 584-591.

Morton, A. and Garcia-del-Pino, F. (2017). Laboratory and field evaluation of entomopathogenic nematodes for control of *Agriotes obscurus* (L.) (Coleoptera: Elateridae). *J. Appl. Entomol.* 141(4), 241-246.

Nakao, T. and Banba, S. (2016). Broflanilide: A meta-diamide insecticide with a novel mode of action. *Bioorg. Med. Chem.* 24(3), 372-377.

Nikoukar, A., Ensafi, P., Lewis, E. E., Crowder, D. W. and Rashed, A. (2021). Efficacy of naturally occurring and commercial entomopathogenic nematodes against sugar beet wireworm (Coleoptera: Elateridae). *J. Econ. Entomol.* 114(5), 2241-2244.

Nikoukar, A. and Rashed, A. (2022). Integrated pest management of wireworms (Coleoptera: Elateridae) and rhizosphere in agroecosystems. *Insects* 13(9), 769.

Osbourn, A. E. (2003). Saponins in cereals. *Phytochemistry* 62(1), 1-4.

Parker, W. E. (1994). Evaluation of the use of food baits for detecting wireworms (*Agriotes* spp., Coleoptera: Elateridae) in fields intended for arable crop production. *Crop Prot.* 13(4), 271-276.

Parker, W. E. (1996). The development of baiting techniques to detect wireworms (*Agriotes* spp., Coleoptera: Elateridae) in the field, and the relationship between bait-trap catches and wireworm damage to potato. *Crop Prot.* 15(6), 521-527.

Parker, W. E. and Howard, J. J. (2001). The biology and management of wireworms (*Agriotes* spp.) on potato with particular reference to the U.K. *Agric. For. Entomol.* 3(2), 85-98.

Parker, W. E. and Seeney, F. M. (1997). An investigation into the use of multiple site characteristics to predict the presence and infestation level of wireworms (*Agriotes* spp., Coleoptera: Elateridae) in individual grass fields. *Ann. Appl. Biol.* 130(3), 409-425.

Pedigo, L. P. (1996). *Entomology and Pest Management* (2nd edn.). New Jersey: Prentice Hall.

Poggi, S., Le Cointe, R., Lehmhus, J., Plantegenest, M. and Furlan, L. (2021). Alternative strategies for controlling wireworms in field crops: A review. *Agriculture* 11(5), 436.

Rashed, A., Etzler, F., Rogers, C. W. and Marshall, J. M. (2015). Wireworms in Idaho cereals: Monitoring and identification. *Univ. Ida. Ext Bull* 898.

Rashed, A., Rogers, C. W., Rashidi, M. and Marshall, J. M. (2017). Sugar beet wireworm *Limonius californicus* damage to wheat and barley: Evaluations of plant damage with respect to soil media, seeding depth, and diatomaceous earth application. *Arthropod Plant Interact.* 11(2), 147-154.

Reddy, G. V., Tangtrakulwanich, K., Wu, S., Miller, J. H., Ophus, V. L., Prewett, J. and Jaronski, S. T. (2014). Evaluation of the effectiveness of entomopathogens for the management of wireworms (Coleoptera: Elateridae) on spring wheat. *J. Invertebr. Pathol.* 120, 43-49.

Rojht, H., Kač, M. and Trdan, S. (2009). Nontarget effect of entomopathogenic nematodes on larvae of twospotted lady beetle (Coleoptera: Coccinellidae) and green lacewing (Neuroptera: Chrysopidae) under laboratory conditions. *J. Econ. Entomol.* 102(4), 1440-1443.

Sadeghi, R., Odubiyi, S., Nikoukar, A., Schroeder, K. L. and Rashed, A. (2021). *Mayetiola destructor* (Diptera: Cecidomyiidae) host preference and survival on small grains with respect to leaf reflectance and phytohormone concentrations. *Sci. Rep.* 11(1), 1-12.

Salt, G. and Hollick, F. S. J. (1949). Studies of wireworm population III. Some effects of cultivation. *Ann. Appl. Biol.* 36(2), 169-186.

Sandhi, R. K., Shapiro-Ilan, D. and Reddy, G. V. P. (2020a). Montana native entomopathogenic nematode species against *Limonius californicus* (Coleoptera: Elateridae). *J. Econ. Entomol.* 113(5), 2104-2111.

Sandhi, R. K., Shapiro-Ilan, D., Sharma, A. and Reddy, G. V. P. (2020b). Efficacy of entomopathogenic nematodes against the sugarbeet wireworm, *Limonius californicus* (Mannerheim)(Coleoptera: Elateridae). *Biol. Control* 143, 104190.

Schallhart, N., Tusch, M. J., Staudacher, K., Wallinger, C. and Traugott, M. (2011). Stable isotope analysis reveals whether soil-living elaterid larvae move between agricultural crops. *Soil Biol. Biochem.* 43(7), 1612-1614.

Seal, D. R., Chalfant, R. B. and Hall, M. R. (1992). Effects of cultural practices and rotational crops on abundance of wireworms (Coleoptera: Elateridae) affecting sweetpotato in Georgia. *Environ. Entomol.* 21(5), 969-974.

Seal, D. R., Chalfant, R. B. and McSorley, R. (1997). Seasonal abundance and mathematical distribution of wireworms and wireworm feeding damage in sweet potato. *J. Entomol. Sci.* 32(3), 311-320.

Serrano, J. M., Collignon, R. M., Zou, Y. and Millar, J. G. (2018). Identification of sex pheromones and sex pheromone mimics for two North American click beetle species (Coleoptera: Elateridae) in the Genus Cardiophorus Esch. *J. Chem. Ecol.* 44(4), 327-338.

Shapiro-Ilan, D. I., Jackson, M., Reilly, C. C. and Hotchkiss, M. W. (2004). Effects of combining an entomopathogenic fungi or bacterium with entomopathogenic nematodes on mortality of *Curculio caryae* (Coleoptera: Curculionidae). *Biol. Control* 30(1), 119-126.

Sharma, A., Sandhi, R. K., Briar, S. S., Miller, J. H. and Reddy, G. V. P. (2019). Assessing the performance of pea and lentil at different seeding densities as trap crops for the management of wireworms in spring wheat. *J. Appl. Entomol.* 143(4), 460-469.

Shirck, F. H. (1945). Crop rotations and cultural practices as related to wireworm control in Idaho12. *J. Econ. Entomol.* 38(6), 627-633.

Shirck, F. H. (1946). Growth of the sugar-beet wireworm on different food plants. *J. Econ. Entomol.* 39(5), 648-651.

Staudacher, K., Pitterl, P., Furlan, L., Cate, P. C. and Traugott, M. (2011). PCR-based species identification of Agriotes larvae. *Bull. Entomol. Res.* 101(2), 201-210.

Sufyan, M., Abbasi, A., Gogi, M. D., Arshad, M., Nawaz, A. and Neuhoff, D. (2017). Efficacy of *Beauveria bassiana* for the management of economically important wireworm species (Coleoptera: Elateridae) in organic farming. *Gesunde Pflanz* 69(4), 197-202.

Toba, H. H., Lindegren, J. E., Turner, J. E. and Vail, P. V. (1983). Susceptibility of the Colorado potato beetle and the sugar beet wireworm to *Steinernema feltiae* and *S. glaseri*. *J. Entomol.* 15(4), 597-601.

Traugott, M., Benefer, C. M., Blackshaw, R. P., van Herk, W. G. and Vernon, R. S. (2015). Biology, ecology, and control of elaterid beetles in agricultural land. *Annu. Rev. Entomol.* 60, 313-334.

Traugott, M., Pázmándi, C., Kaufmann, R. and Juen, A. (2007). Evaluating 15N/14N and 13C/12C isotope ratio analysis to investigate trophic relationships of elaterid larvae (Coleoptera: Elateridae). *Soil Biol. Biochem.* 39(5), 1023-1030.

Traugott, M., Schallhart, N., Kaufmann, R. and Juen, A. (2008). The feeding ecology of elaterid larvae in central European arable land: New perspectives based on naturally occurring stable isotopes. *Soil Biol. Biochem.* 40(2), 342-349.

Van Herk, W., Vernon, B., Bourassa-Tait, G., Tóth, M. and Kovacs, E. (2022). Field evaluation of selected plant volatiles and conspecific pheromones as attractants for *Agriotes obscurus* and *A. lineatus* (Coleoptera: Elateridae). *Insects* 13(2), 173.

Van Herk, W., Vernon, R. S., Vojtko, B., Snow, S., Fortier, J. and Fortin, C. (2015). Contact behaviour and mortality of wireworms exposed to six classes of insecticide applied to wheat seed. *J. Pest Sci.* 88(4), 717-739.

Van Herk, W. G., Labun, T. J. and Vernon, R. S. (2018). Efficacy of diamide, neonicotinoid, pyrethroid, and phenyl pyrazole insecticide seed treatments for controlling the sugar beet wireworm, *Limonius californicus* (Coleoptera: Elateridae), in spring wheat. *J. Entomol. Soc. B.C.* 115, 86-100.

Van Herk, W. G., Lemke, E., Gries, G., Gries, R., Serrano, J. M., Catton, H., Wanner, K., Landolt, P. J., Cooper, W. R., Meers, S., Nikoukar, A., Smith, J. L., Alamsetti, S. K. and Etzler, F. E. (2021a). Limoniic acid and its analog as trap lures for pest Limonius species (Coleoptera: Elateridae) in North America. *J. Econ. Entomol.* 114(5), 2108-2120.

Van Herk, W. G., Vernon, R. S., Goudis, L. and Mitchell, T. (2021b). Broflanilide, a meta-diamide insecticide seed treatment for protection of wheat and mortality of wireworms (*Agriotes obscurus*) in the field. *J. Econ. Entomol.* 114(1), 161-173.

Van Herk, W. G., Vernon, R. S., Labun, T. J., Sevcik, M. H. and Schwinghamer, T. D. (2021c). Distribution of pest wireworm (Coleoptera: Elateridae) species in Alberta, Saskatchewan, and Manitoba (Canada). *Environ. Entomol.* 50(3), 663-672.

Van Herk, W. G. and Vernon, R. S. (2006). Effect of temperature and soil on the control of a wireworm, *Agriotes obscurus* L.(Coleoptera: Elateridae) by flooding. *Crop Prot.* 25(9), 1057-1061.

Van Herk, W. G. and Vernon, R. S. (2007a). Morbidity and recovery of the Pacific Coast wireworm, *Limonius canus*, following contact with tefluthrin-treated wheat seeds. *Entomol. Exp. Appl.* 125(2), 111-117.

Van Herk, W. G. and Vernon, R. S. (2007b). Soil bioassay for studying behavioral responses of wireworms (Coleoptera: Elateridae) to insecticide-treated wheat seed. *Environ. Entomol.* 36(6), 1441-1449.

Van Herk, W. G. and Vernon, R. S. (2013). Categorization and numerical assessment of wireworm mobility over time following exposure to bifenthrin. *J. Pest Sci.* 86(1), 115-123.

Van Herk, W. G., Vernon, R. S., Tolman, J. H. and Ortiz Saavedra, H. O. (2008). Mortality of a wireworm, *Agriotes obscurus* (Coleoptera: Elateridae), after topical application of various insecticides. *J. Econ. Entomol.* 101(2), 375-383.

Vernon, B. and van Herk, W. (2022). Wireworms as pests of potato. In: *Insect Pests of Potato*. Academic Press, pp. 103–148. Cambridge, MA

Vernon, R. and Van Herk, W. (2017). Wireworm and flea beetle IPM in potatoes in Canada: Implications for managing emergent problems in Europe. *Potato Res.* 60(3–4), 269–285.

Vernon, R. S. (2005). Aggregation and mortality of *Agriotes obscurus* (Coleoptera: Elateridae) at insecticide-treated trap crops of wheat. *J. Econ. Entomol.* 98(6), 1999–2005.

Vernon, R. S., Blackshaw, R. P., van Herk, W. G. and Clodius, M. (2014). Mass trapping wild *Agriotes obscurus* and *Agriotes lineatus* males with pheromone traps in a permanent grassland population reservoir. *Agric. For. Entomol.* 16(3), 227–239.

Vernon, R. S., Kabaluk, T. and Behringer, A. (2000). Movement of *Agriotes obscurus* (Coleoptera: Elateridae) in strawberry (Rosaceae) plantings with wheat (Gramineae) as a trap crop. *Can. Entomol.* 132(2), 231–241.

Vernon, R. S., Van Herk, W., Tolman, J., Ortiz Saavedra, H. O., Clodius, M. and Gage, B. (2008). Transitional sublethal and lethal effects of insecticides after dermal exposures to five economic species of wireworms (Coleoptera: Elateridae). *J. Econ. Entomol.* 101(2), 365–374.

Vernon, R. S., Van Herk, W. G., Clodius, M. and Harding, C. (2009). Wireworm management I: Stand protection versus wireworm mortality with wheat seed treatments. *J. Econ. Entomol.* 102(6), 2126–2136.

Vernon, R. S., van Herk, W. G., Clodius, M. and Harding, C. (2013). Crop protection and mortality of Agriotes obscurus wireworms with blended insecticidal wheat seed treatments. *J. Pest Sci.* 86(1), 137–150.

Wakil, W., Yasin, M. and Shapiro-Ilan, D. (2017). Effects of single and combined applications of entomopathogenic fungi and nematodes against *Rhynchophorus ferrugineus* (Olivier). *Sci. Rep.* 7(1), 5971.

Wenninger, E. J., Rashed, A., Rondon, S. I., Alyokhin, A. and Alvarez, J. M. (2020). Insect pests and their management. In: Stark, J. C., Thornton, M. and Nolte, P. (Eds.), *Potato Production Systems*. Cham: Springer Nature, pp. 283–345.

Wilkinson, A. T. (1963). Wireworms of cultivated land in British Columbia. *J. Entomol. Soc. B.C.* 60, 3–17.

Williams, L., Morra, M. J., Brown, P. D. and McCaffrey, J. P. (1993). Toxicity of allyl isothiocyanate-amended soil to *Limonius californicus* (Mann.) (Coleoptera: Elateridae) wireworms. *J. Chem. Ecol.* 19(6), 1033–1046.

Williams, L., Serrano, J. M., Johnson, P. J. and Millar, J. G. (2019). 13-Tetradecenyl acetate, a female-produced sex pheromone component of the economically important click beetle *Melanotus communis* (Gyllenhal) (Coleoptera: Elateridae). *Sci. Rep.* 9(1), 16197.

Wright, R. J., Hein, G. L., Hoback, W. W. and Pavlista, A. D. (2002). Biology and management of potato insects. *University of Nebraska-Lincoln Extension*, EC 1565.

Yates, F. and Finney, D. J. (1942). Statistical problems in field sampling for wireworms. *Ann. Appl. Biol.* 29(2), 156–167.

Zhang, Z., Zhang, X., Zhao, Y., Mu, W. and Liu, F. (2017). Efficacy of insecticidal seed treatments against the wireworm *Pleonomus canaliculatus* (Coleoptera: Elateridae) in China. *Crop Prot.* 92, 134–142.

Part 4

Emerging issues

Chapter 13

Recent invasions of insect pests of wheat and sorghum

Michael J. Brewer and Blake H. Elkins, Texas A&M AgriLife Research, Texas A&M University, USA

1 Introduction

Recent invasions of insect pests onto wheat (*Triticum aestivum*) and other cereals join the growing threat to agriculture from invasive species worldwide (Paini et al., 2016). Invasions by insect pests may disrupt community-scale grain production essential for local food security and larger-scaled cereal production regions that contribute to the global grain distribution and processing system. Recent invasions do not appear to be insect taxa biased as represented by the six example invaders discussed below, although selected taxa appear to be over-represented in some regions such as the periodic aphid invasions into the Great Plains of North America over the last 100 years (Brewer et al., 2019). The case study presented on the invasion of *Melanaphis sorghi* Theobold (Hem: Aphididae) (sugarcane aphid) onto sorghum (*Sorghum bicolor*) is taken from this region.

Regardless of taxa, insect invasions into crop-based agroecosystems pass through general phases (Venette and Morey, 2020). The phases may be modified from the cereal agroecosystem perspective. First, invasive insect entry and establishment into a new area may cause substantial local disruptions, often characterized as outbreaks, that severely impact community based production or segments of larger-scale cereal grain production systems. Next, the extent of regional spread and outbreaks are based on the capacity of insect movement and the extent of a favorable environment of host cereal crops, other host plants, weather, and other relevant landscape features, thereby threatening

http://dx.doi.org/10.19103/AS.2022.0114.18

both multiple small-scale systems and larger-scale production regions to varying degrees across the expansion zone of the invasion. The invasion may progress to a state of episodes of pest population increases and declines as influenced by pest life history and behavior, interactions with natural enemies and other species, abiotic environmental factors, and agroecosystem (natural and manipulated) response. As a result, regional variation may occur in pest population levels, associated agroecosystem services such as pest suppression by natural enemies, and economic impact on cereal crops, most notably in production systems that cross ecological gradients.

This chapter provides a general description of the invasion and management response of six insects that have affected wheat and other cereals (Section 2). A more detailed case study follows: the recent invasion of *M. sorghi* across the sorghum agroecosystem of North America. This near continent-scale invasion provides perspectives on invasion progression (both ecology and management) across ecological and agricultural gradients and on management response appropriate at local and regional scales in the context of environmental (landscape and weather) factors that influence pest-natural enemy-plant dynamics (Section 3). The chapter concludes with a discussion of future trends in light of climate change and lessons learned from these recent insect pest invasions (Section 4).

2 Recent invasions onto wheat and other cereal grains

2.1 Agriphila aeneociliella *Eversmann (Lepid: Crambidae)*

Agriphila aeneociliella is a pest of wheat that has recently expanded its range in China. *A. aeneociliella* is found across Europe and Asia (Shodotova, 2008). It threatens wheat production throughout China (Ji et al., 2014), with the most recent detections from northern (2008) and eastern (2010) China (Shodotova, 2008; Ji et al., 2014). Besides wheat, hosts of this species include smooth brome (*Bromus inermis*) and wild oats (*Avena fatua*) (Shodotova, 2008; Peng et al., 2013; Zeng et al., 2014). Wheat damage is primarily due to larvae boring into stalks. During severe infestations, as many as 20 individuals can infest a single stalk resulting in severe economic loss and a significant population increase of *A. aeneociliella* (Zeng et al., 2014). *A. aeneociliella* is also capable of overwintering in winter wheat (Chi et al., 2016). Current management practices for this pest primarily rely on insecticides (Zeng et al., 2014).

2.2 Metopolophium festucae cerealium *Stroyan (Hemi: Aphididae)*

Metopolophium festucae cerealium has recently expanded its range in the wheat and barley production system of the Pacific Northwest of the United States of America (U.S.) (i.e. Oregon, Washington, and Idaho) (see also Chapter 4 of this volume). Host plants include wheat, oats (*Avena sativa*), barley (*Hordeum vulgare*), and non-crop grasses including *Lolium perenne*, *L. multiflorum*, *Festuca rubra*, *F. pratensis*, *F. arundinacea*, and *Dactylis glomerata* (Dent and Wratten, 1986; Davis et al., 2014). This species is likely a native of the United Kingdom, where it is not considered an economic pest of cereals. *M. festucae* (sensu stricto) was first detected in the U.S. in 1970 in California (Foottit et al., 2006). Halbert and Sandvol (1995) observed this species in Oregon during cereal aphid surveys conducted during the 1990s. There was renewed interest in the early 2010s when aphid surveys in wheat of the Pacific Northwest found it had become ubiquitous (Halbert et al., 2013). Ongoing surveys continue to detect *M. festucae cerealium* as a relatively abundant species that appears to be expanding in the Pacific Northwest cereal systems (S. Eigenbrode and A. Rashed, pers. comm.).

Feeding leads to pigmented and necrotic spots on plants, reducing growth and biomass of wheat but not barley (Sadeghi et al., 2016; Davis et al., 2014). Over a 60% reduction in biomass from *M. festucae cerealium* feeding on wheat has been observed experimentally. It is not a competent vector of barley yellow dwarf virus (Sadeghi et al., 2016). *M. festucae cerealium* does not seem to be causing widespread economic damage to wheat in the U.S. at this time, but its presence and persistence in wheat and barley across the large diverse region of the Pacific Northwest are of concern. This species was introduced into Chile in the 1990s where natural enemies were quick to respond and generally kept populations below economically damaging levels (Starý et al.1994).

2.3 Cephus cinctus *Norton (Hymen: Cephidae) (wheat stem sawfly)*

Cephus cinctus is a major pest of wheat in the northern Great Plains of the U.S. that recently expanded its range of impact onto wheat in Colorado (see also Chapter 9 of this volume) (Irell and Peairs, 2011). This pest has been well established in the U.S. for more than a century, with early identifications coming from Colorado (Norton, 1872), California, and New York (Ainslie, 1920; Comstock, 1889). It is debatable if this species was introduced from Eurasia or is native to North America (Beres et al. 2011) and why substantial wheat damage has expanded into the central Great Plains.

Biology and management of *C. cinctus* was reviewed by Beres et al. (2011). *C. cinctus* utilizes a number of uncultivated grasses in addition to wheat (see Cockrell et al., 2017 for a complete host list). Larvae of *C. cinctus* bore into wheat stems causing severe economic losses as a result of decreased photosynthesis and greater susceptibility to lodging which makes grain more difficult to harvest. Management using conventional contact insecticides has proven difficult as larvae in the stems are not exposed to these insecticides. Solid stem wheat cultivars disrupt stem feeding, but these cultivars generally have lower yield potential and quality in the absence of *C. cinctus* than susceptible cultivars (Beres et al., 2011). These drawbacks in using solid stem wheats have led to mixed adoption by wheat growers. Multiple primary parasitoids have been identified for this pest. However, parasitism rates needed for suppression below economically damaging levels are not often achieved in the field and are subject to cultivar and seasonal variation (Beres et al., 2011; Peirce et al., 2021). Biological control can be integrated with solid stem cultivars, thereby decreasing the amount of parasitism needed to effectively control this pest (Rand et al., 2020).

2.4 Sipha maydis *Passerini (Hemi: Aphididae)*

Sipha maydis is a recent aphid pest of multiple cereals including wheat in the U.S. Hosts include economically important crops such as oats, barley, sorghum, wheat, sugarcane (*Saccharum officinarum*), rye (*Secale cereale*), and corn (*Zea mays*). It also utilizes many weedy and uncultivated grasses (see Puterka et al., 2019 for a host list). This pest was first detected on wheat in the Americas in Argentina in the early 2000s (Corrales et al., 2007). It was also detected in the U.S. in California around the same time period, across the western edge of the Great Plains from New Mexico to Wyoming, and into intermountain production regions of Utah (Puterka et al., 2019). *S. maydis* can utilize a wide range of environments including dry steppe, temperate, and mountain areas, but it has not been detected in subtropical areas (Corrales et al., 2007).

During multistate surveys, aphid densities were generally low, but occasional outbreaks were detected with infestations of 1500 aphids per plant (Puterka et al., 2019). Aphid feeding induces leaf chlorosis in wheat, and *S. maydis* also is a vector of several important plant viruses including barley yellow dwarf virus, zucchini yellow mosaic virus, and cucumber mosaic virus (Blackman and Eastop, 2000; Katis et al., 2006; Corrales et al., 2007). Although it was widely distributed in the western Great Plains and had the capacity to damage multiple cereal crops, outbreaks appeared sporadic (Puterka et al., 2019). Natural enemies of this species have not yet been identified in the U.S. (Puterka et al., 2019), and no parasitism of *S. maydis* was observed in Argentina (Corrales et al., 2007). However, parasitism has been reported from other parts of its range including Europe and Asia, leading to the proposal that resident natural enemies that prey on other cereal aphids in the U.S. Great Plains will adapt to this aphid (Puterka et al., 2019). Corrales et al. (2007) reported that wheat varieties with resistance to the aphids *Schizaphis graminum* (greenbug) and *Diuraphis noxia* (Russian wheat aphid) (Hemi: Aphididae) show no resistance to *S. maydis*.

2.5 Diuraphis noxia *(Kurdjumov) (Hemi: Aphididae) (Russian wheat aphid)*

Diuraphis noxia invaded the wheat production region of Australia and Tasmania in 2016 (see also Chapter 6 of this volume). It quickly spread throughout the warm dry production areas of the southern rim of Australia (Avila et al., 2018). Based on climate matching and its ability to spread via wind-blown alates, the invasion into New Zealand and other island counties of the region is anticipated (Avila et al., 2018). With its previous invasion into cereals of South Africa and North America and its native range in central Asia, it is approaching the world-scale distribution of wheat (Ward et al., 2020).

D. noxia is anholocylic and viviparous with no sexual forms detected, and one biotype has been recovered during its Australian invasion (Ward et al., 2020). Its host plant range includes wheat and barley as the primary cultivated grasses, and another 140 species of the Poaceae (Ward et al., 2020). *D. noxia* is not a significant vector of disease although quick and severe plant response to feeding can be mistaken for disease symptoms. Visual leaf injury is indicated by longitudinal streaking (white to pink-purple in color) and folding of leaves. Aphids feed within the leaf folds, where they are partially protected from contact insecticides (Ward et al., 2020). Wheat response to *D. noxia* feeding is moisture sensitive. In Australia, wheat yield loss exceeded 70% when plants were exposed to *D. noxia* in exclusion cages in dry conditions. In a high rainfall region, the same experimental set up resulted in no yield loss (Ward et al., 2020). Insecticides were available, and decision-making guidelines were adopted from other regions of the world (Botha, 2021). Guidance on insecticide use and delineating regional spread of *D. noxia* helped to minimize control costs. A taxonomically diverse set of aphid natural enemies was seen in Australian wheat production systems, and no aphid-resistant traits have been found during the screening of Australian adapted wheats (Ward et al., 2020). A similar aphid enemy complex and wheat-resistant cultivars have contributed to *D. noxia* suppression in the U.S. (Giles et al., 2017; Brewer et al., 2019). Increasing incidence and parasitism of *D. noxia* along the northwestern edge of the Great Plains were seen over 7 years, while *D. noxia* infestations declined over the same time period and locations of a multistate evaluation (Fig. 1). Parasitism was seen in wheat and barley susceptible and resistance to *D. noxia* (Brewer et al., 1999). For the large wheat agroecosystem of Australia and elsewhere in the region, refinement of decision-making protocols for insecticide use and broadening efforts in breeding for wheat resistance to *D. noxia* are advisable to optimize management and to transition to a more sustainable *D. noxia* management.

2.6 Melanaphis sorghi *Theobold (Hem: Aphididae) (sugarcane aphid)*

M. sorghi invaded the subtropical and southern temperate sorghum agroecosystem of North America in 2013, as first documented along the Gulf Coast of Texas and Mexico. Within 3 years, its range expanded to the

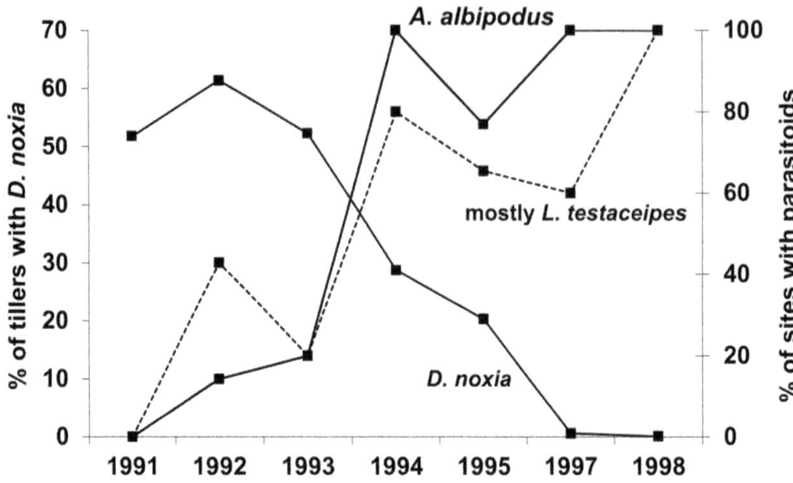

Figure 1 The decline of *Diuraphis noxia* infestation in wheat (scale on left axis) as the incidence of aphelinid (*Aphelinus albipodus*) and braconid (mostly *Lysiphlebus testaceipes*) parasitism increased (scale on right axis) over a 7-year period along the northwestern edge of U.S. Great Plains. *D. noxia* invaded wheat and barley in the region beginning in 1986.Source: Data taken from Brewer et al. (2001, 2005).

northcentral U.S. Great Plains, and from the western Great Plains states to the eastern seaboard (Bowling et al., 2016), with detections seen in California and Arizona toward the end of this expansion period. Worldwide, its host plant range includes many Poaceae genera, and sorghum and sugarcane are the most important cultivated hosts (Singh et al., 2004). Prior to 2013, another biotype was previously found on sugarcane with minimal occurrence on nearby sorghum. The invading *M. sorghi* was found on sorghum and Johnson grass (*Sorghum halepense*) and was designated as the 'superclone' responsible for outbreaks and the range expansion seen on sorghum (Harris-Shultz et al., 2018). Previous publications of *M. sorghi* on sorghum in North America have used the name *Melanaphis sacchari* (Brewer et al., 2019 and references therein) until the invasive biotype was proposed as a separate species based on molecular and morphological evidence (Nibouche et al., 2021).

M. sorghi is fully or predominantly anholocylic and viviparous in North America (Bowling et al., 2016). Populations of *M. sorghi* overwinter on remnant, ratoon, and forage sorghums and the weedy Johnson grass that persist during mild winters in southern regions of the U.S. and Mexico (Esquivel et al., 2021). It is commonly observed on the underside of green leaf tissue. It is also observed in the developing seed heads of sorghum and leaves and stem tissue of Johnson grass (Fig. 2). Rapid colony expansion exceeding 2500 aphids per leaf can occur on susceptible sorghums, particularly in warm dry weather (Gordy et al., 2019, Fig. 2). Long-distance wind-aided dispersal of alates contributes to and is likely the driver of annual re-infestation of grain sorghum planted yearly in more northern (temperate) regions where sorghum and Johnson grass die or senescence during winter (Esquivel et al., 2021). Johnson grass may aid *M. sorghi* overwintering in southern temperate areas where sorghum does not survive the winter. It has been observed on Johnson grass leaves and stem tissue at and just below the soil surface (Fig. 2).

Sorghum injury results from a combination of loss of plant nutrients, leaf chlorosis, and reduction in photosynthetic productivity which is intensified under plant water stress (Singh et al., 2004). Disease vector potential of *M. sorghi* appears low (Bowling et al., 2016). Yield loss can exceed 50%, and loss can be exacerbated by honey dew disrupting mechanical harvest (Bowling et al., 2016). Insecticides were key to managing *M. sorghi* outbreaks early in the invasion, and several insecticide products with low mortality to natural enemies came into common use (Lytle and Huseth, 2021). Predators and parasitoids of *M. sorghi* have been documented throughout North America (Colares et al., 2015; Rodríguez-del-Bosque et al., 2018; Maxson et al., 2019). Host plant resistance traits have been identified in sorghum hybrids previously developed from *S. graminum* resistance, and additional parental lines with *M. sorghi* resistance were identified (Armstrong et al., 2015; Paudyal et al., 2019; Tetreault et al., 2019). Partially *M. sorghi*-resistant hybrids were released and

Figure 2 Photos of *Melanaphis sorghi* (a) on sorghum leaves at early colonization, (b) near peak aphid population growth, (c) as apterae and alatae, (d) on Johnsongrass leaves, (e) on Johnsongrass stems, and (f) with juvenile lady beetles and *Aphelinus nigritus* parasitized mummy present on sorghum leaf.

widely planted, which as a group were able to maintain yield under pressure from field populations of *M. sorghi* (Gordy et al., 2021).

3 Management approaches for a recent invader onto sorghum

The recent invasion of *M. sorghi* across the sorghum agroecosystem of North America provided a contemporary case study, in which pest management and agroecosystem response crossed from subtropical to temperate regions of sorghum cultivation and from small to large production systems. The disruption to sorghum occurred on a near continent scale, yet *M. sorghi* incidence and abundance and severity of sorghum damage varied geographically (Bowling et al., 2016; Brewer et al., 2019). The recent invasion overlapped with prior aphid invasions of wheat and barley in the U.S. Great Plains, where an agroecosystem response to *D. noxia* had been previously documented (Quisenberry and Peairs, 1989; Brewer et al., 2019) (Fig. 1). Partly drawing out of the previous experience of the *D. noxia* invasion on wheat and barley, a mixture of traditional and advanced pest management tools and approaches were developed and implemented as appropriate to local (field) and regional (landscape-scale) management. The management response considered the spatial variation of *M. sorghi* incidence, population growth, and sorghum impact in the context of environmental (landscape and weather) factors that influence pest-natural enemy-plant dynamics. This recent invasion and management response are instructive for developing management approaches for *D. noxia,* a more recent invader of wheat production in Australia (Section 2.5) (Avila et al., 2018; Ward et al., 2020), and other current and future insect pest invasions onto wheat and other cereal grains.

3.1 Local (field-scale) insect pest management framework

Decision-making protocols for insect pest management commonly take the form of economic thresholds supported by in-field sampling methods that guide insecticide use. An economic threshold is defined as the pest population level above which the use of an insecticide or other management tactic is economically justified based on projected economic damage to the crop (Pedigo, 1999). The decision-making protocols are applied to individual fields or occasionally sections of large fields. These protocols are valuable especially in the early expansion phase of a pest invasion when other integrated pest management (IPM) tools are not well developed. Time and cost efficiencies of insect sampling and insecticide use are important considerations to aid adoption by practitioners. As research matures and agroecosystems adapt to new pests, there are benefits to rebalancing the IPM toolbox with more weight given to approaches that are more sustainable (e.g. biological control by resident or introduced natural enemies) and can be readily inserted as part of the existing agronomic practices (e.g. host plant resistance). Adoption of advanced IPM tactics and pest assessment and forecasting tools (see also Chapter 15 of this volume) may be challenged by technology transfer and integration into existing pest management systems.

As in past aphid invasions into cereal agroecosystems, some insecticide products were available in spray formulations to reduce the impact of *M. sorghi* outbreaks that were occurring during the first years of its invasion (Bowling et al., 2016). Newer products (e.g. sulfoxaflor and flupyradifurone) with high efficacy in controlling *M. sorghi* with less disruption to non-targets (such as pollinators and natural enemies) were identified. They became available for use on sorghum via special insecticide registration (use exemption) processes that accelerated use in commercial sorghum production in many states of the U.S. (Bowling et al., 2016). Research on decision-making for insecticide use was stimulated by the availability of these insecticides that tended to be higher in cost than older products and by conditions placed on obtaining special use exemptions. Decision-making at the field-scale (i.e. economic thresholds linked to aphid monitoring) targeted insecticide use when and where *M. sorghi* was threatening the crop. The field-scale approach was appropriate for this pest that varied in population levels across fields of a growing region or even within fields (Elliott et al., 2017; Deleon et al., 2017).

Economic thresholds were based on research that showed an approximate linear decrease in yield as *M. sorghi* density increased from near-zero to hundreds of aphids per leaf. The steady rate of yield decline across increasing aphid densities was relatively stable across the southern U.S. as demonstrated in multi-site replicated experiments using sorghum hybrids commonly planted in the region (Gordy et al., 2019). The yield loss estimates were most applicable under warm dry conditions when aphid population increase can be high (aphid doubling time under 4 days) (Gordy et al., 2021). This consistency in what was termed a high aphid growth environment supported recommendation of a regional density-based economic threshold (i.e. 40 aphids per leaf) to guide insecticide use during vegetative growth through early seed head development (Gordy et al., 2019). Those practitioners with more information on the pace of sugarcane aphid population increase and the economics of using an insecticide (i.e. cost of insecticide use and value of the grain) had opportunity to adjust the threshold (Gordy et al., 2019).

To use economic thresholds, aphid monitoring must be frequent when sorghum is most susceptible to *M. sorghi* feeding (i.e. up to twice per week starting at first *M. sorghi* detection through early seed head development). Monitoring must be efficient in order to implement activities across multiple fields. Monitoring must also provide reliable population estimates (Thomas et al., 2018; Lindenmayer et al., 2020) to gauge whether the population exceeded or was below an economic threshold (Gordy et al., 2019). Density-based economic thresholds and basic sampling methods based on visual counts of aphids were timely and widely promoted (Bowling et al., 2016) but challenging to deploy across the large-scale sorghum production of the U.S. Great Plains and in smaller production areas of the southern U.S. In addition, *M. sorghi* population levels varied between fields and across production regions especially as the invasion matured (Elliott et al., 2017). Therefore, returns on sampling efficiencies per sampling event diminished when placed in the context of visiting multiple fields with variable *M. sorghi* populations in order to make management decisions on an individual field basis.

Adjustments to the density-based threshold of 40 aphids per leaf were explored to increase efficiencies in use. Early decision-making protocols found in outreach literature for growers included several protocols based on % infestation of plants or leaves with *M. sorghi* in order to trigger insecticide use to reduce sampling time (Knutson et al., 2018), but the relationship to yield decline was only known for the density-based thresholds. To stimulate adoption by crop and pest consultants, tally-based thresholds (commonly inserted as part of binomial sampling approaches in the insect pest management literature) were investigated as substitutes for the density-based threshold. A quick visual estimation of whether aphids on a leaf were above or below a tally count may substitute for a more time consuming and variable estimate based on visually counting or estimating the total number of aphids on a leaf (Thomas et al., 2018). Lindenmayer et al. (2020) considered a suite of tallies using linear regression models, resulting in a recommendation of a 50 aphid per leaf tally based on best regression model fit and ease of use of established sampling protocols. Gordy et al. (2021) found that using tallies of 50 aphid or 100 aphid per leaf were satisfactory substitutes for the density-based approach comparing the methods using linear and second-order polynomial regressions (R^2 from 0.74 to 0.94). Visual classification of aphid densities above or below a 50 aphid or 100 aphid per leaf tally was reliable (Lindenmayer et al., 2020, Fig. 2 provides a visual example of a leaf infested with about 100 aphids). Operationally, approximately 15% of leaves infested with 50 aphids or more in a 100-leaf sample was equivalent to the 40 aphids per leaf density-based threshold. Use of this tally-based threshold resulted in the same insecticide use decisions and time savings of about 50% compared with using the density-based threshold (Gordy et al., 2021). Lindenmayer et al. (2020) implemented a binomial sequential sampling plan using a 50 aphids per leaf tally which gained further efficiencies: an average of 22 leaves needed to be inspected to make a reliable decision.

The use of smart phones and automated classification of aphid counts may further reduce sampler variation associated with experience level and sampler fatigue (Thomas et al., 2018). Deng et al. (2020) classified *M. sorghi* infestations on sorghum leaves into four density categories using image processing algorithms applied to images taken by a smart phone. The machine learning algorithm produced average classification accuracies of 85%, including two classifications (0–10 aphids and 11–125 aphids per leaf) that can be modified to support tally-based economic thresholds. Full integration of economic thresholds and sampling protocols may be modeled after the *S. graminum* monitoring system on wheat (sometimes referred to as glance-and-go and rapid sampling in the outreach literature for growers [Royer et al., 2015]) and as recently launched as a smart phone app for *M. sorghi* (Royer et al., 2022).

Advanced methods in detecting aphid stress and the use of aphid-resistant hybrids in areas at risk of *M. sorghi* outbreaks may further optimize management at the local scale, as well as accrue benefits regionally. Remote sensing was considered for in-field use as a surrogate or aid to traditional sampling. Remotely sensed optical data obtained by a multispectral camera mounted on a manned low-altitude aircraft was able to detect plant decline caused by *M. sorghi* (Elliott et al., 2015). In a proof of concept, a commonly used vegetation index (NDVI, normalized differences vegetation index) was calculated using near-infrared and red wavelength bands. Low NDVI values were correlated with *M. sorghi*-induced sorghum injury in fields where *M. sorghi* had exceeded population levels above economic concern preceding the flight. Stanton et al. (2017) extended the application prior to visual detection of plant decline using a commercial camera (three-wavelength bands with a filter to include near-infrared wavelength detection) mounted on a UAS flown at a very low altitude (i.e. agricultural drone). NDVI and plant height maps, generated from structure-from-motion image processing, were constructed for a field where *M. sorghi* densities were manipulated to obtain values below and above economic injury levels of about 100 aphids per leaf. Increasing aphid density was negatively correlated with NDVI, plant height, and grain yield. The variability when *M. sorghi* densities were regressed on NDVI ($R^2 = 0.40$) (Stanton et al., 2017) raises caution in using NDVI as a surrogate of aphid density in decision-making at the individual field level. But NDVI or similar vegetation indices (Zhang et al., 2021) may aid in detecting clusters of fields or regions of sorghum production where *M. sorghi* infestations or other forms of sorghum stress may be occurring, thereby guiding more targeted field monitoring by crop and IPM consultants.

Sorghum resistance to *M. sorghi* is a valuable IPM tool that is well suited to small and community-based sorghum production (if a planting seed distribution system is present) as well as larger production systems. The IPM tactic has the advantage in implementation by 'coming with the seed'. Although it is not without risk in selecting pest biotypes that overcome plant resistance, as well documented for races of the midge *Mayetiola destructor* (Dip: Cecidomyiidae) (Hessian fly) that survive on wheats with different genes for resistance (Hatchett and Gallun, 1970). Some sorghum hybrids in use at the time of the

North American invasion were partially resistant due to a breeding history that included resistance to the aphid *S. graminum* (Armstrong et al., 2015). Additional sources of resistance to *M. sorghi* were identified from parental lines maintained in regional breeding programs (Armstrong et al., 2015). Paudyal et al. (2019) found phenotypic resistance expression varied across sorghum hybrids tested and characterized response of *M. sorghi* as antibiosis, tolerance, and antixenosis depending on the hybrid. In mechanistic and heritability experiments, antixenosis and tolerance traits of parental lines and hybrids were indicated by changes in probing behavior of the piercing-sucking mouthparts into leaves by using electronic penetration graph techniques (Tetreault et al., 2019). Crossing experiments indicated that a single dominant locus was responsible for *M. sorghi* resistance in one parental line used in breeding programs, which may facilitate the rapid development of resistant hybrids (Tetreault et al., 2019). In replicated field experiments conducted in multiple years and states, resistant hybrids developed from several resistance resources reduced *M. sorghi* populations by 80% compared to susceptible hybrids. The population doubling time increased about six-fold on resistant hybrids. A suite of partially *M. sorghi*-resistant hybrids soon became available commercially in North America (Gordy et al., 2021). These resources bode well for further development and deployment of sorghum hybrids resistant to *M. sorghi*, which are especially valuable in regions of moderate to high risk of annual *M. sorghi* infestation.

3.2 Environmental context of insect management

Management options outlined in Section 3.1 are traditionally planned for local (field-scale) use. By considering environmental context, these tactics can be adapted and integrated for improved management of cereal pests at a larger scale beyond individual fields. Ideally, management strategies suitable for regional implementation can be applied across the landscape strategically, such as planting aphid-resistant hybrids where the landscape and weather patterns are conducive to *M. sorghi* colonization and reproduction on annual plantings of grain sorghum. Strategies for best implementation of regional approaches to pest management are challenging due in part to variation of pest-natural enemy-plant interactions across spatial and temporal shifts in crops and landscapes, weather, and scale of production.

Understanding landscape influences on insects in agroecosystems benefits from evaluation using a landscape structure framework, divided into composition (what the landscape consists of: identifiable features organized in patches in the landscape such as a sorghum field, pasture, or riparian area) and configuration (how the patches are arranged in the landscape such as size and distance of wheat fields to grasslands). Host plant availability is an obvious compositional element to consider when investigating landscape relationships to cereal aphid infestations (Brewer et al., 2019). For example, the primary crop (cultivated sorghum) and non-crop (Johnsongrass) hosts of *M. sorghi* contribute to determining aphid outbreak risk within a region. Landscape configuration has more recently been shown to affect insect pests and their natural enemies (Haan et al., 2020). Aspects of configuration include size, shape, and placement of patches within the landscape. The extent of patchiness (e.g. the number of grain fields in a standardized area of production) and interfaces of different patches or habitats (e.g. the arrangement of cereal fields with managed grasslands and waterways) have shown strong relationships to insect pest abundance and natural enemy activity in agroecosystems. For example, edge type within a cereal agroecosystem was found to influence the movement of pest and beneficial insects between cereal and canola fields in Australia (Macfadyen and Muller, 2013). Landscape structure is also dynamic, changing temporally following annual and seasonal cycles of pests, natural enemies, and crop phenology. Wheat and other cereal production occurring in temperate climatic zones are particularly prone to annual cycles of weather and crop cultivation.

Landscape classification schemes are important to understanding landscape influences on insects in agroecosystems. A major classification for landscapes, which can be adapted to different scales using modern geographic information systems (GIS) technologies (see Fig. 3 for an example of a cereal land-use classification), is based on predominant vegetation type. For example, in the grasslands of the U.S. Great Plains modified to support agriculture, the majority of the patches within the landscape are annually planted crops interspersed with managed grasslands used for forage and pasture and unmanaged woody patches (Fig. 3). Landscape heterogeneity, arising from both composition and configuration, may influence the abundance of *M. sorghi*, its natural enemies, and their interaction within agricultural fields and other patches of the landscape. Even within the apparently simplified landscape of the U.S. Great Plains, the heterogeneity offered by the mix of a few crops and managed and unmanaged non-crop vegetation (sometimes labeled as seminatural habitats) contribute to landscape heterogeneity. Landscape heterogeneity has well-documented relationships with pests and beneficial insects in agroecosystems (Veres et al., 2013). For example, *Lysiphlebus testaceipes* (Hymen: Braconidae), an important parasitoid of the aphid *S. graminum* in the southcentral Great Plains, had greater parasitism rates in wheat when grasslands were also present (Elliott et al., 2018). The relationship of specific patches to insect dynamics may also be affected by local abiotic conditions of an area, which has been shown to vary by habitat type within a region (Suggitt et al., 2011). Additionally, climate and weather vary longitudinally and latitudinally over the large-scale production systems where cereal grains are grown around the world, which will modify the influences of landscape composition on pests and their natural enemies.

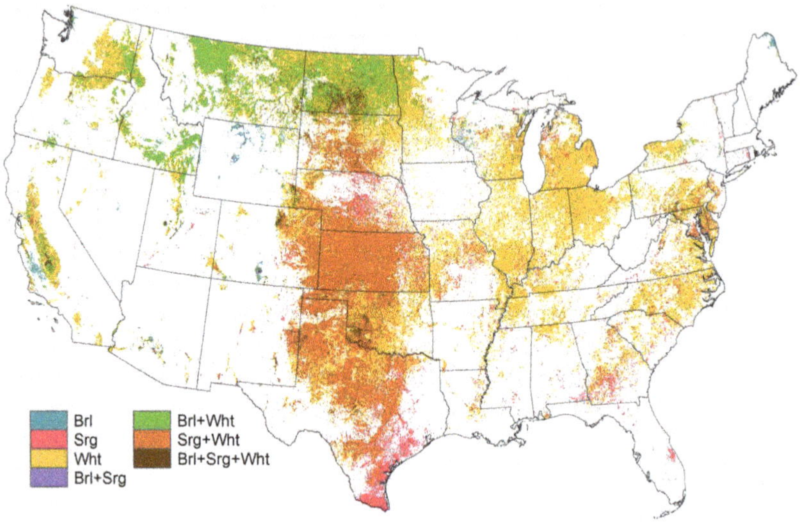

Figure 3 Map of wheat, sorghum, and barley agroecosystems of the U.S. derived from a 2019 cropland data layer product (Han et al., 2012). Different colors correspond to different cereal crops. Many regions, especially in the central Great Plains, show close proximity of multiple cereal crops, as indicated by a 2-km-scale of overlap among fields (depicted here by color-coding). Not shown is the continued expansion of the cereal grain production in the Great Plains of northeastern Mexico and southcentral Canada where sorghum, wheat, and barley are also grown. Legend key: Brl = barley, Srg = sorghum, Wht = wheat; the following are the crops that overlap in the same 2-km radius, Brl+Srg = Barley and sorghum, Brl+Wht = Barley and wheat, Srg+Wht = Sorghum and wheat, Brl+Srg+Wht = Barley and sorghum and wheat.

The abiotic environment will also vary at the larger landscape and regional scales. Abiotic factors, such as temperature and precipitation, can have a direct physiological effect on the growth and development of insects. They have also been found to influence species interactions and alter the species composition (Park, 1954). The variation of temperature and precipitation over large regional scales is the main factor that delineates terrestrial biomes (Jiang et al., 2017), often referred to as climatic or plant hardiness zones when applied to agriculture. However, invasive pests have shown strong adaptability to climactic conditions. For example, *M. sorghi*, a subtropical aphid, was expected to thrive in the subtropical sorghum growing regions of the southern U.S. and Mexico. Yet the aphid was able to rapidly expand to nearly all sorghum-producing regions of North America, including annually into cooler temperate areas of the central Great Plains via seasonal wind-blown movement of winged aphids (Wang et al., 2019). Individual weather events may influence aphid populations and their interaction with parasitoids (Jones, 1979). Less frequent are severe weather events such as hurricanes and extended drought that can lead to mass insect mortality, although the same events may severely damage the crops of these systems. Anthropogenic climate change may be especially concerning to the stability of agroecosystem services in pest management, as supported by data indicating a decoupling of pest-parasitoid interactions (Thomson et al., 2010) (see Section 4.2 for further discussion).

Unfortunately, landscape effects on pest suppression from natural enemies are prone to variability and remain difficult to predict for specific circumstances (Karp et al., 2018). Such variation has made it difficult to extrapolate relationships between systems or use them for pest forecasting purposes at the local scale of individual fields or clusters of fields. Specific habitats may not have consistent effects on pest suppression for agricultural systems separated by space or time due to the specific host, pest, and natural enemy dynamics of the system (Tscharntke et al., 2016). For example, *M. sorghi* is preyed upon by a rich complex of natural enemies in sorghum (Colares et al., 2015; Maxson et al., 2019). But variability in aphid and natural enemy densities resulting from differences in landscape structure may affect pest suppression from natural enemies, making landscape-scale research important in improving forecasting of outbreak risk of *M. sorghi* across regions. Landscape influences on *M. sorghi* and natural enemy abundance in cereal fields can be used to refine or alter management at the field-scale, but there are also avenues for enhancing pest management regionally with this information (sometimes referred to as areawide pest management [Elliott et al., 2008]). Introduced above, planting of *M. sorghi*-resistant sorghum can be targeted for production areas where *M. sorghi* outbreak risk is high or variable. Another application is to improve region-scale forecasts of pest densities and natural enemy response to estimate risk of *M. sorghi* re-infesting sorghum grown in more temperate areas each year, which is affected by wind-aided dispersal of *M. sorghi* and the amount of sorghum grown in different regions (Fig. 3) (Wang et al., 2019, 2020).

Support as well as unintended impediments for implementing areawide pest management fall partly within the realm of land-use (farm) policy. Farm policies designed for improving insect pest management in agroecosystems include landcover regulations and incentives that seek to alter landscape structure to decrease pest abundance or enhance beneficial insects such as natural enemies and pollinators. However, when considered along with traditional pest management decision-making focused on individual fields (see Section 3.1), it becomes challenging to gain support for areawide pest management approaches through grower incentives provided through government-sponsored stewardship program (Brewer and Goodell, 2012). Landscape influences may also be a consideration within field-scale IPM frameworks, which have had partial success in cereal grains. There are multiple ways natural enemies have been incorporated into pest management decision-making. One tactic has been the modification of economic thresholds to include natural enemy presence in the field being monitored for pests, such as done in the *S. graminum* monitoring system in wheat (Royer et al., 2015). Economic thresholds that incorporate natural enemy occurrence and their estimated pest suppression are more economical and sustainable and can set the stage for regional appreciation of conservation biological control tactics that affect local decision-making. Variety selection for planting can be improved by weighing the benefits of plant resistance with negative effects on yield or natural enemies. For example, solid stem cultivars used for the wheat stem sawfly can have antagonistic effects on parasitism (Rand et al., 2020). Therefore, landscapes prone to few parasitoids can prioritize host plant resistance, while landscapes supporting parasitoids should weigh natural enemy suppression against the drawbacks of using solid stem wheats (see Section 3.1). Overall, environment (landscape and weather) affects pest-natural enemy-plant dynamics and should be considered when planning for the management of cereal pests at both local and regional scales.

3.3 *Landscape considerations of natural enemy-based management of* Melanaphis sorghi

Worldwide, *M. sorghi* has a diverse natural enemy complex capable of consuming and suppressing aphid populations in the field (Singh et al., 2004). Natural enemies within North American sorghum agroecosystems quickly responded to *M. sorghi*, where a similar and diverse complex of natural enemies bridged across subtropical and temperate sorghum production regions of the U.S. and Mexico (Colares et al., 2015; Rodríguez-del-Bosque et al., 2018; Maxson et al., 2019). Hinson (2017) studied temperature-dependent development of *M. sorghi*, while consumption rates for several major natural enemy taxa feeding on *M. sorghi* or other aphids were estimated (Table 1). Field-level considerations of natural enemies in pest management decision-making protocols are feasible, as done for *S. graminum* and its parasitoid *L. testaceipes* in wheat (Royer et al., 2015). Regionally, understanding environmental influences on *M. sorghi* can aid strategic use of areawide pest management approaches where they are most useful across regions as discussed in Section 3.2.

Landscape structural effects have been demonstrated for multiple cereal aphids and their natural enemies of the U.S. Great Plains, including *D. noxia* and *S. graminum* on wheat (Brewer et al., 2019). To evaluate landscape influences on *M. sorghi* and its natural enemies, a study was conducted along the Texas Gulf Coast cropping region by the authors (B.H.E. lead investigator). This 2-year study collected data from over 40 sorghum fields to survey *M. sorghi* populations and response of natural enemies. Aphid abundance and natural enemy response were measured in sorghum fields across a range of landscapes differing in composition and configuration. Correlations of temperature and precipitation to *M. sorghi* and natural enemy abundance were also evaluated. Sorghum and cotton were dominant features in the landscape, each accounting for

Table 1 Major natural enemy taxa of *Melanaphis sorghi* in North America with aphid feeding, development time, diversity, and dispersal ability information

Taxa	Aphidophagous life stage	Aphids killed per day	Development days*	N.A. species[i,j,k]	Dispersal ability **
Coccinellidae	Juv+Adult	32.3[a]	19.8[c]	9	High[l]
Chrysopidae	Juvenile	29.9[a]	22.7[d]	6	High[m]
Syrphidae	Juvenile	20.0[a]	21.8[e]	2	Medium[n]
Aphelinidae	Juv+Adult	5.1[a]	11.2[f]	2	Low[o]
Braconidae	Juvenile	7.6[b]	12.0[g]	1	Medium[p]
Anthocoridae	Juv+Adult	8.5[a]	20.0[h]	1	Medium[q]

Citations are indicated by superscript lettering: a. Elliott et al. (2021); b. Rodrigues and Bueno (2001); c. Obrycki and Tauber (1982); d. Liu and Chen (2001); e. Hopper et al. (2011); f. Rohne (2002); g. Hight et al. (1972); h. Isenhour and Yeargan (1981); i. Rodríguez-del-Bosque et al. (2018); j. García-González et al. (2018); k. Maxson et al. (2019); l. Der Werf et al. (2000); m. Duelli (1980); n. Wratten et al. (2003); o. Fauvergue and Hopper (2009); p. Fernandes et al. (1997); q. Zhang and Shipp (1998).
* Number of days to develop from egg to adult based on laboratory assay conducted at 20–25°C.
** Low <10 m; Medium 10–1000 m; High >1000 m are estimated dispersal capabilities per day.

nearly one-third of the landcover, and the remaining was a mix of seminatural vegetation (e.g. shrubland, grassland, and lands of the coastal water system) and developed non-agricultural areas. *Melanaphis sorghi* and natural enemy populations were expected to persist year-round in this region (Esquivel et al., 2021; Maxson et al., 2019), which may affect natural enemy regulation of *M. sorghi* populations.

Analysis showed significant relationships of *M. sorghi* and its natural enemies to landscape structure (Fig. 4). *Melanaphis sorghi* abundance and percent parasitism shifted across landscape configuration, measured by patch edge density. The positive relationship between edge density and *M. sorghi*, especially when combined with the negative relationship of edge density to percent parasitism, indicated landscapes with high edge densities should be targeted for *M. sorghi* monitoring. Also, the strong natural enemy response indicated that more natural enemies were likely to be where *M. sorghi* densities were highest. These results were also relevant to insecticide use strategies, as insecticides with reduced mortality to natural enemies would be the preferred

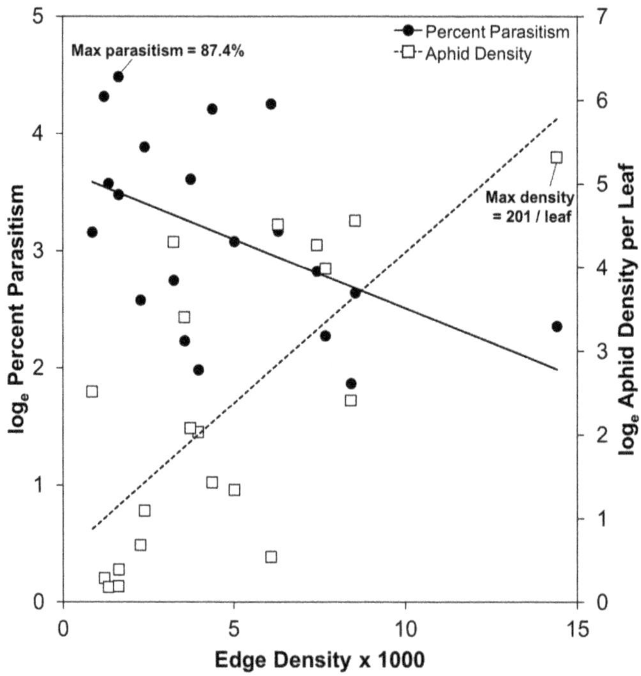

Figure 4 Significant relationships were detected between edge density at a 1-km scale, a measure of landscape configuration, and the insect metrics percent parasitism of *Melanaphis sorghi* (caused by *Aphelinus nigritus*) and *M. sorghi* abundance (aphids per leaf) during 2019 sampling of grain sorghum in South Texas. Sorghum production areas with high edge density may exhibit higher aphid densities and low parasitism rates from *A. nigritus*, thereby increasing risk from *M. sorghi*.

choice for these areas. From this study, the ability to detect landscape structure influences on this recent invader of sorghum was demonstrated, which may guide further studies on how to incorporate natural enemies into IPM practices.

There have also been broader regional scale investigations assessing how *M. sorghi*, its natural enemies, and landscape influences vary across the sub-continent-scale invasion. Elliott et al. (2021) considered *M. sorghi* and the response of natural enemies in distinct geographic regions of the southern Great Plains. Average aphid abundance on sorghum significantly differed across the landscape, but there were strong numeric responses of natural enemies to *M. sorghi* abundance, consistent with the results from the Texas Gulf Coast. A similar study conducted in collaboration with many of the investigators contributing to Elliott et al. (2021) was conducted across four widely separated geographic regions representing important sorghum production regions of the U.S. (Brewer et al., 2022). A numeric response of natural enemies to *M. sorghi* was detected within each region and was strongest along the Texas Gulf Coast, where the aphid and its natural enemies persist year-round. Further, *M. sorghi*-natural enemy dynamics were moderately influenced by landscape structure and weather which varied regionally. These findings are important considerations for areawide pest management, as variable risk from *M. sorghi* appears to be a function of *M. sorghi*-natural enemy interactions that are mediated by landscape and weather conditions (Brewer et al., 2022).

Effects of landscape structure on natural enemies alter how biological control may be incorporated into traditional field-scale management strategies such as economic threshold development. Consumption rates, development time, and dispersal ability are relevant to the response of natural enemies to *M. sorghi* (Table 1). Interspecific differences among members of a taxa are relevant to pest suppression as the species diversity may change from region to region (Snyder, 2019). Such information, given for *M. sorghi* in Table 1, provides a basis for planning and testing specific hypotheses on the integration of natural enemies with other management strategies by allowing cross-taxa consideration of pest suppression.

From the perspective of agroecosystem disturbances from an invasive pest, *M. sorghi* may have shifted the natural enemy complex of winter wheat in Oklahoma. Giles et al. (2021) found *Aphelinus nigritus* (Hymen.: Aphelinidae), a primary parasitoid of *M. sorghi* in sorghum, was more prevalent in parasitizing aphids in winter wheat. This was not surprising considering the overlap of sorghum and wheat agroecosystems, which are concentrated in the central Great Plains (Fig. 3). Possible mechanisms of these effects include ecological spillover (Rand et al., 2006) or shifting host preferences of *A nigritus*.

Simulation of local and regional infestation dynamics of *M. sorghi* in the U.S. Great Plains also may aid efficient allocation of local pest management resources as well as contribute to areawide pest management (Elliott et al.,

2008). A spatially explicit, individual-based model was used to simulate local *M. sorghi* infestations based on temperature-dependent *M. sorghi* growth on sorghum, while an atmospheric sub-model was used to simulate regional aphid migration (Wang et al., 2019). Coupled together, simulation of *M. sorghi* movement at a large spatial scale (i.e. county level in the central U.S.) was used to forecast the likelihood of winged *M. sorghi* re-infesting sorghum fields planted annually, north of where *M. sorghi* overwinters in the far southern Great Plains. These forecasts were relevant to local and regional decision-making using the probability of infestation averages obtained from multiple simulation runs (Koralewski et al., 2020a).

Linking simulation products with local aphid monitoring and weather reporting is valuable to the future development of a smart information network for *M. sorghi* management. Products may include forecasting maps of annual reinvasion at a regional scale as taken from simulation model output of aphid migration and local population growth (Koralewski et al., 2020b). The information network should have data input portals to receive aphid monitoring data and weather data for information sharing and updating simulations with near-real-time data (Koralewski et al., 2020b). Such a network has direct implications for regional or landscape-scale *M. sorghi* management. For example, regions of high risk of annual reinvasion by *M. sorghi* obtained from simulation output (as an indicator of outbreak potential) may be best suited to planting *M. sorghi*-resistant sorghum hybrids. Regions of low risk to *M. sorghi* reinvasion may be best suited to planting sorghums known for local best performance and yield potential when *M. sorghi* is absent. Regions of medium or variable risk may be best suited to planting sorghums with good yield potential for the region using locally adapted *M. sorghi*-resistant sorghum if available and deploying local aphid monitoring as needed. Decisions on the frequency of aphid monitoring can be aided by modeling results using current year weather data and where UAS-enabled remotely sensed data indicate sorghum stress. Feedback of local data (weather and aphid monitoring) into the information network can be shared and used to update same-year simulations. Integration of streams of data into an information network remains in planning and development stages for *M. sorghi* (Koralewski et al., 2020b), but components such as simulation of large-scale infestation dynamics and local in-field decision-making protocols are well advanced (Gordy et al., 2019, 2021, Wang et al., 2019; Koralewski et al., 2020a; Lindenmayer et al., 2020).

4 Future trends and conclusion

4.1 Global forces of change relevant to managing insect pests

Management practices of future pests will have to adapt to changing pest complexes of cereal grains, as they vary between production regions and over time. One reason for the dynamic nature of cereal pests is the continued introduction of invasive species around the world. Paini et al. (2016) analyzed risk and impact the future invasive species may cause on agricultural production globally. Pests of major agricultural commodities, including wheat, are geographically heterogeneous around the world, indicating that continued movement of species through trade poses a major risk of introducing crop pests. Countries with the largest agricultural production and expansive trading networks, such as the U.S. and China, have the greatest threat in terms of total costs from the introduction of invasive pests. These same large producers also have the greatest potential to act as sources of pest species which can become invasive elsewhere, as they also share the greatest proportion of the global pest species pool. However, although total cost of invasions is lower for smaller-scale agriculture in developing nations, they are likely to experience greater economic and community impacts. Preventing the introduction and establishment of invasive insects minimize impacts (Venette and Morey, 2020), but invasions continue as fostered by insect movement and human transport.

Landscape and weather influences may alter the establishment, spread, and impact of an invasion, and these influences may shift as a product of climate change. Climate change may shift the range of invasive species expansion, affect suppression from natural enemies, and increase the susceptibility of a crop to pest feeding. The status of existing insect pests of cereal grains may also change over time. Skvarla et al. (2017) documented detections of novel adventive aphid species in the U.S. that utilized wheat as a host: *Anoecia vagans* (Hemi: Aphididae), *M. festucae cerealium* (see Section 2.2), and *S. maydis* (see Section 2.4). Since their detection, two of the three have expanded in wheat-producing regions. Although there is little evidence that these pests caused broad economic damage to wheat at this time, their status may change as episodes of drought and other climate shifts occur in the future.

Climate change consequences may require modification to current pest management regimes across the world, including cereal production. Specifically, global changes to temperature, precipitation, and carbon dioxide concentrations will physiologically affect crops and their interactions with pest and natural enemies. Climate change impacts on wheat yields may be further affected by climate-induced shifts in pest-natural enemy-plant dynamics. Predictions of climate change on wheat pest management were summarized by Eigenbrode and Macfadyen (2017) and Bajwa et al. (2020). The majority of studies cited indicated negative impacts on wheat production would increase over time. Phloem-feeding insects, such as cereal aphids, benefit from increased carbon dioxide and drought stress, which alter nutrition of the phloem sap. However, populations of *Sitobion avenae* (English grain aphid) (Hemi: Aphididae) have been shown to be suppressed under high temperatures. Range expansions are also predicted for *Rhopalosiphum padi* (bird cherry-oat aphid) (Hemi: Aphididae) and *S. avenae* in Europe due to climate change. Climate change will also alter pest pressure in wheat by altering interspecific interactions of cereal pests and natural enemies that may provide biological control services. Climate change is also expected to increase the susceptibility of natural habitats to invasive species which contain important resources for natural enemies in agroecosystems (Perrings et al., 2010).

4.2 Lessons learned

Agricultural producers around the world will have to meet the increasing demand of an expanding human population. This is especially true for important food commodities such as cereal crops. There is a growing demand for grain in small-scale community-based agriculture and large-scale production systems that supply the global grain distribution system. To meet this demand, pest management will have to adapt over time to counteract the pressure from changing pest complexes, such as the recent insect pest invasions into wheat and sorghum presented in this chapter. The *M. sorghi* case study provided an opportunity to discuss the framework of management response appropriate and essential during the establishment and initial spread of the invasion at the local (field)-scale. The discussion benefited from experiences managing the previous invasion of *D. noxia* on wheat and barley in the same large cereal grain production system of North America (Quisenberry and Peairs, 1989; Brewer et al., 2019). The environmental context of *M. sorghi* spread and variability of impact had regional (landscape-scale) implications to the best approaches to manage *M. sorghi* with a mixture of traditional and advanced pest management tools. Several of the advanced pest management tools are still in the development phase and have yet to be implemented in current management systems, such as advanced forecasting and detection systems enabled by ecological modeling and use of remotely sensed optical data obtained from UAS (agricultural drones). They were presented in the spirit of encouraging the transformation of research and development into operational pest management systems for this or future insect pest invasions onto cereal grains and other agroecosystems. Other management tools have progressed and are ready for use or are being used in pest management systems, including the use of insecticides with reduced kill of natural enemies as guided by tally-based economic thresholds, sorghum hybrids resistant to *M. sorghi*, and recognition of natural enemy suppression of *M. sorghi*.

5 Where to look for further information

Throughout this chapter, journal articles and several book chapters were cited that provided reviews, meta-analyses, and other perspectives on past crop pest invasions, regional scale (areawide) pest management, crop pest and natural enemy dynamics in the context of landscape structure and implications of climate change on agricultural pests. Below is a selection.

5.1 Crop pest invasions

- Perrings et al. (2010) place bioinvasions in the context of globalization.
- Paini et al. (2016) provide a perspective on the global threat to agriculture from invasives.
- Brewer et al. (2019) review a sequence of cereal aphids invading the U.S. Great Plains.

5.2 Crop pest and natural enemy dynamics in the context of landscape structure

- Tscharntke et al. (2012) consider how landscapes moderate biodiversity.
- Macfadyen and Muller (2013) focus on the effects of agricultural edges on natural enemies.
- Veres et al. (2013) review landscape composition effects on pests and natural enemies.

5.3 Regional scale (areawide) pest management

- Elliott et al. (2008) review history and concepts for areawide pest management.
- Brewer and Goodell (2012) focused on pest management adoption at a regional scale.

5.4 Implications of climate change on agricultural pests

- Thomson et al. (2010) present predictions of climate change affecting natural enemies.
- Eigenbrode and Macfadyen (2017) review impact of climate change on wheat insects.

6 References

Ainslie, C. N. (1920). *The Western Grass-Stem Sawfly*. USDA Technical Bulletin No. 841.

Armstrong, J. S., Rooney, W. L., Peterson, G. C., Villenueva, R. T., Brewer, M. J. and Sekula-Ortiz, D. (2015). Sugarcane aphid (Hemiptera: Aphididae): host range and sorghum resistance including cross-resistance from greenbug sources. *J. Econ. Entomol.* 108(2): 576–582.

Avila, G. A., Davidson, M., van Helden, M. and Fagan, L. (2018). The potential distribution of the Russian wheat aphid (*Diuraphis noxia*): an updated distribution model including irrigation improves model fit for predicting potential spread. *Bull. Entomol. Res.* 109(1): 90–101.

Bajwa, A. A., Farooq, M., Al-Sadi, A. M., Nawaz, A., Jabran, K. and Siddique, K. H. M. (2020). Impact of climate change on biology and management of wheat pests. *Crop Prot.* 137: 105304. DOI: 10.1016/j.cropro.2020.105304.

Beres, B. L., Dosdall, L. M., Weaver, D. K., Cárcamo, H. A. and Spaner, D. M. (2011). Biology and integrated management of wheat stem sawfly and the need for continuing research. *Can. Entomol.* 143(2): 105–125.

Blackman, R. L. and Eastop, V. F. (2000). *Aphids on the World's Crops, an Identification and Information Guide* (2nd edn.). Wiley & Sons, Ltd., Chichester, United Kingdom, 466 pp.

Botha, A. M. (2021). Fast developing Russian wheat aphid biotypes remains an unsolved enigma. *Curr. Opin. Insect Sci.* 45: 42–52.

Bowling, R. D., Brewer, M. J., Kerns, D. L., Gordy, J., Seiter, N., Elliott, N. E., Buntin, G. D., Way, M. O., Royer, T. A., Biles, S. and Maxson, E. (2016). Sugarcane aphid (Hemiptera: Aphididae): a new pest on sorghum in North America. *J. Integr. Pest Manag.* 7(1): 12. DOI: 10.1093/jipm/pmw011.

Brewer, M. J., Elliott, N. C., Esquivel, E. L, Jacobson, A. L., Faris, A. M., Szczepaniec, A., Elkins, B. H., Gordy, J. W., Pekarcik, A. J., Wang, H.-H., Koralewski, T. E., Giles, K. L., Jessie, C. N. and Grant, W. E. (2022). Natural enemies, mediated by landscape and weather conditions, shape response of the sorghum agroecosystem of North America to the invasive aphid Melanaphis sorghi. *Front. Insect Sci.* 2: 830997. DOI: 10.3389/finsc.2022.830997.

Brewer, M. J. and Goodell, P. B. (2012). Approaches and incentives to implement integrated pest management that addresses regional and environmental issues. *Annu. Rev. Entomol.* 57: 41–59.

Brewer, M. J., Mornhinweg, D. W. and Huzurbazar, S. (1999). Compatibility of insect management strategies: *Diuraphis noxia* (Hom: Aphididae) abundance on susceptible and resistant barley in the presence of parasitoids. *BioControl* 43(4): 479–491.

Brewer, M. J., Nelson, D. J., Ahern, R. G., Donahue, J. D. and Prokrym, D. R. (2001). Recovery and range expansion of parasitoids (Hymenoptera: Aphelinidae and Braconidae) released for biological control of *Diuraphis noxia* (Homoptera: Aphididae) in Wyoming. *Environ. Entomol.* 30(3): 578–588.

Brewer, M. J., Noma, T. and Elliott, N. C. (2005). Hymenopteran parasitoids and dipteran predators of the invasive aphid *Diuraphis noxia* after enemy introductions: temporal variation and implication for future aphid invasions. *Biol. Control* 33(3): 315–323.

Brewer, M. J., Peairs, F. B. and Elliott, N. C. (2019). Invasive cereal aphids of North America: ecology and pest management. *Annu. Rev. Entomol.* 64: 73–93.

Chi, B., Zheng, X., Liang, X. and Liu, Y. (2016). Temperature-dependent demography of *Agriphila aeneociliella* (Lepidoptera: Crambidae), a new insect pest of wheat in China. *Agric. For. Entomol.* 18(3): 189–197.

Cockrell, D. M., Griffin-Nolan, R. J., Rand, T. A., Altilmisani, N., Ode, P. J. and Peairs, F. (2017). Host plants of the wheat stem Sawfly (Hymenoptera: Cephidae). *Environ. Entomol.* 46(4): 847–854.

Colares, F., Michaud, J. P., Bain, C. L. and Torres, J. B. (2015). Recruitment of aphidophagous arthropods to sorghum plants infested with *Melanaphis sacchari* and *Schizaphis graminum* (Hemiptera: Aphididae). *Biol. Control* 90: 16–24.

Comstock, J. H. (1889). *On a Saw-Fly Borer in Wheat*. Cornell University, Ithaca, NY.

Corrales, C. E., Castro, A. M., Ricci, M. and Dixon, A. F. (2007). *Sipha maydis*: distribution and host range of a new aphid pest of winter cereals in Argentina. *J. Econ. Entomol.* 100(6): 1781-1788.

Davis, T. S., Wu, Y. and Eigenbrode, S. D. (2014). Host settling behavior, reproductive performance, and effects on plant growth of an exotic cereal aphid, *Metopolophium festucae* subsp. *Cerealium* (Hemiptera: Aphididae). *Environ. Entomol.* 43(3): 682-688.

Deleon, L., Brewer, M. J., Esquivel, I. L. and Halcomb, J. (2017). Use of a geographic information system to produce pest monitoring maps for south Texas cotton and sorghum land managers. *Crop Prot.* 101: 50-57.

Deng, X., Thomasson, J. A., Puch, N. A., Chen, J., Rooney, W. L., Brewer, M. J. and Shi, Y. (2020). Estimating the severity of sugarcane aphid infestation on sorghum with machine vision. *Int. J. Precis. Agric. Aviat.* 3(2): 89-96.

Dent, D. R. and Wratten, S. D. (1986). The host-plant relationships of apterous virginoparae of the grass aphid *Metopolophium festucae Cerealium*. *Ann. Appl. Biol.* 108(3): 567-576.

Der Werf, W. V., Evans, E. W. and Powell, J. (2000). Measuring and modelling the dispersal of *Coccinella septempunctata* (Coleoptera: Coccinellidae) in alfalfa fields. *Eur. J. Entomol.* 97(4): 487-493.

Duelli, P. (1980). Adaptive dispersal and appetitive flight in the green lacewing, *Chrysopa cornea*. *Ecol. Entomol.* 5(3): 213-220.

Eigenbrode, S. D. and Macfadyen, S. (2017). The impact of climate change on wheat insect pests: current knowledge and future trends. In Langridge, P. (Ed.), *Burleigh Dodds Series in Agricultural Science. 1: Breeding, Quality Traits, Pests and Diseases*. Burleigh Dodds Science Publishing, Cambridge, UK, 544-567.

Elliott, N., Giles, K., Brewer, M., Szczepaniec, A., Knutson, A., Michaud, J., Jessie, C., Faris, A., Elkins, B., Wang, H., Koralewski, T. and Grant, W. (2021). Recruitment of natural enemies of the invasive sugarcane aphid vary spatially and temporally in sorghum fields in the Southern Great Plains of the USA. *Southwest. Entomol.* 46(2): 357-372.

Elliott, N. C., Backoulou, G. F., Brewer, M. J. and Giles, K. L. (2015). NDVI to detect sugarcane aphid injury to grain sorghum. *J. Econ. Entomol.* 108(3): 1452-1455.

Elliott, N. C., Brewer, M. J. and Giles, K. L. (2018). Landscape context affects aphid parasitism by *Lysiphlebus testaceipes* (Hymenoptera: Aphidiinae) in wheat fields. *Environ. Entomol.* 47(4): 803-811.

Elliott, N. C., Brewer, M., Seiter, N., Royer, T., Bowling, R., Backoulou, G., Gordy, J., Giles, K., Lindenmayer, J., McCornack, B. and Kerns, D. (2017). Sugarcane aphid spatial distribution in grain sorghum fields. *Southwest. Entomol.* 42(1): 27-35.

Elliott, N. C., Onstad, D. W. and Brewer, M. J. (2008). History and ecological basis for areawide pest management. In Koul, O., Cuperus, G. W. and Elliott, N. C. (Eds.), *Area-Wide Pest Management: Theory to Implementation*. CABI Publishing, Oxfordshire, UK, 15-33.

Esquivel, I. L., Faris, A. M. and Brewer, M. J. (2021). Sugarcane aphid (*Melanaphis sacchari* Hemiptera: Aphididae) abundance on sorghum and Johnsongrass in a laboratory and field setting. *Crop Prot.* 148: 105715. DOI: 10.1016/j.cropro.2021.105715.

Fauvergue, X. and Hopper, K. R. (2009). French wasps in the new world: experimental biological control introductions reveal a demographic Allee effect. *Popul. Ecol.* 51(3): 385-397.

Fernandes, O. A., Wright, R. J., Baumgarten, K. H. and Mayo, Z. B. (1997). Use of rubidium to label *Lysiphlebus* testaceipes (Hymenoptera: Braconidae), a parasitoid of greenbugs (Homoptera: Aphididae), for dispersal studies. *Environ. Entomol.* 26(5): 1167-1172.
Foottit, R. G., Halbert, S. E., Miller, G. L., Maw, E. and Russell, L. M. (2006). Adventive aphids (Hemiptera: Aphididae) of America north of Mexico. *Proc. Entomol. Soc. Wash.* 108: 583-610.
García-González, F., Ramírez-Delgado, M., Cortez-Madrigal, H. and de Lourdes Ramírez-Ahuja, MdL. (2018). New reports of parasitoids and hyperparasitoids of *Melanaphis sacchari* (Zehntner) 1 in México. *Southwest. Entomol.* 43(3): 787-790.
Giles, K. L., Elliott, N. C., Butler, H. E., Baum, K. A. and Backoulou, G. F. (2021). Increase in importance of *Aphelinus nigritus* (Howard) on winter crops in Oklahoma coincides with invasion of sugarcane aphid on sorghum in Oklahoma. *Southwest. Entomol.* 46(1): 59-68.
Giles, K. L., McCornack, B. P., Royer, T. A. and Elliott, N. C. (2017). Incorporating biological control into IPM decision making. *Curr. Opin. Insect Sci.* 20: 84-89.
Gordy, J. W., Brewer, M. J., Bowling, R. D., Buntin, G. D., Seiter, N. J., Kerns, D. L., Reay-Jones, F. P. F. and Way, M. O. (2019). Development of economic thresholds for sugarcane aphid (Hemiptera: Aphididae) in susceptible grain sorghum hybrids. *J. Econ. Entomol.* 112(3): 1251-1259.
Gordy, J. W., Seiter, N. J., Kerns, D. L., Reay-Jones, F. P. F., Bowling, R. D., Way, M. O. and Brewer, M. J. (2021). Field assessment of aphid doubling time and yield of sorghum susceptible and partially resistant to sugarcane aphid (Hemiptera: Aphididae). *J. Econ. Entomol.* 114(5): 2076-2087. DOI: 10.1093/jee/toab135.
Haan, N. L., Zhang, Y. and Landis, D. A. (2020). Predicting landscape configuration effects on agricultural pest suppression. *Trends Ecol. Evol.* 35(2): 175-186.
Halbert, S. E. and Sandvol, L. E. (1995). New or unusual pests in Idaho in 1994. In *Research Report: 54th Annual Northwest Insect Manage Conference*, January 1995, Portland, OR. Oregon State University, Corvallis, OR, p. 71.
Halbert, S. E., Wu, Y. and Eigenbrode, S. D. (2013). Metopolophium festucae Cerealium (Hemiptera: Aphididae), a new addition to the aphid fauna of North America. *Insecta Munda* 0301: 1-6. Available at: journals.flvc.org/mundi/article/view/0301/79240 (accessed September 2021).
Han, W., Yang, Z., Di, L. and Mueller, R. (2012). CropScape: a Web service based application for exploring and disseminating US conterminous geospatial cropland data products for decision support. *Comput. Electron. Agric.* 84: 111-123.
Harris-Shultz, K. R., Brewer, M. J., Wadl, P. A., Ni, X. and Wang, H. (2018). A sugarcane aphid 'super-clone' predominates on sorghum and Johnsongrass from four US states. *Crop Sci.* 58(6): 2533-2541.
Hatchett, J. H. and Gallun, R. L. (1970). Genetics of the ability of the Hessian fly, Mayetiola destructor, to survive on wheats having different genes for resistance. *Ann. Entomol. Soc. Am.* 63(5): 1400-1407.
Hight, S. C., Eikenbary, R. D., Miller, R. J. and Starks, K. J. (1972). The greenbug and *Lysiphlebus testaceipes*. *Environ. Entomol.* 1(2): 205-209.
Hinson, P. (2017). *Effect of Temperature on the Development of Sugarcane Aphid, Melanaphis sacchari, on Sorghum*. West Texas A&M University, Canyon, TX.
Hopper, J. V., Nelson, E. H., Daane, K. M. and Mills, N. J. (2011). Growth, development, and consumption by four syrphid species associated with the lettuce aphid, *Nasonovia ribisnigri*, in California. *Biol. Control* 58(3): 271-276.

Irell, B. and Peairs, F. (2011). Wheat stem sawfly: a new pest of Colorado wheat. In *Extension Fact Sheet 5.612*. Colorado State University, Fort Collins, CO.

Isenhour, D. J. and Yeargan, K. V. (1981). Effect of temperature on the development of *Orius insidiosus*, with notes on laboratory rearing. *Ann. Entomol. Soc. Am.* 74(1): 114–116.

Ji, Y. C., Qin, M., Wu, L. F., Chen, X. L. and Zhang, R. Z. (2014). Potential distribution of two wheat pests, *Agriphila aeneociliella* and *A. vulgivagella*, in China. *Chin. J. App. Entomol.* 51: 271-277.

Jiang, M., Felzer, B. S., Nielsen, U. N., Medlyn, B. E. and Fortin, M. (2017). Biome-specific climatic space defined by temperature and precipitation predictability. *Glob. Ecol. Biogeogr.* 26(11): 1270–1282.

Jones, M. G. (1979). Abundance of aphids on cereals from before 1973 to 1977. *J. Appl. Ecol.* 16(1): 1–22.

Karp, D. S., Chaplin-Kramer, R., Meehan, T. D., Martin, E. A., DeClerck, F., Grab, H., Gratton, C., Hunt, L., Larsen, A. E., Martínez-Salinas, A., O'Rourke, M. E., Rusch, A., Poveda, K., Jonsson, M., Rosenheim, J. A., Schellhorn, N. A., Tscharntke, T., Wratten, S. D., Zhang, W., Iverson, A. L., Adler, L. S., Albrecht, M., Alignier, A., Angelella, G. M., Zubair Anjum, M., Avelino, J., Batáry, P., Baveco, J. M., Bianchi, F. J. J. A., Birkhofer, K., Bohnenblust, E. W., Bommarco, R., Brewer, M. J., Caballero-López, B., Carrière, Y., Carvalheiro, L. G., Cayuela, L., Centrella, M., Ćetković, A., Henri, D. C., Chabert, A., Costamagna, A. C., De la Mora, A., de Kraker, J., Desneux, N., Diehl, E., Diekötter, T., Dormann, C. F., Eckberg, J. O., Entling, M. H., Fiedler, D., Franck, P., Frank van Veen, F. J., Frank, T., Gagic, V., Garratt, M. P. D., Getachew, A., Gonthier, D. J., Goodell, P. B., Graziosi, I., Groves, R. L., Gurr, G. M., Hajian-Forooshani, Z., Heimpel, G. E., Herrmann, J. D., Huseth, A. S., Inclán, D. J., Ingrao, A. J., Iv, P., Jacot, K., Johnson, G. A., Jones, L., Kaiser, M., Kaser, J. M., Keasar, T., Kim, T. N., Kishinevsky, M., Landis, D. A., Lavandero, B., Lavigne, C., Le Ralec, A., Lemessa, D., Letourneau, D. K., Liere, H., Lu, Y., Lubin, Y., Luttermoser, T., Maas, B., Mace, K., Madeira, F., Mader, V., Cortesero, A. M., Marini, L., Martinez, E., Martinson, H. M., Menozzi, P., Mitchell, M. G. E., Miyashita, T., Molina, G. A. R., Molina-Montenegro, M. A., O'Neal, M. E., Opatovsky, I., Ortiz-Martinez, S., Nash, M., Östman, Ö, Ouin, A., Pak, D., Paredes, D., Parsa, S., Parry, H., Perez-Alvarez, R., Perović, D. J., Peterson, J. A., Petit, S., Philpott, S. M., Plantegenest, M., Plećaš, M., Pluess, T., Pons, X., Potts, S. G., Pywell, R. F., Ragsdale, D. W., Rand, T. A., Raymond, L., Ricci, B., Sargent, C., Sarthou, J. P., Saulais, J., Schäckermann, J., Schmidt, N. P., Schneider, G., Schüepp, C., Sivakoff, F. S., Smith, H. G., Stack Whitney, K., Stutz, S., Szendrei, Z., Takada, M. B., Taki, H., Tamburini, G., Thomson, L. J., Tricault, Y., Tsafack, N., Tschumi, M., Valantin-Morison, M., Van Trinh, M., van der Werf, W., Vierling, K. T., Werling, B. P., Wickens, J. B., Wickens, V. J., Woodcock, B. A., Wyckhuys, K., Xiao, H., Yasuda, M., Yoshioka, A. and Zou, Y. (2018). Crop pests and predators exhibit inconsistent responses to surrounding landscape composition. *Proc. Natl. Acad. Sci. U. S. A.* 115(33): E7863–E7870. DOI: 10.1073/pnas.1800042115.

Katis, N. I., Tsitsipis, J. A., Lykouressis, D. P., Papapanayotou, A., Margaritopoulos, J. T., Kokinis, G. M., Perdikis, D. C. and Manoussopoulos, I. N. (2006). Transmission of zucchini yellow mosaic virus by colonizing and non-colonizing aphids in Greece and new aphid species vectors of the virus. *J. Phytopathol.* 154(5): 293–302.

Knutson, A., Bynum, E., Kerns, E., Porter, P., Biles, S. and Reed, B. (2018). *Managing Insect and Mite Pests of Texas Sorghum*. ENTO-085, Texas A&M AgriLife Extension, College Station, TX.

Koralewski, T. E., Wang, H.-H., Grant, W. E., Brewer, M. J., Elliott, N. C., Westbrook, J. K., Szczepaniec, A., Knutson, A., Giles, K. L. and Michaud, J. P. (2020a). Integrating models of atmospheric dispersion and crop-pest dynamics: linking detection of local aphid infestations to forecasts of region-wide invasion of cereal crops. *Ann. Entomol. Soc. Am.* 113(2): 79–87.

Koralewski, T. E., Wang, H.-H., Grant, W. E., LaForest, J. H., Brewer, M. J., Elliott, N. C. and Westbrook, J. K. (2020b). Toward near-real-time forecasts of airborne crop pests: aphid invasions of cereal grains in North America. *Comput. Electron. Agric.* 179: 105861. DOI: 10.1016/j.compag.2020.105861.

Lindenmayer, J. C., Giles, K. L., Elliott, N. C., Knutson, A. E., Bowling, R., Brewer, M. J., Seiter, N. J., McCornack, B., Brown, S. A., Catchot, A. L. and Royer, T. A. (2020). Development of binomial sequential sampling plans for sugarcane aphid (Hemiptera: Aphididae) in commercial grain sorghum. *J. Econ. Entomol.* 113(4): 1990–1998.

Liu, T. X. and Chen, T. Y. (2001). Effects of three aphid species (Homoptera: Aphididae) on development, survival and predation of *Chrysoperla carnea* (Neuroptera: Chrysopidae). *Appl. Entomol. Zool.* 36(3): 361–366.

Lytle, A. J. and Huseth, A. S. (2021). Impact of foliar insecticide sprays on *Melanaphis sacchari* (Hemiptera: Aphididae) and natural enemy populations in grain sorghum. *Crop Prot.* 149: 105764. DOI: 10.1016/j.cropro.2021.105764.

Macfadyen, S. and Muller, W. (2013). Edges in agricultural landscapes: species interactions and movement of natural enemies. *PLoS ONE* 8(3): e59659. DOI: 10.1371/journal.pone.0059659.

Maxson, E. L., Brewer, M. J., Rooney, W. L. and Woolley, J. B. (2019). Species composition and abundance of the natural enemies of sugarcane aphid, *Melanaphis sacchari* (Zehnter) (Hemiptera: Aphididae), on sorghum in Texas. *Proc. Wash. Entomol. Soc.* 121(4): 657–680.

Nibouche, S., Costet, L., Medina, R. F., Holt, J. R., Sadeyen, J., Zoogones, A. S., Brown, P. and Blackman, R. L. (2021). Morphometric and molecular discrimination of the sugarcane aphid, *Melanaphis sacchari*, (Zehntner, 1897) and the sorghum aphid *Melanaphis sorghi* (Theobald, 1904). *PLoS ONE* 16(3): e0241881. DOI: 10.1371/journal.pone.0241881.

Norton, E. (1872). Notes on North American Tenthredinidae with descriptions of new species. *Trans. Am. Entomol. Soc.* 4: 77–86.

Obrycki, J. J. and Tauber, M. J. (1982). Thermal requirements for development of *Hippodamia convergence* (Coleoptera: Coccinellidae). *Ann. Entomol. Soc. Am.* 75(6): 678–683.

Paini, D. R., Sheppard, A. W., Cook, D. C., De Barro, P. J., Worner, S. P. and Thomas, M. B. (2016). Global threat to agriculture from invasive species. *Proc. Natl. Acad. Sci. U. S. A.* 113(27): 7575–7579. DOI: 10.1073/pnas.1602205113.

Park, T. (1954). Experimental studies of interspecies competition II. Temperature, humidity, and competition in two species of *Tribolium*. *Physiol. Zool.* 27(3): 177–238.

Paudyal, S., Armstrong, J. S., Giles, K. L., Payton, M. E., Opit, G. P. and Limaje, A. (2019). Categories of resistance to sugarcane aphid (Hemiptera: Aphididae) among sorghum genotypes. *J. Econ. Entomol.* 112(4): 1932–1940.

Pedigo, L. P. (1999). *Entomology and Pest Management* (3rd edn.). Prentice-Hall Publishing, Upper Saddle River, NJ.

Peirce, E. S., Rand, T. A., Cockrell, D. M., Ode, P. J. and Peairs, F. B. (2021). Effects of landscape composition on wheat stem sawfly (Hymenoptera: Cephidae) and its associated braconid parasitoids. *J. Econ. Entomol.* 114(1): 72–81.

Peng, H., Zhang, Y. H. and Li, X. R. (2013). The toxicity and field efficacy of 5 insecticides to *Agriphila aeneociliella*. *Chin. J. Plant Prot.* 39: 184–187.

Perrings, C., Mooney, H. and Williamson, M. (2010). *Bioinvasions and Globalization: Ecology, Economics, Management, and Policy*. Oxford University Press, Oxford, UK.

Puterka, G. J., Hammon, R. W., Franklin, M., Mornhinweg, D. W., Springer, T., Armstrong, S. and Brown, M. J. (2019). Distribution of a new invasive species, *Sipha maydis* (Heteroptera: Aphididae), on cereals and wild grasses in the southern Plains and Rocky Mountain states. *J. Econ. Entomol.* 112(4): 1713–1721.

Quisenberry, S. S. and Peairs, F. B. (Eds.). (1989). *Response Model for an Introduced Pest–The Russian Wheat Aphid*. Thomas Say Publications in Entomology, Entomol. Società Am, Lanham, MD.

Rand, T. A., Richmond, C. E. and Dougherty, E. T. (2020). Modeling the combined impacts of host plant resistance and biological control on the population dynamics of a major pest of wheat. *Pest Manag. Sci.* 76(8): 2818–2828.

Rand, T. A., Tylianakis, J. M. and Tscharntke, T. (2006). Spillover edge effects: the dispersal of agriculturally subsidized insect natural enemies into adjacent natural habitats. *Ecol. Lett.* 9(5): 603–614.

Rodrigues, S. M. M. and Bueno, V. H. P. (2001). Parasitism rates of *Lysiphlebus testaceipes* (Cresson) (Hym.: Aphidiidae) on *Schizaphis graminum* (Rond.) and *Aphis gossypii* Glover (Hem.: Aphididae). *Neotrop. Entomol.* 30(4): 625–629.

Rodríguez-del-Bosque, L. A., Rodríguez-Vélez, B., Sarmiento-Cordero, M. A. and Arredondo-Bernal, H. C. (2018). Natural enemies of *Melanaphis sacchari* on grain sorghum in Northeastern Mexico. *Southwest. Entomol.* 43(1): 277–279.

Rohne, O. (2002). Effect of temperature and host stage on performance of *Aphelinus varipes* Forster (Hym., Aphelinidae) parasitizing the cotton aphid, *Aphis gossypii* Glover (Hom., Aphididae). *J. Appl. Entomol.* 126(10): 572-576.

Royer, T.A., Giles, K. L., and Elliott, N. C. 2022. The Glance-N-Go® sampling system for greenbugs, sorghum aphids, and sorghum headworms: questions and answers. EPP-7191, *Oklahoma Cooperative Extension Service*, Stillwater, OK.

Royer, T. A., Pendleton, B. B., Elliott, N. C. and Giles, K. L. (2015). Greenbug (Hemiptera: Aphididae) biology, ecology, and management in wheat and sorghum. *J. Integr. Pest Manag.* 6(1): 19-19. DOI: 10.1093/jipm/pmv018.

Sadeghi, S. E., Bjur, J., Ingwell, L., Unger, L., Bosque-Pérez, N. A. and Eigenbrode, S. D. (2016). Interactions between *Metopolophium festucae Cerealium* (Hemiptera: Aphididae) and Barley yellow dwarf virus (BYDV-PAV). *J. Insect Sci.* 16(1). DOI: 10.1093/jisesa/iev160.

Shodotova, A. A. (2008). Pyralid moths (Lepidoptera, Pyraloidea) of Buryatia. Family Crambidae. *Entomol. Rev.* 88(3): 286–295.

Singh, B. U., Padmaja, P. G. and Seetharama, N. (2004). Biology and management of the sugarcane aphid, *Melanaphis sacchari* (Zehntner) (Homoptera: Aphididae), in sorghum: a review. *Crop Prot.* 23(9): 739–755.

Skvarla, M. J., Halbert, S. E., Foottit, R. G., Jensen, A. S., Maw, E. and Miller, G. L. (2017). An update to the adventive aphids (Hemiptera: Aphidoidea) of America north of Mexico, with notes on intercepted species. *Proc. Entomol. Soc. Wash.* 119(1): 90–111.

Snyder, W. E. (2019). Give predators a complement: conserving natural enemy biodiversity to improve biocontrol. *Biol. Control* 135: 73-82.

Stanton, C., Starek, M. J., Elliott, N., Brewer, M., Maeda, M. M. and Chu, T. (2017). Unmanned aircraft system-derived crop height and normalized difference vegetation index metrics for sorghum yield and aphid stress assessment. *J. Appl. Remote Sens.* 11(2): 02035. DOI: 10.1117/1.JRS.11.026035.

Starý, P., Rodriguez, A. F., Gerding, P. M., Norambuena, M. H. and Remaudiere, G. (1994). Distribution, frequency, host range and parasitism of two new cereal aphids, *Sitobion fragariae* (Walker) and *Metopolophium festucae cerealium Stroyan* (Homoptera, Aphididae), in Chile. *Agric. Tec. (Santiago)* 54(1): 54–59.

Suggitt, A. J., Gillingham, P. K., Hill, J. K., Huntley, B., Kunin, W. E., Roy, D. B. and Thomas, C. D. (2011). Habitat microclimates drive fine-scale variation in extreme temperatures. *Oikos* 120(1): 1-8.

Tetreault, H. M., Grover, S., Scully, E. D., Gries, T., Palmer, N. A., Sarath, G., Louis, J. and Sattler, S. E. (2019). Global responses of resistant and susceptible sorghum (*Sorghum bicolor*) to sugarcane aphid (*Melanaphis sacchari*). *Front. Plant Sci.* 10: 145. DOI: 10.3389/fpls.2019.00145.

Thomas, J. L., Bowling, R. and Brewer, M. J. (2018). Learning experiences in IPM through concise instructional videos. *J. Integr. Pest Manag.* 9(1): 1-6. DOI: 10.1093/jipm/pmx030.

Thomson, L. J., Macfadyen, S. and Hoffmann, A. A. (2010). Predicting the effects of climate change on natural enemies of agricultural pests. *Biol. Control* 52(3): 296–306.

Tscharntke, T., Karp, D. S., Chaplin-Kramer, R., Batáry, P., DeClerck, F., Gratton, C., Hunt, L., Ives, A., Jonsson, M., Larsen, A., Martin, E. A., Martínez-Salinas, A., Meehan, T. D., O'Rourke, M., Poveda, K., Rosenheim, J. A., Rusch, A., Schellhorn, N., Wanger, T. C., Wratten, S., Zhang, W. (2016). When natural habitat fails to enhance biological pest control – Five hypotheses. *Biol. Conserv.* 204(Part B): 449–458. https://doi.org/10.1016/j.biocon.2016.10.001.

Tscharntke, T., Tylianakis, J. M., Rand, T. A., Didham, R. K., Fahrig, L., Batáry, P., Bengtsson, J., Clough, Y., Crist, T. O., Dormann, C. F., Ewers, R. M., Fründ, J., Holt, R. D., Holzschuh, A., Klein, A. M., Kleijn, D., Kremen, C., Landis, D. A., Laurance, W., Lindenmayer, D., Scherber, C., Sodhi, N., Steffan-Dewenter, I., Thies, C., van der Putten, W. H. and Westphal, C. (2012). Landscape moderation of biodiversity patterns and processes-eight hypotheses. *Biol. Rev. Camb. Philos. Soc.* 87(3): 661–685.

Venette, R. C. and Morey, A. C. (2020). Advances in understanding the ecology of invasive crop insect pests and their impact on IPM, 30 pp. In Kogan, M. and Heinrichs, E. A. (Eds.), *Integrated Management of Insect Pests: Current and Future Developments*. Burleigh Dodds Science Publishing, Cambridge, UK.

Veres, A., Petit, S., Conord, C. and Lavigne, C. (2013). Does landscape composition affect pest abundance and their control by natural enemies? A review. *Agric. Ecosyst. Environ.* 166: 110–117.

Wang, H.-H., Grant, W. E., Elliott, N. C., Brewer, M. J., Koralewski, T. E., Westbrook, J. K., Alves, T. M. and Sword, G. A. (2019). Integrated modelling of the life cycle and aeroecology of wind-borne pests in temporally variable spatially heterogeneous environment. *Ecol. Modell.* 399: 23-38.

Wang, H.-H., Grant, W. E., Koralewski, T. E., Elliott, N. C., Brewer, M. J. and Westbrook, J. K. (2020). Where do all the aphids go? A series of thought experiments within the context of area-wide pest management. *Agric. Syst.* 185: 102957. DOI: 10.1016/j.agsy.2020.102957.

Ward, S., van Helden, M., Heddle, T., Ridland, P. M., Pirtle, E. and Umina, P. A. (2020). Biology, ecology and management of Diuraphis noxia (Hemiptera: Aphididae) in Australia. *Aust. Entomol.* 59(2): 238–252.

Wratten, S. D., Bowie, M. H., Hickman, J. M., Evans, A. M., Sedcole, J. R. and Tylianakis, J. M. (2003). Field boundaries as barriers to movement of hover flies (Diptera: Syrphidae) in cultivated land. *Oecologia* 134(4): 605–611.

Zeng, J., Jiang, Y. Y., Wang, H. Y., et al. (2014). The biological characteristics of *Agriphila aeneociliella* and methods for monitoring and managing this pest. *Chin. J. App. Entomol.* 51: 814–818.

Zhang, J., Maleski, J., Schwartz, B., Dunn, D., Mailhot, D., Ni, X., Harris-Shultz, K., Knoll, J. and Toews, M. (2021). Assessing spatio-temporal patterns of sugarcane aphid (Hemiptera: Aphididae) infestations on silage sorghum yield using unmanned aerial systems (UAS). *Crop Prot.* 146: 105681. DOI: 10.1016/j.cropro.2021.105681.

Zhang, Y. and Shipp, J. L. (1998). Effect of temperature and vapor pressure deficit on the flight activity of *Orius insidiosus* (Hemiptera: Anthocoridae). *Environ. Entomol.* 27(3): 736–742.

Chapter 14

Biotechnology for wheat crop protection: potential and challenges

Anna-Maria Botha, Stellenbosch University, South Africa

1 Introduction

Modern-day bread wheat (*Triticum aestivum* L.) has been cultivated for over 10 000 years and is believed to have originated in the south-eastern part of Turkey, which forms part of the Fertile Crescent (Dubcovsky and Dvorak, 2007). It is also the most widely grown crop with its geographical cultivation range that extends further than any other crop species. This adaptability to a wide range of climates, due to the complex nature of its genome, provides great phenotypic plasticity (Feldman et al., 2012; Suneja et al., 2019). Since the dawn of human civilization some 15 000 years ago (Gill et al., 2004), wheat has had a large impact on human civilization and its development (Shewry, 2009). Bread wheat contributes to roughly 20% of the total calories consumed by humans and therefore crop losses impact global economy and food security. Notwithstanding the fact that wheat production increased steadily since the 'green revolution' in the 1960s, it will have to grow even more to provide nutrition to the estimated 9.8 billion people in 2050 (https://www.un.org/en/desa/world-population-projected-reach-98-billion-2050-and-112-billion-2100). Even though wheat is produced across the world in most countries because of the plants' ability to grow and yield harvest on marginal soils, many

http://dx.doi.org/10.19103/AS.2022.0114.20

of these countries don't produce sufficient wheat to provide for local demands and hence import the grain. Wheat, more than some other crops, is a critical global commodity exacerbated by production shortfalls in many countries and regions. This leaves the international community vulnerable to disruptions in the wheat supply chain (FAO, 2018), most recently like those caused by the conflict in Ukraine in 2022. These conflicts may prove problematic especially if the participating countries (i.e. Russia and Ukraine) are major wheat producers and wheat grain exporters (https://www.sciencedaily.com/releases/2022/06/220627141448.htm) (FOA, 2018, Fig. 1). Hence to address these political uncertainties and secure global bread wheat supply, production must increase beyond the current level: ~770 million tonnes (mt) for the 2021-2 season (~536 mt for human consumption; ~144 mt for fodder) (FAO, 2022).

Regardless of all the progress made in plant protection and plant biotechnology, most advancements in genetic improvement of wheat for sustained production have been made using classical crop breeding (Smith, 2021) and targeting valued agronomic traits such as yield, quality and reduced susceptibility to insect pests and pathogens. In the past, farmers planted these improved pest-resistant varieties supported by well-established integrated pest management (IPM) regimes within cropping systems, which in this context refers to all of the techniques employed by a farmer to produce a crop, such as succession of species cultivated, varieties of each species, soil tillage, dates and means of seeding, planting, pruning, thinning, the dates of application

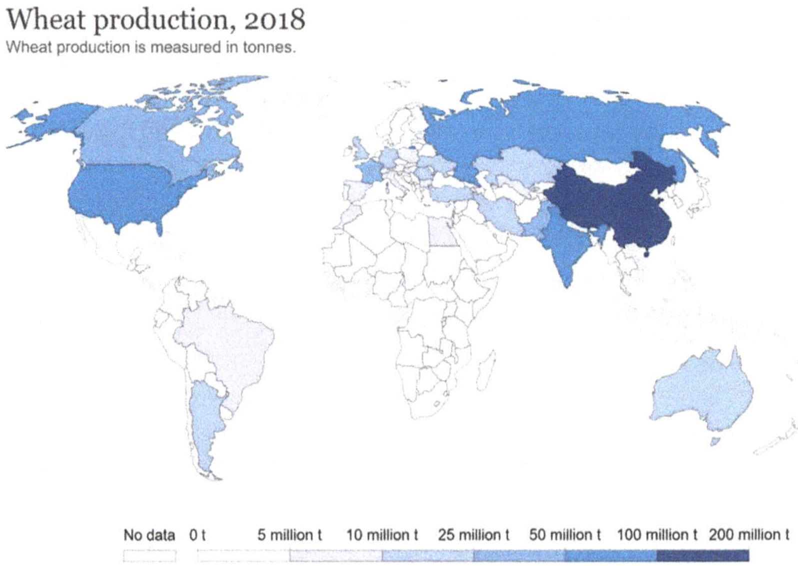

Figure 1 Global wheat production in 2018. The darker the colour the larger the production as indicated by the bar (FAO, 2018).

and doses of mineral or organic fertilizers, growth regulators, dates and doses of irrigation, dates and means of harvesting and chemical control (addition of fungicides, herbicides, or/and insecticides) (Meynard et al., 2003; Schmid et al., 2019; Nikoukar and Rashed, 2022 and references within). Also, breakdown of host resistance is not uncommon (Khan et al., 2015; Simon and Peccoud, 2018).

In a recent review, Denguine et al. (2021, p. 35) concluded that despite all the 'good intentions' displayed over the past 60 years of IPM implementation, successful implementation is rare because of *'quasi-infinite number of definitions and interpretations, IPM roadblocks and adoption barriers'* (p.2) and *'vested interests and corporate responsibility'* (p. 8). In their review, Denguine et al. (2021) proposed an alternative strategy that follows a more holistic systems approach to pest management. They called their approach the 'Agroecological Crop Protection' (ACP) system and argue that it improves on traditionally employed IPM by focusing more on practices that are chemical free, more environmentally friendly and support ecological resilience and sustainability (Ratnadass, 2020). They advocate for ACP that is superior to IPM because it is 'crop-centred' and 'farmer-centred', rather than 'pest-centred' (Denguine et al., 2009; Barrera, 2020). ACP aims to replace the use of chemicals (i.e. pesticides) which have a negative effect on the environment and human health, with ecological services offered by functional diversity both above and below the soil (Sarthou et al., 2017). Examples of ACP practice include the Cansea platform involving six Asian countries in Southeast Asia (http://www.agriculture-biodiversite-oi.org/en/ACP-ACTAE/Agroecological-Crop-Protection#pub1) and funded by the Food and Agriculture Organization of the United Nations (FOA) (Agroecological Crop Protection in South East Asia - Agritrop, https://agritrop.cirad.fr), where weaver ants (*Oecophylla* spp.) are conserved by farmers to control pests in citrus orchards. Implementation of the ACP aligns well with current trends to reduce the use of pesticides as proposed by European Union legislation, which compels breeders and farmers to seek alternatives (Directive 2009/128/EC – EU action to achieve the sustainable use of pesticides, https://eur-lex.europa.eu/EN/legal-content/summary/pesticides-in-the-eu-ensuring-their-sustainable-use.html) that will still enable pest control but that are less harmful to the environment and more aligned with consumer preferences. These changes in policy or preference, however, have significant implications for revenue as was shown in a 2014 study conducted in France; halving pesticide use in wheat fields resulted in additional yield losses ranging from 5% to 13% (Hossard et al., 2014). This 2–3 mt reduction in yield would lead to financial losses of between 800 million and 1200 million USD, calculated at 400 USD per tonne (2022 prices) (FAO, 2022). As with ACP, one of the key objectives with biotechnological research in agriculture is reduction of chemical usage either as fertilizers or pesticides.

Over the years biotechnology tools such as recombinant DNA technology that delivered genetically modified (GM) crops were applied to better the productivity of many crops with promising outcomes. Good examples are *Bacillus thuringiensis* (Bt) maize and Bt cotton (Meissle et al., 2022), but these were underutilized in wheat; and hence few of these resulted in viable or commercialized products in wheat. The underutilization of these technologies in wheat is the consequence of the crop being a high valued food commodity in Europe with its strong directives against planting GM crops for food as published in Regulation (EC) No 1946/2003 of the European Parliament and of the Council of 15 July 2003 on transboundary movements of genetically modified organisms (OJ L 287, 5.11.2003, pp. 1-10', https://eur-lex.europa.eu/EN/legal-content/summary/regulating-gm-crops-eu-countries-rights.html). This review will provide an overview of some of these, highlight the limitations and challenges associated with using most of the technologies in wheat and then propose alternatives focusing on current topical avenues to enhance crop protection in wheat in a sustainable and economically viable manner.

2 New breeding technologies

Genetic engineering offers good opportunities for improving pest resistance of most crops. In the context of this discussion, genetic modification (GM) refers to the genetic engineering of wheat using recombinant DNA technologies to produce improved or novel wheat varieties. After initially experiencing challenges with wheat transformation (Vasil, 2007 and references within), the first stable transformation of wheat was reported in 1992 (Vasil et al., 1992), after which stable transformation of many agronomically important traits followed (Vasil, 2007 and references within). These modifications were made to improve protein and baking quality attributes, enhance resistance to herbicides, resistance to biotic stressors (such as weeds, pests and pathogens) and tolerance to abiotic stressors (Vasil, 2007 and references within; Jones and Shrewy, 2009 and references within). Early examples of GM wheat with resistance to pests include GM wheat with resistance to Angoumois grain moth (*Sitotroga ceralella*, Altpeter et al., 1999) and grain aphid (*Sitobion avenae*, Stoger et al., 1999). However, the first commercial release of GM wheat wasn't aimed at pests, but at drought tolerance. The October 2020 GM wheat released in Argentina and Brazil (Kovak and Blaustein-Rejto, 2022), branded 'EcoWheat®', targets the improvement of drought tolerance, but also contains Rizoderma®, a biofungicide (Bioceres Crop solutions, 2020, https://investors.biocerescrops.com/news/news-details/2020/Bioceres-Crop-Solutions-Corp.-Announces-Regulatory-Approval-of-Drought-Tolerant-HB4-Wheat-in-Argentina/default.aspx). Similarly, the latest GM wheat to be distributed in Australia and New

Zealand is also engineered for drought tolerance but also contains glufosinate-herbicide resistance (Abbott, 2022). Interestingly, despite the fact that the latter two countries changed their policies in favour of the release of GM wheat, neither will allow the planting of the Bioceres Crop Solutions-GMO wheat, which was developed by the Argentinian-based company to be grown by its farmers. To date, declarations that GM-wheat is safe for consumer use/consumption along with the legislation for the release of GM wheat were only granted in Argentina in 2020, Brazil in 2021, and Australia and New Zealand in 2022 (Abbott, 2022).

Despite the significant scientific advances made, the commercial uptake of GM wheat has been slow, with none of the initial products ever commercialized because of consumer resistance (Heller, 2006). The slow release and commercialization of GM wheat in the USA can, in part, be ascribed to negative consumer sentiment which was brought about by claims of the detection of GM wheat escapes. In 2013 a farmer in Oregon State reported GM wheat in his wheat fields, and in 2019 the USDA in Washington State detected GM wheat in unplanted fields (Geimann, 2019).

3 The wheat genome: an obstacle to precise genome editing

Notwithstanding the fact that precise genome editing using the clustered regularly interspaced short palindromic repeats (CRISPR)/CRISPR-associated protein 9 (Cas9)-system (Fig. 2) has been demonstrated in many crop species (Si et al., 2020), genome editing in wheat remains challenging, with most GM wheat developed to date acquired via more traditional methods of genetic engineering (Li et al., 2021 and references within). There are mainly two approaches that are followed whilst engineering plants through CRISPR/Cas9 systems: (i) a knockout approach that generates insertion/deletion (indel) mutations that happen through inaccurate (error-prone) repair of DNA double-strand breaks (DSB) by nonhomologous end joining (NHEJ) and that gives rise to loss-of-function (LOF) alleles and (ii) knock-in approaches directed at introducing defined mutations by engineering an insertion or single-nucleotide polymorphism (SNP). These site-directed base modifications or SNPs are encoded by repair templates that utilize endogenous homology-directed repair (HDR) mechanisms. The NHEJ pathway has been applied to develop gene substitutions and gene knock-ins in rice (Dong et al., 2019). Whilst current available literature on wheat only reports on precise base editing at low efficiency in a limited number of laboratories worldwide (Gil-Humanes et al., 2017; Li and Xia, 2020; Sánchez-Leon et al., 2018; Zhang et al., 2019a, b; Li et al., 2015, 2020a, b), the need is urgent to develop new improved varieties.

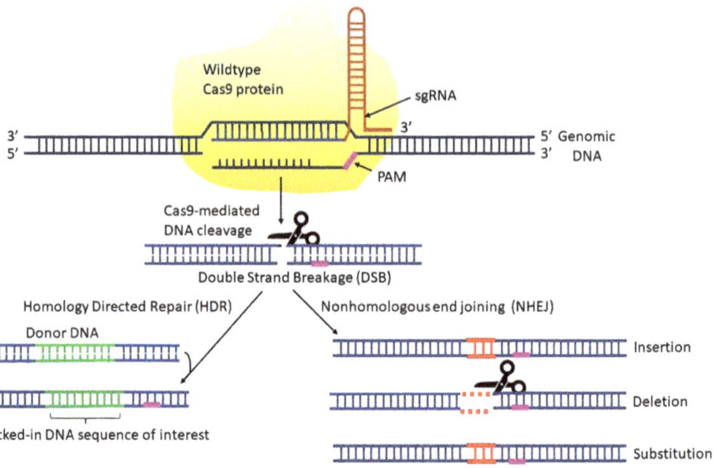

Figure 2 The mechanism of CRISPR-Cas9-mediated genome engineering. The synthetic sgRNA structure directs a Cas9 endonuclease to an arbitrary DNA sequence in the genome and then guides Cas9 around a double-strand break (DSB) in the targeted genomic region. The DSB generated by two distinct Cas9 nuclease domains is repaired by host-mediated DNA repair mechanisms. In the non-homologous end joining (NHEJ) pathway (absence of a repair template), the general error-prone mechanism is activated and causes random insertions and deletions (indels) or even substitutions at the DSB site, often leading to the disruption of gene function. In the homology directed repair (HDR) pathway (error-free), a donor template containing a sequence of interest flanked by homology arms can be initiated to create desired mutations by homologous recombination, providing the basis for the export of precise gene modification (i.e. gene knock-ins, deletions, corrections, or mutagenesis). Abbreviations: PAM, protospacer adjacent motif; sgRNA, single-guide RNA (Adopted from Jiang and Doudna, 2017).

The slow implementation of the CRISPR/Cas9-based technology in wheat can in part be ascribed to the complexity of the bread wheat's genome, which by comparison with those of other crops is larger and more complex. Bread wheat is an allohexaploid (2n = 6x = 42, AABBDD) with three subgenomes inherited from three homoeologous progenitor species (Salamini et al., 2002; Gill et al., 2004; Shewry, 2009). This complex polyploid nature of wheat has hindered the development of functional genomics and breeding, especially compared to other cereal crop species such as rice and maize. Particularly in the case of CRISPR/Cas9 dealing with multiple targets (potentially up to six alleles or two on each of the subgenomes) in the wheat genome mismatch cleavages (DNA restriction at unintended positions, Vicente et al., 2021) may result in unwanted phenotypes. A potential solution to rid the newly developed varieties of off-target mutations is by back-crossing or crossing with wild-type plants, or alternatively, through improved design of targets using web-based design tools (Li et al., 2021 and references within). The improved understanding of the genome, and recent release of the high-quality IWGSC RefSeq v1.0 genome (International Wheat Genome Sequencing Consortium, 2018), will greatly assist in revealing targets for fast breeding of adaptive and sustainable pest-resistant bread wheat varieties.

4 Epibreeding

Epibreeding refers to the exploration of methylated DNA sites, which originate during the activation of the epigenetic pathway in response to environmental stressors, as sources of new genetic variation. These environmentally induced DNA modifications can be applied as a breeding tool during the selection of new varieties after exposure of set varieties to specific stress such as insect infestation. These methylation changes in the DNA can then be followed, and if associated with a phenotype, they can be used for the selection of set trait. During exposure to, for example, a specific pest, covalent changes to DNA molecules and histones are induced that influence the transcriptional activity of the chromatin without changing the sequence of the DNA (Iwasaki and Paszkowski, 2014). Chromatin structure consists of nucleosomes that result in DNA packaging by the interacting DNA and the histone proteins. There are two types of chromatin, namely euchromatin - less condensed and associated with transcriptional active regions and heterochromatin - condensed and associated with transcriptionally silenced regions, specific modification of histones, and during epigenetics, the hypermethylation of DNA (Vaillant and Paszkowski, 2007). Since the expression of genes is dependent on access to DNA, the level of chromatin condensation dictates regulation. These induced changes are 'hallmarks' of induced epigenetic changes.

Several studies highlighted the hallmarks of epigenetics in plants as DNA methylation and histone modification, with in many instances small RNAs that prompt DNA methylation and modification of the chromatin (Meyer, 2015). When plants are exposed to environmental stressors, DNA gets methylated through the enzyme methyltransferase and related mechanisms such as RNA interference (RNAi) (Bewick and Schmitz, 2017), which may modify the plant genomic DNA rapidly and reversibly. This DNA methylation occurs mainly in three contexts in plants, namely CG, CHG and CHH, where H indicates any base other than G (Law and Jacobsen, 2010). Recent evidence demonstrated that the plants' somatic cell 'memorizes' the methylation of mostly GC-rich sequences which acts as the focal point for most epigenetic mechanisms to ensure post-generational transfer to progeny resulting in the formation of epialleles and heritable phenotypes (Matzke and Mosher, 2014; Quadrana and Colot, 2016) which may have potential value for breeding applications.

The modification of DNA methylation in response to pathogen attack has also been shown with significant host genome hypomethylation (less methylated than normal) and net chromocenter de-condensation (less packed and more open to be transcriptionally active) which directly influences the expression of defence genes. It is known that some epigenetic marks can be removed or retained and transferred to progeny, and hence they may provide plasticity or adaptability to new arising pests without the need to create genomic lesions (Alvarez et al., 2010 and references within; Furci et al., 2019). Such an example comes after analysis of data generated from screening 104 landraces of common wheat with 110-Mb exon capture probes, which revealed that the observed local adaptation may be underpinned by newly developed SNPs because of 5-methylcytosine deamination of inherited methylation states (Gardiner et al., 2015, 2018). Similarly, while analysing DNA methylation patterning in *Ae. tauschii* upon fungal infection, it was found that DNA methylation status - particularly methylation at CHH trinucleotides (where H indicates any nucleotide other than C), was involved in fungal defence and that this mechanism may be fixed in common wheat (Geng et al., 2018).

Recent studies have further shown that there exists a direct link between epigenetic and phenotypic variation which is typically modulated through the induction of gene expression, albeit not fully understood (Seymor and Becker, 2017). In wheat, it was found that interphase nuclei form regions called 'intergenic condensed spacers' (ICS), which separates gene-rich regions, where the DNA is less methylated but enriched with histones H3K9ac and H3K36me3, from folded and condensed segments enriched in cytosine methylation and enriched with histone H3K27me3 (Concia et al., 2020; Flavell, 2022 and references within). Less condensed regions can form loops at the boundaries of the ICS which enables interaction with promotor regions to activate and regulate transcription. Since these loop complexes are associated with RNA polymerase II, it was concluded that they form 'transcriptional factories' (TFs), which drive coordinated expression associated with these loops (Concia et al., 2020). Any perturbations of these interactions may be a source of phenotypic variations (Zhang et al., 2021), which can be defined by quantitative genetic maps as eQTLs (Furci et al., 2019; Zhao et al., 2019). Recently, a genome-wide atlas of the epigenetic states of genetic elements (i.e. genes/promotors and enhancers) all varying in activity were compiled based on chromatin and the histone marks H3K9ac and H3K36me3 and repressive H3K27me3 (Wang et al., 2021). This atlas of gene activity correlates well with the physical loop linkages reported by Concia et al. (2020) and has significant implications for haplotype breeding approaches (Brinton et al., 2020). Collectively, when the epigenetic status and relative activities of genes can be combined in new wheat hybrids, it may revolutionize breeding outcomes.

5 Host plant-insect pest interaction

Understanding host plant-insect interaction underpins the successful application of crop pest management. Current research shows that host plants recognize insect attacks through damage-associated molecular patterns (DAMPs) and herbivore-associated molecular patterns (HAMPs) (Acevedo et al., 2015), whereupon a diverse set of defensive mechanisms are activated to deter (antixenosis) or intoxicate (antibiosis) the invader. The ability of the host plant to recognize the invasion from signals such as cellular damage during grazing herbivory or hemipteran stylet penetration during feeding affects its capacity to respond. During feeding, hemipteran pests secrete salivary 'effector' proteins, a decoy strategy to manipulate the host's metabolism, suppressing the host defence and benefitting the insect (Hogenhout and Bos, 2011; Atamian et al., 2013; Naessens et al., 2015; Kaloshian and Walling, 2016; Mugford et al., 2016; Schimmel et al., 2017; Rodriguez et al., 2013, 2017). Hence, understanding how plants recognize herbivores and how herbivores manipulate their hosts will underpin genetic technology for pest resistance.

Wheat is under attack by a suite of insects, as reviewed in this volume. Some of the most damaging pests are members of the Coleoptera such as wireworms (i.e. larval state of the click beetle, *Limonius californicus*, Reddy et al., 2014; Nikoukar and Rashed, 2022), Diptera such as Hessian fly (*Maytiola destructor*, Schmid et al., 2019) and wheat stem maggot (*Meromyze americana,* Knodel et al., 2009), Lepidoptera such as fall armyworm (*Spodoptera frugiperda*, Schmid et al., 2019), Acari such as leaf curl mite (*Aceria tosichella*, Schmid et al., 2019) and Hymenoptera such as wheat stem sawfly (*Cephus cinctus*) (Knodel et al., 2009). Aphids (Hemiptera, Aphididae) can be especially damaging to wheat due to their nature of feeding (Dedryver et al., 2010), because they are vectors for about 50% of insect-transmitted plant viruses (Brunt et al., 1996; Nault, 1997). Thus, more attention will be given to aphids in the following section.

6 The importance of understanding the pest beyond pest biology

Analysis of the salivary proteome from a variety of aphid species revealed that it consists of a complex mixture of proteins (Miles, 1999; Harmel et al., 2008; Carolan et al., 2009, 2011; Cooper et al., 2010, 2011; Rao et al., 2013; Sharma et al., 2014; Nicholson and Puterka, 2014; Vandermoten et al., 2014; Chaudhary et al., 2015; Yang et al., 2018; MacWilliams et al., 2020). Putative effectors from this proteome differ between aphid species in a plant-species-specific manner, indicating that effectors are determining factors of aphid host range (Djamei et al., 2011; Pitino et al., 2011; Pitino and Hogenhout, 2013; Rodriguez et al., 2017). Nonetheless, specific effectors occur in multiple aphid species which suggests that there is a conserved repertoire of effectors required for infestation (Bos et al., 2010; Carolan et al., 2011; Atamian et al., 2013; Rao et al., 2013; Thorpe et al., 2016, 2018, 2020; Zhang et al., 2017; Boulain et al., 2018, 2019; Lemus et al., 2020; MacWilliams et al., 2020). Currently information on effector proteins is limited, with only a few effector proteins functionally characterized (Mutti et al., 2008; Bos et al., 2010; Pitino and Hogenhout, 2013; Furch et al., 2015; Rodriguez et al., 2014, 2017; Elzinga et al., 2014; Mugford et al., 2016; Atamian et al., 2013; Naessens et al., 2015; Wang et al., 2015; Chaudhary et al., 2014, 2015, 2019; Kettles and Kaloshian, 2016; Rodriguez and Bos, 2013; Cui et al., 2019; Escudero-Martinez et al., 2020) and effector functional predictions are hampered by their diversity (Boulain et al., 2018; Dommel et al., 2020). Despite these challenges, more focus should be given to effector proteins for plant protection as they are key determinants of pest-host interaction. Needed are (i) a better catalogue of effector proteins from different aphid species to help us understand their diversity and appreciate their evolution and (ii) an improved mechanistic understanding of their action which includes their function and structure.

Good opportunities exist to explore this field of pest biology using different modern 'omics' approaches such as genomics, transcriptomics and proteomics. However, to make full use of such big data to fully capitalize on functional host plant-pest interaction studies for biopesticide development, a particularly well-curated genome with comprehensive gene sets is required. Unfortunately, these are not yet available. Although there are at least 5000 different hemipteran species (Blackman et al., 2007) causing extensive crop yield losses through their feeding either directly (35–40%) by probing and sucking of plant's phloem or indirectly (20–80%) through the transmission of viral and fungal diseases (Ng and Perry, 2004), only a limited number of genome drafts are available (International Aphid Genomics Consortium, 2010; Biello et al., 2021; Mathers et al., 2021, 2022), with even fewer of relevance to wheat (Nicholson et al., 2015; Burger and Botha, 2017; Mathers et al., 2022) (Table 1, Fig. 3). A recent review explored the use of 'Effectoromics' in this field as a novel approach to enable the cataloguing of candidate effectors which then can be applied to specify their corresponding host resistance and susceptibility genes (Kanja and Hammond-Kosack, 2020 and references within). In the review, several features of effector proteins are highlighted as 'signatures' or 'markers' that can be applied during this process such as for example the identification of a signal peptide-specific amino acid sequence followed by a restriction-amenable site commonly found at the N-terminus of secreted proteins (Nielsen and Krogh, 1998; Torto et al., 2003) or the presence of the LysM domain that targets chitin (de Jonge and Thomma, 2009; Lee et al., 2014). Kanja and Hammond-Kosack (2020) argued in favour of the value of developed effector databases that contain repertoires of effectors from different host-invader systems but note

Table 1 All available genome assemblies (excluding transcriptome and targeted genome sequencing) of wheat-feeding aphids as obtained from the NCBI's Assembly database (https://www.ncbi.nlm.nih.gov/assembly/?term=txid27482%20%5BORGN%5D) along with select assembly metrics

Aphid species and common name	GenBank assembly accession	Date submitted	Sequencing platforms	Genome sequencing coverage	Total assembly length gapped in bp (ungapped)	Number of Scaffolds (N50 in bp)	Number or contigs (N50 in bp)
Diuraphis noxia Russian wheat aphid	GCA_001186385.1	2015/07/23	Illumina HiSeq	104 x	395 073 589 (296 540 817)	5637 (397 774)	50 913 (13 141)
	GCA_001465515.1	2015/12/14	Illumina HiSeq + IonTorrent + 454	44.8 x	399 654 885 –	– –	190 686 (5727)
Rhopalosiphum padi Bird cherry-oat aphid	GCA_019425515.1	2021/07/30	Illumina HiSeq	200 x	315 402 387 (314 608 461)	9628 (2 027 176)	10 981 (420 804)
	GCA_020882245.1	2021/11/16	PacBio Sequel + Illumina NovaSeq	21 x	338 184 925 (338 066 425)	35 (92 921 235)	272 (14 338 858)
Schizaphis graminum Greenbug	GCA_003264975.1	2018/06/25	Illumina HiSeq	100 x	385 232 317 (359 705 026)	7859 (1 292 312)	15 967 (54 502)
	GCA_019425635.1	2021/07/30	Illumina HiSeq	184 x	401 374 768 (394 173 158)	43 470 (849 019)	53 000 (67 785)
	GCA_020882235.1	2021/11/16	PacBio Sequel + Illumina NovaSeq	25 x	499 221 988 (499 172 187)	22 (104 490 323)	300 (15 350 123)
Sitobion avenae English grain aphid	GCA_019425605.1	2021/07/30	Illumina HiSeq+ PacBio Sequel	242 x	393 018 683 (390 965 188)	27 478 (807 389)	30 179 (244 199)
					(60 277 968)	833 (14 626)	37 500 (1 900)

(*Continued*)

Table 1 (*Continued*)

Aphid species and common name	GenBank assembly accession	Date submitted	Sequencing platforms	Genome sequencing coverage	Total assembly length gapped in bp (ungapped)	Number of Scaffolds (N50 in bp)	Number or contigs (N50 in bp)
Sitobion miscanthi	GCA_008086715.1	2019/08/28	PacBio Sequel	87.0 x	397852643	653	1145
Indian grain aphid					(397803443)	(36263045)	(1613763)
Metopolophium dirhodum	GCA_019925205.1	2021/09/14	PacBio Sequel + Illumina NovaSeq + Complete Genomics	85.31x	480691620	67	307
Rose-grain aphid					(480691620)	(41762923)	(7032305)

Figure 3 (a) *Rhopalosiphum padi*, commonly known as the Bird cherry oat aphid. Note the rusty band near the end of its abdomen. Photo courtesy of Adam Varenhorst (https://extension.sdstate.edu/scouting-winter-wheat-aphid-pests); (b) *Sitobion avenae*, also known as the English grain aphid, displaying its black cornicles and alternating black and green segments of the antennae and legs. Photo by Adam Varenhorst (https://extension.sdstate.edu/scouting-winter-wheat-aphid-pests); (c) *Sitobion miscanthi* or Indian grain aphid – are elongate ovate and are variably coloured greenish, yellowish or reddish-brown to dark brown, with shiny black siphunculi and a pale cauda (https://influentialpoints.com/Gallery/Sitobion_miscanthi_indian_grain_aphid.htm); (d) *Schizaphis graminum* also known as Greenbug. Hosts display discoloration at their site of the feeding (Photo by Alton N. Spark, Jr. University of Georgia, Bugwood.org, https://extension.sdstate.edu/scouting-winter-wheat-aphid-pests); (e) *Metopolophium dirhodum* or Rose-grain aphid is a medium-sized spindle-shaped aphid which ranges from green to yellowish green, with a brighter green longitudinal mid-dorsal stripe (https://influentialpoints.com/Gallery/Metopolophium_dirhodum_Rose-grain_aphid.htm); and (f) *Diuraphis noxia* or Russian wheat aphid (RWA). Symptoms of feeding include leaf rolling and chlorotic streaking, with the latter often associated with viral infection, although RWA doesn't serve as a virus vector.

that corresponding expression data following host-invader interactions must also be available to fast-track the development of pest control strategies. A current example of such a platform in the field of plant/microbe pathology is the Pathogen–Host Interactions database (PHI-base, www.phi-base.org) (Urban et al., 2020). Upon release, the PHI-base version 4.8 (September 2019) contained 3454 manually curated references; provided information on 6780 genes from 268 pathogens that were tested on 210 hosts in 13 801 interactions with almost equal representation between prokaryotic and eukaryotic pathogens; while 60% of the host species are plants (split 50:50 between cereal and non-cereal plants) the rest were of value to the field of medicine/epidemiology.

Recently such an effectoromics approach was followed to catalogue the putative effectors of an economically important pest of wheat, the Russian wheat aphid (RWA, *Diuraphis noxia*). In this study, the putative effectors were predicted from data sets of whole aphid transcriptome and the RWA genome using a bioinformatic pipeline (Nicolis et al., 2022). While a previous analysis of the salivary gland proteome identified 156 proteins of RWA, among which are proteins sharing similarity to known effectors as well as the first report of ornithine decarboxylase antizyme 1 in aphid saliva, analysis of the transcriptome predicted 831 transcripts encoding putatively secreted proteins and included in this collection were 24 transcripts specific to RWA. The authors selected ten of the predicted effector proteins and tested their preferential expression in aphid head rather than the whole body with an accuracy of 95% but confirmed the presence of a further three other predicted effector proteins at a 99% probability to be preferentially expressed in salivary gland, rather than the whole body, providing the basis for future functional characterization studies to gain further insight into this aphid-plant interaction.

It is thus proposed that the availability of comprehensive effector databases coupled with knowledge of crop and pest genomes and their interactions, could facilitate modern approaches such as machine/deep learning (i.e. artificial intelligence devoted to understanding and building methods that improve performance and inform predictions) to enable the development of prediction models of interacting genes that confers resistance or susceptibility in these interacting systems. The genes that codes for these interacting proteins can then be applied as targets in RNA interference studies. Such an approach will eventually make sustainable solutions to crop pest control a reality.

7 RNA interference

RNA interference, a highly conserved cellular mechanism that occurs in all eukaryotic organisms - also reported in plants as post-transcriptional gene silencing, was first described in the nematode *Caenorhabditis elegans* (Fire et al., 1998). The first insect documentation to follow was the fruit fly (*Drosophila melanogaster*) (Kennerdell and Carthew, 1998). Subsequently, RNAi has been observed in many insect species (Mao et al., 2007; Mao and Zeng, 2014; Coleman et al., 2015; de Souza Pacheco et al., 2020) including the pea aphid (Mutti et al., 2006, 2008) and *Sitobion avenae* (Abdellatef et al., 2015), with three documented RNAi pathways, namely the small interfering RNA (siRNA), microRNA and Piwi-interacting RNA pathways (Zhu and Palli, 2020). RNAi has been widely applied as a common reverse genetics tool for determining gene function. The principle of the technique lies in the delivery of siRNA or double-stranded RNA (dsRNA) capable of target-specific recognition of sequence followed by the breakdown of the mRNA targets (Fire et al., 1998; Zamore et al., 2000; Marques et al., 2010; Pratt and MacRae, 2009), which are cleaved into siRNAs upon cell entry by an enzyme called Dicer (Bernstein et al., 2001; Hannon, 2002). Dicer is a member of the RNase III family that produce 5′-phosphorylated termini after restriction of dsRNA (Hannon, 2002). When siRNA is produced by Dicer (about 21–25 nucleotides in length) the resulting element contains 5′-phosphorylated termini and a 3′ overhang of 2 nucleotides. This siRNA binds to an Argonaute (Ago) protein, to form a new multiprotein complex known as the RNA-induced silencing complex (RISC) (Preall and Sontheimer, 2005). The siRNA/RISC complex binds the complementary sequence of the target mRNA resulting in post-transcriptional gene silencing (PTGS) through degradation of the mRNA or inhibition of its translation. But transcriptional gene silencing (TGS) can also occur in the nucleus via chromatin modifications (Liu et al., 2021). In fungi and nematodes, the silencing can be maintained by the action of the RNA-dependent RNA polymerase (RdRP) for siRNA secondary intensification (Liu et al., 2021). Although the presence of RdRP was confirmed in Diptera (Pinzón et al., 2019), its function as a secondary amplifier of siRNA must still be demonstrated.

In the context of crop pest management, the RNAi system is viewed as a potential approach to sustainable and eco-friendly insect pest management by avoiding the excessive application of chemicals and resulting chemical residues (Zotti and Smagghe, 2015; Zhang et al., 2015a; Cai et al., 2018). Despite the theoretical potential and many proposed delivery systems, few have been successfully implemented in practice. There are several reasons for this. During a gene silencing experiment, the design of the RNAi molecule is critical, since for successful execution of gene silencing, it must be sequence-specific; a base differences as small as one nucleotide base could cause the silencing effect to disappear (Joseph and Osman, 2012). This, however, can be advantageous as high specificity where many genes with high similarity are present in the same targeted organism. The efficiency of the outcome of such silencing experiments may also be determined by siRNA molecule size. In insects, the efficacy differs from species to species, but also it was demonstrated that in the case of dsRNA molecules, ideal sizes range between 50 and 200 nucleotides in length in insects (Huvenne and Smagghe, 2010; Koch and Wassenegger, 2021). Insect species also differ in their responsiveness to dsRNA and siRNA, for example, the mechanism works well in Coleoptera but results prove inconsistent in some species and these require longer dsRNA rather than the smaller siRNA for any response (Saleh et al., 2006; Bolognesi et al., 2012).

The two most used delivery systems are host-induced gene silencing (HIGS) and spray-induced gene silencing (SIGS). HIGS entails the delivery of dsRNA by expressing the targets in GM crops and was achieved with commercialization of some products (Zhang et al., 2017). This approach makes use of the expression of dsRNA derived from pest or pathogen genes *in planta*. Silencing will occur in the pest provided that the insect ingests enough of the dsRNA or sRNA to activate the process (Liu et al., 2021; Koch and Wassenegger, 2021). Table 2 provides a list of HIGS-delivered RNAi target genes reported in pests, pathogens and nematodes. While GM crops are expensive and tedious to develop and may face many regulatory obstacles, non-transformative (NT) delivery strategies for pest management are favoured because of their short development time, low development cost, low risk of consumer/market-related resistance, simple regulatory process and feasibility and viability for all crops (Cagliari et al., 2019). Applications of these technologies, however, requires that RNAi efficiency of these NT-products is improved to achieve better success rates.

Delivery through topical application (i.e. spraying), also known as SIGS, has been widely demonstrated (Pridgeon et al., 2008; El-Shesheny et al., 2013; San Miguel and Scott, 2016; Zhang et al., 2015b; Dias et al., 2019) and may offer a suitable and affordable alternative means of application. Since RNAi is highly dependent on the sequence specificity, it poses little risk to non-target organisms (pests or microbes) (Liu et al., 2021) and despite this, off-target effects have been documented for highly conserved region sequences such as in the case of the use of *vATPAseA* and *vATPAseE* from *Leptinotarsa decemlineata* (Colorado potato beetle, CPB) that share high sequence similarities to those from the Western corn rootworm, resulting in reduced fitness of the CPB in a bioassay (Baum et al., 2007). Cuticle penetration of dsRNA is still a limiting factor for the successful execution of SIGS. Once the penetration efficiency-problem is solved stability of the dsRNA and siRNA remains a concern especially with SIGS application, as rapid degradation because of low or high pH in the gut lumens or some pests is a possibility (Cooper et al., 2019). Similarly, degradation of dsRNA to below detectable levels after application on different soil types within 7 days has also been reported (Albright et al., 2017; Fischer et al., 2017). Like with other biopesticide applications, resistance may also occur if not managed properly (Liu et al., 2021). Lastly, mass production for large-scale applications poses another challenge as conventional laboratory methods will be too costly and impractical and even though microbial-based dsRNA production has been demonstrated (Parsons et al., 2018; Ahn et al., 2019), more research focusing on upscaling during production is required to be feasible. As these technical challenges are met, SIGS for pest management may become feasible.

Recent promising solutions to SIGS-delivery of dsRNA for pest control may involve nanoparticles or nanosheets. Nanoparticles are usually defined by their

Table 2 List of reported RNAi target genes tested in organisms using host-induced RNAi silencing (HIGS) modified from Liu et al. (2022)

Organism	Target gene(s)	Organismal response	Reference
Insects			
Sitobion avenae	Salivary sheath protein (*SHP*)	Mortality/fecundity/transgenerational gene silencing	Abdellatef et al. (2015)
Sitobion avenae	Lipase maturation factor 2-like gene	Mortality/fecundity	Xu et al. (2017)
Sitobion avenae	Zinc finger protein (*SaZFP*)	Mortality/transgenerational gene silencing	Sun et al. (2019)
Sitobion avenae	Chitin synthase 1 (*CHS1*)	Mortality/fecundity	Zhao et al. (2018)
Nematodes			
Meloidogyne incognita	Heat-shock protein 90, isocitrate lyase, *Mi-cpl-1*	Reduced reproduction	Lilley et al. (2007)
Pathogens			
Fusarium graminearum	Cytochrome P450 lanosterol C-14a-demethylase (*CYP51*)	Inhibiting fungal mycelium formation	Koch et al. (2013)
Fusarium graminearum	*Chs3b*	Restriction of fungal growth	Cheng et al. (2015)
Blumeria graminis	Virulence effector (*Avra10*)	Reduced fungal development	Nowara et al. (2010)
B. graminis f. sp. hordei	Ribonuclease-like protein, Virulence effector, Glucanase, Metalloprotease, Virulence effector	Reduced virulence	Pliego et al. (2013)
Fusarium culmorum	Secreted lipase (*Fgl1*) Mitogen-activated protein (MAP) kinase (*Fmk1*), Beta 1,3-Glucan synthase (*Gls1*)	Reduced virulence	Chen et al. (2016)
Blumeria graminis f.sp. tritici	Three virulence effectors (*SvrPm3a1/f1*)	Reduced virulence	Schaefer et al. (2020)
Puccinia striiformis f. sp. tritici	Glycine-serine-rich effector (*PstGSRE1*)	Reduced virulence and increased H_2O_2 accumulation	Qi et al. (2019)

sizes which range between 1 and 100 nm and consists of different chemistries (Miele et al., 2012; Kumar et al., 2019; Kunte et al., 2020). Nanoparticles, such as chitosan, liposomes and cationic dendrimers, can protect dsRNA/siRNA molecules from enzymatic breakdown. They are also amenable to cross cell membranes transport and have been demonstrated to be excellent carriers during administering treatment in the field of medicine (Ahmadzada et al., 2018; Chalbatani et al., 2019). In recent years, several types of nanoparticles have been tested for nucleic acid delivery in insects with the objective to increase the stability and cellular uptake of the dsRNA/siRNA (Dong et al., 2014; Zhang et al., 2015a, b, c, d; Li et al., 2019; Yan et al., 2021), with clay particle-based systems gaining more attention (Mitter et al., 2017). However, before they can be applied broadscale in field pest protection systems, several issues need to be addressed, such as the cost of large-scale production, the risk they pose to human health, and the environment being an intentionally diffuse source.

8 Conclusion and future trends

Notwithstanding the benefits that new breeding technologies such as genetically modified wheat and precise genome edited wheat varieties may offer, their development remains difficult and costly, and hence, it may be worthwhile to focus on other pest management alternatives such as effectoromics and RNAi. With the advances in next-generation sequence technologies that improved the quality of the available insect genomes and other genomic resources (i.e. proteomics, transcriptomics, metabolomics, effectoromics), researchers in the field can now tap these to enable better pest control strategies. Machine/deep learning approaches for big data management should also be integrated to enable optimization of such informative effectoromics platforms to enable the identification and prediction of effector protein-host proteins interactions. This will not only increase our understanding of the mechanistic action that underpins these complex interactions but also assist in the identification of suitable gene targets for pest control in the future.

9 References

Abbott, C. (2022). GMO wheat approved for consumption in Australia and New Zealand. Available at: https://www.agriculture.com/news/business/gmo-wheat-approved-for-consumption-in-australia-and-new-zealand. Downloaded 12 July 2022.

Abdellatef, E. T., Koch, W. A., Imani, J., Vilcinskas, A. and Kogel, K. H. (2015). Silencing the expression of the salivary sheath protein causes transgenerational feeding suppression in the aphid *Sitobion avenae*. *Plant Biotechnol. J.* 13, 849–857.

Acevedo, F. E., Rivera-Vega, L. J., Chung, S. H., Ray, S. and Felton, G. W. (2015). Cues from chewing insects - The intersection of DAMPs, HAMPs, MAMPs and effectors. *Curr. Opin. Plant Biol.* 26, 80–86.

Ahmadzada, T., Reid, G. and McKenzie, D. R. (2018). Fundamentals of siRNA and miRNA therapeutics and a review of targeted nanoparticle delivery systems in breast cancer. *Biophys. Rev.* 10(1), 69–86. doi: 10.1007/s12551-017-0392-1.

Ahn, S. J., Donahue, K., Koh, Y., Martin, R. R. and Choi, M. Y. (2019). Microbial based double-stranded RNA production to develop cost effective RNA interference application for insect pest management. *Int. J. Insect Sci.* 11, 1179543319840323. doi: 10.1177/1179543319840323.

Albright, V. C. 3rd, Wong, C. R., Hellmich, R. L. and Coats, J. R. (2017). Dissipation of double-stranded RNA in aquatic microcosms. *Environ. Toxicol. Chem.* 36(5), 1249–1253. doi: 10.1002/etc.3648.

Altpeter, F., Diaz, I., McAuslane, H., Gaddour, K., Carbonero, P. and Vasil, I. K. (1999). Increased insect resistance in transgenic wheat stably expressing trypsin inhibitor CMe. *Mol. Breed.* 5(1), 53–63.

Alvarez, A. E., Nota, F. and Cambiagno, D. A. (2010). Epigenetic control and plant immunity. *Mol. Plant Pathol.* 11, 563–576. doi: 10.1111/J.1364-3703.2010.00621.

Atamian, H. S., Chaudhary, R., Cin, V. D., Girke, T. and Kaloshian, I. (2013). *In planta* expression or delivery of potato aphid *Macrosiphum euphorbiae* effectors Me10 and Me23 enhances aphid fecundity. *Mol. Plant Micro. Interac.* 26, 67–74. doi: 10.1094/MPMI-06-12-0144-FI.

Barrera, J. F. (2020). Holistic pest management. In: *Aera-Wide Management of Fruit Fly Pests*. Boca Raton: CRC Press, 275–293.

Baum, J. A., Bogaert, T., Clinton, W., Heck, G. R., Feldmann, P., Ilagan, O., Johnson, S., Plaetinck, G., Munyikwa, T., Pleau, M., Vaughn, T. and Robert, J. (2007). Control of coleopteran insect pests through RNA interference. *Nat. Biotechnol.* 25(11), 1322–1326. doi: 10.1038/nbt1359.

Bernstein, E., Caudy, A. A., Hammond, S. M. and Hannon, G. J. (2001). Role for a bidentate ribonuclease in the initiation step of RNA interference. *Nature* 409(6818), 363–366.

Bewick, A. J. and Schmitz, R. J. (2017). Gene body DNA methylation in plants. *Curr. Opin. Plant Biol.* 36, 103–110. doi: 10.1016/j.pbi.2016.12.007.

Biello, R., Singh, A., Godfrey, C. J., Fernández, F. F., Mugford, S. T., Powell, G., Hogenhout, S. A. and Mathers, T. C. (2021). A chromosome-level genome assembly of the woolly apple aphid, *Eriosoma lanigerum* Hausmann (Hemiptera: Aphididae). *Mol. Ecol. Resour.* 21(1), 316–326.

Blackman, R. L., Eastop, V. F., Van Emden, H. F. and Harrington, R. (2007). Aphids as crop pests': Taxonomic issues. In: van Emden, H. F. and Harrington, R. (Eds.), 1–29.

Bolognesi, R., Ramaseshadri, P., Anderson, J., Bachman, P., Clinton, W., Flannagan, R., Ilagan, O., Lawrence, C., Levine, S., Moar, W., Mueller, G., Tan, J., Uffman, J., Wiggins, E., Heck, G. and Segers, G. (2012). Characterizing the mechanism of action of double-stranded RNA activity against western corn rootworm (*Diabrotica virgifera* LeConte). *PLoS ONE* 7(10), e47534.

Bos, J. I., Prince, D., Pitino, M., Maffei, M. E., Win, J. and Hogenhout, S. A. (2010). A functional genomics approach identifies candidate effectors from the aphid species *Myzus persicae* (green peach aphid). *PLoS Genet.* 6(11), e1001216. doi: 10.1371/journal.pgen.1001216.

Botha, A.-M., Kunert, K. J., Malinga, J. M. and Foyer, C. H. (2019). Defining biotechnological solutions for insect control in Sub-Saharan Africa. *Food Ener. Sec.* e191. doi: 10.1002/fes3.191.

Boulain, H., Legeai, F., Guy, E., Morliere, S., Douglas, N. E., Oh, J., Murugan, M., Smith, M., Jaquiéry, J., Peccoud, J., White, F. F., Carolan, J. C., Simon, J. C. and Sugio, A. (2018). Fast evolution and lineage-specific gene family expansions of aphid salivary effectors driven by interactions with host-plants. *Genome Biol. Evol.* 10(6), 1554-1572. doi: 10.1093/gbe/evy097.

Boulain, H., Legeai, F., Jaquiéry, J., Guy, E., Morlière, S., Simon, J. C. and Sugio, A. (2019). Differential expression of candidate salivary effector genes in pea aphid biotypes with distinct host plant specificity. *Front. Plant Sci.* 10, 1301. doi: 10.3389/fpls.2019.01301.

Brinton, J., Ramirez-Gonzalez, R. H., Simmonds, J., Wingen, L., Orford, S., Griffiths, S., 10 Wheat Genome Project, Haberer, G., Spannagl, M., Walkowiak, S., Pozniak, C. and Uauy, C. (2020). A haplotype-led approach to increase the precision of wheat breeding. *Commun. Biol.* 3(1), 712. doi: 10.1038/s42003-020-01413-2.

Brunt, A. A., Crabtree, K., Dallwitz, M. J., Gibbs, A. J., Watson, L. and Zurcher, E. J. (1996). Plant viruses online: Descriptions and lists from the VIDE database. *Version: 20 August 1996*. Available at: http://biology.anu.edu.au/Groups/MES/vide.

Burger, N. F. V. and Botha, A.-M. (2017). Genome of Russian wheat aphid an economically important cereal aphid. *Stand. Genom. Sci.* 12, 1.

Cagliari, D., Dias, N. P., Galdeano, D. M., dos Santos, E. Á, Smagghe, G. and Zotti, M. J. (2019). Management of pest insects and plant diseases by non-transformative RNAi. *Front. Plant Sci.* 10, 1319. doi: 10.3389/fpls.2019.01319.

Cai, Q., He, B., Kogel, K. H. and Jin, H. (2018). Cross-kingdom RNA trafficking and environmental RNAi – Natures blueprint for modern crop protection strategies. *Curr. Opin. Microbiol.* 46, 58–64. doi: 10.1016/j.mib.2018.02.003.

Carolan, J. C., Caragea, D., Reardon, K. T., Mutti, N. S., Dittmer, N., Pappan, K., Cui, F., Castaneto, M., Poulain, J., Dossat, C., Tagu, D., Reese, J. C., Reeck, G. R., Wilkinson, T. L. and Edwards, O. R. (2011). Predicted effector molecules in the salivary secretome of the pea aphid (*Acyrthosiphon pisum*): A dual transcriptomic/proteomic approach. *J. Proteome Res.* 10(4), 1505–1518. doi: 10.1021/pr100881q.

Carolan, J. C., Fitzroy, C. I., Ashton, P. D., Douglas, A. E. and Wilkinson, T. L. (2009). The secreted salivary proteome of the pea aphid *Acyrthosiphon pisum* characterised by mass spectrometry. *Proteomics* 9(9), 2457–2467. doi: 10.1002/pmic.200800692.

Chalbatani, M. G., Dana, H., Gharagouzloo, E., Grijalvo, S., Eritja, R., Logsdon, C. D., Memari, F., Miri, S. R., Rad, M. R. and Marmari, V. (2019). Small interfering RNAs (siRNAs) in cancer therapy: A nano-based approach. *Int. J. Nanomed.* 2(14), 3111-3128. doi: 10.2147/IJN.S200253.

Chaudhary, R., Atamian, H. S., Shen, Z., Briggs, S. P. and Kaloshian, I. (2014). GroEL from the endosymbiont *Buchnera aphidicola* betrays the aphid by triggering plant defense. *Proc. Natl. Acad. Sci. U. S. A.* 111(24), 8919–8924. doi: 10.1073/pnas.1407687111.

Chaudhary, R., Atamian, H. S., Shen, Z., Briggs, S. P. and Kaloshian, I. (2015). Potato aphid salivary proteome: Enhanced salivation using resorcinol and identification of aphid phosphoproteins. *J. Proteome Res.* 14(4), 1762–1778. doi: 10.1021/pr501128k.

Chaudhary, R., Peng, H. C., He, J., MacWilliams, J., Teixeira, M., Tsuchiya, T., Chesnais, Q., Mudgett, M. B. and Kaloshian, I. (2019). Aphid effector Me10 interacts with tomato

TFT 7, a 14-3-3 isoform involved in aphid resistance. *New Phytol.* 221(3), 1518–1528. doi: 10.1111/nph.15475.

Chen, W., Kastner, C., Nowara, D., Oliveira-Garcia, E., Rutten, T., Zhao, Y., Deising, H. B., Kumlehn, J. and Schweizer, P. (2016). Host-induced silencing of Fusarium culmorum genes protects wheat from infection. *J. Exp. Bot.* 67(17), 4979–4991. doi: 10.1093/jxb/erw263.

Cheng, W., Song, X. S., Li, H. P., Cao, L. H., Sun, K., Qiu, X. L., Xu, Y. B., Yang, P., Huang, T., Zhang, J. B., Qu, B. and Liao, Y. C. (2015). Host-induced gene silencing of an essential chitin synthase gene confers durable resistance to Fusarium head blight and seedling blight in wheat. *Plant Biotechnol. J.* 13(9), 1335–1345. doi: 10.1111/pbi.12352.

Coleman, A. D., Wouters, R. H. M., Mugford, S. T. and Hogenhout, S. A. (2015). Persistence and transgenerational effect of plant-mediated RNAi in aphids. *J. Exp. Bot.* 66(2), 541–548.

Concia, L., Veluchamy, A., Ramirez-Prado, J. S., Martin-Ramirez, A., Huang, Y., Perez, M., Domenichini, S., Rodriguez Granados, N. Y., Kim, S., Blein, T., Duncan, S., Pichot, C., Manza-Mianza, D., Juery, C., Paux, E., Moore, G., Hirt, H., Bergounioux, C., Crespi, M., Mahfouz, M. M., Bendahmane, A., Liu, C., Hall, A., Raynaud, C., Latrasse, D. and Benhamed, M. (2020). Wheat chromatin architecture is organized in genome territories and transcription factories. *Genom. Biol.* 21(1), 104. doi: 10.1186/s13059-020-01998-1.

Cooper, W. R., Dillwith, J. W. and Puterka, G. J. (2010). Salivary proteins of Russian wheat aphid (Hemiptera: Aphididae). *Environ. Entomol.* 39(1), 223–231. doi: 10.1603/EN09079.

Cooper, W. R., Dillwith, J. W. and Puterka, G. J. (2011). Comparisons of salivary proteins from five aphid (Hemiptera: Aphididae) species. *Environ. Entomol.* 40(1), 151–156. doi: 10.1603/EN10153.

Cooper, A. M., Silver, K., Zhang, J., Park, Y. and Zhu, K. Y. (2019). Molecular mechanisms influencing efficiency of RNA interference in insects. *Pest Manag. Sci.* 75(1), 18–28. doi: 10.1002/ps.5126.

de Jonge, R. and Thomma, B. P. (2009). Fungal LysM effectors: Extinguishers of host immunity? *Trends Microbiol.* 17(4), 151–157.

de Souza Pacheco, IdS., Galdeano, D. M., Maluta, N. K. P., Lopes, J. R. S. and Machado, M. A. (2020). Gene silencing of *Diaphorina citri* candidate effectors promotes changes in feeding behaviors. *Sci. Rep.* 10(1), 1–14. doi: 10.1038/s41598-020-62856-5.

Dedryver, C. A., Le Ralec, A. and Fabre, F. (2010). The conflicting relationships between aphids and men: A review of aphid damage and control strategies. *C. R. Biol.* 333(6–7), 539–553.

Denguine, J.-P., Aubertot, J.-N., Flor, R. J., Lescourrett, F., Wyckhuys, K. A. G. and Ratnadas, A. (2021). Integrated pest management: Good intentions, hard realities. A review. *Agron. Sustain. Dev.* 41(38), 1–35.

Denguine, J.-P., Ferron, P. and Russell, D. (2009). *Crop Protection: From Agrochemistry to Agroecology*. Enfield, NH: Science Publishers.

Deutsch, C. A., Tewksbury, J. J., Tigchelaar, M., Battisti, D. S., Merrill, S. C., Huey, R. B. and Naylor, R. L. (2018). Increase in crop losses to insect pests in a warming climate. *Science* 361(6405), 916–919.

Dias, N., Cagliari, D., Kremer, F. S., Rickes, L. N., Nava, D. E., Smagghe, G. and Zotti, M. (2019). The South American fruit fly: An important pest insect with RNAi-sensitive larval stages. *Front. Physiol.* 10, 794.

Djamei, A., Schipper, K., Rabe, F., Ghosh, A., Vincon, V., Kahnt, J., Osorio, S., Tohge, T., Fernie, A. R., Feussner, I., Feussner, K., Meinicke, P., Stierhof, Y. D., Schwarz, H., Macek, B., Mann, M. and Kahmann, R. (2011). Metabolic priming by a secreted fungal effector. *Nature* 478(7369), 395–398. doi: 10.1038/nature10454.

Dommel, M., Oh, J., Huguet-Tapia, J. C., Guy, E., Boulain, H., Sugio, A., Murugan, M., Legeai, F., Heck, M., Smith, C. M. and White, F. F. (2020). Big genes, small effectors: Pea aphid cassette effector families composed from miniature exons. *Front. Plant Sci.* 11, 1230. doi: 10.3389/fpls.2020.01230.

Dong, Y., Siegwart, D. J. and Anderson, D. G. (2019). Strategies, design, and chemistry in siRNA delivery systems. *Adv. Drug Deliv. Rev.* 144, 133–147.

Dubcovsky, J. and Dvorak, J. (2007). Genome plasticity a key factor in the success of polyploid wheat under domestication. *Science* 316(5833), 1862–1866.

El-Shesheny, I., Hajeri, S., El-Hawary, I., Gowda, S. and Killiny, N. (2013). Silencing abnormal wing disc gene of the Asian citrus psyllid, *Diaphorina citri* disrupts adult wing development and increases nymph mortality. *PLoS ONE* 8(5), e65392.

Elzinga, D. A., De Vos, M. and Jander, G. (2014). Suppression of plant defenses by a *Myzus persicae* (green peach aphid) salivary effector protein. *Mol. Plant. Microbe Interact.* 27(7), 747–756. doi: 10.1094/MPMI-01-14-0018-R.

Escudero-Martinez, C., Rodriguez, P. A., Liu, S., Santos, P. A., Stephens, J. and Bos, J. I. B. (2020). An aphid effector promotes barley susceptibility through suppression of defence gene expression. *J. Exp. Bot.* 71(9), 2796–2807. doi: 10.1093/jxb/eraa043.

Feldman, M., Levy, A., Chalhoub, B. and Kashkush, K. (2012). Genomic plasticity in polyploid wheat. In: Soltis, P. and Soltis, D. (Eds.), *Polyploidy and Genome Evolution*. Berlin, Heidelberg: Springer, 109–135. doi: 10.1007/978-3-642-31442-1_7.

Fire, A., Xu, S., Montgomery, M. K., Kostas, S. A., Driver, S. E. and Mello, C. C. (1998). Potent and specific genetic interference by double-stranded RNA in *Caenorhabditis elegans*. *Nature* 391(6669), 806–811.

Fischer, J. R., Zapata, F., Dubelman, S., Mueller, G. M., Uffman, J. P., Jiang, C., Jensen, P. D. and Levine, S. L. (2017). Aquatic fate of a double-stranded RNA in a sediment-water system following an over-water application. *Environ. Toxicol. Chem.* 36(3), 727–734. doi: 10.1002/etc.3585.

Flavell, R. B. (2022). Wheat breeding, transcription factories, and genetic interactions: New perspectives. *Front. Plant Sci.* 13, 807884. doi: 10.3389/fpls.2022.807884.

Food and Agriculture Organization of the United Nations (FAO). (2021). *World Food and Agriculture - Statistical Yearbook 2021*. Rome: Food and Agriculture Organization of the United Nations (FAO).

Food and Agriculture Organization of the United Nations (FAO) (2022). *Food Outlook, Biannual Report on Global Food Markets*. Rome: Food and Agriculture Organization of the United Nations (FAO).

Furch, A. C., van Bel, A. J. and Will, T. (2015). Aphid salivary proteases are capable of degrading sieve-tube proteins. *J. Exp. Bot.* 66(2), 533–539. doi: 10.1093/jxb/eru487.

Furci, L., Jain, R., Stassen, J., Berkowitz, O., Whelan, J., Roquis, D., Victoire Baillet, V., Colot, V., Johannes, F. and Ton, J. (2019). Identification and characterisation of hypomethylated DNA loci controlling quantitative resistance in Arabidopsis. *eLife* 8, e40655. doi: 10.7554/eLife.40655.

Gardiner, L. J., Joynson, R., Omony, J., Rusholme-Pilcher, R., Olohan, L., Lang, D., Bai, C., Hawkesford, M., Salt, D., Spannagl, M., Mayer, K. F. X., Kenny, J., Bevan, M., Hall, N. and Hall, A. (2018). Hidden variation in polyploid wheat drives local adaptation. *Genome Res.* 28(9), 1319–1332.

Gardiner, L. J., Quinton-Tulloch, M., Olohan, L., Price, J., Hall, N. and Hall, A. (2015). A genome-wide survey of DNA methylation in hexaploid wheat. *Genome Biol.* 16, 273.

Geimann, S. (2019). Genetically engineered wheat found in unplanted Washington field. *Bloomberg.com*. Retrieved 9 June 2019. Bloomberg.

Geng, S., Kong, X., Song, G., Jia, M., Guan, J., Wang, F., Qin, Z., Wu, L., Lan, X., Li, A. and Mao, L. (2019). DNA methylation dynamics during the interaction of wheat progenitor *Aegilops tauschii* with the obligate biotrophic fungus *Blumeria graminis* f. sp. *tritici*. *New Phytol.* 221(2), 1023–1035.

Gil-Humanes, J., Wang, Y., Liang, Z., Shan, Q., Ozuna, C. V., Sanchez-Leon, S., Baltes, N. J., Starker, C., Barro, F., Gao, C. and Voytas, D. F. (2017). High efficiency gene targeting in hexaploid wheat using DNA replicons and CRISPR/Cas9. *Plant J.* 89(6), 1251-1262. doi: 10.1111/tpj.13446.

Gill, B. S., Appels, R., Botha-Oberholster, A.-M., Buell, C. R., Bennetzen, J. L., Chalhoub, B., Chumley, F., Dvořák, J., Iwanaga, M., Keller, B. and Li, W. (2004). Workshop report: A workshop report on wheat genome sequencing: International genome research on wheat consortium. *Genetics* 168, 1087–1096.

Hannon, G. J. (2002). RNA interference. *Nature* 418(6894), 244–251.

Harmel, N., Létocart, E., Cherqui, A., Giordanengo, P., Mazzucchelli, G., Guillonneau, F., De Pauw, E., Haubruge, E. and Francis, F. (2008). Identification of aphid salivary proteins: A proteomic investigation of *Myzus persicae*. *Insect Mol. Biol.* 17(2), 165–174. doi: 10.1111/j.1365-2583.2008.00790.x.

Heller, K. J. (Eds.) (2006). *Genetically Engineered Food: Methods and Detection*. Retrieved 12 May 2013. John Wiley and Sons. ISBN: 9783527606450, 36.

Hogenhout, S. A. and Bos, J. I. (2011). Effector proteins that modulate plant-insect interactions. *Curr. Opin. Plant Biol.* 14(4), 422–428.

Hossard, L., Philibert, A., Bertrand, M., Colnenne-David, C., Debaeke, P., Munier-Jolain, N., Jeuffroy, M. H., Richard, G. and Makowski, D. (2014). Effects of halving pesticide use on wheat production. *Sci. Rep.* 4(1), 4405.

Huvenne, H. and Smagghe, G. (2010). Mechanisms of dsRNA uptake in insects and potential of RNAi for pest control: A review. *J. Insect Physiol.* 56(3), 227–235.

International Aphid Genomics Consortium. (2010). Genome sequence of the pea aphid *Acyrthosiphon pisum*. *PLoS Biol.* 8(2), e1000313.

International Wheat Genome Sequencing Consortium (IWGSC). (2018). Shifting the limits in wheat research and breeding using a fully annotated reference genome. *Science* 361(6403), eaar7191.

Jiang, F. and Doudna, J. A. (2017). CRISPR–Cas9 structures and mechanisms. *Annu. Rev. Biophys.* 46, 505–529. doi: 10.1146/annurev-biophys-062215-010822.

Jones, H. D. and Shrewy, P. R. (Eds.). (2009).*Transgenic Wheat, Barley and Oats: Production and Characterisation Protocols*. New York, NY. Humana Press. ISBN: 9781588299611.

Joseph, T. T. and Osman, R. (2012). Convergent transmission of RNAi guide-target mismatch information across Argonaute internal allosteric network. *PLoS Comput. Biol.* 8(9), e1002693.

Kaloshian, I. and Walling, L. L. (2016). Hemipteran and dipteran pests: Effectors and plant host immune regulators. *J. Integr. Plant Biol.* 58(4), 350–361.

Kanja, C. and Hammond-Kosack, K. E. (2020). Proteinaceous effector discovery and characterisation in filamentous plant pathogens. *Mol. Plant Pathol.* 21(10), 1353-1376. doi: 10.1111/mpp.12980.

Kennerdell, J. R. and Carthew, R. W. (1998). Use of dsRNA mediated genetic interference to demonstrate that *frizzled* and *frizzled 2 act* in the wingless pathway. *Cell* 95(7), 1017-1026.

Kettles, G. J. and Kaloshian, I. (2016). The potato aphid salivary effector Me47 is a glutathione-S-transferase involved in modifying plant responses to aphid infestation. *Front. Plant Sci.* 7, 1142. doi: 10.3389/fpls.2016.01142.

Khan, Md. A., Khan, Z., Ahmad, W., Paul, B., Paul, S., Aggarwal, C. and Sayeed Akhtar, M. (2015). Insect pest resistance: An alternative approach to crop protection. In: Hakeem, K. R. (Ed.), *Crop Production and Global Environmental Issues*, 257-282. doi: 10.1007/978-3-319-23162-4_11.

Knodel, J. J., Beauzay, P. B., Eriksmoen, E. D. and Pederson, J. D. (2009). Pest management of wheat stem maggot (Diptera: Chloropidae) and wheat stem sawfly (Hymenoptera: Cephidae) using Insecticides in spring wheat. *J. Agric. Urban Entomol.* 26(4), 183-197.

Koch, A., Kumar, N., Weber, L., Keller, H., Imani, J. and Kogel, K. H. (2013). Host-induced gene silencing of cytochrome P450 lanosterol C14α-demethylase-encoding genes confers strong resistance to Fusarium species. *Proc. Natl. Acad. Sci. U. S. A.* 110(48), 19324-19329. doi: 10.1073/pnas.1306373110.

Koch, A. and Wassenegger, M. (2021). Host-induced gene silencing - Mechanisms and applications. *New Phytol.* 231, 54-59.

Kovak, E. and Blaustein-Rejto, D. (2022). The World's first genetically engineered wheat is here. Argentina's new genetically engineered, drought-tolerant wheat could have large environmental benefits. Available at: https://thebreakthrough.org/issues/food/the-worlds-first-genetically-engineered-wheat-is-here. Downloaded 12 July 2022.

Kumar, S., Nehra,M., Dilbaghi, N., Marrazza, G., Hassan, A. A. and Kim, K. H. (2019). Nano-based smart pesticide formulations: Emerging opportunities for agriculture. *J. Control. Release* 294, 131-153.

Kunte, N.,McGraw, E., Bell, S., Held, D. and Avila, L. A. (2020). Prospect, challenges and current status of RNAi through insect feeding. *Pest Manag. Sci.* 76(1), 26-41.

Law, J. A. and Jacobsen, S. E. (2010). Establishing, maintaining and modifying DNA methylation patterns in plants and animals. *Nat. Rev. Genet.* 11(3), 204-220. doi: 10.1038/nrg2719.

Lee, W. S., Rudd, J. J., Hammond-Kosack, K. E. and Kanyuka, K. (2014). *Mycosphaerella graminicola* LysM effector-mediated stealth pathogenesis subverts recognition through both CERK1 and CEBiP homologues in wheat. *Mol. Plant. Microbe Interact.* 27(3), 236-243.

Lemus, L. P., Tricard, J., Duclercq, J., Coulette, Q., Giron, D., Hano, C., Huguet, E., Lamblin, F., Cherqui, A. and Sallé, A. (2020). Salivary proteins of *Phloeomyzus passerinii*, a plant-manipulating aphid, and their impact on early gene responses of susceptible and resistant poplar genotypes. *Plant Sci.* 294, 110468. doi: 10.1016/j.plantsci.2020.110468.

Li, F., Zhao, X., Li, M., He, K., Huang, C., Zhou, Y., Li, Z. and Walters, J. R. (2019). Insect genomes: Progress and challenges. *Insect Mol. Biol.* 28(6), 739-758.

Li, J., Jiao, G., Sun, Y., Chen, J., Zhong, Y., Yan, L., Jiang, D., Ma, Y. and Xia, L. (2021). Modification of starch composition, structure and properties through editing of

TaSBEIIa in both winter and spring wheat varieties by CRISPR/Cas9. *Plant Biotechnol. J.* 19(5), 937–951.

Li, J., Li, Y. and Ma, L. (2021). Recent advances in CRISPR/Cas9 and applications for wheat functional genomics and breeding. *Abiotech* 2(4), 375–385. doi: 10.1007/s42994-021-00042-5.

Li, J., Zhang, S., Zhang, R., Gao, J., Qi, Y., Song, G. and Li, W. (2020). Efficient multiplex genome editing by CRISPR/Cas9 in common wheat. *Plant Biotechnol. J.* 19, 427–429.

Li, S. Y. and Xia, L. Q. (2020). Precise gene replacement in plants through CRISPR/Cas genome editing technology: Current status and future perspectives. *Abiotech* 1(1), 58–73. doi: 10.1007/s42994-019-00009-7.

Li, Y. F., Wu, Y., Wang, T., Li, L. R., Lu, L., Zhang, C. Y., Li, J. M., Zhang, L., Liu, Z. H. and Zheng, S. G. (2015). Polyphenol oxidase activity and yellow pigment content in *Aegilops tauschii*, *Triticum turgidum*, *Triticum aestivum*, synthetic hexaploid wheat and its parents. *J. Cereal Sci.* 65, 192–201.

Lilley, C. J., Bakhetia, M., Charlton, W. L. and Urwin, P. E. (2007). Recent progress in the development of RNA interference for plant parasitic nematodes. *Mol. Plant Pathol.* 8(5), 701–711. doi: 10.1111/j.1364-3703.2007.00422.x.

Liu, S., Geng, S., Li, A., Mao, Y. and Mao, L. (2021). RNAi technology for plant protection and its application in wheat. *Abiotech* 2(4), 365–374. doi: 10.1007/s42994-021-00036-3.

MacWilliams, J. R., Dingwall, S., Chesnais, Q., Sugio, A. and Kaloshian, I. (2020). AcDCXR is a cowpea aphid effector with putative roles in altering host immunity and physiology. *Front. Plant Sci.* 11, 605. doi: 10.3389/fpls.2020.00605.

Mao, J. and Zeng, F. (2014). Plant-mediated RNAi of a gap gene-enhanced tobacco tolerance against the *Myzus persicae*. *Transgen. Res.* 23(1), 145–152.

Mao, Y. B., Cai, W. J., Wang, J. W., Hong, G. J., Tao, X. Y., Wang, L. J., Huang, Y. P. and Chen, X. Y. (2007). Silencing a cotton bollworm P450 monooxygenase gene by plant-mediated RNAi impairs larval tolerance of gossypol. *Nat. Biotechnol.* 25(11), 1307–1313.

Marques, J. T., Kim, K., Wu, P. H., Alleyne, T. M., Jafari, N. and Carthew, R. W. (2010). Loqs and R2D2 act sequentially in the siRNA pathway in Drosophila. *Nat. Struct. Mol. Biol.* 17(1), 24–30.

Mathers, T. C., Mugford, S. T., Wouters, R. H. M., Heavens, D. and Botha, A.-M. (2022). Aphidinae comparative genomics resource (Version v1) [Data set]. *Zenodo*. doi: 10.5281/zenodo.5908005.

Mathers, T. C., Wouters, R. H. M., Mugford, S. T., Swarbreck, D., Van Oosterhout, C. and Hogenhout, S. A. (2021). Chromosome-scale genome assemblies of aphids reveal extensively rearranged autosomes and long-term conservation of the X chromosome. *Mol. Biol. Evol.* 38(3), 856–875.

Matzke, M. A. and Mosher, R. A. (2014). RNA-directed DNA methylation: An epigenetic pathway of increasing complexity. *Nat. Rev. Genet.* 15(6), 394–408. doi: 10.1038/nrg3683.

Meissle, M., Naranjo, S. E. and Romeis, J. (2022). Does the growing of Bt maize change abundance or ecological function of non-target animals compared to the growing of non-GM maize? A systematic review. *Environ. Evid.* 11(1), 21. doi: 10.1186/s13750-022-00272-0.

Meyer, P. (2015). Epigenetic variation and environmental change. *J. Exp. Bot.* 66(12), 3541–3548. doi: 10.1093/jxb/eru502.

Meynard, J. M., Doré, T. and Lucas, P. (2003). Agronomic approach: Cropping systems and plant diseases. *C. R. Biol.* 326(1), 37–46. doi: 10.1016/S1631-0691(03)00006-4.

Miele, E., Spinelli, G. P., Miele, E., Fabrizio, E. D., Ferretti, E., Tomao, S. and Gulino, A. (2012). Nanoparticle-based delivery of small interfering RNA: Challenges for cancer therapy. *Int. J. Nanomed.* 7, 3637–3657.

Miles, P. W. (1999). Aphid saliva. *Biol. Rev. Cam. Phil. Soc.* 74(1), 41–85.

Mitter, N., Worrall, E. A., Robinson, K. L., Peng, L., Jain, R. G., Taochy, C., Fletcher, S. J., Carroll, B. J., Lu, G. Q. and Xu, Z. P. (2017). Clay nanosheets for topical delivery of RNAi for sustained protection against plant viruses. *Nat. Plants* 3, 16207. doi: 10.1038/nplants.2016.207. Available at: www.nature.com/natureplants.

Mugford, S. T., Barclay, E., Drurey, C., Findlay, K. C. and Hogenhout, S. A. (2016). An immuno-suppressive aphid saliva protein is delivered into the cytosol of plant mesophyll cells during feeding. *Mol. Plant. Microbe Interact.* 29(11), 854–861. doi: 10.1094/MPMI-08-16-0168-R.

Mutti, N. S., Louis, J., Pappan, L. K., Pappan, K., Begum, K., Chen, M. S., Park, Y., Dittmer, N., Marshall, J., Reese, J. C. and Reeck, G. R. (2008). A protein from the salivary glands of the pea aphid, *Acyrthosiphon pisum*, is essential in feeding on a host plant. *Proc. Natl. Acad. Sci. U. S. A.* 105(29), 9965–9969. doi: 10.1073/pnas.0708958105.

Mutti, N. S., Park, Y., Reese, J. C. and Reeck, G. R. (2006). RNAi knockdown of a salivary transcript leading to lethality in the pea aphid, Acyrthosiphon pisum. *J. Insect Sci.* 6, 1–7.

Naessens, E., Dubreuil, G., Giordanengo, P., Baron, O. L., Minet-Kebdani, N., Keller, H. and Coustau, C. (2015). A secreted MIF cytokine enables aphid feeding and represses plant immune responses. *Curr. Biol.* 25(14), 1898–1903. doi: 10.1016/j.cub.2015.05.047.

Nault, L. R. (1997). Arthropod transmission of plant viruses–A new synthesis. *Ann. Entomol. Soc. Am.* 90(5), 521–541.

Ng, J. C. and Perry, K. L. (2004). Transmission of plant viruses by aphid vectors. *Mol. Plant Pathol.* 5(5), 505–511.

Nicholson, S. J., Nickerson, M. L., Dean, M., Song, Y., Hoyt, P. R., Rhee, H., Kim, C. and Puterka, G. J. (2015). The genome of *Diuraphis noxia*, a global aphid pest of small grains. *BMC Genomics* 16(1), 429.

Nicholson, S. J. and Puterka, G. J. (2014). Variation in the salivary proteomes of differentially virulent greenbug (*Schizaphis graminum* Rondani) biotypes. *J. Proteomics* 105, 186–203. doi: 10.1016/j.jprot.2013.12.005.

Nicolis, V. F., Burger, N. F. V. and Botha, A. M. (2022). Whole-body transcriptome mining for candidate effectors from *Diuraphis noxia*. *BMC Genomics* 23(1), 493. doi: 10.1186/s12864-022-08712-4.

Nielsen, H. and Krogh, A. (1998). Prediction of signal peptides and signal anchors by a hidden Markov model. *Proc. Int. Conf. Intell. Syst. Mol. Biol.* 6, 122–130.

Nikoukar, A. and Rashed, A. (2022). Integrated pest management of wireworms (Coleoptera: Elateridae) and the rhizosphere in agroecosystems. *Insects* 13(9), 769. doi: 10.3390/insects13090769.

Nowara, D., Gay, A., Lacomme, C., Shaw, J., Ridout, C., Douchkov, D., Hensel, G., Kumlehn, J. and Schweizer, P. (2010). HIGS: Host-induced gene silencing in the obligate biotrophic fungal pathogen *Blumeria graminis*. *Plant Cell* 22(9), 3130–3141. doi: 10.1105/tpc.110.077040.

Parsons, K. H., Mondal, M. H., McCormick, C. L. and Flynt, A. S. (2018). Guanidinium-functionalized interpolyelectrolyte complexes enabling RNAi in resistant insect pests. *Biomacromolecules* 19(4), 1111-1117. doi: 10.1021/acs.biomac.7b01717.

Pinzón, N., Bertrand, S., Subirana, L., Busseau, I., Escrivá, H. and Seitz, H. (2019). Functional lability of RNA-dependent RNA polymerases in animals. *PLoS Genet.* 15(2), e1007915. doi: 10.1371/journal.pgen.1007915.

Pitino, M., Coleman, A. D., Maffei, M. E., Ridout, C. J. and Hogenhout, S. A. (2011). Silencing of aphid genes by dsRNA feeding from plants. *PLoS ONE* 6(10), e25709.

Pitino, M. and Hogenhout, S. A. (2013). Aphid protein effectors promote aphid colonization in a plant species-specific manner. *Mol. Plant. Microbe Interact.* 26(1), 130–139. doi: 10.1094/MPMI-07-12-0172-FI.

Pliego, C., Nowara, D., Bonciani, G., Gheorghe, D. M., Xu, R., Surana, P., Whigham, E., Nettleton, D., Bogdanove, A. J., Wise, R. P., Schweizer, P., Bindschedler, L. V. and Spanu, P. D. (2013). Host-induced gene silencing in barley powdery mildew reveals a class of ribonuclease-like effectors. *Mol. Plant. Microbe Interact.* 26(6), 633–642. doi: 10.1094/MPMI-01-13-0005-R.

Pratt, A. J. and MacRae, I. J. (2009). The RNA-induced silencing complex: A versatile gene-silencing machine. *J. Biol. Chem.* 284(27), 17897-17901.

Preall, J. B. and Sontheimer, E. J. (2005). RNAi: RISC gets loaded. *Cell* 123(4), 543–545.

Pridgeon, J. W., Zhao, L., Becnel, J. J., Strickman, D. A., Clark, G. G. and Linthicum, K. J. (2008). Topically applied *AaelAP1* double-stranded RNA kills female adults of *Aedes aegypti*. *J. Med. Entomol.* 45(3), 414–420.

Qi, T., Guo, J., Liu, P., He, F., Wan, C., Islam, M. A., Tyler, B. M., Kang, Z. and Guo, J. (2019). Stripe rust effector *PstGSRE1* disrupts nuclear localization of ROS-Promoting transcription factor *TaLOL2* to defeat ROS-Induced defense in wheat. *Mol. Plant* 12(12), 1624–1638. doi: 10.1016/j.molp.2019.09.010.

Quadrana, L. and Colot, V. (2016). Plant transgenerational epigenetics. *Annu. Rev. Genet.* 50, 467–491. doi: 10.1146/annurev-genet-120215-035254.

Rao, S. A., Carolan, J. C. and Wilkinson, T. L. (2013). Proteomic profiling of cereal aphid saliva reveals both ubiquitous and adaptive secreted proteins. *PLoS ONE* 8(2), e57413. doi: 10.1371/journal.pone.0057413.

Ratnadass, A. (2020). Crop protection for agricultural intensification systems in Sub-Saharan Africa. In: Lichtfouse, E. (Ed.), *Agric. Rev. 39. Sustainable Agriculture Reviews, Sustainable*. Cham: Springer 39, 1–34. doi: 10.1007/978-3-030-38881-2_1.

Reddy, G. V. P., Tangtrakulwanich, K., Wu, S., Miller, J. H., Ophus, V. L., Prewett, J. and Jaronski, S. T. (2014). Evaluation of the effectiveness of entomopathogens for the management of wireworms (Coleoptera: Elateridae) on spring wheat. *J. Invertebr. Pathol.* 120, 43–49, doi: 10.1016/j.jip.2014.05.005.

Rodriguez, P. A. and Bos, J. I. (2013). Toward understanding the role of aphid effectors in plant infestation. *Mol. Plant. Microbe Interact.* 26(1), 25–30.

Rodriguez, P. A., Escudero-Martinez, C. and Bos, J. I. (2017). An aphid effector targets trafficking protein VPS52 in a host-specific manner to promote virulence. *Plant Physiol.* 173(3), 1892–1903. doi: 10.1104/pp.16.01458.

Rodriguez, P. A., Stam, R., Warbroek, T. and Bos, J. I. (2014). Mp10 and Mp42 from the aphid species *Myzus persicae* trigger plant defenses in *Nicotiana benthamiana* through different activities. *Mol. Plant. Microbe Interact.* 27(1), 30–39. doi: 10.1094/MPMI-05-13-0156-R.

Salamini, F., Ozkan, H., Brandolini, A., Schafer-Pregl, R. and Martin, W. (2002). Genetics and geography of wild cereal domestication in the near east. *Nat. Rev. Genet.* 3(6), 429–441.

Saleh, M. C., van Rij, R. P., Hekele, A., Gillis, A., Foley, E., O'Farrell, P. H. and Andino, R. (2006). The endocytic pathway mediates cell entry of dsRNA to induce RNAi silencing. *Nat. Cell Biol.* 8(8), 793–802.

San Miguel, K. and Scott, J. G. (2016) The next generation of insecticides: dsRNA is stable as a foliar-applied insecticide. *Pest Manag. Sci.* 72(4), 801–809.

Sánchez-Leon, S., Gil-Humanes, J., Ozuna, C. V., Gimenez, M. J., Sousa, C., Voytas, D. F. and Barro, F. (2018). Low-gluten, nontransgenic wheat engineered with CRISPR/Cas9. *Plant Biotechnol. J.* 16(4), 902–910.

Sarthou, J.-P., Jacquot, M. and Deguine, J. P. (2017). Functional soil and aerial biodiversity. In: Deguine, J.-P., Gloanec, C., Laurent, P., Ratnadass, A. and Aubertot, J. N. (Eds.), *Agroecological Crop Protection.* Dordrecht: Springer Science+Business Media B.V., 182–187. doi: 10.1007/978-94-024-1185-0. ISBN: 978-94-024-1184-3.

Schaefer, L. K., Parlange, F., Buchmann, G., Jung, E., Wehrli, A., Herren, G., Müller, M. C., Stehlin, J., Schmid, R., Wicker, T., Keller, B. and Bourras, S. (2020). Cross-kingdom RNAi of pathogen effectors leads to quantitative adult plant resistance in wheat. *Front. Plant Sci.* 11, 253. doi: 10.3389/fpls.2020.00253.

Schimmel, B. C. J., Ataide, L. M. S., Chafi, R., Villarroel, C. A., Alba, J. M., Schuurink, R. C. and Kant, M. R. (2017). Over compensation of herbivore reproduction through hyper-suppression of plant defenses in response to competition. *New Phytol.* 214(4), 1688–1701.

Schmid, R. B., Knutson, A., Giles, K. L. and McCormack, B. P. (2019). Hessian fly (Diptera: Cecidomyiidae) biology and management in wheat. *J. Intgr Pest Manag.* 9(1), 1–12.

Seymor, D. K. and Becker, C. (2017). The causes and consequences of DNA methylome variation in plants. *Curr. Opin. Plant Biol.* 36, 56–63. doi: 10.1016/j.pbi.2017.01.005.

Sharma, A., Khan, A. N., Subrahmanyam, S., Raman, A., Taylor, G. S. and Fletcher, M. J. (2014). Salivary proteins of plant-feeding hemipteroids-implication in phytophagy. *Bull. Entomol. Res.* 104(2), 117–136. doi: 10.1017/S0007485313000618.

Shewry, P. R. (2009). Wheat. *J. Exp. Bot.* 60(6), 1537–1553.

Simon, J. C. and Peccoud, J. (2018). Rapid evolution of aphid pests in agricultural environments. *Curr. Opin. Insect Sci.* 26, 17–24.

Smith, C. M. (2021). Conventional breeding of insect-resistant crop plants still the best way to feed the world population. *Curr. Opin. Insect Sci.* 45, 7–13.

Stoger, E., Williams, S., Christou, P., Down, R. E. and Gatehouse, J. A. (1999). Expression of the insecticidal lectin from snowdrop (*Galanthus nivalis* agglutinin; GNA) in transgenic wheat plants: Effects on predation by the grain aphid *Sitobion avenae*. *Mol. Breed.* 5(1), 65–73.

Sun, Y., Sparks, C., Jones, H., Riley, M., Francis, F., Du, W. and Xia, L. (2019). Silencing an essential gene involved in infestation and digestion in grain aphid through plant-mediated RNA interference generates aphid-resistant wheat plants. *Plant Biotechnol. J.* 17(5), 852–854. doi: 10.1111/pbi.13067.

Suneja, Y., Gupta, A. K. and Bains, N. S. (2019). Stress adaptive plasticity: *Aegilops tauschii* and *Triticum dicoccoides* as potential donors of drought associated morpho-physiological traits in wheat. *Front. Plant Sci.* 10, 211. doi: 10.3389/fpls.2019.00211.

Thorpe, P., Cock, P. J. and Bos, J. (2016). Comparative transcriptomics and proteomics of three different aphid species identifies core and diverse effector sets. *BMC Genomics* 17, 172. doi: 10.1186/s12864-016-2496-6.

Thorpe, P., Escudero-Martinez, C. M., Cock, P. J. A., Eves-van den Akker, S. and Bos, J. I. B. (2018). Shared transcriptional control and disparate gain and loss of aphid parasitism genes. *Gen. Biol. Evol.* 10(10), 2716–2733.

Thorpe, P., Escudero-Martinez, C. M., Eves-van den Akker, S. and Bos, J. I. B. (2020). Transcriptional changes in the aphid species *Myzus cerasi* under different host and environmental conditions. *Insect Mol. Biol.* 29(3), 271–282.

Torto, T. A., Li, S., Styer, A., Huitema, E., Testa, A., Gow, N. A., van West, P. and Kamoun, S. (2003). EST mining and functional expression assays identify extracellular effector proteins from the plant pathogen Phytophthora. *Genome Res.* 13(7), 1675–1685.

Urban, M., Cuzick, A., Seager, J., Wood, V., Rutherford, K., Venkatesh, S. Y., De Silva, N., Martinez, M. C., Pedro, H., Yates, A. D., Hassani-Pak, K. and Hammond-Kosack, K. E. (2020). PHI-base: The pathogen–host interactions database. *Nucleic Acids Res.* 48(D1), D613–D620. doi: 10.1093/nar/gkz904.

Vaillant, I. and Paszkowski, J. (2007). Role of histone and DNA methylation in gene regulation. *Curr. Opin. Plant Biol.* 10(5), 528–533. doi: 10.1016/j.pbi.2007.06.008.

Vandermoten, S., Harmel, N., Mazzucchelli, G., De Pauw, E., Haubruge, E. and Francis, F. (2014). Comparative analyses of salivary proteins from three aphid species. *Insect Mol. Biol.* 23(1), 67–77. doi: 10.1111/imb.12061.

Vasil, I. K. (2007). Molecular genetic improvement of cereals: Transgenic wheat (*Triticum aestivum* L.). *Plant Cell Rep.* 26(8), 1133–1154. doi: 10.1007/s00299-007-0338-3.

Vasil, V., Castillo, A. M., Fromm, M. E. and Vasil, I. K. (1992). Herbicide resistant fertile transgenic wheat plants obtained by microprojectile bombardment of regenerable embryogneic callus. *Biotechnology* 10, 667–674.

Vicente, M. M., Chaves-Ferreira, M., Jorge, J. M. P., Proença, J. T. and Barreto, V. M. (2021). The off-targets of clustered regularly interspaced short palindromic repeats gene editing. *Front. Cell Dev. Biol.* 9, 718466. doi: 10.3389/fcell.2021.718466.

Wang, W., Dai, H., Zhang, Y., Chandrasekar, R., Luo, L., Hiromasa, Y., Sheng, C., Peng, G., Chen, S., Tomich, J. M., Reese, J., Edwards, O., Kang, L., Reeck, G. and Cui, F. (2015). Armet is an effector protein mediating aphid-plant interactions. *FASEB J.* 29(5), 2032–2045. doi: 10.1096/fj.14-266023.

Wang, M., Li, Z., Zhang, Y., Zhang, Y., Xie, Y., Ye, L., Zhuang, Y., Lin, K., Zhao, F., Guo, J., Teng, W., Zhang, W., Tong, Y., Xue, Y. and Zhang, Y. (2021). An atlas of wheat epigenetic regulatory elements reveals subgenome divergence in the regulation of development and stress responses. *Plant Cell* 33(4), 865–881. doi: 10.1093/plcell/koab028.

Xu, L., Hou, Q., Zhao, Y., Lu, L., Li, B., Ni, Z. and Liang, R. (2017). Silencing of a lipase maturation factor 2-like gene by wheat-mediated RNAi reduces the survivability and reproductive capacity of the grain aphid *Sitobion avenae*. *Arch. Insect Biochem. Physiol.* 95(3), e21392. doi: 10.1002/arch.21392.

Yan, S., Ren, B. Y. and Shen, J. (2021). Nanoparticle-mediated double-stranded RNA delivery system: A promising approach for sustainable pest management. *Insect Sci.* 28(1), 21–34. doi: 10.1111/1744-7917.12822.

Yang, Z., Ma, L., Francis, F., Yang, Y., Chen, H., Wu, H. and Chen, X. (2018). Proteins identified from saliva and salivary glands of the Chinese gall aphid *Schlechtendalia chinensis*. *Proteomics* 18(9), e1700378. doi: 10.1002/pmic.201700378.

Zamore, P. D., Tuschl, T., Sharp, P. A. and Bartel, D. P. (2000). RNAi: Double-stranded RNA directs the ATP-dependent cleavage of mRNA at 21 to 23 nucleotide intervals. *Cell* 101(1), 25–33.

Zhang, J., Khan, S. A., Hasse, C., Ruf, S., Heckel, D. G. and Bock, R. (2015a). Pest control. Full crop protection from an insect pest by expression of long double-stranded RNAs in plastids. *Science* 347(6225), 991–994.

Zhang, H., Li, H., Guan, R. and Miao, X. (2015b). Lepidopteran insect species-specific, broad-spectrum, and systemic RNA interference by spraying dsRNA on larvae. *Entomol. Exp. Appl.* 155, 218–228.

Zhang, X., Mysore, K., Flannery, E., Michel, K., Severson, D. W., Zhu, K. Y. and Duman-Scheel, M. (2015c). Chitosan/interfering RNA nanoparticle mediated gene silencing in disease vector mosquito larvae. *J. Vis. Exp.* 97, 181–191.

Zhang, Q., Hua, G. and Adang, M. J. (2015d). Chitosan/DsiRNA nanoparticle targeting identifies AgCad1 cadherin in *Anopheles gambiae* larvae as an *in vivo* receptor of Cry11Ba toxin of *Bacillus thuringiensis* subsp. *jegathesan*. *Insect Biochem. Mol. Biol.* 60, 33–38.

Zhang, J., Khan, S. A., Heckel, D. G. and Bock, R. (2017). Next generation insect-resistant plants: RNAi-mediated crop protection. *Trends Biotechnol.* 35(9), 871–882.

Zhang, S., Zhang, R., Gao, J., Gu, T., Song, G., Li, W., Li, D., Li, Y. and Li, G. (2019a). High efficient and heritable targeted mutagenesis in wheat via the *Agrobacterium tumefaciens*-mediated CRISPR/Cas9 system. *Int. J. Mol. Sci.* 20(17), 4257.

Zhang, Z., Hua, L., Gupta, A., Tricoli, D., Edwards, K. J., Yang, B. and Li, W. (2019b). Development of an *Agrobacterium*-delivered CRISPR/Cas9 system for wheat genome editing. *Plant Biotechnol. J.* 17(8), 1623–1635.

Zhang, Y., Fan, J., Sun, J., Francis, F. and Chen, J. (2017). Transcriptome analysis of the salivary glands of the grain aphid, *Sitobion avenae*. *Sci. Rep.* 7(1), 15911. doi: 10.1038/s41598-017-16092-z.

Zhang, Y., Li, Z., Lin, K., Peng, Y., Ye, L., Zhuang, Y., Wang, M., Xie, Y., Guo, J., Teng, W. and Tong, Y. (2021). Evolutionary rewiring of the wheat transcriptional regulatory network by lineage-specific transposable elements. *Genom. Res.* 31. doi: 10.1101/gr.275658.

Zhao, L., Wang, S., Cao, Z., Ouyang, W., Zhang, Q., Xie, L., Zheng, R., Guo, M., Ma, M., Hu, Z., Sung, W. K., Zhang, Q., Li, G. and Li, X. (2019). Chromatin loops associated with active genes and heterochromatin shape rice genome architecture for transcriptional regulation. *Nat. Commun.* 10(1), 3640. doi: 10.1038/s41467-019-11535-9.

Zhao, Y., Sui, X., Xu, L., Liu, G., Lu, L., You, M., Xie, C., Li, B., Ni, Z. and Liang, R. (2018). Plant-mediated RNAi of grain aphid *CHS1* gene confers common wheat resistance against aphids. *Pest Manag. Sci.* 74(12), 2754–2760. doi: 10.1002/ps.5062.

Zhu, K. Y. and Palli, S. R. (2020). Mechanisms, applications, and challenges of insect RNA interference. *Annu. Rev. Entomol.* 65, 293–311.

Zotti, M. J. and Smagghe, G. (2015). RNAi technology for insect management and protection of beneficial insects from diseases: Lessons, challenges and risk assessments. *Neotrop. Entomol.* 44(3), 197–213. doi: 10.1007/s13744-015-0291-8.

Chapter 15

Online decision support systems, remote sensing and artificial intelligence applications for wheat pest management

Daniel J. Leybourne, Leibniz University Hannover, Germany and RSK ADAS Ltd, UK; Mark Ramsden and Sacha White, RSK ADAS Ltd, UK; Rujing Wang, He Huang and Chengjun Xe, Institute of Intelligent Machines, Chinese Academy of Sciences, China; and Po Yang, University of Sheffield, UK

1 Introduction

Infestation with herbivorous invertebrates ('pests') can be extremely damaging to wheat production, potentially resulting in up to 80% yield loss (Nancarrow et al., 2021). Reducing the damage caused by these pests is a central component of crop protection practices. Farmers often use plant protective products, such as pesticides, to manage pest infestations. With increasing environmental, ecological, and public concerns around the use of pesticides (Leather, 2018), there is a growing need to use these in a more environmentally conscious manner and to reduce overall pesticide use and promote greater uptake of alternative management options.

Integrated pest management (IPM) aims to reduce reliance on plant protection products through the implementation of combinations of pest

http://dx.doi.org/10.19103/AS.2022.0114.21

management techniques, including the uptake of alternative management options, the development of less environmentally damaging pesticides, and sustainable pesticide application where necessary. There are eight key IPM principles that underpin these aims (European Union, 2009; Barzman et al., 2015), these are:

1 Prevention and suppression;
2 Pest identification and monitoring;
3 Decision based on monitoring and thresholds;
4 Non-chemical methods;
5 Pesticide selection;
6 Reduced pesticide use;
7 Anti-resistance strategies; and
8 Evaluation.

These eight principles allow the development of effective, practicable, and site-specific IPM programmes that increase the resilience of cropping systems and account for the knowledge-intensive, dynamic, and systemic nature of IPM (Barzman et al., 2015). They provide a logical, stepwise process to assist growers in their decision-making. The first step relates to actions that prevent or minimise the severity of pest outbreaks (Principle 1). This is followed by steps to monitor pest populations (Principle 2) and decide on the need to carry out intervention(s) based on treatment thresholds (Principle 3). If an intervention is required, non-chemical control methods should be considered first (Principle 4). Where pesticides are deemed necessary, consideration should be given to select pesticides with minimal non-target impacts (Principle 5), minimise pesticide use (e.g. through reduced dose and/or application frequency, or spot-spraying) to that which provides effective control (Principle 6), and implement anti-resistance strategies where necessary (Principle 7). Finally, the IPM programme should be evaluated to assess its effectiveness and consider improvements (Principle 8).

Decision support systems (DSS) are interactive systems (usually software-based) that help users identify and solve problems and make decisions as part of an IPM strategy. DSS play an important role in pest management, especially in relation to treatment application and interventions. DSS can be used as an umbrella term for any interactive support system that helps farmers make management and production decisions. The majority of DSS currently available for wheat pest management provide precautionary, supportive, and preventative measures to achieve effective pest management (Elliott et al., 2004) as well as resources and techniques that support tactical decision-making through continuous pest monitoring and the integration of pest population data into decision-making processes. Figure 1 shows a simple graphical

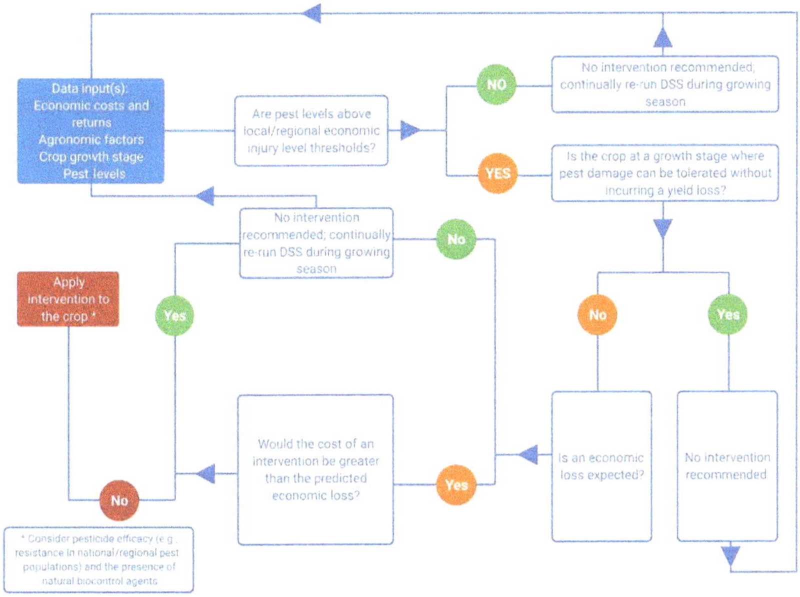

Figure 1 Representative flowchart showing key processes involved in a DSS-based decision-making process for pest management. Image created with BioRender.com.

representation of the various factors that are taken into account when farmers and their advisors are deciding whether an intervention should be applied to the crop to manage a pest. DSS can contribute either specific components in the decision-making process (e.g. pest prediction models, economic damage forecasts) or comprehensively assess pest risk (e.g. a fully integrated DSS for a specified pest that accounts for all variables).

A good DSS will enable the user to make strategic, tactical, and operational decisions with confidence and in full knowledge of the associated risks and assumptions. In order to be successful, DSS need to be fully integrated into IPM plans and, therefore, need to align with the eight IPM principles (Barzman et al., 2015). DSS can help growers achieve effective decision-making through the selection and implementation of:

1 Strategic pest management techniques: The application of precautionary, supportive, and preventative measures to achieve effective pest management. This includes all aspects of pest control from pesticide application to cultural control.
2 Tactical management techniques: The continuous monitoring of pests and the use of pest population data for the effective prescriptive application of pest control treatments, including smart pesticide use.

3 Operational techniques: The use of operational tools to monitor and assess the success of deployed IPM measures (including strategic and tactical techniques).

Generally, DSS that provide pest management support are often based on computational and statistical models that predict, forecast, or confirm pest abundance, migration, activity, and crop-associated crop risk. Many such DSS have been developed to support farmers and play integral roles in the development of IPM plans, including:

- Pest forecasting tools that are used pre-sowing to predict seasonal pest populations to determine annual risk (e.g. determining wheat bulb fly risk, Leybourne et al., 2022; predicting aphid population dynamics, Fabre et al., 2006).
- Pest forecasting tools that are used in-season to predict crop risk and advise on treatment application in real-time (e.g. the Agriculture and Horticultre Development Board barley yellow dwarf virus (BYDV) tool, accessible: https://ahdb.org.uk/bydv).
- Computational resources to aid with pest identification (e.g. Liu et al., 2019).
- Computational resources that estimate pest infestation levels (e.g. using soundscapes to predict infestation levels; Mishra et al., 2018).

The primary purpose of a DSS is to enable users to make more effective decisions by presenting the likelihood of various outcomes (Rose et al., 2016). The use of in-crop monitoring, forecasting models, and thresholds of economic injury are good foundational tools for inclusion in an overall IPM plan (Ramsden et al., 2017), and many DSS build on these tools to enable users to optimise their decision-making (Leybourne et al., 2022). In general, DSS should not be viewed as a replacement for farmer experience or industry support but should be seen to complement the farmer's and/or their advisor's own expertise with other sources of information to improve the decision-making process.

The chapter will focus on providing an overview of DSS targeting the primary wheat pests covered in this book, including cereal aphids, sunn pest, hessian fly, frit fly, gout fly, wheat bulb fly, yellow cereal fly, orange wheat blossom midge, saddle gall midge, leather jackets, wheat stem sawfly, wireworms, cereal leaf beetle, wheat curl mite, slugs, and common stored grain pests. First, in Section 2, we will present an overview of the forecasting and prediction tools available for wheat pest management. In Section 3, we will provide an overview of advances in remote sensing and artificial intelligence (AI) technology and discuss the future role of these technologies in DSS development. Section 4 will provide an overview of potential barriers to DSS uptake among farmers. Sections 5 and 6 will cover future research and further reading.

2 Pest forecasting, observation and support tools for wheat pest management

DSS confer numerous benefits for growers. Regular monitoring of large areas of wheat crops is labour intensive, and while continuous monitoring is advisable, the time and labour intensity create a bottleneck that restricts effective monitoring for most farmers (Ramsden et al., 2017). DSS for pest management broadly fall into three categories: (1) computational pest forecasting models that provide pest population and crop risk estimations (e.g. models that compute annual risk for a defined pest; Leybourne et al., 2022, models that estimate pest activity; Young et al., 1993); (2) pest observation tools that support users in understanding the likelihood of a pest being present in their crop (e.g. pest monitoring networks such as the Rothamsted Suction Trap Network); and (3) support systems that aid the overall decision-making process (e.g. economic threshold and computer-aided resources for pest identification; Liu et al., 2019, 2021). In this section, we will provide a brief overview of each DSS type, cover case studies for forecasting and observation DSS, and highlight computational DSS that have been developed to support wheat pest management.

2.1 Pest forecasting tools

Pest forecasting tools are DSS that predict the timings of pest emergence or transition to the next life stage, pest migration into crops, estimate overall populations, or estimate pest activity. Generally, these DSS can support the farmer before and during the growing season. Pre-season DSS can include tools that predict pest populations and, by association, provide estimations of annual crop risk before crop drilling starts (e.g. annual prediction of wheat bulb fly risk, Leybourne et al., 2022; predicting cereal aphid population dynamics, Fabre et al., 2006); these DSS can help farmers estimate the annual crop risk for a given pest and will enable farmers to implement early interventions if crop risk is predicted to be high, such as cultural control methods, insecticide seed treatment, or altering the crop rotation so that a less susceptible crop is planted (e.g. substituting a cereal crop for oilseed rape if DSS predicts high wheat bulb fly risk for a specific field). In-season DSS support decision-making throughout the crop growing season and can include tools that estimate pest activity (e.g. slug activity (Young et al., 1993)), predict when a phenological event will occur that increases crop risk (e.g. predicting the emergence of orange wheat blossom midge adults (Jacquemin et al., 2014)), or predict pest distribution within the field (e.g. for slugs (Petrovskii et al., 2022)). The aim of these in-season DSS is to identify the optimal timing and positioning of interventions so that insect management coincides with peak periods or locations of pest activity. For example, targeting molluscicide application during periods of slug activity on the soil surface (Young et al., 1993) or targeting applications to locations within the field where pest density is high (Petrovskii et al., 2022). DSS that predict migration or emergence can then be used to guide in-field trapping efforts.

Pest forecasting DSS are often based on computational or statistical models that take climatic and meteorological data and use these to make predictions for the phenology and behaviour of the target pest (Gözüaçik et al., 2016; Leybourne et al., 2022). These systems have many benefits: climate and meteorological data are often open-access and available on a local level, so estimations can be based on local inputs; these tools can also replace onerous and time-intensive monitoring, saving farmers money and resources; and estimations on pest populations can be linked with regional thresholds to estimate whether an economic loss is likely based on predicted pest numbers. However, these systems also have limitations. For example, the systems are often based on generic assumptions, which can limit their accuracy thereby affecting the confidence farmers have in the predictions.

2.1.1 Pest forecasting decision support system case study: wheat bulb fly decision support system

A recent example of a pest forecasting DSS is a predictive model developed in the UK to estimate annual crop risk to wheat bulb fly (Leybourne et al., 2022), an important pest in eastern parts of the UK that can cause yield loss of 4 t/ha (Rogers et al., 2015). This predictive model uses pest population (wheat bulb fly egg counts) and meteorological data from 1971 to 2019 to estimate the annual crop risk to wheat bulb fly (Leybourne et al., 2022). This pest forecasting tool represents an important DSS as currently the only other options available to predict wheat bulb fly risk are through time- and labour-intensive soil sampling and manual egg counting (Ramsden et al., 2017). However, predicting pest population levels alone is only one component of an effective DSS. In order for DSS uptake to be high, it is important that DSS represent an integrated system that can estimate the risk to the crop and provide farmers with additional management advice. In order to develop a comprehensive DSS for wheat bulb fly management, this population prediction model was combined with accepted wheat bulb fly thresholds and a further shoot number prediction model (i.e. crop tolerance estimation model) to develop an integrated DSS for wheat bulb fly management. Figure 2 shows the decision workflow for this DSS involving: (1) pest population prediction; (2) economic thresholds; (3) crop tolerance prediction; (4) crop risk; and (5) damage mitigation.

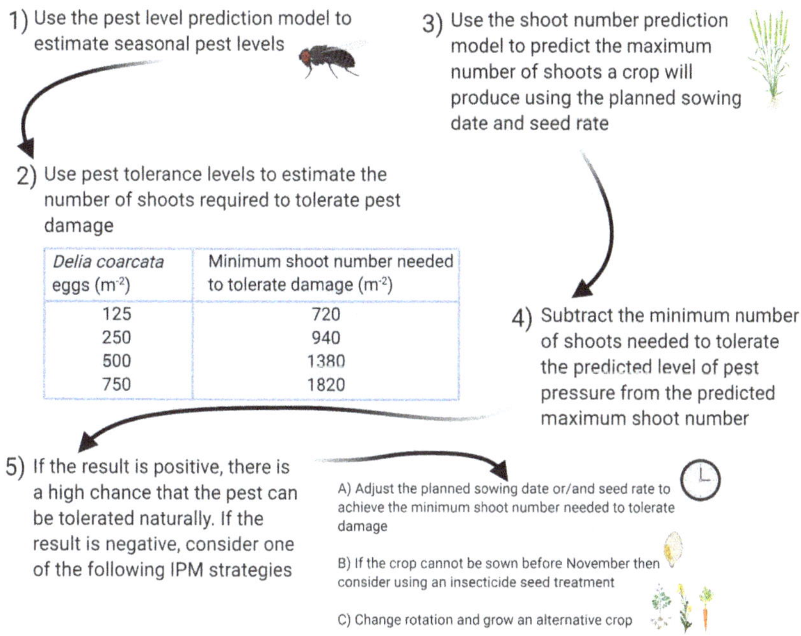

Figure 2 A graphical overview of a DSS system for wheat bulb fly (*Delia coarctata*); image from Leybourne et al., 2022, reproduced with permission. Image created with BioRender .com.

This pest forecasting tool highlights some of the limitations that often affect these predictive tools - the predictions are generic and unable to provide farmers with precise, farm-scale, bespoke risk predictions. In the case of the wheat bulb fly model, predictions are currently restricted to two geographic regions in the UK (Leybourne et al., 2022). While this provides farmers with an overall risk category for their growing region, advancing the tool to provide farmers with estimations at a greater resolution will improve farmer confidence in the DSS and stimulate end-user uptake and usage.

Currently, the wheat bulb fly DSS is not available as an online tool. However, the DSS is being integrated into the IPM Decisions platform (https://www.ipmdecisions.net/) where it will be publicly available. A central barrier in DSS development is the lack of post-project resources available to many applied research projects to facilitate the development of user interfaces and eventual computational tools, this often results in DSS remaining as inaccessible research and development tools. Barriers to DSS uptake will be discussed in Section 4.

2.2 Pest observation networks

Pest observation networks represent a DSS of growing importance, especially as farmers are under increasing pressure to justify any chemical-based interventions. Observation networks offer similar functionality to forecasting models, but these DSS provide farmers with actual data on pest occurrence rather than modelled predictions. Historically, observation networks have comprised fixed infrastructure to monitor pest migration and abundance. Pest observation networks have been established in many countries, with suction trap networks forming a backbone for aphid monitoring and management (Miao et al., 2011; Lagos-Kutz et al., 2020). In the UK, the Rothamsted Insect Survey (RIS) network of suction traps has been running for 58 years (Bell et al., 2015) and is used to provide weekly information on the timing and scale of the migration of aphid pests of many crops, including wheat. The network has also been used to monitor changes in pesticide resistance status and of viruliferous aphid populations (Harrington et al., 2012), and as a source of input values for DSS used in the management of barley yellow dwarf virus (BYDV) (Morgan, 2000). Recent developments in remote sensing technology have enabled observation networks to expand into short-term, real-time monitoring tools. These systems use high-resolution imaging to identify pest presence in real-time. Recent examples of this include real-time monitoring of locust swarms over Africa (for an overview of these applications, see Klein et al. (2021). With regards to wheat pests, remote sensing observation networks can include the use of soundscapes to detect stored grain pests (Mishra et al., 2018) and the use of hyperspectral imaging to identify Hessian fly infestation (Bhattarai et al., 2019). Remote sensing will be discussed in more detail in Section 3.

Fixed monitoring networks remain the most widely used type of pest observation network. These networks are often managed and maintained by government bodies, research institutions, or university extension services. There are several benefits to pest observation networks: these networks are an invaluable tool that can easily monitor pest migratory patterns and, alongside the monitoring of established pests, can also act as early warning systems for invasive species (Teulon and Scott, 2006). If used over long periods, the long-term data produced can provide important insights into climate change effects on pest migration patterns (Bell et al., 2015), and the data can also be used to produce predictive DSS that estimate the timings of annual migrations; this approach has been used to estimate aphid migration and abundance across multiple countries and climates (Harrington et al., 1991, 2012; Howling et al., 1993; Lankin-Vega et al., 2008; Thackray et al., 2009; Klueken et al., 2009; Jarošová et al., 2019). These networks can also provide farmers with an early warning system for the migration of virus vectors in their local area (Thackray et al., 2009), allowing them to install on-farm traps and implement a pest management strategy if needed. A key downside of observation networks is that efficient monitoring requires a comprehensive and reliable network that is sufficiently distributed across agriculturally important regions, and the processing of trap contents relies on dedicated entomologists with a high level of taxonomic expertise.

2.2.1 Observation network case study: suction trap monitoring for aphid management

Suction trap networks have been installed for aphid monitoring in multiple countries, with fixed monitoring stations installed in the UK (https://insectsurvey.com/aphid-bulletin), France (Luquet et al., 2019), the Czech Republic (Jarošová et al., 2019), the USA (Lagos-Kutz et al., 2020), and China (Miao et al., 2011). Mobile units are also being used to establish short-term mobile monitoring networks in Australia (https://imappests.com.au/). These monitoring networks comprise suction traps of varying heights, usually between 6 m and 12 m, that sample alate aphids in-flight during key migratory periods. This information provides farmers with near real-time monitoring data on the movement of their local or regional aphid population, which is subsequently used to estimate crop risk (Miao et al., 2011; Lagos-Kutz et al., 2020). Most monitoring networks provide this information in publicly accessible online tools, which can be easily accessed in most countries, for example, suction trap networks in the UK (https://insectsurvey.com/aphid-bulletin) and the USA (https://suctiontrapnetwork.org/data/) provide pest abundance data across the network during key periods of pest risk within the growing season. In the UK, farmers can subscribe to their local trap to receive weekly text and email bulletins during the growing season.

The long-term data produced by these networks have been used to develop additional DSS that estimate pest migration. In the UK, an aphid forecast DSS has been developed that uses mean temperature in January and February to predict the date of 'first capture' and the total number of aphids that will be caught from migration up until July for each suction trap in the network (https://insectsurvey.com/aphid-forecast). In the Czech Republic, models have been developed to estimate population peaks for autumn and summer migrations (Jarošová et al., 2019)

Although the suction trap networks provide data that are invaluable across the growing season, including the start of aphid migration and frequent pest abundance monitoring, there are still improvements that could be made to the networks. One avenue that is currently being explored in the RIS network in the UK is to combine aphid monitoring with molecular methods of virus monitoring by sub-sampling the weekly aphid catch and determining the percentage virus incidence (including the aphid-borne cereal virus BYDV) within the sampled aphid population. This additional information enables farmers to make more informed management decisions as not all aphids carry the same level of risk: If aphid abundance is high but virus infection rate is low, then crop risk will be relatively low, whereas if aphid abundance is at a moderate level but virus presence is high, crop risk would be higher.

2.3 Systems that support decision-making

Alongside pest forecasting tools and observation networks, a suite of additional DSS support farmers in using the forecasting and observation data to make grounded and informed decisions. Although not a focal point of this chapter, they will be briefly described as they are central components of integrated support systems.

Economic thresholds are DSS that are integral for IPM. These thresholds detail the pest population, usually per unit area (e.g. per plant, per year, per soil sample), over which the farmer can be confident that an economic loss will be observed if no pest intervention strategy is deployed (Ramsden et al., 2017); i.e. if a threshold is exceeded a farmer should consider taking an action to manage the pest population. Economic thresholds have been developed for many economically damaging wheat pests (Bahrami et al., 2003; Kindler et al., 2003; Larsson, 2005; Steinger et al., 2020) and farmers should take care to ensure they observe national or regional thresholds. Although thresholds have been described for many wheat pests, and these thresholds are usually followed by farmers, several thresholds are outdated and oversimplified and are therefore not considered sufficiently reliable. For example, a recent sensitivity analysis of the wheat bulb fly threshold suggested that the current UK threshold of 250 eggs/ m^2 soil is too stringent when crop density was included in the threshold calculation (Leybourne et al., 2022). As some thresholds were devised as long ago as the 1950s (Gough et al., 1961), it is likely that the majority of pest thresholds need updating to account for environmental and climate change as well as changes in agronomic practice and the development of new crop varieties. Many thresholds used in the UK were found to have no basis in published evidence and almost none took account of the differential ability of crops to tolerate damage (Ramsden et al., 2017).

Other tools that can help farmers make effective decisions include image recognition tools that assist farmers with carrying out pest identification with greater confidence. Correct taxonomic identification is an important component of IPM; in theory, pesticide application should only occur if (1) a correct pest identification has been achieved and (2) the regional threshold for that pest has been exceeded. Several mobile apps have been developed that can help farmers with the identification of pests and beneficial species and, therefore, enable the farmers to identify insect pests with greater confidence and justify any chemical-based intervention. Recent advances in AI and machine-learning technology have increased the confidence with which researchers can make taxonomic identifications using image recognition techniques (Liu et al., 2019, 2021); these advanced methods enable a greater number of data inputs to be introduced into the image-recognition process that can help with pest identification. The applications of advanced technology for image-recognition development will be discussed in more detail in Section 3.

2.4 An overview of currently available decision support systems for pest forecasting and population prediction

Table 1 provides an overview of computational-based pest prediction and forecasting DSS developed for the management of wheat pests.

3 Remote sensing and artificial intelligence image analysis to inform pest identification and promote decision-making

Alongside pest forecasts and pest observations (Section 2), farmers also need to be able to confidently identify pest infestation or pest damage within their crop in order to make informed pest management decisions. While manual inspections achieve some success, a bottleneck for the effectiveness of these efforts is the taxonomic challenges associated with correct pest identification and the labour- and time-intensive processes associated with manual inspection and species identification (Ramsden et al., 2017). Furthermore, other external factors can also influence the effectiveness of these efforts, including the weather and climate, inappropriate use and/or deployment of sensing or monitoring devices, high time and/or monetary costs associated with installing pest monitoring facilities, and increased labour and transportation costs. These issues reduce the extent to which farmers carry out in-crop observations for pest infestation. Remote sensing and image analysis technology are two technologies that can support farmers with their pest monitoring and identification efforts.

3.1 Overview of the Decision support systems' potential of remote sensing technology

Recent advances in remote sensing technology mean that it is possible to receive and interpret large-scale, multi-level information of the earth's surface, including environmental data and multispectral imaging data. Remote sensing technology can use this data, often collected through aviation (e.g. planes and drones) or via satellite feeds, at a variety of spatial and temporal scales to monitor pest migration, damage, and outbreak patterns (Abd El-Ghany et al.,

Table 1 An overview of DSS that has been developed for the management of internationally important wheat pests, where no DSS has been developed this is indicated.

Pest name	Pest	Purpose of DSS; country of development	DSS inputs	Reference	Comments on availability
Aphids	Virus vectors during crop risk stage: *Sitobion avenae, Rhopalosiphum padi*	Predicts aphid activity and virus risk; UK	Monitoring data, climate data, crop data, agronomic data	Kendall et al. (1992), Morgan (2000)	Web-based tool available, AHDB BYDV tool: https://ahdb.org.uk/bydv This tool is currently being updated by incorporating Kendall and Morgan models to increase tool efficiency
		Predicts aphid arrival, BYDV incidence, and yield loss; Australia	Monitoring data, climate data, crop data, agronomic data	Thackray et al. (2009)	Not publicly available
		Provides various management options; USA	Monitoring data, climate data, agronomic data, pest phenological and population data	Walls et al. (2016)	Not publicly available
		Estimates the proportion of plants infected with BYDV; France	Monitoring data, climate data	Gillet et al. (1990)	Not publicly available
		Estimates secondary virus spread; France	Frequent aphid infestation assessments	Leclercq-Le Quillec et al. (2000)	Not publicly available

		Predicts autumn aphid abundance; New Zealand	Monitoring data, climate data, degree-days	Lankin-Vega et al. (2008)	Not publicly available
	Populations causing direct damage: Sitobion avenae, Metopolophium dirhodum, Schizaphis graminum, *Diuraphis noxia*	Predicts population dynamics; UK	Aphid infestation assessment	Fabre et al. (2006)	Not publicly available
		Predicts migration; UK	Monitoring data, climate data	Harrington et al. (1991), Howling et al. (1993)	Predictions reported on the aphid forecast website: https://insectsurvey.com/aphid-forecast
		Predicts economic impact of aphid infestation; USA	Climate data, economic data, agronomic data	Elliott et al. (2004)	Web-tool available at Oklahoma State University: http://entoplp.okstate.edu/gbweb/tutorial2.htm
	Economic loss models: *Sitobion avenae*	Estimates potential yield loss; France	Percentage of stems infested with aphids	Fabre et al. (2003)	Not publicly available
		Estimates aphid damage from in-field infestation level and predicts potential of a yield loss, provides a spray recommendation; UK	Economic costs (spray costs), expected yield, market price, growth stage, percentage of stems infested with aphids	Mann et al. (1986)	This system is outdated and no longer used
Sunn pest	*Eurygaster integriceps*	Provides various management options; Iran	Agronomy data, economic data, climate data	Atri and Amir-maafi (2006)	Not publicly available
		Predicts critical phenological stages; Turkey	T-sum degree-days	Gözüaçik et al. (2016)	Not publicly available

(Continued)

Table 1 (*Continued*)

Pest name	Pest	Purpose of DSS; country of development	DSS inputs	Reference	Comments on availability
Hessian fly	*Mayetiola destructor*	Detecting infestation; USA	Geographical data, imaging data	Bhattarai et al. (2019)	Not publicly available
Frit fly	*Oscinella frit*	Predicts infestation; Sweden	Environmental data	Lindblad (2001)	Not publicly available
		Predicts first occurrence; Sweden	T-sum degree-days	Lindblad and Sigvald (1996)	Not publicly available
Gout fly	*Chlorops pumilionis*	Predicts summer emergence; Switzerland	Climate data, agronomy data	Derron and Goy (1990)	Not publicly available
Wheat bulb fly	*Delia coarctata*	Predicts egg abundance; UK	Climate data	Leybourne et al. (2022)	Not publicly available
Yellow cereal fly	*Opomyza florum*	No DSS available	N/A		
Orange wheat blossom midge	*Sitodiplosis mosellana*	Predicts emergence; Belgium	Climate data	Jacquemin et al. (2014)	Not publicly available
Saddle gall midge	*Haplodiplosis marginata*	Predicts emergence; UK	Climate data	Rowley et al. (2016)	Not publicly available
Leather jackets	*Tipula spp.*	No DSS available	N/A		
Wheat stem sawfly	*Cephus cinctus*	Predicts potential pest damage; Canada	Precipitation	Beres et al. (2017)	Model available from authors

Slugs	*Deroceras reticulatum*	Predicts slug activity; UK	Climate data	Young et al. (1993)	Widely used
		Predicts slug distribution in-field; UK		Petrovskii et al. (2022)	Not publicly available
Wireworms	*Agriotes spp.*	No DSS available	N/A		
Cereal leaf beetle	*Oulema melanopus*	Predicts egg and larvae populations; USA	T-sum degree-days	Philips et al. (2012)	Not publicly available
Wheat curl mite	*Aceria tosichella*	No DSS available	N/A		
Stored grain pests	*Rhyzopertha dominica, Sitophilus oryzae, Cryptolestes ferrugineus, Tribolium castaneum, and Oryzaephilus surinamensis*	Predicts risk of grain damage; USA	Insect density, grain temperature, grain moisture	Flinn et al. (2007)	Stored Grain Advisor system is available to download from USDA
	Sitophilus granarius	Predicts infestation levels; India	Sound	Mishra et al. (2018)	Not publicly available

This is a non-exhaustive list and might not include all available DSS or reported models; monitoring networks are not included in this DSS list; any weblinks are accurate at the time of publication.

2020). Remote sensing technology has been used to monitor the occurrence and range of important pests and to identify and distinguish the occurrence and severity of pest infestations (Sishodia et al., 2020). Recent examples of the successful use of remote sensing technology in wheat pest management include the use of multispectral normalised difference vegetation index data to detect Hessian Fly infestation in the USA (Bhattarai et al., 2019) and the use of remote sensing techniques to identify aphid presence and aphid damage on wheat (Luo et al., 2013). Here, the authors showed that hyperspectral imaging was able to discriminate aphid damage in a similar manner to manual observation when damage was scored into a defined Leaf Aphid Damage Index.

While remote sensing technologies have the potential to be used in pest management processes (Luo et al., 2013; Bhattarai et al., 2019) various limitations restrict the extent to which these technologies can be used on a large scale. The primary limitation here is a lack of understanding on the efficacy of remote sensing technologies by sensor and platform type, for example, Bhattarai et al. (2019) showed contrasting effectiveness between high-resolution image data collected from aircrafts and lower resolution satellite data, with lower resolution satellite data providing a better estimation of Hessian Fly infestation. Furthermore, significant consideration needs to be given to the techno-economic benefit of remote sensing technology. As most remote sensing imaging technology will involve hundreds of narrow spectral bands per image, there is significant complexity and data storage requirements associated with the collection, processing, and analysis of these data which need to be considered; if high computational power is required to source, process, and analyse the data, this can impact the usability of any developed DSS for the end-user. AI algorithms can supplement remote sensing technology to develop remote sensing-based smart DSS tools. These AI algorithms can address the complexities and uncertainties associated with remote sensing data, as well as overcoming the sparsity of remote sensing data by supporting scalable and accurate real-time pest data mining methods. As with standard predictive model-based DSS, remote sensing-based DSS also require visualisation tools in order to provide user-friendly interfaces that can reduce and refine large complex remote sensing datasets to simple graphs, maps, or tables in order to facilitate the delivery of actionable decisions.

For example, image recognition technology is suitable for automatic identification and monitoring of pests living on the surface of crops. Insect radar (Lidar or Entomological radar) technology is particularly suitable for automatic identification and monitoring of flying pests, while audio recognition technology is capable of identifying and monitoring hidden pests (Blaschke, 2010), including in the detection of stored grain pests (Mishra et al., 2018).

3.2 Role of artificial intelligence and image analysis in decision-making

One important aspect of successful IPM is the correct identification of pest species in order to ensure appropriate management and treatment options are selected and implemented. Accurate identification of insect pests is challenging for most farmers, and low confidence in their ability to correctly identify pests usually results in the application of insurance sprays. One avenue for improving this decision-making process is the use of AI algorithms to optimise image analysis processes to support on-farm insect identification and prescriptive treatment.

Image analysis techniques involve the use of mathematical models and image processing algorithms to analyse and extract specific features from an image object. For example, the models and algorithms will identify and extract information on the colour, shape, and size of an object as well as the presence or absence of specific identifiable traits, with these features used as inputs in decision-making trees (Fomalont, 1999), the process detects and measures the objects of interest in the image to obtain their objective information, thereby establishing a description of the image. The main process of image analysis is input, segmentation, recognition, and interpretation. The AI image analysis decision-making process follows four key steps:

1. Data and information acquisition. Information collection is a prerequisite for image analysis. Various signals are converted into electrical signals and then the required information and data are extracted from them.
2. Data pre-processing. This stage is to denoise, scale, calibrate, smooth, and transform the image and the purpose is to highlight the important information of the image itself.
3. Feature extraction and selection. This is the key element of image analysis technology, especially in the recognition mode, which directly determines whether the image can be successfully recognised and whether the extracted features can be stored.
4. Design of decision-making algorithms, which is the last step of AI combined with image analysis tasks such as detection, classification, or counting. This link mainly formulates decision-making rules based on operating procedures and analyses images in accordance with standards (Das, 2016).

At the fundamental level, AI-based image recognition processes work through a workflow in a similar way a human taxonomist would work through a taxonomic key.

3.2.1 Overview of computational techniques used in artificial intelligence image analysis algorithms

AI-based decision-making is a process relying on 'big data'. In other words, machines summarise laws from a large amount of data, which can be applied to real-life scenarios. All the knowledge available to the AI comes from real-world data and real scenes (Jarrahi, 2018). Compared with traditional decision-making models, it can deal with various complex and detailed situations and make more accurate judgments to optimise the system. The ability of AI machines, and by association AI-based computational models, to deal with multi-modal complex data and detailed situations enables these models to make more robust estimations that are more representative of the real world than standard computational models. These AI-based processes can be further enhanced by integrating a machine-learning component into these systems.

Machine-learning algorithms, including deep-learning processes, have recently become a powerful tool for image analysis and modelling of agricultural conditions. A number of deep-learning approaches, such as Single Shot Detector (Liu et al., 2016), Faster Region-based Convolutional Neural Networks (R-CNN) (Ren et al., 2015), Feature Pyramid Network (FPN) (Lin et al., 2017), and other extended variations of these networks (Dai et al., 2016), have led to highly promising progress in the field of generic object recognition. To extract features on a single-scale feature map, Faster R-CNN employs Region-of-Interest pooling. However, when it comes to tiny object detection, FPN outperforms Microsoft's large-scale object detection dataset COCO (Lin et al., 2014) with a mean average precision of 56.9%. FPN allows a model to identify all objects in different scales over a wide variety of sizes and pyramid levels by constructing a pyramid of multi-scale feature maps. Furthermore, the Convolutional Neural Network (CNN) feature pyramid structure has become a popular choice since it simultaneously includes low-level object characteristics and high-level semantic features, which is beneficial for detecting little objects, such as pest detection. Deep learning does not require manual extraction and annotation of agricultural information in images, but its effectiveness relies on massive databases. Currently, deep-learning-based agricultural applications are widely used, such as disease and pest diagnosis and identification, plant identification and detection, remote sensing area classification and monitoring, agricultural product classification, and fruit carrier detection (Steele-Dunne et al., 2017). Particularly, towards diverse wheat pest detection, the combination of CNN and Region Proposal Network (RPN) has been proposed to identify and count wheat mites from digital images (Chen et al., 2021); furthermore, Li et al. have improved the method integrating a CNN of Zeiler and Ferus model and RPN with non-maximum suppression to remove overlapping detections with enhanced mean accuracy precision of 88.5% (Li et al., 2019); also DeepPest suggests a novel two-stage mobile vision-based multi-projection pest detection approach using multi-scale contextual information towards large-scale multiple species of wheat pests (Wang et al., 2020). Recently, AgriPest dataset is published as the benchmark containing over 49.7K annotated pest images of 4 main crops with 14 species of wheat pests. The scalability and diversity of AgriPest will offer great chances to researchers for evaluating the performance of deep learning technologies in pest detection (Wang et al., 2021).

3.3 Device-driven artificial intelligence as a 21st-century integrated pest management tool

Decision-making as part of IPM is a process that combines various technologies and information including control methods, production practices, and pest biology and phenology. IPM is a comprehensive discipline based on decision science, systematics, ecology, cultivation, economics, genetics, chemistry, and sociology (Kogan, 1998). Modern IPM practices are centred on the interactions between pest populations with other biological populations and non-biological elements present within the ecosystem across spatial and temporal scales. This creates a complex system with the phenology, distribution, and occurrence of the pest population a dynamic process within this system. Therefore, as these systems are dynamic and complex, the computer and AI-based tools will become an effective decision-making tool for manipulating this complex system (Deguine et al., 2021). In recent years, the application of computers and computer decision-making technology in pest management research has rapidly increased, showcasing the application of using this technology in pest management processes.

In the field of pest management, its application involves many aspects such as the diagnosis of pests and diseases, the prediction of pest pressure, the selection of pesticides, and the training of personnel, which greatly broadens the scope of IPM technology (Isman, 2019). Device-driven AI IPM reduces the time for non-professional managers to acquire domain knowledge and frees them from the limitations of failing to make timely decisions and failing to make predictions and forecasts due to lack of domain experts. AI is a computer program that simulates the reasoning process of human experts, and many systems themselves combine models, Geographic Information System technology, and multimedia technology, etc., so the system can greatly reduce human effort during operation and decision-making reasoning (Coulson, 1992). Therefore, it can give full play to its advantages in the comprehensive management of ecosystems, including ecological problems such as accounting for a range of natural enemies, environmental factors, and complex interrelationships.

3.3.1 Integrating remote sensing and device-driven artificial intelligence for on-farm pest monitoring

Advanced pest sensing systems enable the transmission of multi-modal and heterogeneous pest data (e.g. images, remote sensing data, environmental information) to computational servers in real-time, supporting automated data analysis and management. This can include systems that install fixed pest trapping devices in the field to collect and transmit images of trap contents in real-time, and then utilise AI technology, such as machine learning image analysis models, to precisely detect and classify pests present in these images (Liu et al., 2021; Wang et al., 2021). These models can then estimate the severity of the pest infestation to support intelligent prediction and prevention (Liu et al., 2021; Wang et al., 2021). Figure 3 provides a graphical representation of these systems.

These systems can also utilise historical information of crop pests and diseases, combine this with additional real-time remote sensing data such as soil temperature, weather change trends, specific remote sensing images, and other influencing factors, and through machine-learning algorithms simulate the occurrence of pests and diseases (Durgabai et al., 2018) or detect the intrusion of pests in a specific area. By counting the number of corresponding pests, combined with relevant information such as time and location, the trend chart of the occurrence of pests can be displayed visually. The long-term accumulated data can be used to mine valuable information such as the outbreak period and pest living habits, which is convenient for forecasting the future trends of the pests, and formulating prevention and control plans in time to effectively respond to the occurrence of major pest epidemics (Damos, 2015). Through the preliminary identification of the pests and the judgement and confirmation of the symptoms that cause harm, the system can ultimately give the prevention and control opinions.

Advances in this area are focused on developing tools and functionality in order to 'scale down' the process in order to provide farmers the capability, the convenience, and the precision of the AI-based DSS on a mobile phone. Through the AI programme, the situation of pests and diseases is scanned, and professional prevention and control advice is obtained to quickly and efficiently adopt pest prevention and control measures to reduce farmers' losses. Advances in the field mean that in the future it will be possible for a farmer to conduct AI and machine learning enhanced pest monitoring during routine crop walks in order to make rapid management decisions in real-time.

4 The uptake of digital support systems amongst farmers

While numerous DSS have been developed, relatively few are used routinely (Rossi et al., 2012). Various factors can influence the uptake of DSS amongst farmers. Generally, DSS uptake is thought to be constrained by the following (Gent et al., 2010; Shtienberg, 2013):

1. A lack of end-user engagement during the DSS development process resulting in unwanted or unsuitable DSS;
2. Development of DSS is often derived from academic initiatives rather than end-user requirements, so availability does not necessarily match end-user demand;
3. Lack of quantification of the economic, environmental, and consumer benefits of using DSS - often limited by access to sufficiently large datasets across a wide range of sites and seasons;
4. DSS which are insufficiently risk-averse to meet farmer needs;

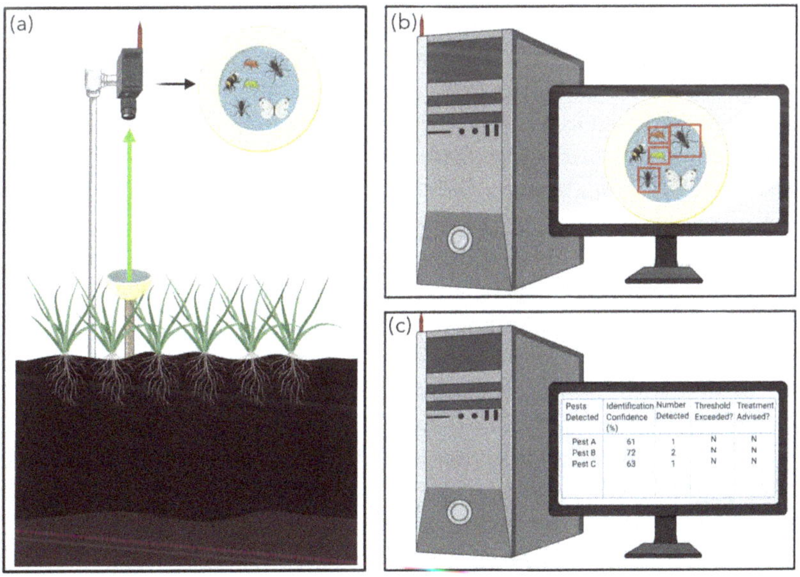

Figure 3 Graphical overview of a potential integrated pest management station for automated detection and identification of insect pests. (a) A remote sensing station captures images from in-field traps and transmits images to the computer server. (b) AI machine-learning algorithms identify potential pest species present in the image. (c) An end-user output provides the user information that can be used to guide decision-making processes. Image created with BioRender.com.

5 Inadequate testing of DSS outputs in target regions for accuracy and predictive value (also data limited);
6 Short-term funding curtailing long-term updating and user support;
7 DSS limited in focus to single pests, whilst farmer decisions need to account for multiple pests that often occur concurrently;
8 Regional fragmentation of DSS development and user communities, limiting awareness of and access to available DSS; and
9 Limitations in the accessibility, usability, or publicity of the DSS tool. This is usually the most prominent factor that limits DSS uptake; if end-users are unable to access or use the DSS tools, or are unaware of its existence, uptake will be low.

These constraints limit the rate of uptake of DSS and should be considered during the development of new systems or addressed to facilitate greater uptake of current systems. A lack of coordination across DSS development has led to individual systems being created without a route to long-lasting uptake and support. Systems developed as part of research initiatives are largely bound by short-term funding cycles; when the project ends, the system is no longer updated, promoted, or supported. Funded research often prioritises publications over practical impact, resulting in high-impact papers describing the resulting models and potential application without a lasting user interface to make the results accessible and implementable by farmers and advisors. DSS developers should incorporate both behavioural and socio-economic driver factors in the design and function of the models and user interface, such as farm size and type, and user experience; younger farmers, farmers with higher levels of education, and farmers on relatively larger farms are, for example, more likely to use DSS (Rose et al., 2016). DSS need to be designed with farmers since their use of DSS is influenced by their farming styles and usual practices (Kanatas et al., 2020). One international research project, the EU Horizon 2020 IPM Decisions Project, aims to remove the access barrier by providing a holistic DSS environment, hosting DSS for crops with a common interface, such as the wheat bulb fly DSS (Leybourne et al., 2022). The IPM Decisions Platform automatically integrates the climate data required to run each system, for any given location in Europe, and provides the user with the essential information needed to support decisions; the current predicted risk level based on underlying DSS models, guidance on how to interpret the risk, any assumptions or limitations on the model, and a description of how the model works. This project is initially focused on DSS developed for EU, but the framework developed in this project could be expanded globally.

4.1 Factors determining the success of decision support systems: gaining end-user trust

DSS are developed as a result of various initiatives and ambitions. The target end-users are normally farmers and/or agricultural advisors. The most successful DSS are presented in very simple formats to encourage wider use (e.g. the AHDB BYDV tool; https://ahdb.org.uk/bydv). Simplification has a benefit in making DSS outputs readily accessible; however, it can create barriers to use where outcomes are perceived as untrustworthy due to oversimplification of the recommendations and a lack of transparency. Conversely, if the system is too complex, users may be put off by the time taken to provide the required inputs and interpret the outputs. Designing sophisticated tools with simple user interfaces can also miss opportunities to support farmers with specific challenges, such as supporting adherence to legislative requirements (Rose et al., 2016). DSS are most effective where they are used in tandem with supplementary information from peers and technical experts. Ideally, the system will have been developed with direct input from representative target users ensuring it is fit for its purpose. Creating and involving a network of informed potential end-users able to help promote trust in, and support the use of, the final system should also help stimulate uptake (Jakku and Thorburn, 2010).

Primarily, most DSS provide quantifiable information about the likelihood and/or scale of pest infestations, based on validated data appropriate to the target user. In addition, DSS directly, or through supporting materials, provide information about the pest and its ecology that further enables improved pest management by the user. This can occur either by improving their ability to interpret and apply the DSS outputs and recommendations to specific cropping situations or by providing insight into combinations of techniques which may improve overall management.

Generally, the success of a specific DSS is determined by end-user buy-in; i.e. how willing farmers and advisers are to trust the output (Jakku and Thorburn, 2010). Trust is dictated by:

1 how engaged farmers were in DSS development (i.e. can end-users trust that their needs were adequately accounted for and considered in the development process?);
2 the evidence available on validation exercises (i.e. has the data the DSS is based on been sufficiently and transparently provided to end-users via case studies and/or peer-to-peer demonstration/discussion); and
3 risk aversion (i.e. can the end-users trust that the DSS outputs adequately and accurately account for any potential economic loss?).

Ultimately, in order for DSS uptake to be successful, end-users trust and confidence in the outputs provided by the system need to be high. From the end-user perspective, the repercussions of applying treatments where no pest risk was present are low (in the short term), whereas the repercussions of not treating a damaging infestation can lead to considerable financial losses. Therefore, DSS need to be sufficiently reliable to provide farmers and advisers with enough confidence to follow the associated guidance, to trust that any guidance leading to a 'do not treat' decision is sufficiently justified and to be satisfied that the system is suitably risk averse.

4.2 Factors determining the success of decision support systems: end-user involvement and validation

Early engagement with end-users is key in developing successful DSS. Early engagement inspires greater trust in the DSS and promotes greater uptake and use once the system is available.

In order to convince farmers and advisors of the value of a DSS, validation testing must demonstrate a reliable outcome, in terms of both improved pest management and economic returns (Jørgensen et al., 2020). Ideally, validation takes place alongside DSS development, together with direct engagement with potential users (Jakku and Thorburn, 2010). This establishes a relationship between the developer and the relevant stakeholders from the start of DSS development, allowing both to be involved in the initial concept and providing an avenue for end-users to highlight potential user requirements. Technical development of the DSS algorithms and associated identification of assumptions and limitations can be completed, and the system can be validated on farms within the initial end-user group. Further testing in the wider community ensures the DSS is working correctly, and then the user interface can be developed (Oliver et al., 2012). Research-led DSS development is often the result of short-term projects and tends to focus on the development of technical aspects of the algorithm behind the system. These projects often engage well with stakeholders early on, but ultimately leave the resultant DSS largely unsupported, often without any form of user interface. This can lead to good systems being lost due to atrophy, and ultimately losing their validity, since peer-to-peer community validation diminishes over time (Gent et al., 2010). Frequently, DSS have been developed by technical experts, focusing on the biological factors associated with pest distribution, impact, and management; as such the assumptions made about the end goal of management may differ between the DSS developer (targeting long-term sustainable management) which conflicts with a farmer's ambition to have immediate solutions to the problem of an infestation (Wilson et al., 2009). For example, where a DSS is built on an assumed understanding that weed dispersal is predominantly driven by human activity, the outputs may not be seen as feasible by users who do not agree with this assumption (Wilson et al., 2009).

Development of DSS that focuses too heavily on the user experience can also limit uptake. While it is critical that DSS development is user-centric (Parker and Sinclair, 2010), rapid advances in technology can render equipment and software obsolete in a short period of time. Even if past versions are still functional, they may no longer be supported by the company they were purchased from (Kanatas et al., 2020). The repeated and/or frequent changes in the DSS user interface and/or content can be off-putting or distracting to the user (Kanatas et al., 2020).

In general, barriers to DSS uptake include a lack of end-user engagement during the development process, lack of economic and environmental quantification, a disregard of consumer benefits to using DSS, and limited end-user trust in DSS recommendations. These can broadly be alleviated by including end-users during the DSS development process and providing end-users access to the validation and testing processes used during DSS development. Validation is especially important for pest forecast DSS, as in these cases it is important that the DSS is validated in multiple regions as pest biology and phenology can differ geographically (Jørgensen et al., 2020). For example, if end-users can see that a DSS developed, tested, and validated in the UK has also been validated for use in other geographical areas, confidence in the DSS will be greater and uptake will increase. DSS uptake can also be increased by ensuring end-users can trust the outputs and recommendations, with trust increasing if these recommendations have been subjected to robust economic risk assessments.

5 Conclusion and future trends in research

DSS should only ever be used to support pest management decisions and should not be used to replace agronomic expertise and on-farm pest monitoring experience. This is because DSS cannot be expected to predict all situations all of the time, and farmers/advisors will have the best understanding of specific field characteristics, history, etc. that may affect final management decisions. In addition, many DSS have clear assumptions and limitations, and these must be considered on a case-by-case basis when applying the outputs on the farm. For example, DSS that predict pest pressure should be followed by in-field inspection. Such an in-field inspection need not be as onerous as crop monitoring without the assistance of DSS but only enough to check that the recommendation is broadly correct. DSS should help farmers make effective decisions, especially as crop protection practices are becoming increasingly complex. In an industry of growing regulation and supervision, both by government and processors, regarding the use of pesticides and other inputs, DSS can play a vital role in effective decision-making. Increased reliance on pesticides has gone hand in hand with a delegation of responsibility for making hard decisions in crop protection. The pressures applied to farmers and their agronomists to improve yield and quality have in the past made it all but impossible to reduce pesticide use, with high levels of inputs made easy by the accessibility of relatively cheap and easy-to-use pesticides. As agricultural priorities shift towards long-term economic and environmental sustainability, the responsibility for pest management is shifting back onto farmers, along with the associated risks. The role of DSS in both IPM plans and as a component in the standard crop production process is likely to increase, with this in mind we have highlighted some key areas for future DSS research and development.

One focus for DSS research and development should be to develop more holistic systems. Developing integrative systems will require a knowledge base far wider than a few individuals can be expected to retain; instead, local stakeholders need to harmonise their vision and strategies to engender long-term sustainability, with the input of relevant experts where necessary (Swiergiel et al., 2019). DSS should be used as part of this process but must be used only as a conduit for improving understanding of key decision points and aiding in drawing appropriate conclusions in combination with appropriate monitoring data from the targeted crop area. In the future, we are going to see more systemic approaches to decision support, incorporating a range of stakeholders accessing DSS through shared platforms. These will engender much greater trust in systems, providing assurance that effective pest management can be maintained (Bailey et al., 2009; Geiger et al., 2010; Lechenet et al., 2017). It will also support innovation in the development of new DSS across all aspects of crop production.

A second focus should be on developing holistic multi-DSS platforms that are accessible and user-friendly. A key benefit of multi-DSS platforms is the ability to standardise DSS output formats, enabling the relative assessment of different pest risks during the cropping season, via a range of multimedia tools and supportive materials (text, graphs, maps, photos, and video; Rossi et al., 2012). VIPS (vips-landbruk.no) is a prime example of a successful holistic multi-DSS platform and is the basis of the pan-European IPM Decisions platform. VIPS is an Open Source platform for IPM DSS in Norway. The structure of the system makes all available IPM tools for pests and diseases within a cropping system open for further development and validation and provides a single point of access for users looking to use DSS. VIPS, therefore, brings farmers, advisors, researchers, software developers, and other IPM stakeholders together within one system. This also facilitates the rapid release of new tools, without any delays or reprogramming of models when research and development are completed. The structure behind VIPS makes it possible to connect it to any online source of input data, such as weather stations, external forecast systems, and automated insect traps. It can also replace missing weather data where necessary and use locally sourced information from observation networks where required by models within the platform. Model output is typically presented with colour codes in red, yellow, or green to show the risk of infection/infestation. Each model is also explained and presented in more detail, providing any assumptions or conditions that should be considered when interpreting the outputs. Currently, the VIPS system includes DSS forecasting models for four cereal pests: pollen beetle, oat leafminer fly, grain aphid, and bird cherry-oat aphid. A current project, IPM Decisions project, aims to develop a similar holistic platform at a larger pan-European scale. Future research should aim to replicate this in other geographical regions. As a minimum, these projects should also aim to provide an environment where all available DSS can be accessed, DSS models updated where needed, DSS validated and tested in new regions, and user interfaces simplified.

Future work should also focus on adapting pre-existing DSS to new regions and diversifying the range of pests covered. A significant number of current DSS for wheat pest management focus on aphid control and BYDV management (Mann et al., 1986; Kendall et al., 1992; Morgan, 2000; Fabre et al., 2003, 2006; Elliott et al., 2004; Thackray et al., 2009); this is justifiable as aphids and aphid-vectored viruses can cause substantial yield loss, recent estimates suggest up to 84% (Nancarrow et al., 2021). However, it is important that researchers realise that although aphids are important pests they are not the only pest of significance. DSS have been developed for most economically important wheat pests (Table 1) although there are still pest species (yellow cereal fly and wheat curl mite) where no DSS are available. Going forward it is important that these DSS gaps are plugged and that, where possible, DSS are integrated into multi-pest systems as this will increase end-user uptake and usability. Nonetheless, the fact that various aphid DSS exist and were developed and validated in multiple regions showcases the need for a region-specific DSS for each pest. Therefore, future research should also focus on validating pre-existing DSS for use in new regions (or adapting the underlying models and assumptions) as an initial step in developing pest-specific DSS on a regional basis. Where pre-existing models cannot be adapted, then new models should be developed on a regional basis and, where possible, integrated into multi-pest DSS environments. Finally, future research and development should focus on integrating remote sensing, AI models, and other smart systems to provide integrated real-time smart pest monitoring systems.

In summary, future research should focus on:

- Developing integrative systems that encompass the entire crop production aspect into a single DSS, instead of multiple DSS for individual crop production components;

- Developing user-friendly holistic multi-DSS platforms;
- Developing DSS for a greater range of pest species and adapting/validating pre-existing DSS to new geographical regions;
- Developing DSS that target multiple wheat pests, or promoting the use of holistic platforms that provide end-users easy access to multiple single-pest DSS tools; and
- Incorporating remote sensing and AI to promote real-time smart pest management, including automated pest detection/monitoring and instant risk determination.

6 Where to look for further information

The availability of DSS for wheat pest management is indicated in Table 1, where available links to the associated online DSS are also provided. We have highlighted sources for further information in the following section.

6.1 Further reading

- An overview of the role of DSS in developing effective IPM strategies: (Damos, 2015).
- A case-study for the use of cloud-based support systems: (Rupnik et al., 2019).
- An overview of the potential of AI and deep-learning models in pest management science: (Clark, 2020).

7 Funding

Production of this chapter was supported in part by the Alexander von Humboldt Foundation through an Alexander von Humboldt Postdoctoral Research Fellowship to DJL. Production of this chapter was also supported in part by the IPM Decisions project, which has received funding from the European Research Council (ERC) under the European Union's Horizon 2020 research and innovation programme (grant agreement 817617).

8 References

Abd El-Ghany, N. M., Abd El-Aziz, S. E. and Marei, S. S. 2020. A review: application of remote sensing as a promising strategy for insect pests and diseases management. *Environmental Science and Pollution Research International* 27(27), 33503-33515.

Atri, A. and Amir-maafi, M. 2006. A decision support system for management of Sunn pest (*Eurygaster integriceps* puton.) in wheat. In: *Computers in Agriculture and Natural*

Resources, 23-25 July 2006. Orlando, FL: American Society of Agricultural and Biological Engineers, p. 414.

Bahrami, N., Radjabi, G. H., Rezabeygi, M. and Kamali, K. 2003. Study on economic injury level of sunn pest (*Eurygaster integriceps* put.) on wheat in rainfed regions of Kermanshah province. *Applied Entomology and Phytopathology* 70, 29-45.

Bailey, A. S., Bertaglia, M., Fraser, I. M., Sharma, A. and Douarin, E. 2009. Integrated pest management portfolios in UK arable farming: results of a farmer survey. *Pest Management Science* 65(9), 1030-1039.

Barzman, M., Bàrberi, P., Birch, A. N. E., Boonekamp, P., Dachbrodt-Saaydeh, S., Graf, B., Hommel, B., Jensen, J. E., Kiss, J., Kudsk, P., Lamichhane, J. R., Messéan, A., Moonen, A., Ratnadass, A., Ricci, P., Sarah, J. and Sattin, M. 2015. Eight principles of integrated pest management. *Agronomy for Sustainable Development* 35(4), 1199-1215.

Bell, J. R., Alderson, L., Izera, D., Kruger, T., Parker, S., Pickup, J., Shortall, C. R., Taylor, M. S., Verrier, P. and Harrington, R. 2015. Long-term phenological trends, species accumulation rates, aphid traits and climate: five decades of change in migrating aphids. *Journal of Animal Ecology* 84(1), 21-34.

Beres, B. L., Hill, B. D., Cárcamo, H. A., Knodel, J. J., Weaver, D. K. and Cuthbert, R. D. 2017. An artificial neural network model to predict wheat stem sawfly cutting in solid-stemmed wheat cultivars. *Canadian Journal of Plant Science* 97, 329-336.

Bhattarai, G. P., Schmid, R. B. and McCornack, B. P. 2019. Remote sensing data to detect Hessian fly infestation in commercial wheat fields. *Scientific Reports* 9(1), 6109.

Blaschke, T. 2010. Object based image analysis for remote sensing. *ISPRS Journal of Photogrammetry and Remote Sensing* 65(1), 2-16.

Chen, P., Li, W., Yao, S., Ma, C., Zhang, J., Wang, B., Zheng, C., Xie, C. and Liang, D. 2021. Recognition and counting of wheat mites in wheat fields by a three-step deep learning method. *Neurocomputing* 437, 21-30.

Clark, R. D. 2020. Putting deep learning in perspective for pest management scientists. *Pest Management Science* 76(7), 2267-2275.

Coulson, R. N. 1992. Intelligent geographic information systems and integrated pest management. *Crop Protection* 11(6), 507-516.

Dai, J., Li, Y., He, K. and Sun, J. 2016. R-FCN: object detection via region-based fully convolutional networks. *Advances in Neural Information Processing Systems* 29.

Damos, P. 2015. Modular structure of web-based decision support systems for integrated pest management: a review. *Agronomy for Sustainable Development* 35(4), 1347-1372.

Das, T. K. 2016. Intelligent techniques in decision making: a survey. *Indian Journal of Science and Technology* 9(12), 1-6.

Deguine, J.-P., Aubertot, J.-N., Flor, R. J., Lescourret, F., Wyckhuys, K. A. G. and Ratnadass, A. 2021. Integrated pest management: good intentions, hard realities: a review. *Agronomy for Sustainable Development* 41(3), 38.

Derron, J. O. and Goy, G. 1990. The gout-fly (*Chlorops pumilionis* Bjerk.): biology, injuriousness, control methods. *Revue Suisse d'Agriculture* 22, 101-105.

Durgabai, R. P. L., Bhargavi, P. and Joyothi, S. 2018. Pest management using machine learning algorithms: a review. *International Journal of Computer Science Engineering and Information Technology Research* 8(1), 13-22.

Elliott, N. C., Royer, T. A., Giles, K. L., Kindler, S. D., Porter, D. R., Elliott, D. T. and Waits, D. A. 2004. A web-based decision support system for managing greenbugs in wheat. *Crop Management* 3(1), 1-6.

European Union. 2009. *Regulation (EC) No 1107/2009 of the European Parliament and of the Council of 21 October 2009 Concerning the Placing of Plant Protection Products on the Market and Repealing Council Directives 79/117/EEC and 91/414/ EEC.*

Fabre, F., Dedryver, C. A., Leterrier, J. L. and Plantegenest, M. 2003. Aphid abundance on cereals in autumn predicts yield losses caused by barley yellow dwarf virus. *Phytopathology* 93(10), 1217-1222.

Fabre, F., Pierre, J. S., Dedryver, C. A. and Plantegenest, M. 2006. Barley yellow dwarf disease risk assessment based on Bayesian modelling of aphid population dynamics. *Ecological Modelling* 193(3-4), 457-466.

Flinn, P. W., Hagstrum, D. W., Reed, C. R. and Phillips, T. W. 2007. Stored grain Advisor Pro: decision support system for insect management in commercial grain elevators. *Journal of Stored Products Research* 43(4), 375-383.

Fomalont, E. B. 1999. Image analysis. *Synthesis Imaging in Radio Astronomy* II, 301.

Geiger, F., Bengtsson, J., Berendse, F., Weisser, W. W., Emmerson, M., Morales, M. B., Ceryngier, P., Liira, J., Tscharntke, T., Winqvist, C., Eggers, S., Bommarco, R., Pärt, T., Bretagnolle, V., Plantegenest, M., Clement, L. W., Dennis, C., Palmer, C., Oñate, J. J., Guerrero, I., Hawro, V., Aavik, T., Thies, C., Flohre, A., Hänke, S., Fischer, C., Goedhart, P. W. and Inchausti, P. 2010. Persistent negative effects of pesticides on biodiversity and biological control potential on European farmland. *Basic and Applied Ecology* 11(2), 97-105.

Gent, D. H., De Wolf, E. and Pethybridge, S. J. 2010. Perceptions of risk, risk aversion, and barriers to adoption of decision support systems and integrated pest management: an introduction. *Phytopathology* 101, 640-643.

Gillet, H., Defryver, C. A., Robert, Y., Gamon, A. and Pierre, J. S. 1990. Assessing the risk of primary infection of cereals by barley yellow dwarf virus in autumn in the Rennes basin of France. *Annals of Applied Biology* 117(2), 237-251.

Gough, H. C., Woods, A., Maskell, F. E. and Towler, M. J. 1961. Field experiments on the control of wheat bulb fly, *Leptohylemyia coarctata* (Fall.). *Bulletin of Entomological Research* 52(3), 621-634.

Gözüaçik, C., Yiğit, A. and Şimşek, Z. 2016. Predicting the development of critical biological stages of Sunn pest, *Eurygaster integriceps* Put. (Hemiptera: Scutelleridae), by using sum of degree-days for timing its chemical control in wheat. *Turkish Journal of Agriculture and Forestry* 40, 577-582.

Harrington, R., Howling, G. G., Bale, J. S. and Clark, S. 1991. A new approach to the use of meteorological and suction trap data in predicting aphid problems. *EPPO Bulletin* 21(3), 499-505.

Harrington, R., Taylor, M. S., Shortall, C. R., Alderson, L., Mallott, M. and Verrier, P. J. 2012. The Rothamsted Insect Survey: old traps, new tricks. *Aspects of Applied Biology* 117, 157-164.

Howling, G. G., Harrington, R., Clark, S. J. and Bale, J. S. 1993. The use of multiple regression via principal components in forecasting early season aphid (Homoptera: Aphididae) flight. *Bulletin of Entomological Research* 83(3), 377-381.

Isman, M. B. 2019. Challenges of pest management in the twenty first century: new tools and strategies to combat old and new foes alike. *Frontiers in Agronomy* 1, 2.

Jacquemin, G., Chavalle, S. and De Proft, M. 2014. Forecasting the emergence of the adult orange wheat blossom midge, *Sitodiplosis mosellana* (Géhin) (Diptera: Cecidomyiidae) in Belgium. *Crop Protection* 58, 6-13.

Jakku, E. and Thorburn, P. J. 2010. A conceptual framework for guiding the participatory development of agricultural decision support systems. *Agricultural Systems* 103(9), 675-682.

Jarošová, J., Żelazny, W. R. and Kundu, J. K. 2019. Patterns and predictions of barley yellow dwarf virus vector migrations in Central Europe. *Plant Disease* 103(8), 2057-2064.

Jarrahi, M. H. 2018. Artificial intelligence and the future of work: human-AI symbiosis in organizational decision making. *Business Horizons* 61(4), 577-586.

Jørgensen, L. N., Matzen, N., Ficke, A., Nielsen, G. C., Jalli, M., Ronis, A., Andersson, B. and Djurle, A. 2020. Validation of risk models for control of leaf blotch diseases in wheat in the Nordic and Baltic countries. *European Journal of Plant Pathology* 157(3), 599-613.

Kanatas, P., Travlos, I. S., Gazoulis, I., Tataridas, A., Tsekoura, A. and Antonopoulos, N. 2020. Benefits and limitations of decision support systems (DSS) with a special emphasis on weeds. *Agronomy* 10(4), 548.

Kendall, D. A., Brain, P. and Chinn, N. E. 1992. A simulation model of the epidemiology of barley yellow dwarf virus in winter sown cereals and its application to forecasting. *Journal of Applied Ecology* 29(2), 414-426.

Kindler, S. D., Elliott, N. C., Giles, K. L. and Royer, T. A. 2003. Economic injury level for the greenbug, *Schizaphis graminum*, in Oklahoma winter wheat. *Southwestern Entomologist* 28, 163-166.

Klein, I., Oppelt, N. and Kuenzer, C. 2021. Application of remote sensing data for locust research and management–a review. *Insects* 12(3), 233.

Klueken, A. M., Hau, B., Ulber, B. and Poehling, H.-M. 2009. Forecasting migration of cereal aphids (Hemiptera: Aphididae) in autumn and spring. *Journal of Applied Entomology* 133(5), 328-344.

Kogan, M. 1998. Integrated pest management: historical perspectives and contemporary developments. *Annual Review of Entomology* 43, 243-270.

Lagos-Kutz, D., Voegtlin, D. J., Onstad, D., Hogg, D., Ragsdale, D., Tilmon, K., Hodgson, E., Difonzo, C., Groves, R., Krupke, C., Laforest, J., Seiter, N. J., Duerr, E., Bradford, B. and Hartman, G. L. 2020. The soybean aphid suction trap network: sampling the aerobiological "soup". *American Entomologist* 66(1), 48-55.

Lankin-Vega, G., Worner, S. P. and Teulon, D. A. J. 2008. An ensemble model for predicting *Rhopalosiphum padi* abundance. *Entomologia Experimentalis et Applicata* 129(3), 308-315.

Larsson, H. 2005. A crop loss model and economic thresholds for the grain aphid, *Sitobion avenae* (F.), in winter wheat in southern Sweden. *Crop Protection* 24(5), 397-405.

Leather, S. R. 2018. "Ecological Armageddon" - more evidence for the drastic decline in insect numbers. *Annals of Applied Biology* 172(1), 1-3.

Lechenet, M., Dessaint, F., Py, G., Makowski, D. and Munier-Jolain, N. 2017. Reducing pesticide use while preserving crop productivity and profitability on arable farms. *Nature Plants* 3, 17008.

Leclercq-Le Quillec, F., Plantegenest, M., Riault, G. and Dedryver, C. A. 2000. Analyzing and modeling temporal disease progress of barley yellow dwarf virus serotypes in barley fields. *Phytopathology* 90(8), 860-866.

Leybourne, D. J., Storer, K. E., Berry, P. and Ellis, S. 2022. Development of a pest threshold decision support system for minimising damage to winter wheat from wheat bulb fly, *Delia coarctata*. *Annals of Applied Biology* 180(1), 118-131.

Li, W., Chen, P., Wang, B. and Xie, C. 2019. Automatic localization and count of agricultural crop pests based on an improved deep learning pipeline. *Scientific Reports* 9(1), 7024.

Lin, T.-Y., Dollár, P., Girshick, R., He, K., Hariharan, B. and Belongie, S. 2017. Feature pyramid networks for object detection. In: *Proceedings of the IEEE Conference on Computer Vision and Pattern Recognition*, pp. 2117–2125.

Lin, T.-Y., Maire, M., Belongie, S., Hays, J., Perona, P., Ramanan, D., Dollár, P. and Zitnick, C. L. 2014. Microsoft COCO: common objects in context. In: *Lecture Notes in Computer Science: European Conference on Computer Vision*. Springer, pp. 740–755.

Lindblad, M. 2001. Development and evaluation of a logistic risk model: predicting frit fly infestation in oats. *Ecological Applications* 11(5), 1563–1572.

Lindblad, M. and Sigvald, R. 1996. A degree-day model for regional prediction of first occurrence of frit flies in oats in Sweden. *Crop Protection* 15(6), 559–565.

Liu, L., Wang, R., Xie, C., Yang, P., Wang, F., Sudirman, S. and Liu, W. 2019. PestNet: an end-to-end deep learning approach for large-scale multi-class pest detection and classification. *IEEE Access* 7, 45301–45312.

Liu, L., Xie, C., Wang, R., Yang, P., Sudirman, S., Zhang, J., Li, R. and Wang, F. 2021. Deep learning based automatic multiclass wild pest monitoring approach using hybrid global and local activated features. *IEEE Transactions on Industrial Informatics* 17(11), 7589–7598.

Liu, W., Anguelov, D., Erhan, D., Szegedy, C., Reed, S., Fu, C.-Y. and Berg, A. C. 2016. SSD: single shot multibox detector. In: *Lecture Notes in Computer Science: European Conference on Computer Vision*. Springer, pp. 21–37.

Luo, J., Huang, W., Guan, Q., Zhao, J. and Zhang, J. 2013. Hyperspectral image for discriminating aphid and aphid damage region of winter wheat leaf. *International Geoscience and Remote Sensing Symposium - (IGARSS)* 2013, 3726–3729.

Luquet, M., Hullé, M., Simon, J. C., Parisey, N., Buchard, C. and Jaloux, B. 2019. Relative importance of long-term changes in climate and land-use on the phenology and abundance of legume crop specialist and generalist aphids. *Insect Science* 26(5), 881–896.

Mann, B. P., Wratten, S. D. and Watt, A. D. 1986. A computer-based advisory system for cereal aphid control. *Computers and Electronics in Agriculture* 1(3), 263–270.

Miao, L., Jian-Feng, Z., Qing-Quan, C., Jia, Z. L., Wang, H. T., Liang, H. B., Zhang, H., Li, X., Zhang, J. H., Jiang, L. Y. and Qin, Q. L. 2011. Construction of a preliminary network of suction traps to monitor the migration of alate aphids in China. *Chinese Journal of Applied Entomology* 48, 1874–1878.

Mishra, G., Srivastava, S., Panda, B. K. and Mishra, H. N. 2018. Prediction of *Sitophilus granarius* infestation in stored wheat grain using multivariate chemometrics & fuzzy logic-based electronic nose analysis. *Computers and Electronics in Agriculture* 152, 324–332.

Morgan, D. 2000. Population dynamics of the bird cherry-oat aphid, *Rhopalosiphum padi* (L.), during the autumn and winter: a modelling approach. *Agricultural and Forest Entomology* 2(4), 297–304.

Nancarrow, N., Aftab, M., Hollaway, G., Rodoni, B. and Trębicki, P. 2021. Yield losses caused by barley yellow dwarf virus-PAV infection in wheat and barley: a three-year field study in South-Eastern Australia. *Microorganisms* 9(3), 645.

Oliver, D. M., Fish, R. D., Winter, M., Hodgson, C. J., Heathwaite, A. L. and Chadwick, D. R. 2012. Valuing local knowledge as a source of expert data: farmer engagement and the design of decision support systems. *Environmental Modelling and Software* 36, 76–85.

Parker, C. and Sinclair, M. 2010. User-centred design does make a difference: the case of decision support systems in crop production. *Behaviour and Information Technology* 20(6), 449–460.

Petrovskii, S., Ellis, J., Forbes, E., Petrovskaya, N. and Walters, K. F. A. 2022. A predictive model and a field study on heterogeneous slug distribution in arable fields arising from density dependent movement. *Scientific Reports* 12(1), 2274.

Philips, C. R., Herbert, D. A., Kuhar, T. P., Reisig, D. D. and Roberts, E. A. 2012. Using degree-days to predict cereal leaf beetle (Coleoptera: Chrysomelidae) egg and larval population peaks. *Environmental Entomology* 41(4), 761-767.

Ramsden, M. W., Kendall, S. L., Ellis, S. A. and Berry, P. M. 2017. A review of economic thresholds for invertebrate pests in UK arable crops. *Crop Protection* 96, 30–43.

Ren, S., He, K., Girshick, R. and Sun, J. 2015. Faster R-CNN: towards real-time object detection with region proposal networks. *Advances in Neural Information Processing Systems* 28.

Rogers, C. D., Guimarães, R. M. L., Evans, K. A. and Rogers, S. A. 2015. Spatial and temporal analysis of wheat bulb fly (*Delia coarctata*, Fallén) oviposition: consequences for pest population monitoring. *Journal of Pest Science* 88(1), 75–86.

Rose, D. C., Sutherland, W. J., Parker, C., Lobley, M., Winter, M., Morris, C., Twining, S., Ffoulkes, C., Amano, T. and Dicks, L. V. 2016. Decision support tools for agriculture: towards effective design and delivery. *Agricultural Systems* 149, 165–174.

Rossi, V., Caffi, T. and Salinari, F. 2012. Helping farmers face the increasing complexity of decision-making for crop protection. *Phytopathologia Mediterranea* 51, 457–479.

Rowley, C., Cherrill, A., Leather, S., Nicholls, C., Ellis, S. and Pope, T. 2016. A review of the biology, ecology and control of saddle gall midge, *Haplodiplosis marginata* (Diptera: Cecidomyiidae) with a focus on phenological forecasting. *Annals of Applied Biology* 169(2), 167–179.

Rupnik, R., Kukar, M., Vračar, P., Košir, D., Pevec, D. and Bosnić, Z. 2019. AgroDSS: a decision support system for agriculture and farming. *Computers and Electronics in Agriculture* 161, 260–271.

Shtienberg, D. 2013. Will decision-support systems be widely used for the management of plant diseases? *Annual Review of Phytopathology* 51, 1–16.

Sishodia, R. P., Ray, R. L. and Singh, S. K. 2020. Applications of remote sensing in precision agriculture: a review. *Remote Sensing* 12(19), 3136.

Steele-Dunne, S. C., McNairn, H., Monsivais-Huertero, A., Judge, J., Liu, P.-W. and Papathanassiou, K. 2017. Radar remote sensing of agricultural canopies: a review. *IEEE Journal of Selected Topics in Applied Earth Observations and Remote Sensing* 10(5), 2249–2273.

Steinger, T., Klötzli, F. and Ramseier, H. 2020. Experimental assessment of the economic injury level of the cereal leaf beetle (Coleoptera: Chrysomelidae) in winter wheat. *Journal of Economic Entomology* 113(4), 1823–1830.

Swiergiel, W., Manduric, S., Rämert, B., Porcel, M. and Tasin, M. 2019. Development of sustainable plant protection programs through multi-actor co-innovation: an 8-year case study in Swedish apple production. *Journal of Cleaner Production* 234, 1178–1191.

Teulon, D. A. J. and Scott, I. A. W. 2006. The use of suction traps for detection of unwanted invasive insects and other invertebrates. *New Zealand Plant Protection* 59, 125–131.

Thackray, D. J., Diggle, A. J. and Jones, R. A. C. 2009. BYDV predictor: a simulation model to predict aphid arrival, epidemics of Barley yellow dwarf virus and yield losses in wheat crops in a Mediterranean-type environment. *Plant Pathology* 58(1), 186–202.

Walls, J. T., Caciagli, P., Tooker, J. F., Russo, J. M., Rajotte, E. G. and Rosa, C. 2016. Modeling the decision process for barley yellow dwarf management. *Computers and Electronics in Agriculture* 127, 775–786.

Wang, F., Wang, R., Xie, C., Yang, P. and Liu, L. 2020. Fusing multi-scale context-aware information representation for automatic in-field pest detection and recognition. *Computers and Electronics in Agriculture* 169, 105222.

Wang, R., Liu, L., Xie, C., Yang, P., Li, R. and Zhou, M. 2021. AgriPest: a large-scale domain-specific benchmark dataset for practical agricultural pest detection in the wild. *Sensors* 21(5), 1601.

Wilson, R. S., Hooker, N., Tucker, M., LeJeune, J. and Doohan, D. 2009. Targeting the farmer decision making process: a pathway to increased adoption of integrated weed management. *Crop Protection* 28(9), 756–764.

Young, A. G., Port, G. R. and Green, D. B. 1993. Development of a forecast of slug activity: validation of models to predict slug activity from meteorological conditions. *Crop Protection* 12(3), 232–236.

Index

Ingram Content Group UK Ltd.
Milton Keynes UK
UKHW020604050523
421277UK00001B/25

9 781801 461139